Polymer-Based Materials and Composites—Synthesis, Assembly, Properties and Applications

MATERIALS RESEARCH SOCIETY
SYMPOSIUM PROCEEDINGS VOLUME 1312

Polymer-Based Materials and Composites–Synthesis, Assembly, Properties and Applications

Symposium held November 29–December 3, Boston, Massachusetts, U.S.A.

EDITORS

Zhongyang Cheng
Auburn University
Auburn, Alabama, U.S.A.

Vivek Bharti
3M Company
St. Paul, Minnesota, U.S.A.

Zhuo Xu
Xian Jiaotong University
Xian, China

Debra A. Wrobleski
Los Alamos National Laboratory
Los Alamos, New Mexico, U.S.A.

Mircea Chipara
The University of Texas Pan American
Edinburg, Texas, U.S.A.

Pulickel M. Ajayan
Rice University
Houston, Texas, U.S.A.

Ali Nasar
CNC Coatings
Lancashire, United Kingdom

Alan Kin-Tak Lau
University of Southern Queensland
Toowoomba
Queensland, Australia

Padma Gopalan
University of Wisconsin-Madison
Madison, Wisconsin, U.S.A.

Teruaki Hayakawa
Tokyo Institute of Technology
Tokyo, Japan

Hilmar Koerner
Air Force Research Laboratory
Wright-Patterson AFB, Ohio, U.S.A.

D. Venkataraman
University of Massachusetts Amherst
Amherst, Massachusetts, U.S.A.

Qing Wang
The Pennsylvania State University
University Park, Pennsylvania, U.S.A.

MRS
Materials Research Society
Warrendale, Pennsylvania

CAMBRIDGE
UNIVERSITY PRESS

CAMBRIDGE UNIVERSITY PRESS
Cambridge, New York, Melbourne, Madrid, Cape Town,
Singapore, São Paulo, Delhi, Tokyo, Mexico City

Cambridge University Press
32 Avenue of the Americas, New York, NY 10013-2473, USA

www.cambridge.org
Information on this title: www.cambridge.org/9781605112893

Materials Research Society
506 Keystone Drive, Warrendale, PA 15086, USA
http://www.mrs.org

First published 2011

Single article reprints from this publication are available through University
Microfilms Inc., 300 North Zeeb Road, Ann Arbor, MI 48106, USA

CODEN: MRSPDH

ISBN: 978-1-60511-289-3 Hardback

CONTENTS

*Invited Paper

*Invited Paper

*Invited Paper

*Invited Paper

*Invited Paper

NANOSTRUCTURED POLYMERIC MATERIALS—SYNTHESIS AND ASSEMBLY

*Invited Paper

PREFACE

Symposium HH, "Polymer-Based Smart Materials—Process, Properties, and Application," Symposium II, "Polymer-Based Nanocomposites," and Symposium JJ, "Nanostructured Polymeric Materials—Synthesis and Assembly," were held Nov. 29–Dec. 3 at the 2010 MRS Fall Meeting in Boston, Massachusetts.

Polymeric materials pervade all walks of life. Recent advances in polymerization chemistry, polymer assembly and polymer processing have resulted in a rich array of polymer-based materials with tailored properties through precise control of molecular structures and architecture. This symposium proceedings volume represents the recent advances in polymerization chemistry, polymer assembly and processing. The papers are divided into three sections: (1) Polymer-based Smart Materials, (2) Polymer-based Nanocomposites and (3) Nanostructured Polymeric Materials. Each paper in this volume provides a glimpse of the exciting recent developments occurring in polymeric materials such as new and efficient drug delivery vehicles, novel polymer composite materials, smart and self-healing materials, and novel optoelectronic and dielectric applications. We hope that these papers convey the breadth of exciting advancements happening in the area of polymeric materials.

Vivek Bharti
Mircea Chipara
Dhandapani Venkataraman

February 2011

ACKNOWLEDGMENTS

The papers published in this volume result from three MRS Fall 2010 symposia—HH, II, and JJ. We sincerely thank all of the oral and poster presenters of the symposia who contributed to this proceedings volume. We also thank the reviewers of these manuscripts, who provided valuable feedback to the editors and to the authors. It is an understatement to say that the symposia and the proceedings would not have happened without the organizational help of the Materials Research Society and its staff, particularly the Publications staff for guiding us smoothly through the submission/review process and constantly nudging us to move forward. The organizers of Symposium JJ thank the Air Force Research Laboratory for its financial support.

Materials Research Society Symposium Proceedings

Volume 1275— Structural and Chemical Characterization of Metals, Alloys and Compounds, R. Pérez Campos, A. Contreras Cuevas, R.A. Esparza Muñoz, 2011, ISBN 978-1-60511-252-7
Volume 1276— Advanced Structural Materials—2010, H.A. Calderon, A. Salinas Rodriguez, H. Balmori-Ramirez, 2010, ISBN 978-1-60511-253-4
Volume 1277E—Biomaterials—2010, S.E. Rodil, A. Almaguer-Flores, K. Anselme, 2010, ISBN 978-1-60511-254-1
Volume 1278E—Composite, Hybrid Materials and Ecomaterials, R. Bernal, C. Cruz Vazquez, L.E. Rendon Diaz Miron, V.M. Castaño, 2010, ISBN 978-1-60511-255-8
Volume 1279— New Catalytic Materials, J.A. Wang, J. Manuel Domínguez, 2010, ISBN 978-1-60511-256-5
Volume 1280E—Nanomaterials for Biomedical Applications, L. Zhang, T.J. Webster, A. Salinas Rodriguez, 2010, ISBN 978-1-60511-257-2
Volume 1282— Diamond Electronics and Bioelectronics—Fundamentals to Applications IV, P. Bergonzo, J.E. Butler, C.E. Nebel, M. Nesladek, A.T.S. Wee, 2011, ISBN 978-1-60511-259-6
Volume 1283E—Carbon-Based Electronic Devices—Processing, Performance and Reliability, M. Chhowalla, R.R. Keller , M. Meyyappan, W.J. Ready, 2011, ISBN 978-1-60511-260-2
Volume 1284— Fundamentals of Low-Dimensional Carbon Nanomaterials, J.J. Boeckl, L. Dai, W. Lu, M.H. Rümmeli, J. Warner, 2011, ISBN 978-1-60511-261-9
Volume 1285E—Challenges in Roll-to-Roll (R2R) Fabrication for Electronics and Other Functionalities, T. Blaudeck, G. Cho, J.H. Daniel, M.R. Dokmeci, 2011, ISBN 978-1-60511-262-6
Volume 1286E—Molecular and Hybrid Materials for Electronics and Photonics, J. Liu, 2011, ISBN 978-1-60511-263-3
Volume 1287E—Low-Temperature-Processed Thin-Film Transistors, E. Fortunato, 2011, ISBN 978-1-60511-264-0
Volume 1288E—Novel Fabrication Methods for Electronic Devices, P. Andrew, 2011, ISBN 978-1-60511-265-7
Volume 1289E—Controlling Material Properties and Charge-Carrier Interactions with Quantum-Dot Coupling, 2011, ISBN 978-1-60511-266-4
Volume 1290E—Magnetism and Correlated Electronic Structure of Nitrides—Rare-Earth and Transition Metals as Constituents and Dopants, W.R.L. Lambrecht, A. Ney, K. Smith, H.J. Trodahl, 2011, ISBN 978-1-60511-267-1
Volume 1291E—Integrated Nonreciprocal Photonics—Materials, Phenomena and Devices, V. Fratello, M. Levy, B. Stadler, M. Vanwolleghem, 2011, ISBN 978-1-60511-268-8
Volume 1292— Oxide Nanoelectronics, H. Hwang, J. Levy, P. Makysymovych, G. Medeiros-Ribeiro, R. Waser, 2011, ISBN 978-1-60511-269-5
Volume 1293E—Liquid-Crystal Materials—Beyond Displays, N.L. Abbott, D.J. Broer, T. Kato, T.J. White, 2011, ISBN 978-1-60511-270-1
Volume 1294E—Resonant Optical Antennas—Sensing and Shaping Materials, K.B. Crozier, N. Engheta, G. Ju, R. Quidant, R. Zia, 2011, ISBN 978-1-60511-271-8
Volume 1295— Intermetallic-Based Alloys for Structural and Functional Applications, M. Palm, B. Bewlay, S. Kumar, K. Yoshimi, 2011, ISBN 978-1-60511-272-5
Volume 1296E—New Methods in Steel Design—Steel Ab Initio, Y. Adachi, R. Dronskowski, D. Raabe, P.E.A. Turchi, 2011, ISBN 978-1-60511-273-2
Volume 1297— Deformation Mechanisms, Microstructure Evolution and Mechanical Properties of Nanoscale Materials, J.R. Greer, D.S. Gianola, B.G. Clark, T. Zhu, A.H.W. Ngan, 2011, ISBN 978-1-60511-274-9
Volume 1298— Advanced Materials for Applications in Extreme Environments, T.S. Byun, R. Smith, M. Li, 2011, ISBN 978-1-60511-275-6
Volume 1299— Microelectromechanical Systems—Materials and Devices IV, M.P. de Boer, F.W. DelRio, C. Eberl, E.P. Gusev, 2011, ISBN 978-1-60511-276-3
Volume 1300E—Bulk Metallic Glasses and their Applications, K.F. Yao, 2011, ISBN 978-1-60511-277-0

MATERIALS RESEARCH SOCIETY SYMPOSIUM PROCEEDINGS

Volume 1301— Soft Matter, Biological Materials and Biomedical Materials—Synthesis, Characterization and Applications, A.J. Nolte, K. Shiba, R. Narayan, D. Nolte, 2011, ISBN 978-1-60511-278-7
Volume 1302E—Nanowires—Growth and Device Assembly for Novel Applications, 2011, ISBN 978-1-60511-279-4
Volume 1303— Nanomaterials Integration for Electronics, Energy and Sensing, D. E. Perea, Y. Jung, J. B. Hannon, M. A. Reed, S. T. Picraux, 2011, ISBN 978-1-60511-280-0
Volume 1304E—Hierarchical Materials and Composites—Combining Length Scales from Nano to Macro, J.H. Moon, G.M. Odegard, M.S.P. Shaffer, B.L. Wardle, 2011, ISBN 978-1-60511-281-7
Volume 1305E—Group IV Semiconductor Nanostructures and Applications, L. Dal Negro, 2011, ISBN 978-1-60511-282-4
Volume 1306E—Aerogels and Aerogel-Inspired Materials, S. Brock, G. Gould, A. Roig, D. Rolison, 2011, ISBN 978-1-60511-283-1
Volume 1307E—Boron and Boron Compounds—From Fundamentals to Applications, M. Dudley, J.H. Edgar, M. Kuball, 2011, ISBN 978-1-60511-284-8
Volume 1308E—Artificially Induced Crystalline Alignment in Thin Films and Nanostructures, A.T. Findikoglu, R. Huehne, T. Shimada, J.Z. Wu, 2011, ISBN 978-1-60511-285-5
Volume 1309— Solid-State Chemistry of Inorganic Materials VIII, K-S. Choi, S.J. Clarke, P.S. Halasyamani, D.G. Mandrus, 2011, ISBN 978-1-60511-286-2
Volume 1310E—Magneto Calorics and Magnetic Cooling, A. Fujita, K. Gschneidner Jr., O. Gutfleisch, K.G. Sandeman, A. Yan, 2011, ISBN 978-1-60511-287-9
Volume 1311— Next-Generation Fuel Cells—New Materials and Concepts, T. He, K. Swider-Lyons, B. Park, P.A. Kohl, 2011, ISBN 978-1-60511-288-6
Volume 1312— Polymer-Based Materials and Composites—Synthesis, Assembly, Properties and Applications, V. Bharti, M. Chipara, D. Venkataraman, 2011, ISBN 978-1-60511-289-3
Volume 1313— Materials for Advanced Lithium Batteries, G.-A. Nazri, J-M Tarascon, D. Guyomard, A. Yamada, 2011, ISBN 978-1-60511-290-9
Volume 1314E—Thermoelectric Materials for Solid-State Power Generation and Refrigeration, Y. Grin, G.S. Nolas, J. Sharp, T.M. Tritt, 2011, ISBN 978-1-60511-291-6
Volume 1315— Transparent Conducting Oxides and Applications, J.J. Berry, E. Fortunato, J. Medvedeva, Y. Shigesato, 2011, ISBN 978-1-60511-292-3
Volume 1316E—Nanofunctional Materials, Nanostructures and Nanodevices for Biomedical Applications II, R. Rao, 2011, ISBN 978-1-60511-293-0
Volume 1317E—Interdisciplinary Approaches to Safe Nanotechnologies, C. Chaneac, S. Harper, G.V. Lowry, R.I. MacCuspie, 2011, ISBN 978-1-60511-294-7
Volume 1318— Advances in Spectroscopy and Imaging of Surfaces and Nanostructures, J. Cumings, J. Guo, F.M. Granozio, O.V. Kolosov, 2011, ISBN 978-1-60511-295-4
Volume 1319— Materials Issues in Art and Archaeology IX, P.B. Vandiver, C.L. Reedy, J.L. Ruvalcaba Sil, W. Li, 2011, ISBN 978-1-60511-296-1
Volume 1320— Materials Education Development and Outreach—From K–Grad, D. Bahr, K. Jones, M. Glass, E. Allen, 2011, ISBN 978-1-60511-297-8

Prior Materials Research Society Symposium Proceedings available by contacting Materials Research Society

Polymer-Based Smart Materials—Process, Properties, and Application

Mater. Res. Soc. Symp. Proc. Vol. 1312 © 2011 Materials Research Society
DOI: 10.1557/opl.2011.668

Hydrostatically Coupled Dielectric Elastomer Actuators: New Opportunities for Haptics

Federico Carpi, Gabriele Frediani and Danilo De Rossi

Interdepartmental Research Centre 'E. Piaggio', University of Pisa,
School of Engineering,Via Diotisalvi, 2 - 56100 Pisa, Italy.

ABSTRACT

Dielectric elastomer actuators (DEAs) have been demonstrated to represent today a high-performance technology for electromechanical transducers based on electroactive polymers. As a means to improve versatility and safety of DEAs for several fields of application, so-called 'hydrostatically coupled' DEAs (HC-DEAs) have recently been described. HC-DEAs are based on an incompressible fluid that mechanically couples a DE-based active part to a passive part interfaced to the load, so as to enable hydrostatic transmission. This paper presents ongoing developments of bubble-like HC-DEAs and their promising potential application in the field of haptics. In particular, the first part of the paper describes a static and dynamic characterization of a prototype actuator made of two pre-stretched membranes (20 mm wide, 1.8 mm high, and 60 μm thick) of 3M VHB acrylic elastomer, coupled via silicone grease. The actuator exhibited a maximum stress of 1.3 kPa at 4.4 kV, a relative displacement of -80% at 4.4 kV, a -3dB bandwidth of 3 Hz, and a resonance frequency of 160 Hz. The second part of the paper presents possible applications of the tested actuator configuration for haptic interfaces. Two specific examples are considered. The first deals with a wearable tactile/haptic display used to provide users with tactile feedback during electronic navigation in virtual environments. The display consists of HC-DEAs arranged in contact with finger tips. As a second example of usage, an up-scaled prototype version of an 8-dots refreshable cell for dynamic Braille displays is shown. Each Braille pin consists of a miniature HC-DEA, with a diameter lower than 2 mm. Both types of applications clearly show the potential of the new technology and the prospective opportunities for haptics.

INTRODUCTION

Electromechanical actuators based on ElectroActive (or Electromechanically Active) Polymers (EAPs) represent today a well established and promising scientific field of research and development [1-7]. EAPs form a broad family of smart materials capable of transducing energy from the electrical to the mechanical form, and vice versa. As such, they are studied for electromechanical actuation and mechanoelectrical sensing, as well as mechanical energy harvesting to generate electricity, thus gaining the nickname of 'Artificial Muscles' [1-7]. EAPs are commonly classified in two major families: ionic EAPs, activated by an electrically-induced transport of ions and/or solvent, and electronic EAPs, activated by electrostatic forces. Ionic EAPs include polymer gels, ionic polymer metal composites, conjugated polymers, and carbon nanotubes. Electronic EAPs include piezoelectric polymers, electrostrictive polymers, dielectric elastomers, liquid crystal elastomers, and carbon nanotube aerogels [1-7].

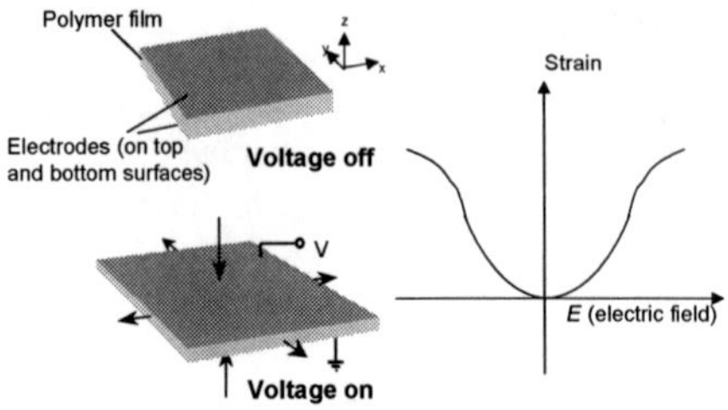

Figure 1. Fundamental working principle of dielectric elastomer actuators, and typical electromechanical quadratic response (see equation 1) with tendency to 'saturation' which usually shows up at high fields.

Figure 2. Examples of actuating configurations for dielectric elastomer actuators (from [4]).

While several EAP materials and their properties have been known for many decades, they have found very limited applications so far. Such a trend is currently going to change, as a result of an effective synergy of at least three main factors: key scientific breakthroughs have been achieved in some of the existing EAP technologies; unprecedented electromechanical properties have been discovered in materials previously developed for different purposes; and industrial efforts have been focused on real exploitation of EAP materials. As an outcome, after several years of basic research, today the EAP field is just starting to undergo transition from academia into commercialization, with significant investments from large companies [8].

This is further demonstrated by the fact that a network entirely focused on EAPs, called "European Scientific Network for Artificial Muscles" has just been established among all of the major research and industrial organizations active in Europe in the EAP field [9].

Electromechanical performance of EAPs heavily depends on the considered type of class and materials [1-7]. Among them, dielectric elastomer (DE) transducers, pioneered by the Stanford Research Institute [10], are today widely recognised to represent the most performing technology emerging from EAPs.

This technology relies on rubber-like insulating materials (typically silicone, acrylic and polyurethane elastomers) which combine low elastic modulus and high dielectric strength, so as to exhibit significant actuation when subject to electric fields. In particular, DE actuators are deformable capacitors, obtained by processing the material as a thin film, coated with compliant electrodes (typically made of metal particles, graphite powder, carbon black, or carbon nanotubes, applied by brushing, spray coating, or mixing with an elastomeric matrix). An applied voltage charges the electrodes, causing a thickness reduction and a surface expansion of the film (figure 1), owing to electrostatic interactions among the electrode free charges. The achievable thickness strain S can be modelled as follows [10]:

$$S = -\frac{1}{Y}\varepsilon_0\varepsilon_r\left(\frac{V}{d}\right)^2 = -\frac{1}{Y}\varepsilon_0\varepsilon_r E^2 \tag{1}$$

where ε_0 is the free-space dielectric permittivity, ε_r is the relative dielectric constant of the material, Y is its Young's modulus, V is the applied voltage, d is the film thickness and E is the electric field. This effect can produce significant output mechanical work [1,4,10].

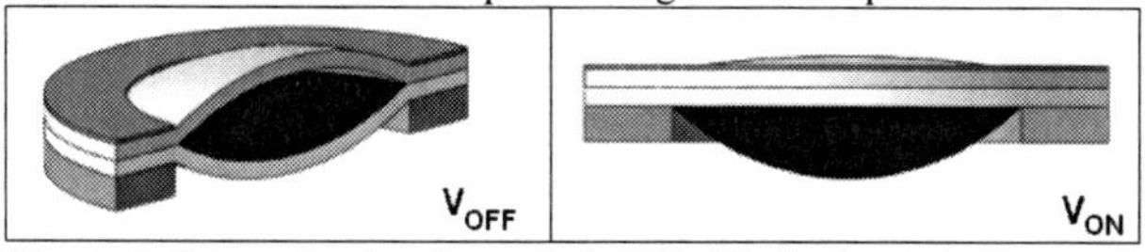

Figure 3. Schematic drawings of a bubble-like hydrostatically coupled dielectric elastomer actuator: section view with no applied voltage (left); lateral view with applied voltage (right).

On the other hand, an imposed deformation of the structure can be detected as a change of electrical capacitance, making the device able to work as a piezo-capacitive strain sensor. Similarly, by maintaining a given charge stored on the electrodes, the structure can work as a mechanical energy harvester, producing electrical energy when an external source deforms it, varying the capacitance [1,4,10].

As compared to other EAP technologies, DE transducers uniquely combine high transduction performance (such as high strains, stresses, energy density, response speed, and efficiency) with other key functional properties (such as low power consumption, no generation of acoustic noise or heat, and long lifetime), physical properties (such as low specific weight, high resilience, shock tolerance, and environmental stability), and processing properties (such as ease of material processing, device shaping, dimensional scaling, as well as low cost) [1,4,10].

Having no equals in terms of overall performance within the whole EAP field, DE transducers show a great potential as 'artificial muscle materials' for many potential fields of applications. In fact, so far a huge variety of devices has been developed for experimental purposes, such as planar, benders, extenders, stacks, folded, helical, rolls, push-pull, diaphragm, and many others, as shown in the examples of figure 2 [4]. The DE technology has been shown to be suitable for applications which so far have been precluded to conventional drive mechanisms. Possible usage spans from the micro- to the macro- scale in different sectors, such as medical and haptic devices, consumer electronics, and automation and robotic systems [1,3,4,8,11].

As a means to improve versatility and safety of DEAs for several fields of application, so-called 'hydrostatically coupled' DEAs (HC-DEAs) have recently been described [12]. HC-DEAs are based on an incompressible fluid that mechanically couples a DE-based active part to a passive part interfaced to the load, so as to enable hydrostatic transmission. Such an elementary general concept can be exploited in several ways, by designing devices with different structure, shape and intended functionality [12]. Figure 3 shows the structure of a bubble-like HC-DEA [13], which includes the following parts: an electromechanically active membrane, made of a DE film coated with compliant electrodes; an electromechanically passive membrane, working as the end effector in contact with the load; an incompressible fluid contained between the two membranes. Both membranes are radially constrained by bonding them to a hollow stiff support. Actuation of the active membrane through a voltage difference applied between its electrodes exerts onto the fluid a pressure, which is hydrostatically transmitted to the passive membrane. The active membrane buckles outwards, owing to its pre-curvature, while the passive membrane follows inwards, driven by hydrostatic transmission. This principle allows for safe transmission and delivery of actuation from the active membrane to the load, without any direct contact between them.

General features and advantages of HC-DEAs include directionability of transmission, remote allocability and scalability of the end effector, modulability of displacements and forces, maximisation of electromechanical coupling, self-compensation against load-induced deformations, and electrical safety, as discussed in details elsewhere [12].

The following sections present ongoing developments of bubble-like HC-DEAs and their potential for two applications in the field of haptics.

MATERIALS AND METHODS

Prototype actuators were manufactured using for each membrane a widely studied [1,4,10] acrylic elastomer film (VHB 4910, 3M, USA). During manufacturing, the film was bi-axially pre-stretched by 4 times, causing a thickness reduction from the rest value of 1 mm down to a pre-stretched value of about 60 μm. One pre-stretched membrane was made active by coating both sides of it with carbon grease (Carbon Conductive Grease 846, M.G. Chemicals, Canada). The transmission medium between the two membranes was silicone grease (8462, M.G. Chemicals, Canada).

To assemble the actuator, the passive and active membranes were sandwiched between two plastic frames (0.15 and 2 mm thick) having a circular hole of radius 10 mm. The frames provided the membranes with both support and in-plane radial constraint. Adhesion between membranes and frames was easily achieved owing to the inherent adhesiveness of the VHB film. While combining the two membranes, the volume between them was filled with the coupling medium, so that the pressurized passive membrane reached an apical height of 1.78 mm from the base plane.

Electromechanical performance of prototype actuators was assessed by measuring displacement and force at the apex of the bubble actuator, in response to d.c. and a.c. voltages. Electrical driving was delivered by means of a high voltage amplifier (10/10B, Trek, USA). In particular, two testing phases were performed. In a static testing phase, free stroke and blocking force were measured (using a two-column dynamometer), at the apex of the passive membrane while driving the active membrane at different voltages, up to 4.4 kV. In a dynamic testing phase, the actuator was characterised in terms of free stroke (optically measured with a LASER displacement transducer) induced by monopolar voltage waveforms, varying sinusoidally from 0 to 5 kV, at a frequency included in the range 0.1-300 Hz; the dynamic data were used to obtain the actuator frequency response.

RESULTS

Static electromechanical performance of a prototype actuator is presented in figures 4 and 5, which plot relative values of the apical displacement (figure 4) and apical force (figure 5), as a function of the applied voltage.

Dynamic electromechanical performance of the actuator is presented in figure 6, which shows the frequency response in terms of *minimum* and *maximum* relative displacement for an oscillating voltage at given frequencies. A resonance peaks at 160 Hz is clearly visible. To estimate the –3dB bandwidth, figure 7 reports the spectrum of the *maximum* relative displacement normalized by its value at low frequency (0.1 Hz) and expressed in decibels (i.e.

20·log(normalized displacement)). The bandwidth is shown to be equal to 3 Hz. Such a rather limited value of the bandwidth is due to the viscosity of the used acrylic elastomer.

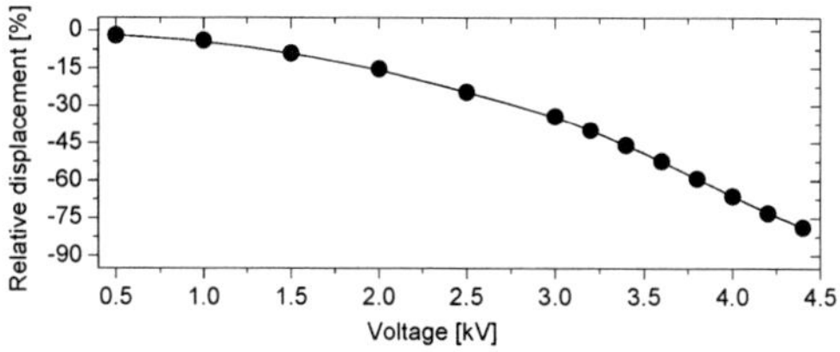

Figure 4. Voltage dependence of the apical relative displacement of the developed bubble-like hydrostatically coupled dielectric elastomer actuator.

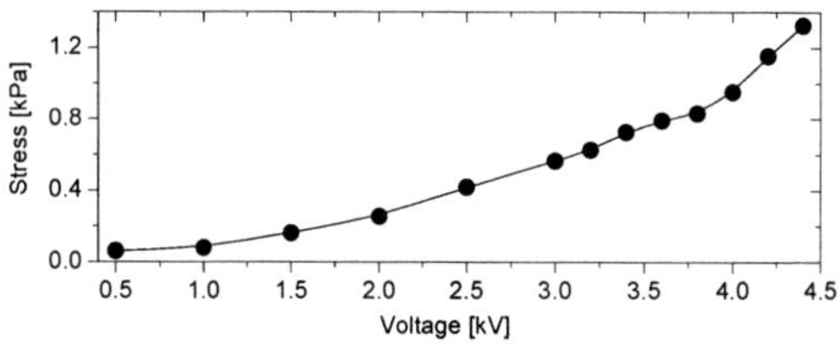

Figure 5. Voltage dependence of the apical stress of the developed bubble-like hydrostatically coupled dielectric elastomer actuator.

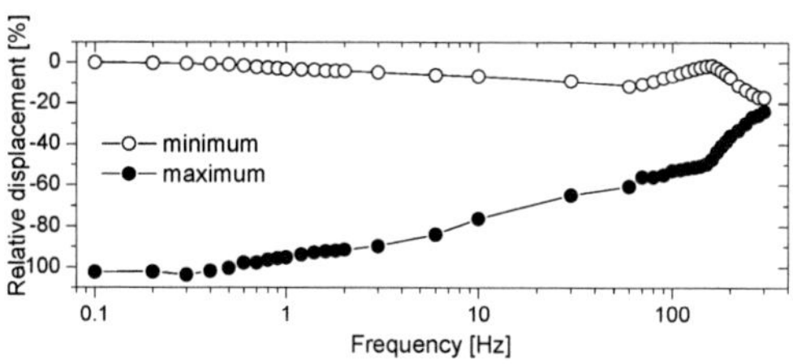

Figure 6. Frequency response: spectrum of the minimum and maximum relative displacement for the powder-filled (a) and the grease-filled (b) actuators.

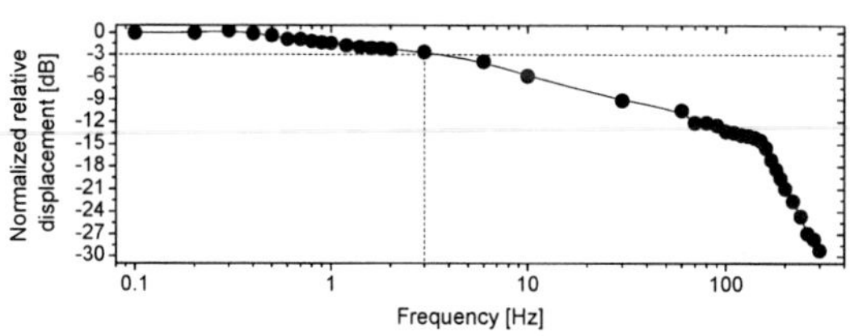

Figure 7. Spectrum of the normalized maximum relative displacement for the powder-filled and grease-filled actuators.

ENVISAGED APPLICATIONS

This technology might be considered for applications that could take specific advantage of actuators having a safe and compliant interface with the load, low specific weight, no acoustic noise, and low cost. The following sub-sections present a couple of possible uses that we are currently exploring at different developmental stage.

Wearable haptic/tactile displays for virtual reality

We are using the HC-DE actuation technology to develop a wearable haptic/tactile display, intended to provide users with tactile feedback during electronic navigation in virtual environments. This work is part of a European research project called "CEEDS: The Collective Experience of Empathic Data Systems" (information available on the ICT research website of the European Commission). Since the project has just started, no data are available yet. Nevertheless, we describe here the considered design. The display consists of an HC-DEA arranged in contact with the finger tip, as shown in figure 8.

In particular, a bubble-like HC-DEA is integrated within a plastic case, so that the passive membrane is in contact with the finger, while the active membrane protected by a plastic chamber. The active membrane is electrically driven by a miniaturized (about one cube centimetre) high-voltage converter, which is equipped within the plastic case, as shown in figure 8. The converter is fed with a 0-5V signal to generate a 0-4 kV input for the actuator. The converter will safely work at low maximum electrical power (less than 1 W) and the user will never be exposed to high voltage parts of the device.

The overall system is aimed at providing the user with haptic/tactile feedback, controlled by a variety of inputs coming from a central controller which manages navigation in the virtual environment. The first goal to be achieved in the project is the creation of the sensation of contact with virtual objects. Furthermore, vibro-tactile stimulation will be implemented and studied, as a confirmation feedback upon virtual commands made by the user with fingers (e.g. pressure of keys on virtual control panels).

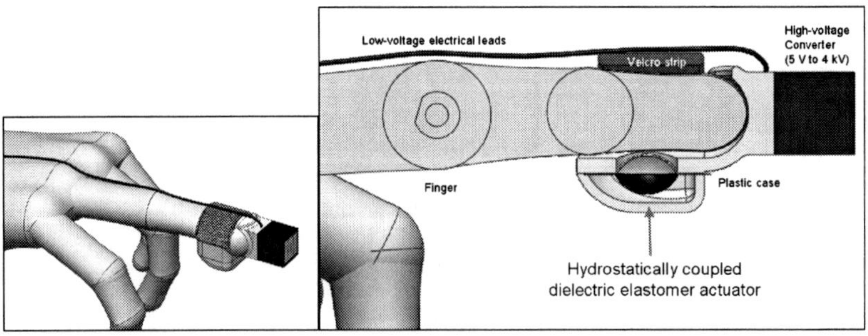

Figure 8. Design of a wearable device to provide a finger tip with haptic/tactile feedback, using an hydrostatically coupled dielectric elastomer actuator.

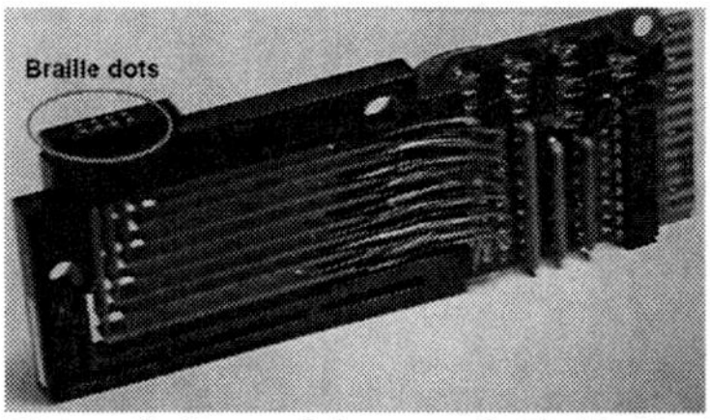

Figure 9. State-of-the-art refreshable Braille cell.

Refreshable Braille displays

The current reference technology for refreshable Braille displays is based on piezoceramic bimorph cantilever actuators (figure 9). At best, such displays can not be made with more than two lines, due to the considerable encumbrance of cantilevers (figure 9) which are necessary to amplify the small stroke of piezoceramics. Actually, assembling two lines of Braille cells requires putting two series of actuators nose-to-nose, with their cantilevers pointing away from the cells, laterally [14].

To reduce encumbrance and costs, new actuation technologies are being searched for next generations of refreshable Braille displays [14]. New actuators for Braille cells should comply with stringent safety issues and allow for high versatility in design, ease of miniaturization at low production costs, so as to enable cost-effective products with proper lay-out and high-density packaging.

We are currently working towards the development of refreshable Braille cells based on the HC-DE actuation concept. The basic idea is that an electrically controllable Braille dot can be based on the structure sketched in figure 10: a top passive DE membrane covers a plastic support which has an internal channel filled with an incompressible fluid, and is closed at the bottom by an active (electroded) DE membrane. The internal fluid is pressurized during manufacturing, so as to provide each membrane with the shape of a roughly spherical cap. The pressurized top membrane works as the Braille cell dot (passive interface with the user). The pressurized bottom membrane behaves as a buckling DE actuator. The latter buckles outwards as a voltage difference is applied between its electrodes, while the passive membrane passively follows

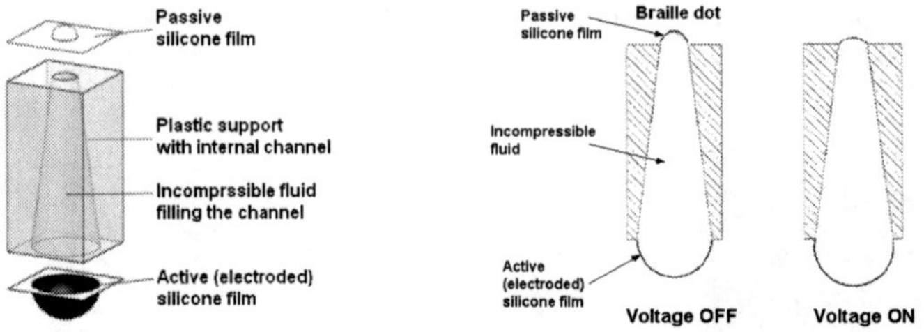

Figure 10. Exploded view (left) and lateral sectional view (right) of the structure of the proposed concept: Braille dot driven by an hydrostatically coupled dielectric elastomer actuator.

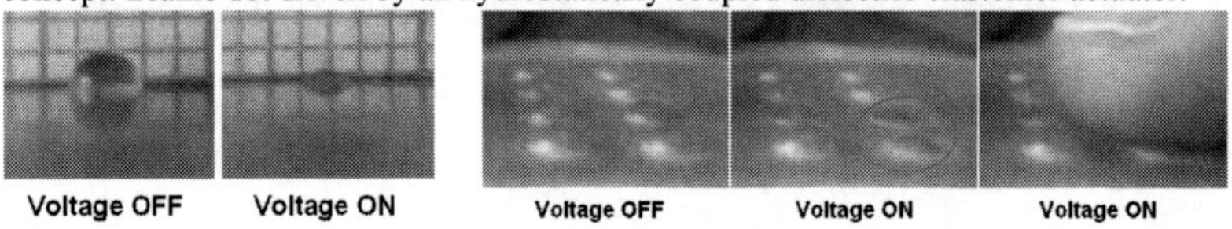

Figure 11. Proof-of-concept demonstrators: single Braille dot (left) and full Braille cell (right) driven by hydrostatically coupled dielectric elastomer actuators. Note: in these prototypes the dots and their spacing are oversized (up-scaled) with respect to Braille standards.

inwards, according to the fluid-enabled hydrostatic transmission (figure 10). Therefore, the cell dot is lowered or raised as a voltage is applied or removed, respectively. This basic structure can be replicated to assemble a standard 8-dots Braille cell.

Braille cells of such kind could provide enormous advantages over the state of the art, as they would be:

1) compact: the design versatility enabled by hydrostatic coupling favours a very compact packaging;
2) lightweight: the low density of the constitutive materials has no equals in the state of the art;
3) shock-tolerant: the constitutive materials are able to sustain high mechanical shocks;
4) portable: the above-mentioned properties favour portability;
5) acoustically silent: this property will make the device comfortable for the user;
6) cheap: the device will simply consist of two elastomer films, an intermediate fluid and structural plastic.

So far we have manufactured and tested preliminary prototypes. To achieve proof-of-concept demonstration more easily, the Braille dots and their spacing were oversized with respect to Braille standards. The dots had diameters of about 2 mm. Both single-dot and full-cell demonstrators were successfully driven by means of square-wave voltages, generating motion between two possible states: dot raised (voltage off) and dot lowered (voltage on), as shown in figure 11. Sighted volunteers tested the devices, reporting ability of discrimination of state between raised and lowered dots while touching but not looking at them.

CONCLUSIONS AND FUTURE DEVELOPMENTS

Bubble-like HC-DE actuators made of two pre-stretched membranes of acrylic elastomer coupled via silicone grease were manufactured and characterized. Two examples of applications currently under development, at different stage, were presented.

The picture emerging from preliminary results suggests that HC-DE actuators might be a promising new technology for a variety of haptic devices. These evidences open new perspectives to develop very compact, lightweight, low-cost, shock-tolerant, portable, silent and cheap haptic systems. No alternative actuation technology seems to offer today comparable combinations of performance with such features. This makes HC-DE actuators worthy of further investigations.

Such activities will significantly benefit of a fundamental improvement, needed by the whole field of DE actuation, in general, which is represented by new elastomer films that can be driven at lower voltages. This topic, which has not been addressed in this paper, is the greatest challenge in the field today. The possibility of ideally driving DE actuators at the same voltage range of piezoelectric actuators would dramatically reduce costs for electronics and, thus, would boost industrial exploitation.

To solve the problem of lowering the activation voltages, without reducing actuation performance, equation 1 clearly shows that either the film thickness d should be decreased, or the ratio ε_r/Y should be increased. The first approach is purely 'geometric' and deals with technological capabilities for thin-film processing of very soft elastomers (usually with elastic modulus of 0.1-1 MPa). The second approach deals with developing new materials able to intrinsically offer superior electromechanical properties. Increasing the dielectric constant, while maintaining the elastic modulus (as well as the dielectric breakdown strength) adequately limited, raises a fundamental scientific problem. While a number of encouraging achievements are being reported at present [15-21], much research has still to be done and solving the problem represents the greatest challenge for material scientists engaged in this compelling and emerging field of science and technology.

ACKNOWLEDGMENTS

Partial financial support for this work was provided by the European Commission, in the framework of the European research projects "CEEDS: The Collective Experience of Empathic Data Systems" (FP7-ICT-2009.8.4, Grant 258749) and "VIACTORS: Variable Impedance Actuation Systems Embodying Advanced Interaction Behaviours" (FP7-ICT-2007-3, Grant 231554).

REFERENCES

1. P. Brochu, Q. Pei, *Macromol Rapid Comm* **31(1)**, 10 (2010).
2. F. Carpi, *Polymer International*, **59(3)**, 277 (2010).
3. F. Carpi and E. Smela, Editors, *Biomedical applications of electroactive polymer actuators* (Wiley, 2009).
4. F. Carpi, D. De Rossi, R. Kornbluh, R. Pelrine and P. Sommer-Larsen, Editors, *Dielectric Elastomers as Electromechanical Transducers* (Elsevier, 2008).
5. T. Mirfakhrai et al., *Mater Today* **10(4)**, 30 (2007).
6. J. Madden et al., *IEEE J. Oceanic Eng* **29(3)**, 706 (2004).

7. Y. Bar-Cohen, Editor, *Electroactive Polymer (EAP) Actuators as Artificial Muscles* (SPIE, 2004).
8. F. Carpi, H.-E. Kiil, R. Kornbluh, P. Sommer-Larsen, G. Alici in *Proc. of ACTUATOR 2010*, edited by H. Borgmann, 12th International Conference on New Actuators, Bremen, Germany, 14–16 June 2010.
9. Online. "European Scientific Network for Artificial Muscles (ESNAM)": www.esnam.eu.
10. R. Pelrine, R. Kornbluh, Q. Pei and J. Joseph, **287**, 836 (2000).
11. J. Biggs, R. Hitchcock in *Proc. SPIE*, **7642**, 764201-1 (2010).
12. F. Carpi, G. Frediani and D. De Rossi, *IEEE/ASME T. Mechatronics*, **15(2)**, 308 (2010).
13. F. Carpi, G. Frediani, S. Tarantino, and D. De Rossi, *Polymer International*, **59(3)**, 407 (2010).
14. N. H. Runyan, *WW-EAP Newsletter*, **11(1)**, 9 (2009).
15. H. Stoyanov, M. Kollosche, D. N. McCarthy and G. Kofod, *Journal of Materials Chemistry*, **20**, 7558 (2010).
16. D.N. McCarthy, D. Rychkov, H. Ragusch, S. Risse, M. Melzer, G. Kofod, in *Proc. of ACTUATOR 2010*, edited by H. Borgmann, 12th International Conference on New Actuators, Bremen, Germany, 14–16 June 2010.
17. G. Gallone, F. Galantini and F. Carpi, *Polymer International*, **59(3)**, 400 (2010).
18. D. Opris, D. Crespy, C. Löwe, M. Molberg, F. Nüesch, *Proc. SPIE*, **7287**, 72870L (2009).
19. F. Carpi, G. Gallone, F. Galantini and D. De Rossi, *Advanced Functional Materials*, **18(2)**, 235 (2008).
20. F. Carpi, G. Gallone, F. Galantini and D. De Rossi, "Enhancing the dielectric permittivity of elastomers", In *Dielectric elastomers as electromechanical transducers.,* edited by F. Carpi, D. De Rossi, R. Kornbluh, R. Pelrine and P Sommer-Larsen, (Elsevier, 2008) pp. 51-68.
21. F. Carpi and D. De Rossi, *IEEE T. Dielectrics and Electrical Insulation*, **12(4)**, 835 (2005).

Mater. Res. Soc. Symp. Proc. Vol. 1312 © 2011 Materials Research Society
DOI: 10.1557/opl.2011.669

Piezoelectric Multimaterial Fibers

Noémie Chocat[1], Zheng Wang[2,3], Shunji Egusa[2], Zachary M. Ruff[1], Alexander M. Stolyarov[4], Dana Shemuly[1], Fabien Sorin[1,2], Peter T. Rakich[2], John D. Joannopoulos[2,3] and Yoel Fink[1,2]
[1]Department of Materials Science and Engineering, Massachusetts Institute of Technology, 77 Massachusetts Avenue, Cambridge, MA 02139, U.S.A.
[2]Research Laboratory of Electronics, Massachusetts Institute of Technology, 77 Massachusetts Avenue, Cambridge, MA 02139, U.S.A.
[3]Department of Physics, Massachusetts Institute of Technology, 77 Massachusetts Avenue, Cambridge, MA 02139, U.S.A.
[4]School of Engineering and Applied Sciences, Harvard University, 29 Oxford Street, Cambridge, MA 02138, U.S.A.

ABSTRACT

Here we report on the design, fabrication, and characterization of fiber containing an internal crystalline non-centrosymmetric phase enabling piezoelectric functionality over extended fiber lengths [1]. A ferroelectric polymer layer of 30 μm thickness is spatially confined and electrically contacted by internal viscous electrodes and encapsulated in an insulating polymer cladding hundreds of microns in diameter. The structure is thermally drawn in its entirety from a macroscopic preform, yielding tens of meters of piezoelectric fiber. Electric fields in excess of 50V/μm are applied through the internal electrodes to the ferroelectric layer leading to effective poling of the structure. To unequivocally establish that the internal copolymer layer is macroscopically poled we adopt a two-step approach. First, we show that the internal piezoelectric modulation indeed translates to a motion of the fiber's surface using a heterodyne optical vibrometer at kHz frequencies. Second, we proceed to an acoustic wave measurement at MHz frequencies: a water-immersion ultrasonic transducer is coupled to a fiber sample across a water tank, and frequency-domain characterizations are carried out using the fiber successively as an acoustic sensor and actuator. These measurements establish the broadband piezoelectric response and acoustic transduction capability of the fiber. The potential to modulate sophisticated optical devices is illustrated by constructing a single-fiber electrically-driven device containing a high-quality-factor Fabry-Perot optical resonator and a piezoelectric transducer.

INTRODUCTION

In recent years, a unique process has emerged that allows a multiplicity of solid materials with disparate electrical, optical, and mechanical properties to be arranged into a single fiber material. Multimaterial fibers have extended the responsivity and functionality of fibers from the traditional optical transmission domain to encompass optoelectronic properties. Applications such as fiber reflectors, thermal detectors, photodetectors, surface-emitting fiber lasers, and fiber diodes have recently been realized using this process [2]. However, multimaterial fibers similar to their traditional single-material fiber counterparts have been static devices, incapable of controllably changing their properties over a wide range of frequencies. Enabling electrically-modulated fibres could pave the way for unprecedented technological capabilities including sensitive pressure and flow measurements in capillary blood vessels, low-cost large-scale fabric-to-fabric communications, *in vivo* endovascular imaging and acoustic microscopy inside acoustically-opaque organs, and minimally-perturbative sparse sensor meshes for large-area studies of pressure/velocity fields in fluid flows. Previous approaches to realizing time-dependent variations in fibers focused on refractive index modulation [3-5], non-linear optical mechanisms in silica glass fibers [6-8], and electroactively modulated polymer fibers [9]. However, the inert nature of traditional glassy fiber materials has limited these approaches to short fiber lengths, simple geometries, and high driving fields. The aim of this study is to lift these limitations through the incorporation of a crystalline non-centrosymmetric phase in a fiber, enabling piezoelectric functionality over extended lengths.

EXPERIMENT

The piezoelectric fiber fabrication process starts with the construction of a preform, which is identical in its composition and geometry to the final fiber, but is thirty to sixty times larger in its cross-sectional dimensions and shorter in length. The preform is then thermally drawn into extended lengths of fiber using a tower draw procedure. To be compatible with this thermal draw process, materials must flow at a common temperature and have a high viscosity to prevent capillary break-up mechanisms. Poly(vinylidene fluoride) (PVDF) seems at first like a good choice for the piezoelectric material. It has a low melting temperature (160-170°C) that allows it to be co-drawn with plastics such as polycarbonate (PC), and the stress during the draw should in principle induce a transition from the non-polar alpha to the ferroelectric beta phase [10]. Assembling a preform made of a layer of PVDF with metal electrodes on either side in a parallel plate capacitor configuration and clad with insulating PC, and drawing it down into fibers appears like a reasonable approach. Upon careful examination, however, a number of challenges emerge. With such a design, the PVDF and metal layers would turn into three adjacent low-viscosity domains of high aspect ratio at the draw temperature, and would be prone to capillary break-up and flow instabilities. These could lead to layer thickness variations that would cause premature breakdown during the electrical poling step. Moreover, the stress and strain conditions necessary to induce the thermodynamic phase transition in PVDF cannot be sustained during the draw, leading to a non-polar phase.

To overcome these challenges, we select alternate materials with special properties. For the piezoelectric material, we choose a copolymer of PVDF and trifluoroethylene (P(VDF-TrFE), 70:30 molar ratio) that has been shown to spontaneously solidify from the melt into the beta phase [11]. For the electrodes, we need a material that is simultaneously conductive and viscous during the draw. We decide to use carbon-loaded polycarbonate (CPC), which exhibits both high viscosity (10^5-10^6 Pa.s) at the draw temperature and sufficient conductivity (1-10^4 Ω.m) to allow charge transport on length scales associated with the fiber cross-section (hundreds of micrometers). A thin filament of indium is used to ensure long-range electrical transport along the length of the fiber.

The preform is constructed by assembling the materials as shown in figure 1. A 700-μm thick layer of P(VDF-TrFE) is melt-pressed from pellets and is sandwiched between two 250-μm thick layers of CPC, each contacted by a filament of indium. A PC cladding surrounds the whole structure. The viscous state of the fiber draw enables the departure from classical round cross sections to ones of lower symmetry thus opening interesting opportunities for shaping acoustic wave fronts (figure 1). The preform is consolidated under vacuum at 185°C for 15 minutes to remove trapped air. Subsequently, it is drawn in a three-zone vertical tube furnace with a middle-zone temperature of 230°C. A laser micrometer and a tension meter are used to monitor the outer dimensions of the drawn fiber and the stress. The last step of the process is electrical poling. The metallic electrodes in the fiber are contacted to an external circuit and an electric field larger than 50 $MV.m^{-1}$ is applied to the copolymer layer.

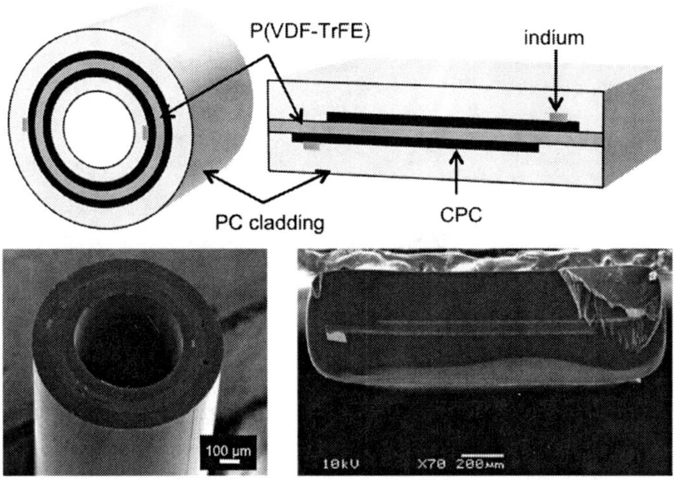

Figure 1. Schematics and SEM micrographs of piezoelectric fibers with round and rectangular cross-sections.

P(VDF-TrFE) samples harvested from the preform and the fiber were analyzed using a wide-angle X-ray diffractometer (Rigaku high power powder diffractometer). The XRD patterns showed that the polymer is in the desired ferroelectric beta phase after the thermal draw [1]. Fiber sample cross-sections were polished using an ion-beam polisher and imaged with a JEOL 6060 Scanning Electron Microscope (SEM). The SEM micrographs show that the structure of the fiber is well conserved during the draw (figure 1).

DISCUSSION

To unequivocally establish the piezoelectric nature over a wide range of frequencies, segments of piezoelectric fiber were tested using two measurement techniques. This section details these characterization measurements.

Measurement using a laser Doppler vibrometer

To demonstrate piezoelectricity in the kHz range, a heterodyne optical vibrometer was used that characterized the speed and frequency of the fiber's surface vibration. The fiber sample was electrically driven by a sine wave of amplitude 10V and frequency ω_D. The fiber surface served as a vibrating reflector and Doppler-shifted the frequency of the reflected light, yielding a frequency-modulated signal. In the Fourier domain, this translates into side bands whose spacing corresponds to the modulation frequency ω_D and whose amplitude is proportional to the velocity amplitude of the vibration.

Figure 2a shows the observed side bands for the cylindrical fiber of figure 1. The presence of these side bands demonstrated a macroscopic piezoelectric response from the fiber. Figure 2b shows the same side bands for a rectangular fiber. The flat fiber surface has a more directional reflectivity, thus leading to an increase in the signal-to-noise ratio. We observe both a 20 dB increase in the side-band amplitude with respect to the heterodyne carrier, and an increase in the side-band signal compared with its immediate background.

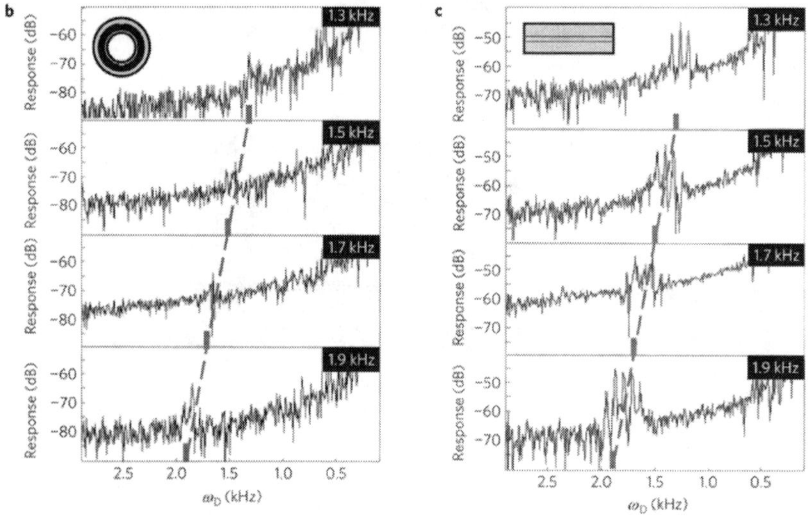

Figure 2. Measured frequency-modulation side-bands for piezoelectric fibers with cylindrical (a) and rectangular (b) cross-sections.

Direct acoustic measurement in a water-tank

To characterize the piezoelectric response in the MHz range, a direct acoustic measurement was performed in a water-tank (figure 3a). A water-immersion ultrasonic transducer (Olympus Panametrics-NDT 1.0MHz-centered) was coupled to a fiber sample across a water tank, and frequency-domain characterizations were carried out using the fiber successively as an acoustic sensor and actuator. To filter out the electromagnetic interference, which can be significant at MHz frequencies, the measurement used a pulsed excitation and time-gated the received signals.

Frequency domain characterizations of the rectangular fiber are presented in figure 3b. The results show an excellent agreement between the spectral features in the fiber-to-transducer and the transducer-to-fiber configurations. The fiber's piezoelectric response essentially follows the intrinsic frequency profile of the ceramic transducer. The finer spectral features are due to the shape of the water tank and are also present when performing the same measurement with a commercial film transducer.

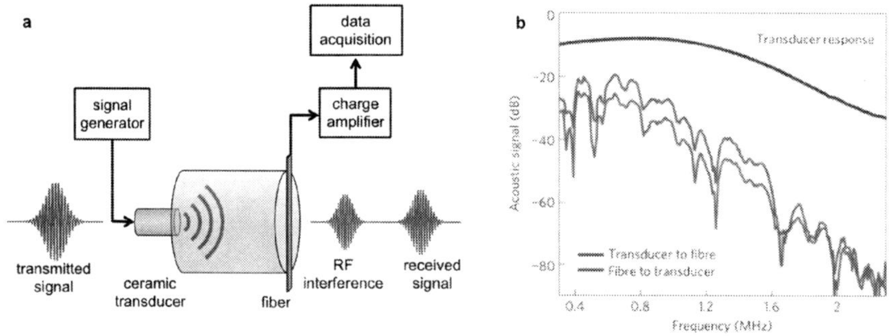

Figure 3. a) Schematic of the experimental set-up for the direct acoustic characterization of piezoelectric fibers (shown in the transducer-to-fiber configuration). b) Acoustic signal detected (top curve) and emitted (bottom curve) by the piezoelectric fiber.

<u>Integration with a Fabry-Pérot cavity</u>

One of the main interests of the thermal drawing approach to fabricating fibers is the possibility to achieve intra-fiber integration. One could in principle fabricate a preform that comprises many devices, for example a thermal-sensing device, a photo-detecting device, and a piezoelectric device, and draw it down into a fiber that would be capable of multiple functionalities (figure 4a). To illustrate this potential, we fabricated a fiber containing a piezoelectric transducer and a high quality Fabry-Perot optical resonator (figure 4b). The SEM image shows that the structure is well conserved during the draw. Acoustic measurements and Fourier transform infrared microscopy revealed that both the piezoelectric transducer and the optical cavity are functional, making this a truly multifunctional fiber. In the future, one could imagine fibers in which the piezoelectric element would be able to modulate an adjacent optical device.

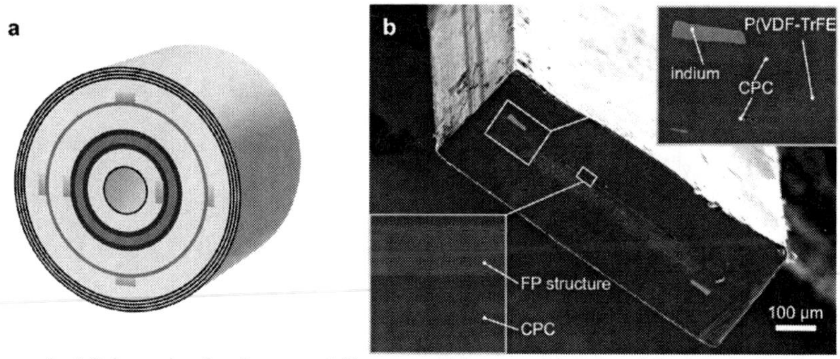

Figure 4. a) Schematic of an integrated fiber comprising a multiplicity of devices. b) SEM micrograph of an actual integrated Fabry-Perot piezoelectric fiber.

CONCLUSIONS

A piezoelectric fiber composed of a P(VDF-TrFE) active layer, CPC polymer electrodes, indium filaments, and a PC cladding, was fabricated using a thermal drawing method. SEM micrographs show that this materials system allows excellent conservation of the fiber structure during the draw. The fiber was characterized using a laser Doppler vibrometer and an acoustic transmission set-up in a water tank, which demonstrated the fiber's broadband piezoelectric response and acoustic transduction capability. An integrated piezoelectric Fabry-Perot fiber was constructed that illustrates the potential of this approach to achieve more sophisticated fibers.

ACKNOWLEDGMENTS

The authors acknowledge A. F. Abouraddy, G. Benoit, M. Spencer, J. Rigling, M. Thompson, S. Griggs, and J. F. Viens for their critical help and for discussions; S. A. Speakman for assistance with the XRD measurements; and E. L. Thomas for guidance. This work was supported in part by the Materials Research Science and Engineering Program of the US National Science Foundation under award number DMR-0819762, DARPA/Griggs and also in part by the US Army Research Office through the Institute for Soldier Nanotechnology under contract number W911NF-07-D-0004.

REFERENCES

1. Egusa, S., Wang, Z., Chocat, N., Ruff, Z.M., Stolyarov, A.M., Shemuly, D., Sorin, F., Rakich, P.T., Joannopoulos, J.D., & Fink, Y., *Nature Materials* **9** (8), 643-648 (2010).
2. Abouraddy, A.F., Bayindir, M., Benoit, G., Hart, S.D., Kuriki, K., Orf, N., Shapira, O., Sorin, F., Temelkuran, B., & Fink, Y., *Nature Materials* **6** (5), 336-347 (2007).
3. Larsen, T.T., Bjarklev, A., Hermann, D.S., & Broeng, J., *Optics Express* **11** (20), 2589-2596 (2003).
4. Benoit, G., Kuriki, K., Viens, J.F., Joannopoulos, J.D., & Fink, Y., *Optics Letters* **30** (13), 1620-1622 (2005).
5. Kerbage, C., Hale, A., Yablon, A., Windeler, R.S., & Eggleton, B.J., *Applied Physics Letters* **79** (19), 3191-3193 (2001).
6. Bergot, M.V., Farries, M.C., Fermann, M.E., Li, L., Poyntzwright, L.J., Russell, P.S.J., & Smithson, A., *Optics Letters* **13** (7), 592-594 (1988).
7. Townsend, P.D., Poustie, A.J., Hardman, P.J., & Blow, K.J., *Optics Letters* **21** (5), 333-335 (1996).
8. Fokine, M., Nilsson, L.E., Claesson, A., Berlemont, D., Kjellberg, L., Krummenacher, L., & Margulis, W., *Optics Letters* **27** (18), 1643-1645 (2002).
9. Carpi, F., & De Rossi, D., *Ieee Transactions on Information Technology in Biomedicine* **9** (3), 295-318 (2005).
10. Matsushige, K., Nagata, K., Imada, S., & Takemura, T., *Polymer* **21** (12), 1391-1397 (1980).
11. Yagi, T., Tatemoto, M., & Sako, J., *Polymer Journal* **12** (4), 209-223 (1980).

Mater. Res. Soc. Symp. Proc. Vol. 1312 © 2011 Materials Research Society
DOI: 10.1557/opl.2011.900

Electromechanical Response of Multilayered Polymer Films for High Energy Density Capacitors.

Mason A. Wolak,[1] James S. Shirk,[1] Matt Mackey,[2] Joel Carr,[2] Ann Hiltner,[2] and Eric Baer[2]

[1]Optical Sciences Division, U.S. Naval Research Laboratory, Washington, DC 20375. USA.
[2]Department of Macromolecular Science and Center for Applied Polymer Research, Case Western Reserve University, Cleveland, OH 44106-7202. USA.

ABSTRACT

Multilayered films comprising alternating layers of polycarbonate (PC) and poly(vinylidene fluoride-hexafluoropropylene) (P[VDF-HFP]) show an enhanced dielectric strength (E_B> 750 kV/mm) and an increased energy storage density ($U_d \sim$ 13.5 J/cm^3) compared to monolithic PC and P[VDF-HFP] films. Here the role of electromechanical effects in the breakdown of multilayer films is explored both by imaging the changes in the layer structure caused by electrical fields below the breakdown field and by a direct measurement of the strain in multilayer PC/ P[VDF-HFP] films subjected to similar fields. Focused Ion Beam (FIB)/ Scanning Electron Microscopy (SEM) images of the layer structure in films subjected to repeated cycles at near-breakdown fields showed local changes in the thickness of individual layers, suggesting that mechanical forces arising from field-induced compression may play a role in the steps preceding the breakdown. The directly measured field induced strain showed evidence for both an elastic and a flow component to the strain. The mechanical responses of films with $\leq$ 50 vol% P[VDF-HFP] were modeled as simply the sum of an elastic and viscous flow. The observed electromechanical properties vary with the layer structure. This suggests that multilayering polymers may provide a means to mitigate deleterious electromechanical effects in low modulus, high dielectric materials.

INTRODUCTION

It was recently reported that multilayer films comprising alternating layers of polycarbonate (PC) and poly[vinylidene fluoride-hexafluoropropylene] (P[VDF-HFP]) exhibit superior dielectric properties including high dielectric strength, low dielectric loss, and high deliverable energy density.[1] Films containing 256 layers of 50 vol% PC and 50 vol% P[VDF-HFP] have a dielectric strength, E_B, of > 750 kV/mm and an average maximum energy density, U_d, of 13.5 J/cm^3. The measured energy density represents a 100% increase relative to parallel plate capacitors fabricated from PC control films, and a 60% increase relative to similar devices fabricated from monolithic P[VDF-HFP] control films.[2] The enhanced dielectric strength (relative to single component films of PC and P[VDF-HFP]) appears to originate from a barrier effect,[3] whereby the interfaces between layers serve to impede the propagation of a breakdown tree through the film in the direction of an applied field. We have reported on the use of a Focused Ion Beam (FIB)/ Scanning Electron Microscopy (SEM) technique to image the layer structure both before and after the application of a field resulting in electrical breakdown.[4] The FIB/SEM technique revealed that the breakdown processes in layered composite films differ from those in single component films. In layered films, voids were observed extending 30 μm or more from the breakdown site along layer interfaces, resulting in considerable layer delamination. In single component films such extended features were not observed. The images are consistent with the idea that layer boundaries provide barriers that deflect the propagation of an electrical breakdown across a multilayer polymer.

The present report describes initial studies designed to explore the importance of electromechanical effects in the breakdown of multilayer films. The studies include imaging changes in the layer structure caused by electrical fields just below the breakdown field and by measuring the mechanical response in multilayer PC/ P[VDF-HFP] films subjected to similar fields. Recent studies of PVDF derivatives suggest that the breakdown mechanism is influenced by electrically-induced mechanical stress.[5] In this study we examine such effects in multilayer films. In the pre-breakdown imaging studies, multilayer films were subjected to repeated charge/discharge cycles at fields of roughly 70 to 80% of the measured dielectric strength. The applied field was sufficient to cause some electrical stress but not enough to cause a catastrophic breakdown. Following electrical cycling, the films are sectioned with the FIB and imaged with SEM to reveal any changes to the layer structure. The goal is to correlate changes in the layer structure with the applied mechanical stresses.

EXPERIMENT

Film Fabrication

32- and 256-layer films containing alternating layers with varying volume fractions of PC (Makrolon 2207) and P[VDF-HFP] (Solef 21508 $[-(CH_2\text{-}CF_2)_{85}\text{-}(CFCF_3CF_2)_{15}\text{-}]$) were prepared using the melt coextrusion method previously reported.[6] Single component (homogeneous) films were also prepared by the same melt coextrusion method. Typical sample thicknesses were approximately 13±1 μm as measured with a Measure-It-All LE1000-2 thickness gauge.

Pre-Breakdown Imaging

First, top electrodes (25 nm Au) were sputter coated on the samples through a shadow mask (d=100 to 200 μm) using a Bal-Tec SD-005 sputter coater. The films were then placed on a ground plane electrode (brass panel), submersed in Fluorinert FC-40, and electrical contact was made to the top electrode with a needle probe (tip d=40 μm). Samples were cycled 400 times at a frequency of 10 Hz, using a Trek 610D amplifier as the voltage source. A single electrical cycle was defined as application of a field up to 0.7^*E_B kV/mm, followed by reversal of the polarity to -0.7^*E_B kV/mm, followed by final release to zero field. After electrical cycling, a thin 5 nm Al contact layer was applied to the surface using a Temescal E-beam Evaporator. The contact layer was necessary for SEM imaging, and is sufficiently thin that the 25 nm thick Au electrodes can be observed through it. The film volume directly below the Au electrodes was then milled and sequentially imaged using an FIB/SEM technique that has been described in some detail.[4]

Measurement of Electromechanical Strain

The strain vs. applied field was measured for both 32- and 256-layer films at different PC/P[VDF-HFP] compositions. E-beam evaporation was used to deposit Al electrodes (d=4.7 mm) through shadow masks on both film surfaces. The sample was then placed between two spherical electrodes (d=6 mm) connected to a Trek 610D amplifier. A micrometer was used to adjust the sphere positions so that they made contact to the film without compressing it.

The displacement of the sample was monitored by a linear variable differential transformer (LVDT), interfaced to a Stanford Research SR830 DSP lock-in amplifier.[7] When the thickness of the film changes as a result of an applied field, a voltage is induced in the LVDT. The lock-in amplifier converts the AC signal from the LVDT to DC while filtering random fluctuations of a different time constant (1 ms). Labview was used to collect the DC voltage and convert it into the displacement using a scaling factor (1V = 1000 nm) derived from the time constant (1 ms) of the lock-in amplifier. The applied voltage was sinusoidal with a period of 3 seconds. The maximum of the sinusoidal field was increased in steps of 50 kV/mm until breakdown was achieved. Samples were cycled 3 times at each voltage step. The measured strains (fractional change in thickness) as a function of the variation in the applied field presented here are the averages of those measured for 6 film samples.

RESULTS AND DISCUSSION

Pre-Breakdown Imaging

Figure 1 shows a cross-section of a 32 layer control film that has not been subjected to high field and two different 32 layer films that have been subjected to 'pre-breakdown' electrical stress. The layer structure of the control film is uniform with discrete continuous layers (Figure 1A). The layer structure of both films subjected to high field is distorted. Regions with a periodic buckling are seen in the 10% PC / 90% P[VDF-HFP] film in Figure 1B. Changes in layer thickness are also seen in the 50% PC / 50% P[VDF-HFP] film in Figure 1C. In addition, small voids (d<500 nm) were found at the interfaces between layers in a small fraction (<3%) of the cross sectional images of the films subjected to pre-breakdown fields. The pre-breakdown images show changes to the layer structure that are similar to, but smaller than the deformations of the layer structure and voids observed in similar films after a complete electrical breakdown. It is reasonable that these features are precursors to the larger scale changes reported in a FIB/SEM study of films driven to electrical breakdown.[4]

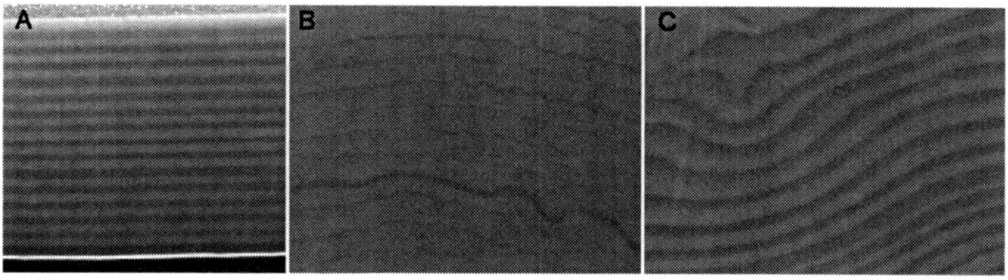

Figure 1A. SEM micrograph of the layer structure of a 32-layer 50 vol% PC / 50 vol% P[VDF-HFP] control film that has not been subjected to high field. Lighter layers are P[VDF-HFP] and darker layers are PC. **1B and 1C** are SEM micrographs showing the changes to the layer structures of 32-layer 10% PC/90% P[VDF-HFP] (**1B**) and 50% PC/50 % P[VDF-HFP] (**1C**) films after 400 successive charge/discharge cycles. The maximum applied field in the 10/90 film was 320 kV/mm, for the 50/50 film it was 600 kV/mm.

The local changes in P[VDF-HFP] layer thickness are consistent with material flow. The periodic variations are reminiscent of the buckling known to occur as a result of mechanical stress in two-component systems whose component materials have disparate moduli.[8] This apparent flow, seen in both breakdown and pre-breakdown images, may play a role in the breakdown mechanism and the dielectric strength of the composites. The images motivated us to quantify the electromechanical response of the composites, with an emphasis on assessing how the layer structure influences strain under high field.

Electromechanical Properties

Some typical graphs of the measured thickness strain vs. applied field are shown in Figure 2. Compressive strain under high fields is easily observed. The magnitude of the strain depends on the composition of the film. Figure 2A shows the result for 32-layer and 256 layer 50% PC / 50% P[VDF-HFP] films. They show a maximum strain between -0.4% to -0.5% at fields of E_{max}=±500 kV/mm. The observed strain at a given field increases with increasing P[VDF-HFP] content. The strains in the 70% PC / 30% P[VDF-HFP] films in Figure 2B are generally smaller. For pure P[VDF-HFP] control films, (not shown) the maximum strain is 2.8% at E_{max}=±400 kV/mm.

The strain vs. field plots show a hysteresis and an offset between the initial film thickness and the final film thickness after a full electrical cycle (from E=0 to $\pm E_{max}$, back to E=0). The offset can be interpreted as arising from a viscoelastic and/or a flow component, where there is some flow in a direction perpendicular to the applied field. The plots in Figure 2 are consistent with both an elastic and a flow component to the mechanical response on this time scale. The magnitude of the offset increases as the vol% of P[VDF-HFP] increases, indicating that it is probably responsible for much of the flow.

An estimate of the effect of the layer structure on the elastic and flow components can be made by fitting the observed data to a simple model. In equation 1, the strain is modeled as the sum of the compressive elastic (first term) and a flow response (second term).

$$strain(V,t,\phi) = c_1\left(V_{\max}\sin\left(\frac{2\pi}{t_0}+\phi\right)\right)^2 + c_2\int_0^t\left(V_{\max}\sin\left(\frac{2\pi}{t_0}+\phi\right)\right)^2 dt \qquad (1)$$

Where V is applied field, t is time, ϕ is the phase delay between the applied voltage and the strain, c_1 is the coefficient of the elastic response, and c_2 is the coefficient of the viscous flow response.

This simple model appears to describe the electromechanical response of films containing 50 vol% P[VDF-HFP] or less. However, the model is not accurate at higher P[VDF-HFP] concentrations perhaps because it does not include electrostrictive and piezoelectric effects that may be more important as the P[VDF-HFP] concentration increases. Efforts to model films with higher P[VDF-HFP] concentration are underway.

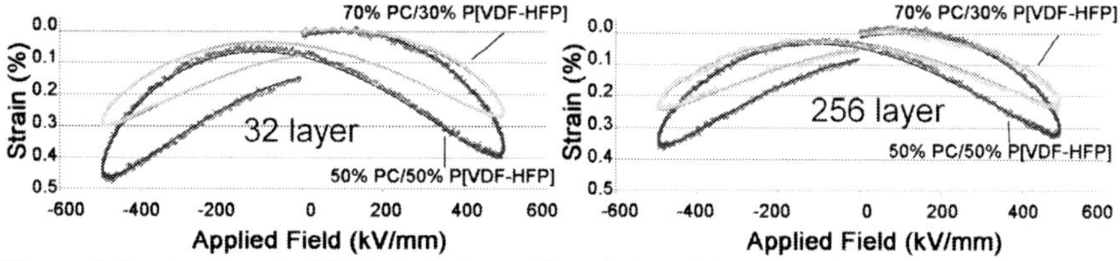

Figure 2. Strain vs. applied field for 32-layer films (left) and 256-layer films (right). The PC/P[VDF-HFP] volume concentrations are indicated. Solid lines are fits to Equation 1.

Table I displays the relative coefficients of the elastic and viscous flow response for 32- and 256-layer films of both 50% PC / 50% P[VDF-HFP] and 70% PC / 30% P[VDF-HFP] films. The c_1 coefficient is related to the modulus and the c_2 coefficient is related to the viscosity. For a given number of layers, both the apparent modulus and viscosity increase as the P[VDF-HFP] content is decreased, consistent with the fact that P[VDF-HFP] has a low bulk modulus (480 MPa), roughly 1/5th that of PC.[9]

	50/50	50/50	70/30	70/30
	32 Layer	256 Layer	32 Layer	256 Layer
c_1 (modulus)	1.00	1.19	1.43	1.67
c_2 (viscosity)	1.00	1.88	1.88	2.96

Table I. Coefficients of the elastic (c_1) and viscous flow (c_2) response for both 32- and 256-layer 50/50 and 70/30 films.

Elastic Response of P[VDF-HFP]

The strain as a function of applied field was also measured for pure P[VDF-HFP] control samples. The magnitude of the elastic component of the strain, estimated by subtracting the apparent flow response, was approximately ~2.8% at a field of E_{max} = 400 kV/mm. The estimated force exerted on P[VDF-HFP] control films due to Coulomb attraction between the top and bottom Al electrodes is approximately 150 N as calculated from the Eq. 2.

$$F = \frac{\varepsilon_0 \varepsilon_r A V^2}{2t^2} \qquad (2)$$

Where ε_0 is vacuum permittivity, ε_r is relative permittivity (12), A is the area of the electrodes (.172 cm^2), V is the applied voltage (5200 V), and t is the film thickness (13 μm). A force of 150 N yields a stress of 8.5 MPa (Stress = F/A), which predicts an elastic compression of 1.8%. The measured strain at a field of E_{max} = 400 kV/mm (2.8%) is higher than the predicted strain. The estimated Coulomb compression accounts for a major portion of the strain, but may not account for all of it. It is possible that electrostriction and/or piezoelectric effects may also contribute.

CONCLUSIONS

FIB/SEM images of cross sections of multilayered PC/P[VDF-HFP] films exposed to electric fields just short of a breakdown show layer deformations and local changes in the P[VDF-HFP] layer thicknesses. These features are consistent with mechanical deformations at these fields. The electromechanical response of multilayered films, measured directly, shows a hysteresis and an offset consistent with both an elastic and a viscoelastic and/or flow component to the strain. The mechanical response of films with ≤ 50 vol% P[VDF-HFP] was modeled as simply the sum of an elastic and viscous flow. The apparent modulus and viscosity of these composites varies with the layer structure; both increased with decreasing the P[VDF-HFP] content and with decreasing the layer thickness. While a significant component of the response is due to Coulomb attraction between the top and bottom electrodes, it is possible that piezoelectric or electrostrictive effects may also play a role.

The results here, while mostly qualitative, imply that multilayering polymers with differing mechanical properties can provide a means to mitigate deleterious electromechanical effects in low modulus, high dielectric materials.

ACKNOWLEDGEMENTS

The authors wish to thank the Office of Naval Research for funding this work.

REFERENCES

1 M.A. Wolak, M.-J. Pan, A. Wan et al., Applied Physics Letters 92 (11), 113301 (**2008**); M. Mackey, A. Hiltner, E. Baer et al., J. Phys. D: Appl. Phys. 42, 175304 (**2009**).

2 The breakdown strength of homogeneous P[VDF-HFP] films in parallel plate capacitors with Al electrodes (E_B = 400 kV/mm) was substantially lower than that of non-metalized native films (E_B = 575 kV/mm). The inability to drive these devices to high field, coupled with considerable polarization hysteresis, limited their measured energy density to U_d = 8 J/cm^3.

3 R. Vogelsang, T. Farr, and K. Frohlich, IEEE Trans. Dielect. El. Insul. 13 (1), 373 (**2006**); S.M. Lebedev, O.S. Gefle, and Y.P. Pokholkov, IEEE Trans. Dielect. El. Insul. 12 (3), 537 (**2005**).

4 M.A. Wolak, A.S. Wan, J.S. Shirk et al., J. of Appl. Polym. Sci. (in press).

5 J. Claude, Y. Y. Lu, K. Li et al., Chem. Mater. 20 (6), 2078 (**2008**).

6 M. Ma, K. Vijayan, A. Hiltner et al., J. Mater. Sci. 25, 2039 (**1990**).

7 D. Viehland and Y.-H. Chen, J. Appl. Phys. 88 (11), 6696 (**2000**).

8 S. Faulhaber, C. Mercer, M. W. Moon et al., J. Mech. Phys. Solids 54 (5), 1004 (**2006**).

9 Solef 21508 and Makrolon 2207 datasheets.

Mater. Res. Soc. Symp. Proc. Vol. 1312 © 2011 Materials Research Society
DOI: 10.1557/opl.2011.901

Influence of Process Condition on the Dielectric Properties of CCTO-P(VDF-TrFE) 0-3 Composites

Xiaobing Shan, Lin Zhang and Z.-Y. Cheng
Materials Research and Education Center
Auburn University, Auburn, AL 36849

Abstract

A high dielectric constant was observed in 0-3 composites using P(VDF-TrFE) copolymer as matrix and $CaCu_3TiO_4$ (CCTO) ceramic powders as fillers with a composition of 50 vol% CCTO. CCTO-P(VDF-TrFE) 0-3 composites with a CCTO content from 0 to 50 vol% were prepared and their dielectric properties, including frequency dependence and temperature dependence of the dielectric constant and loss, were characterized. It was experimentally observed that the hot-press process, inlcuding the configuration and hot-press time, has a strong influence on the dielectric properties of the composites. Additionally, the influence of the CCTO partice size on the dielectric properties of the composites was studied.

Introduction

Dielectric materials with a high dielectric constant and high dielectric strength (electric breakdown field) are highly desirable for use in miniaturizing electronic circuits and components, as well as the development of capacitors with high energy storage density. From an application point of view, flexible dielectric materials are very desirable for device fabrication and dielectric materials with a low process temperature are needed to integrate electronic components. In general, a high dielectric constant is observed in brittle inorganic materials and high dielectric strength is obtained in flexible polymers. Therefore, flexible ceramic-polymer 0-3 composites are interesting for the development of flexible dielectric materials with a high dielectric constant and strength. This has been a hot topic over last three decades.

For the development of ceramic-polymer 0-3 composites with a high dielectric constant, the ceramics used are usually ferroelectric-based materials, such as relaxor ferroelectrics, due to the fact that these materials exhibit a very high dielectric constant [1,2]. Unfortunately, the ferroelectric-based materials exhibit a strong electromechanical effect, which is desirable for electromechanical devices, such as transducers and actuators, but not desirable for dielectric applications, such as capacitors. Additionally, the dielectric constant of the ferroelectric-based materials exhibits a strong dependence on temperature [1]. Regarding the polymers used in the development of composites, a polymer with a high dielectric constant is highly desirable. PVDF-based polymers, which exhibit a dielectric constant of about 10 that is much higher than ~5 obtained in most of the polymers, are widely used [2].

In the last decade, ceramics with giant dielectric constants and weak electromechanical coupling effects have been developed. For example, $CaCu_3TiO_4$ (CCTO), which does not have a piezoelectric effect, exhibits a giant dielectric constant, $10^4 \sim 10^5$ [3-4]. It was also been found that the dielectric constant of the CCTO is almost independent of temperature over a broad temperature range, including room temperature.

Therefore, the discovery of the CCTO provides a unique opportunity for developing high-dielectric-constant ceramic-polymer 0-3 composites [5]. We reported that the dielectric constant obtained in CCTO-P(VDF-TrFE) composites with 50 vol% CCTO can reach more than 1000 [5,6]. After that, a great deal of research has been conducted on the development of composites using CCTO as filler [7-10]. In this paper, the dielectric properties of CCTO-P(VDF-TrFE) 0-3 composites with different compositions and process conditions are reported.

Experiments

The CCTO ceramics were prepared using the conventional powder processing method. High purity metal oxide powders of calcium carbonate ($CaCO_3$, 99.5 %), copper oxide (CuO, 99.7 %), and titanium dioxide (TiO_2, 99.8 %), all from Alfa Aesar, were used. The mixture was calcined at 900~1075 °C in an Al_2O_3 crucible, then the ceramic powders were sintered at 1075 °C. Two types of CCTO powders were used: micro-powders with an average size of ~ 5 μm, prepared using a regular ball miller and nano-powers with an average size of ~0.8 μm, prepared using a high-energy miller [11].

P(VDF-TrFE) 55/45 mol% copolymer was used as the polymer matrix. This copolymer was selected due to the fact that it exhibits a high dielectric constant and a weak piezoelectric effect. The copolymer was first dissolved in dimethyl formamide (DMF) to make the solution. Then, the CCTO powders were added into the solution, stirred for 8 hours and followed by ultrasonic treatment for about 20 mins. The final CCTO-P(VDF-TrFE) suspension/solution was cast onto a glass plate at 70 °C for 8 hours to form the as-cast composite film.

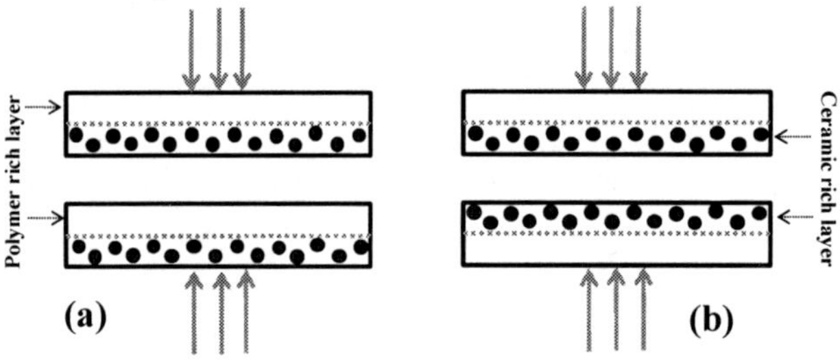

Figure 1. Two configurations of the stack of the as-cast composite film used in HP process: (a) in which the polymer-rich side faces the non-polymer-rich side (or ceramic rich side) of the as-cast composite film; (b) in which the polymer-rich side faces the polymer-rich side (i.e. ceramic-rich side faces the cramic-rich side)

To improve the uniformity of the composites, a hot press (HP) technique, in which the as-cast composite films were stacked together and then hot-pressed at 200 °C to form the final composition, was used. As described in Ref. [5], there is a polymer-rich layer on the bottom of the as-cast composite film. During the HP process, the as-cast composite films were stacked with two configurations: the polymer-rich side of the as-cast film facing the non polymer-rich side (or ceramic-rich side) as shown in **Figure 1(a)** and the polymer-rich side of the as-cast film facing the polymer-rich side of the as-cast film as shown in **Figure 1(b)**. The CCTO-P(VDF-TrFE) composites with different concentrations, 0 to 50 vol.% of CCTO, were prepared. The dielectric properties of the composites were characterized. In the experiments, gold was sputtered on both surfaces of the composite samples to serve as the electrodes. The impedance of the samples at frequencies from 100 Hz to 1 MHz was measured using an Agilent Impedance Analyser (4291A). The dielectric constant of the composites was calculated from the impedance data using a parallel plate capacitor model.

Results and Discussion

For the CCTO-P(VDF-TrFE) composites with CCTO powders in micrometers, it was found that the wettability between the CCTO and P(VDF-TrFE) is poor, which results in a non-uniform morphology in the as-cast composites. In order to solve this problem, an HP process has been introduced [5]. As shown in **Figure 2** and **Figure 3**, when more layers were used in the HP process, the dielectric constant obtained in the final composites became higher. For the composites shown in **Figure 2**, the configuration of the as-cast composite film used in the HP process is a configuration shown in **Figure 1(a)**. When the number of layers reaches to six, the dielectric constant is about to reach its saturated value. Additionally, it was experimentally found that the configuration of the as-cast composite film stacks used in the HP process has a strong influence on the dielectric properties of the final composites. It was found that the as-cast composite film stacks with a configuration shown in **Figure 1(b)** result in a higher dielectric constant.

For CCTO-P(VDF-TrFE) composites with 50 vol% CCTO, a dielectric constant of about 500 [5,6] is obtained in the composites at room temperature, as shown in **Figure 3**. This is much higher than the dielectric constant observed in other 0-3 composites. For this composite, the configuration of the as-cast composite film stack used in the HP process is shown in **Figure 1(b)**. When the samples were heated to about 90 °C, the dielectric constant was much higher (>1000) due to the high dielectric constant of the polymer.

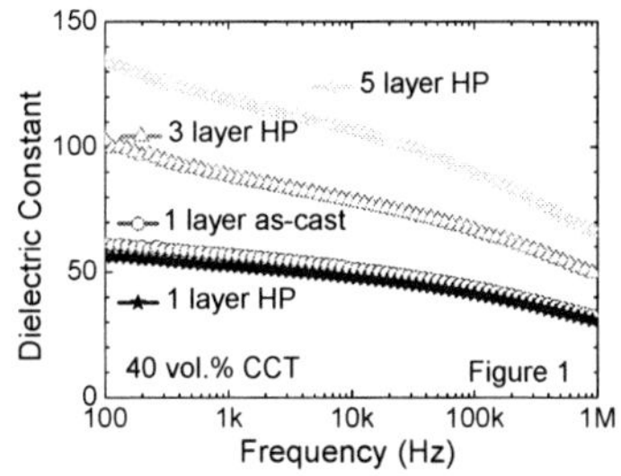

Figure 2. Dielectric constant of 40 vol.% CCTO composites: as-cast, and after hot-press (HP) of different (1, 3, and 5) layers together.

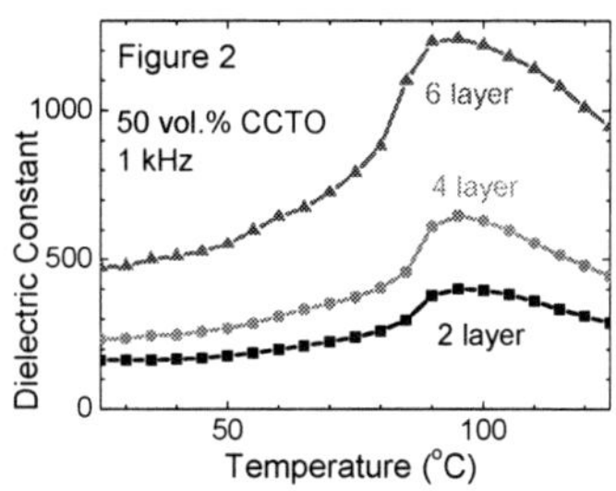

Figure 3. Dielectric constant at 1 kHz of 50 vol.% CCTO composites versus temperatures: for configuration in which the ceramic-rich layer faces the ceramic-rich layer.

It is also found that the HP process time has a strong influence on the dielectric response observed in the final composites, as shown in **Figure 4**, where the HP time is 10 seconds and 30 seconds, respectively. For the composites shown in **Figure 4**, the configuration shown in **Figure 1(b)** was used to form the stack of the as-cast composite film for the HP process.

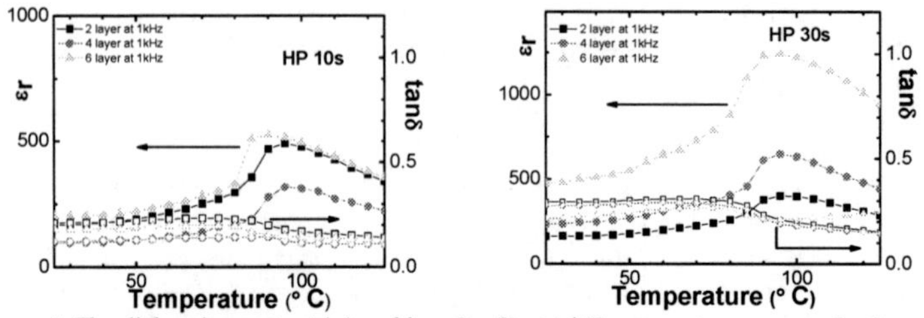

Figure 4. The dielectric constant (ε_r) and loss (tanδ) at 1 kHz versus temperature for the composites with 50 vol% CCTO. During the HP process, different numbers of as-cast composite film were used: 2 layers, 4 layers, and 6 layers. On the left, the data were from the samples with an HP time of 10 seconds, while on the right an HP time of 30 seconds.

Regarding the composition dependence of the dielectric constant, the dielectric constant at 1 kHz for composites at room temperature is presented in **Figure 5**, where the dielectric constant is the highest among different HP conditions used for composites with the same composition.

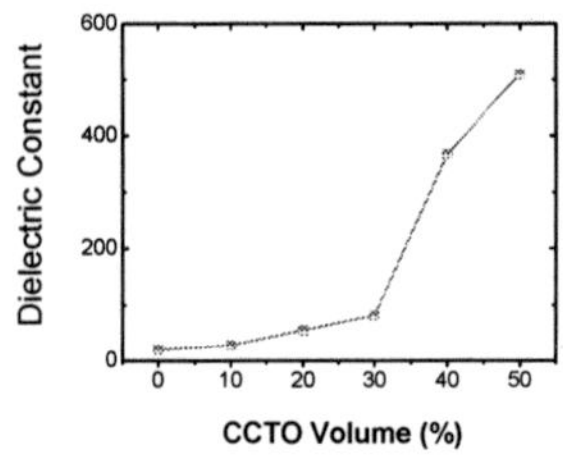

Figure 5. The dielectric constant at 1kHz for the composites at room temperature versus the composition of the composites.
For each composition, the composites were prepared using different HP condition, including configuration of the stack used in the HP and HP times. The dielectric constant reported here is the highest among all dielectric constants obtained from the composites with the same composition and the HP process, but different HP times.

The results shown in **Figure 6** are the frequency dependence of the dielectric properties, dielectric constant and loss, of the composites with 40 vol% CCTO at room temperature, where three samples prepared under the same process condition were used. The results indicate that the composites reported are relatively uniform.

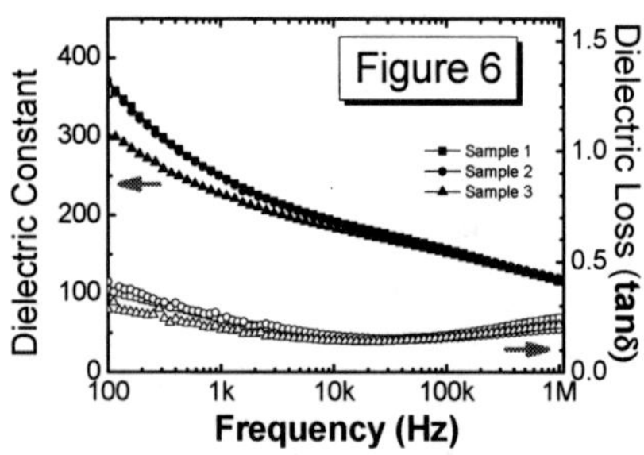

From the data shown in **Figure 6**, it is found that the dielectric loss of the composite can be higher than that obtained in the pure polymer matrix and the pure CCTO ceramics; however, based on dielectric physics, the dielectric loss of a composite should be smaller than the higher dielectric loss observed in the polymer matrix and filler material. That is, the dielectric loss of the composites reported here can not be explained/understood based on the two-phase dielectric composite. It is believed that the dielectric loss in the composites is strongly related to the interfacial layers between the ceramic powders and polymer matrix within the composites [11].

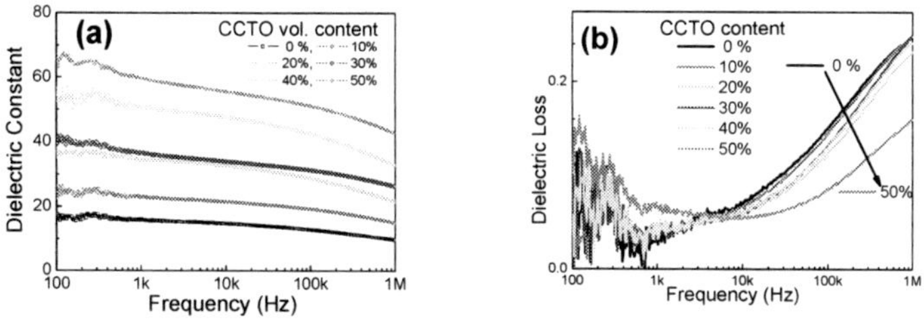

Figure 7. Dielectric properties, dielectric constant and dielectric loss, of nano-sized CCTO composites at room temperature. The composition of the composites is given in the figure.

To study the influence of CCTO particle size on the dielectric properties (i.e. dielectric constant and dielectric loss) of the composites, the composites using nanosized CCTO were prepared. The frequency dependence of the dielectric property of the composites at room temperature is shown in **Figure 7** for composites with different compositions, where the composites were prepared using HP process with six layers of as-cast composite film in the stack using the configuration shown in **Figure 1(b)**.

Comparing composites using microsized CCTO particles, the data shown in **Figure 7** clearly indicates the composites using nanosized CCTO particles have a lower dielectric constant and a smaller dielectric loss. More importantly, the dielectric loss of the composites is smaller than that of the polymer matrix, which is desirable for applications such as energy storage and capacitors.

The physics behind the difference in the dielectric properties of the composites using the micro- and nano-sized CCTO is under investigating.

For the composites using nano-sized CCTO, it was also been experimentally found that the HP process condition has an influence on the dielectric property of the final composites. Compared with composites using microsized CCTO, the influence of the HP process on the dielectric properties of the final composites using nanosized CCTO is much weaker, as shown in **Figure 8**, where the dielectric constant and loss at 1 kHz for the composites at room temperature is plotted against the composition of the composite. These composites were prepared using HP process with a configuration shown in **Figure 1(b)**.

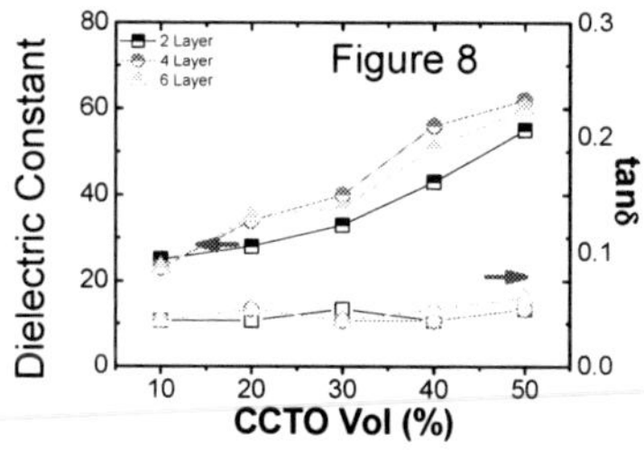

Conclusion

The ceramic-polymer 0-3 composites using CCTO powders as filler and P(VDF-TrFE) 55/45 as polymer matrix exhibit a higher dielectric constant than other systems. It was found that the preparation process, especially the HP process, has a strong influence on the dielectric property of the composites. The composites using nano-sized CCTO powders exhibit a small dielectric loss.

References

[1] Z.-Y. Cheng, R.S. Katiyar, X. Yao, A.S. Bhalla, Phys. Rev.B 57, 8166 (1998).

[2] Y. Bai, Z.-Y. Cheng, V. Bharti, H.S. Xu, and Q.M. Zhang, Appl. Phys. Lett. 76, 3804 (2000).

[3] C.C. Homes, T. Vogt, S.M. Shapiro, S. Wakimoto, A.P. Ramirez, Science 293, 293-673 (2001).

[4] M.A. Subramanian, L. Dong, N. Duan, B.A. Reisner, A.W. Sleight, J. Solid State Chem. 151, 323-325 (2000).

[5] M. Arbatti, X.B. Shan, Z.-Y. Cheng, Advanced Materials 19, 1369 (2007).

[6] M. Arbatti, MS Thesis, Auburn University (2004).

[7]. Y.P. Shen, A.J. Gu, G.Z. Liang, L. Yuan, Composites A – Applied Science and Manufacturing 41(11), 1668 (2010).

[8]. E.A. Stefanescu, X.L. Tan, Z.Q. Lin, N. Bowler, M.R. Kessier, Polymer 51 (24), 5823 (2010).

[9]. P. Thomad, S. Statapathy, K. Dwarakanath, K.B.R. Varma, Express Polymer Letters 4(10), 632(2010).

[10] Z.M. Dang, T. Zhou, S.H. Yao, J.K. Yuan, J.W. Zha, Song, H.T. J.Y. Li, Q. Chen, W.T. Yang, J. Bai, Advanced Materials 21(20), 2077 (2009).

[11] X.B. Shan, PhD Thesis, Auburn University (2009).

Mater. Res. Soc. Symp. Proc. Vol. 1312 © 2011 Materials Research Society
DOI: 10.1557/opl.2011.902

Ionizing Radiation Total Dose Detectors Using Oligomer Organic Semiconductor Material and Devices

Harshil N. Raval and V. Ramgopal Rao
Centre of Excellence in Nanoelectronics, Department of Electrical Engineering,
Indian Institute of Technology, Bombay, Mumbai – 400 076, India

ABSTRACT

Organic semiconducting oligomer – Pentacene, as a material and organic electronic devices based on it, are proposed here as total dose detectors for ionizing radiation. Pentacene, when exposed to ionizing radiation of γ – rays using Cobalt – 60 (^{60}Co) radiation source, shows increase in the conductivity of the material which can be used as a sensing phenomenon for determining the dose of ionizing radiation. The change in material property was also verified using UV-visible (UV-VIS) spectrum for pentacene thin-films with rising absorption peaks at the oxidized positions in the wavelength. A pentacene resistor can be used as a detector, as the change in the conductivity of the pentacene film can be easily quantified by measuring the change in resistance of the pentacene resistor after different total radiation dose exposures. The experiments resulted in a sensitivity of 340 kΩ/Gy for a total 100 Gy radiation dose for the pentacene resistor. Furthermore, employing this simple electrical measurement technique for determining the dose of ionizing radiation and to improve the sensitivity of the sensor by transistor action, a pentacene based organic field effect transistor (OFET) was exposed to ionizing radiation. Change in OFF current (I_{OFF}) of the OFET sensor with W/L = 19350 μm/100 μm, suggests a sensitivity of 21 nA/Gy for 100 Gy dose. Also, changes in various other parameters like threshold voltage, subthreshold swing, field effect mobility, number of interface states etc. can be extracted from the electrical characterizations which prove their usefulness as a detector for ionizing radiation.

INTRODUCTION

Organic semiconducting polymer materials are being explored for various organic electronic applications [1-3] because of various advantages, such as, large area coverage on flexible substrates and cost effective fabrication process. These polymer semiconducting materials have also found their applications in different sensing applications due to their chemical affinity to different atmospheric conditions [4-11]. However, with the highest carrier mobility and stability, organic semiconducting oligomer – Pentacene is the most engineered and known to be the best material of choice for numerous organic electronic applications such as displays, RFID tags, and flexible circuits [12-13]. Moreover, pentacene was also explored for various sensing applications, such as, gas and vapor sensing [14], humidity sensing [15], and pH monitoring [16]. Here, we report the use of the Pentacene for determining total dose of ionizing radiation using the change in conductivity of the material as a measure of ionizing radiation dose. Pentacene resistor and a pentacene OFET are proposed here as the devices for determining ionizing radiation.

EXPERIMENTS

Pentacene degradation and passivation

Pentacene is the most widely used oligomer organic semiconductor for various organic electronic and circuit applications. So, it is extensively studied in various operating and environmental conditions. Pentacene films are known to degrade in the atmosphere due to oxygen and moisture if not encapsulated [17], resulting in degraded field effect mobility and current ratio for a pentacene OFET. To overcome this degradation, various types of encapsulations were also reported [18]. A thin layer of hot-wire CVD (HWCVD) silicon nitride, processed at low temperature, is proposed here as a passivation layer to avoid such degradation, as the change in conductivity of the oligomer - Pentacene is used as a measure of radiation dose. An OFET was fabricated on a highly doped n-type silicon wafer (Si wafer) acting as the gate electrode. A 100 nm silicon dioxide (SiO_2) was thermally grown using dry oxidation of the Si wafer to serve as the gate dielectric and titanium/gold (Ti/Au) (10nm/70nm) layers were deposited by RF sputtering and patterned using a lift-off process with optical lithography for source and drain (S/D) electrodes. A 100nm thick layer of pentacene was deposited by thermal evaporation to form a pentacene OFET. To study the effect of the passivation layer, a set of OFETs were not encapsulated while another set had the passivation layer. A thin layer of silicon nitride (~20 nm) was deposited by HWCVD process for encapsulation of the pentacene film, as the silicon nitride is considered to be a good barrier for oxygen and moisture. In HWCVD process, the sample was kept at room temperature to avoid any effect of high temperature processing on the organic semiconducting layer [19]. The study of the passivation layer was carried out by measuring the drain current (I_{DS}) of the OFET with varying gate to source voltage (V_{GS}) at a fixed drain to source voltage (V_{DS}) (I_{DS}-V_{GS} characteristics) at room temperature in the normal ambient conditions after different time intervals. The electrical characterization was performed on a probe station using Keithley 2602 and Keithley 236 source – measure units (SMUs) for applying voltages and measuring currents from all the three terminals of an OFET.

UV-VIS spectrum analysis for Pentacene films with ionizing radiation dose

The effect of ionizing radiation on the thin film of pentacene was observed by studying its UV-VIS spectrum after different total radiation dose. For the experiment, a thin film (100 nm) of pentacene was deposited on four-transparent glass samples by thermal evaporation. Three of these samples were irradiated by γ – rays using ^{60}Co radiation source for different radiation dose by varying the time for exposure. The radiation source had a radiation rate of 7 Gy/min. UV-VIS spectrum was measured and analyzed for a non-irradiated sample and three samples exposed to ionizing radiation for different radiation doses.

Pentacene resistor and OFET radiation sensors

Effect of ionizing radiation on the conductivity/resistivity of the pentacene film was studied using a pentacene film resistor. For sample preparation, thick (> 500 nm) SiO_2 was grown using wet oxidation process on a Si wafer, this SiO_2 layer isolates the Ti/Au interdigitated electrodes patterned with 100 μm distance between them for 19050 μm length of the electrodes, from the Si wafer. The pentacene film and silicon nitride passivation layer were deposited as described in the previous section. Change in the resistance was measured for the samples for different total dose ionizing radiation on the probe station using a Keithley 2602 SMU. Significant changes in the resistance of the pentacene resistor with increasing dose of ionizing radiation prove it a potential candidate for acting as a sensor. These changes were found encouraging for a study of the effect of ionizing radiation on the well-known device - pentacene OFET. A pentacene OFET was fabricated with a passivation layer as described in the previous section. The $I_{DS}V_{GS}$ characteristics for the OFET were measured and changes were studied with increasing dose of ionizing radiation. Also, a stability study was performed on a pentacene OFET to observe if the changes due to ionizing radiation are permanent in nature.

RESULTS & DISCUSSION

Pentacene degradation and passivation

Figure 1 shows the I_{DS}-V_{GS} characteristics of the OFETs wherein figure 1(a) and 1(b) show characteristics for OFETs without the passivation layer and OFETs with silicon nitride passivation layer respectively. As can be observed from the plots, devices with silicon nitride passivation layer show a 2.5X less degradation compared to devices without any passivation.

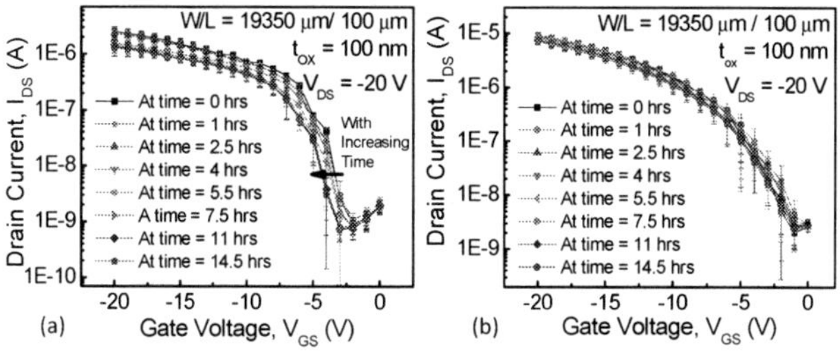

Figure 1. I_{DS}-V_{GS} characteristics for (a) OFETs without any passivation layer, and (b) OFETs with 20 nm silicon nitride passivation layer showing the improvement of 2.5X as compared to (a)

UV-VIS spectrum analysis for Pentacene films with ionizing radiation dose

The normalized UV-VIS spectra for pentacene thin films on transparent glass substrate with different dose of ionizing radiation are shown in figure 2 (a). The prominent absorption peaks for pentacene can be observed at 580 nm and 670 nm wavelengths as reported [20, 21]. At the same time, absorption peaks observed at 225 nm, 270 nm and 300 nm wavelengths are due to the oxidized pentacene molecules in the film [20, 21]. With increasing dose of ionizing radiation, higher absorption was found at these wavelengths, indicating increased concentration of oxidized pentacene molecules in the film.

Pentacene resistor and OFET device sensors

The change in resistance of the samples was measured after subsequent dose of ionizing radiation and the variation in the resistance of the film is plotted in figure 2 (b). As can be observed, the film resistance was 51 MΩ before exposure to γ – radiation and it was found decreasing with subsequent total dose exposure to the ionizing radiation. At 100 Gy total dose of ionizing radiation the resistance was decreased to 17 MΩ resulting in a sensitivity of 340 kΩ/Gy. With such a significant change in resistance for increasing dose of ionizing radiation, the pentacene resistor is proposed here as a detector for total dose ionizing radiation.

The $I_{DS}V_{GS}$ characteristics for the OFET without any radiation and with subsequent total-dose of ionizing radiation are shown in figure 3 (a). The OFET was irradiated for 10 Gy, 50 Gy and 100 Gy total-dose values and after every exposure to γ – radiation; its $I_{DS}V_{GS}$ characteristics were measured. As can be observed from figure 2 (b), the pentacene film becomes more conducting with increasing dose of ionizing radiation, the same can be observed from the increase in I_{OFF} for the OFET which indicates increasing conductivity of the film with increasing dose of ionizing radiation. However, the ON current (I_{ON}) of the OFET behaves differently due to various mechanisms affecting I_{ON} differently. The increasing conductivity of the material contributes to a proportional increase in the I_{ON}, and at the same time, as commonly observed in MOS-like devices, increasing interface traps and defects in the gate oxide due to the ionizing radiation degrade the I_{ON} due to an increased scattering and mobility degradation [22]. As a combined effect of these two mechanisms, the change in I_{ON} with increasing total-dose of radiation is quite different as compared to I_{OFF} as can be seen in figure 3 (a). Figure 3 (b) shows the change in I_{OFF} for the pentacene OFET for ionizing radiation dose upto 100 Gy. For a 100 Gy dose of ionizing radiation the change in I_{OFF} was found 200X compared to the unexposed value, resulting in a sensitivity of 21 nA/Gy for an OFET with W/L = 19350 μm/100 μm. Thus, the I_{OFF} sensitivity was found to be 1.1 pA/Gy/1-μm width of the pentacene OFET sensor for 100 Gy dose of ionizing radiation. Thus, the width of the OFET can be used here as a tuning parameter to tune the current sensitivity of the sensor [11].

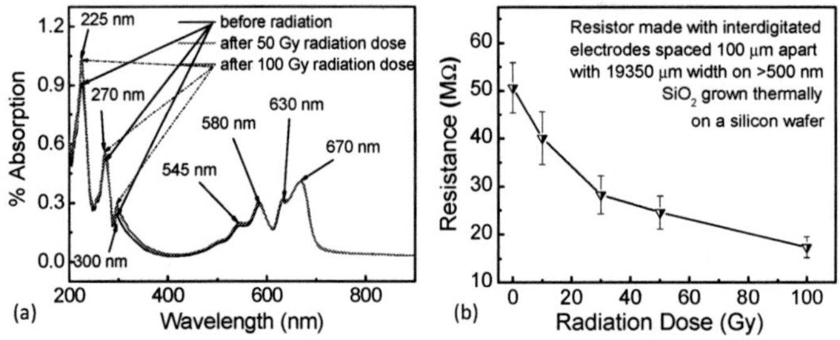

Figure 2. (a) UV-VIS spectrum of Pentacene film on transparent glass substrate showing increase in % absorption peaks for the film due to oxidation of the material with higher radiation dose, and (b) Change in resistance of a Pentacene resistor with increasing dose of ionizing radiation

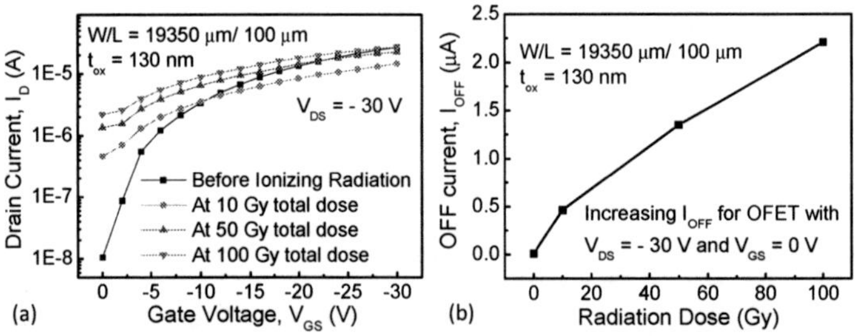

Figure 3. (a) $I_{DS}V_{GS}$ characteristics for a Pentacene OFET with increasing dose of ionizing radiation, and (b) Change in I_{OFF} for the pentacene OFET sensor extracted out of (a)

From the I_{DS}-V_{GS} characteristics in figure 3 (a), other characteristic parameters such as shift in the threshold voltage, change in subthreshold swing which in turn is proportional to the increasing number of interface traps due to high energy exposure [9], degradation in field effect mobility etc. can be extracted and analyzed with respect to total dose effect. The change in threshold voltage (V_{TH}) is shown in figure 4 (a) with increasing dose of ionizing radiation. Considering the shift in threshold voltage as a measure of ionizing radiation, the sensitivity observed for a dose of 100 Gy for the OFET is 0.22 V/Gy. Experiments were also done to see if the effects in Pentacene film due to ionizing radiation are permanent in nature. Figure 4 (b) shows the I_{DS}-V_{GS} characteristics of an OFET sensor before and after 100 Gy dose of ionizing radiation as a function of time after the ionizing radiation. It can be observed that the changes are permanent in nature.

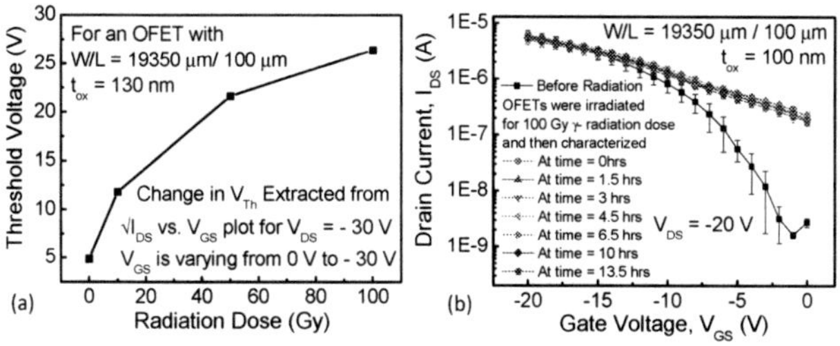

Figure 4. (a) Shift in threshold voltage extracted from $I_{DS}V_{GS}$ characteristics in figure 3. (a), and (b) Time study after irradiation of 100 Gy dose to an OFET sensor, to observe the permanent nature of the effect

CONCLUSIONS

In summary, the change in conductivity of pentacene film can be used as a measure of ionizing radiation using simple electrical measurement techniques in pentacene resistor sensor and pentacene OFET sensor. A thin layer of silicon nitride deposited using HWCVD process has been shown to be a good passivation layer for pentacene OFET sensor for ionizing radiation applications, without sacrificing on the sensitivity. Silicon nitride acts as a good barrier for oxygen and moisture and 250 % improved atmospheric stability was observed with these passivated OFETs. With pentacene resistor a sensitivity of 340 kΩ/Gy was achieved with 100 Gy radiation dose. However, pentacene OFET was proved to be better sensor with various electrical characteristics parameters changing with increasing dose of ionizing radiation. OFF current sensitivity of 1.1 pA/Gy/1-μm width was achieved with pentacene OFET as a sensor. Pentacene based total dose radiation detectors will also retain all advantages of organic electronic devices such as lower cost, large area deposition etc. as shown in this work.

REFERENCES

1. C. D. Dimitrakopoulos, P. R. L. Malenfant, Adv. Mater. **14** (2), 99 (2002)
2. Kazuhito Tsukagoshi, Jun Tanabe, Iwao yagi, Kunji Shigeto, Keiichi Yanagisawa, Yoshinobu aoyagi, J. Appl. Phys. **99**, 064506 (2006)

3. H. N. Raval, S. P. Tiwari, R. R. Navan, Subodh G. Mhaisalkar, V. R. Rao, IEEE Elect. Dev. Lett. **30** (5), 484-486 (2009)
4. G. Guillaud, J. Simon, J. P. Germain, Coord. Chem. Rev. **178** (2), 1433 (1998)
5. R. S. Dudhe, J. Sinha, A. Kumar, V. R. Rao, Sens. and Actu. B: Chem. **148** (1), 158 (2010)
6. T. Agostinelli, M. Campoy-Quiles, J. C. Blakesley, R. Speller, D. D. C. Bradley, J. nelson, Appl. Phys. Lett. **93** (20), 0203305 (2008)
7. J. W. Kingsley, S. J. Weston, D. G. Lidzey, IEEE Journal of selected topics in quantum electronics, **PP** (99), 1-6 (2010)
8. E. A. B. Silva, J. F. Borin, P. Nicolucci, C. F. O. Graeff, T. G. Netto, R. F. Bianchi, Appl. Phys. Lett. **86** (13), 131902 (2005)
9. H. N. Raval, S. P. Tiwari, R. R. Navan, V. R. Rao, Appl. Phys. Lett. **94** (12), 123304 (2009)
10. J. M. Lobez, and T. M. Swager, Chem. Int. Ed. **49** (1), 95-98 (2010)
11. H. N. Raval and V. R. Rao, IEEE Elect. Dev. Lett. **31** (12), 1482 - 1484 (2010)
12. P. F. Baude, D. A. Ender, M. A. Haos, T. W. Kelley, D. V. Muyres, S. D. Theiss, Appl. Phys. Lett. **82** (22), 3964-3966 (2003)
13. F. Eder, H. Klauk, M. Halik, U. Zschieschang, G. Schmid, C. Dehm, Appl. Phys. Lett. **84** (14), 2673-2675 (2004)
14. B. Crone, A. Dodabalapur, A. Gelperin, L. Torsi, H. E. Katz, A. J. Lovinger, Z. Bao, Appl. Phys. Lett. **78** (15), 2229-2231 (2001)
15. T. Zhu, J. T. Mason, R. Dieckermann, G. G. Malliaras, Appl. Phys. Lett. **81** (24), 4643 (2002)
16. C. Bartic, B. Palan, A. Campitelli, G. Borghs, Sens. and Actu. B:Chemical, **83**, 115 (2002)
17. C. R. Kagan, A. Afzali, T. O. Graham, Appl. Phys. Lett. **86**, 193504 (2005)
18. Ho Nyeon Lee, Young Gu Lee, Ik Hwan Ko, Eok Chai Hwang, Sung Kee Kang, Current Applied Physics **8**, 626-630 (2008)
19. S. P. Tiwari, P. Srinivas, S. Shriram, N. S. Kale, S. G. Mnaisalkar, and V. R. Rao, Thin Solid Films, **516** (5), 770-772 (2008)
20. Takashi Minakata, Ichiro Nagoya, Masaru Ozaka, J. Appl. Phys. **69** (10), 7354-7356 (1991)
21. Chun-Lan Tao, Xu-Hui Zhang, Fu-Jia Zhang, Yi-Yang Liu, Hao-Li Zhang, Materials Science and Engineering B **140**, 1-4 (2007)
22. R. van Langevelde, F. M. Klaassen, IEEE Tran. On Elect. Dev. **44** (11), 2044-2052 (1997)

Mater. Res. Soc. Symp. Proc. Vol. 1312 © 2011 Materials Research Society
DOI: 10.1557/opl.2011.903

The Study about the Control of Defect Factors to Improve Properties of P(VDF-TeFE) Thin Film

Jong-Hyeon Jeong, Daiki Terashima, Chiharu Kimura and Hidemitsu Aoki
Department of Electrical, Electronic and Information Engineering, Osaka University, Suita, Osaka, 565-0871, Japan

ABSTRACT

In order to apply P(VDF-TeFE) piezoelectric polymer to micro-generator as a membrane, the polymer is deposited on a substrate by spin-coating method. Since a solvent affects the film properties and surface stability, we have carried out the thermal process at a temperature higher than melting point. In the annealing process of a P(VDF-TeFE) thin film, electrical properties of the film was improved by an application of an electric field. The established study was the investigation of variations in characteristics by an application of a certain electric field. We have attempted to measure about the critical intensity of an electric field and investigate what the influences of the application is caused using a XPS spectrum in this study. Moreover, in order to control the intrusion of impurities, we have used the vacuum chamber to carry out the annealing process in it.

INTRODUCTION

For the development of portable electronic devices and wireless interface, new energy-storage technologies or self-powered energy sources are needed to drive the systems for longer time. Since conventional energy-storage technologies based on chemicals have disadvantages such as finite life-time, instability and large scale, these disadvantages are blocking a miniaturization of the device. Moreover, the conventional self-powered source has also generated a little power in comparison with a scale. With the equipment of a self-power generator or an additional electric-power compensator on a fundamental battery, it is necessary to extend the life-time of the devices [1-4].

In all of energy harvesting technologies, we have focused on human voice to extend the battery time of portable electronic devices. A frequency of human voice is 80~4k Hz. However, the bandwidth of 90~400 Hz is used to speak ordinarily. Then, high conversion efficiency can be expected by realizing a resonant frequency as low as 100 Hz for receiving parts of the devices. Since a human voice has irregular frequency and energy as an acoustic wave, we have conceived a membrane structure formed by piezoelectric material which has low resonant frequency and low density material. In order to fabricate a micro generator that can generate electricity through low-frequency acoustic waves, we have studied the fabrication of the device with micro electromechanical system (MEMS) technology and improved the properties of a piezoelectric polymer. Owing to the absorption of an acoustic wave with a frequency lower than 100 Hz, a polymer with low mechanical properties is more efficient than lead zirconate titanate (PZT) at the same resonant frequency [5-8]. In this study, poly vinylidene difluoride and tetrafluoro-ethylene (P(VDF-TeFE)) (supplied by Daikin Co. Ltd.) was used as a PVDF series.[9,10] Although electrical properties of this polymer are lower than those of PVDF, this polymer is the same with PVDF as a copolymer in all aspects [11-14].

TABLE I. Comparison of mechanical properties among piezoelectric materials

Parameters	Unit	PZT	PVDF stretched	P(VDF-TeFE) unstretched
Density	10^3 kg/m^3	7.5	1.8	1.9
Elastic modulus	10^9 N/m^2	110	3.0	1.2
Electromechanical coupling coefficient [Kt]		0.51	0.2	0.21

*This table is referred to a research report of Daikin Co. Ltd.

As summarized in Table I, for average mechanical parameters of P(VDF-TeFE) and PVDF film, a resonant frequency of PVDF film is almost double compared with that of P(VDF-TeFE) film because elastic modulus of PVDF is higher than that of P(VDF-TeFE). For example, if a thickness of membrane formed by self-standing P(VDF-TeFE) film is 1um and a radius is 500um, the resonant frequency will be estimated to be about 50 Hz from a calculation of resonant frequency [7]. Since a stretching method is impossible to apply to MEMS process, the selection of casting solvents and processing method are very important. Moreover, in order to form a membrane which has about 300 nm thickness, the stability of the film surface is an important factor. For this reason, we have carried out a thermal treatment at a temperature higher than melting point. As a result, a polarity of the film is lowered seriously.

In the established our study, we studied about the new thermal treatment method for restoring the polarization properties of the P(VDF-TeFE) thin film annealed at a temperature higher than its melting point. To assist the rearrangement of monomers in the film, we have applied an electric field during the thermal treatment. In this study, for the application of an electric field, we have investigated the variation of the film properties as a function of an electric field during the annealing process.

EXPERIMENT

The film forming process of P(VDF-TeFE) was carried out by a sol-gel method. As a solvent for dissolving a P(VDF-TeFE) granule, methyl-ethyl-ketone (MEK) has been used generally [15]. MEK can dissolve the P(VDF-TeFE) granule as shown Fig. 1, however, the probability on the recombination of the monomer is very low. In this study, we used two solvents; one is MEK 100% solvent and the other is one that adds n, n-dimethylacetamide (DMAc) (27%) in MEK(73%) solvents. Although, in the case of the mixing a DMAc solvent, reacted material can be possible to remain in the solvent, we used the two solvents because of the benefit which the physical properties of the film is improved.

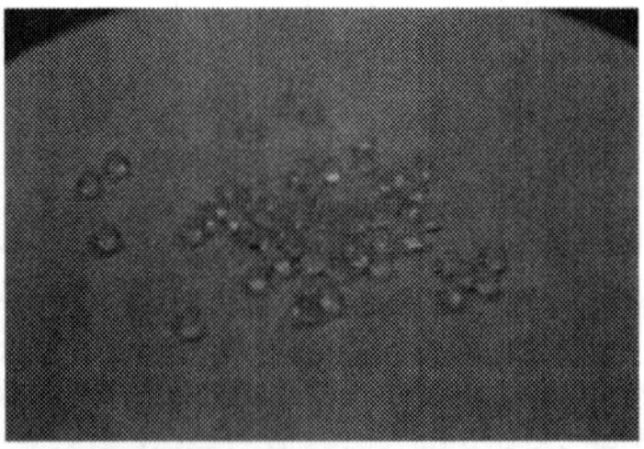

Figure 1. Granule of Poly-Vinylidene DiFluoride - TetraFluoroEthylene

Surface roughness is different from the concentration of the solution used. When the film is formed using a solution of low concentration, the surface roughness of the film becomes more flat. Also, humidity affects the adhesion between the film and a substrate [16]. Therefore, we used a solution of 4% concentration. As a result, the number of pin holes decreased as 85% when using a solution of 4% concentration compared with the case of using a solution of 10% concentration.

The substrate used was Pt/Ta-deposited Si. Ta was used as a buffer layer, and Pt was used to reduce the resistance of electrodes. Using a spin coater, the P(VDF-TeFE) film was deposited to have a thickness of 500 nm from an initial speed of 1,000 rpm to a terminal speed of 5,000 rpm. To prevent impurities from contaminating the film, the chamber of the spin coater was filled with N_2 gas.

In order to improve the adhesion and uniformity of the film, an annealing process was carried out at 195 °C after coating process. As the thermal properties of P(VDF-TeFE) are similar to those of PVDF, the Curie point is about 140 °C and the melting point is about 170 °C. In a previous study about the relationship between the annealing temperature and the film surface, the surface condition of the film annealed at 175 °C is similar to that of the film annealed at a temperature lower than 175 °C [17]. Therefore, the annealing process was carried out at 195 °C, since film properties changed at the temperature higher than 200 °C. The electric field was applied by 0, 0.5, 1 and 2 MV/m for 2 mm distance between electrodes during the annealing process as shown in Fig.2. In addition, the annealing process was carried out in a vacuum chamber to prevent the invasion of impurities.

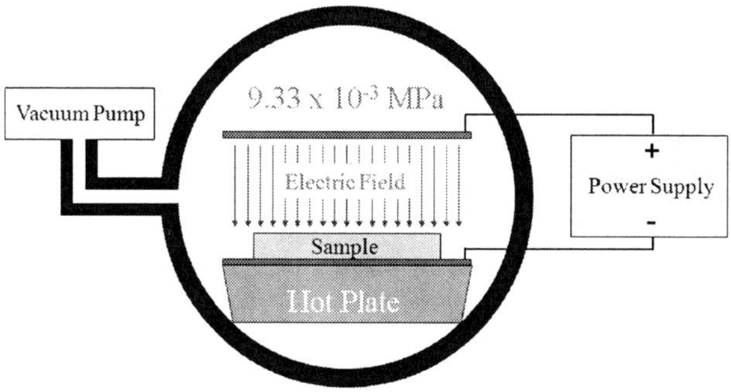

Figure 2. Diagram of the annealing process with an electric field

RESULTS AND DISCUSSION

XPS spectra in C 1s area

In order to analyze effects of an electric field during an annealing process visually, X-ray photoelectron spectroscopy measurements were carried out for the samples. All spectra are modified to prevent an influence of charge-up state through comparing with CF_2 binding energy.

Fig. 3 shows XPS spectra of C 1s for variation of the electric field. In the region of 284 ~ 286.5 eV, CH_n, –C–O–C– and CC bond are detected [18, 19]. In Fig. 3-(a), the detection of $-CF_2-\underline{C}H_2-CF_2-$ and $-CH_2-\underline{C}F_2-CH_2-$ bonds for 0.5 and 2 MV/m are smaller than the others.

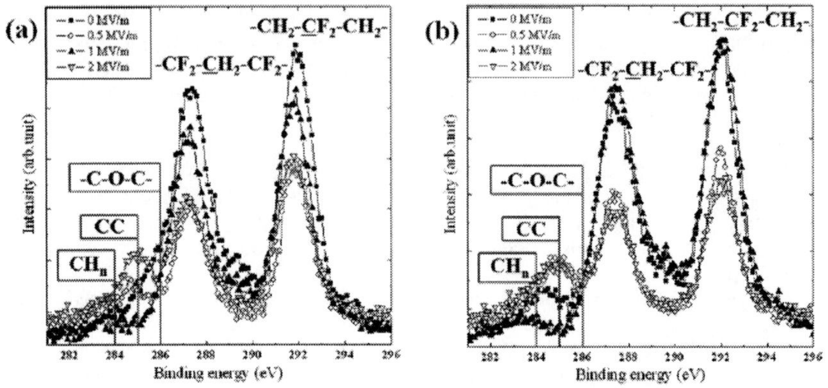

Figure 3. XPS spectra of C 1s core for (a) single solution and (b) mixed solution

Relatively, CC bonds are detected largely. It means to be increased the damaged chain. This phenomenon is also happened in a mixed solution as shown in Fig. 3-(b). On the other hand, CH_n bonds are detected except for the case of 1 MV/m. Since the detection of CH_n bond means the probability of the recombination of monomers, the probability is not good except for the case of 1 MV/m. CH_n bonds are divided into two cases; one is CH_2 that is contained in the monomers and MEK solvent, and the other is CH_3 that is contained only in the MEK solvent. The CH_n bond of the ionic state indicates that the bond of the monomers or the solvent is broken. Thus, $–CH_2$ bonds exist in the end of the chain or in the cutoff of the chain. –C–O–C– bonds mean the invasion of O atom or the remaining solvent. As for O atoms, it can be two cases; one is the –C–O–C– bond that is formed by the substitution of O atoms for C atoms in the –C–C–C– chain, and the other is the C=O bond that is formed by the cutoff of the chain structure by O atoms. Unless there is the invasion of O atom from the air [17], it is quite probable that the remaining solvent disturbs the recombination of monomers of P(VDF-TeFE). CC bonds are –C=C– bond not –C–C– bond, since –C–C– bond are presented by $–CF_2–CH_2–CF_2–$ bond or $–CH_2–CF_2–CH_2–$ bond. In other words, the detection of CH_n, –C–O–C– and CC bond means the invasion of impurities and the damage to a chain structure.

XPS spectra in O 1s area

In order to investigate the bonds related to O atoms, the region of O 1s is observed, as shown in Fig. 4, there are many differences for an electric field and a type of solvents. Since there is not O atom in the P(VDF-TeFE) polymer in principle, any peaks have not to be detected in this area. However, O atom in a lattice, –OH, H_2O and –C–O–C– bonds are detected in this area. O atom in a lattice is possible to be a remaining solvent. –OH bond and H_2O are residues which are caused by a mixing MEK and DMAc. –C–O–C– bonds are also detected in spectra of C 1s. On the whole, peaks which are detected in the case of 1 MV/m are smaller than the others. Only –OH bonds are detected in a single solution, and any peaks are not detected in a mixed solution. In contrast, numbers of peaks are detected in the case of 2 MV/m for a single and a mixed solution. By the electric field, an atom or a molecule which is not related to the chain structure of P(VDF-TeFE) is possible to vanish from the film.

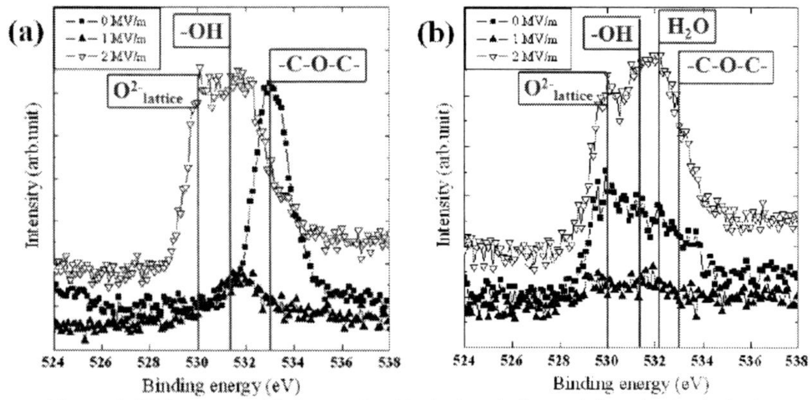

Figure 4. XPS spectra of O 1s core for (a) single solution and (b) mixed solution

Electric field and chain structure

P(VDF-TeFE) has a chain structure which is constructed by $-CF_2-\underline{C}H_2-CF_2-$ and $-CH_2-\underline{C}F_2-CH_2-$ bond. If the chain structure is broken, a function as a piezoelectric material will be lost. Fig. 5-(a) shows a relationship between bonds by which a chain structure is constructed and an electric field which is applied during an annealing process. In a case of 0.5 and 2 MV/m, the ratio is low. It means to be decrease the remanent polarization of the film. The case of 1 MV/m is lower than the case of 0 MV/m, however, the difference is a little. Moreover, a case of a mixed solution is more sensitive than a single solution for an electric field. A ratio of mixed solution in 1 MV/m is almost same with that of single solution in 0 MV/m. The recombination probability of monomers can be affected by the application of an electric field.

Fig. 5-(b) shows the ratio of defects such as CH_n, $-\underline{C}-O-C-$ and CC bonds. On the whole, the ratio of the impurities is higher than that of 1 MV/m. The case of 0.5 and 2 MV/m is high, but rather higher than that of 0 MV/m. While an annealing process progresses at the temperature higher than the melting point, monomers that are spread out with a chain structure in the film lose their preferred orientation.

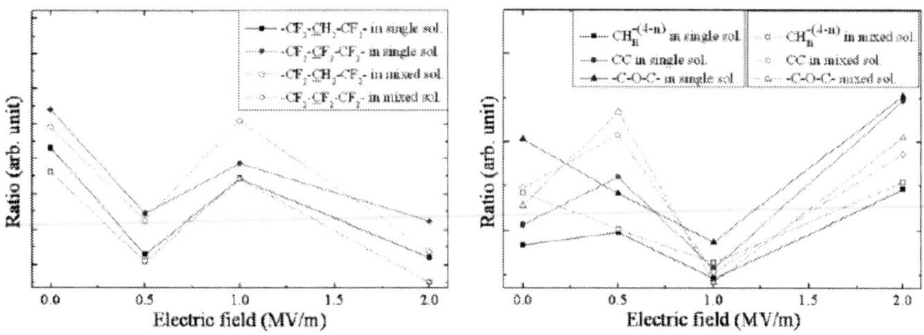

Figure 5. Relationship between bonds and an intensity of an electric field

However, owing to the polarity of the monomers, the monomers are rearranged by an electric field. When the annealing process is ended and the temperature of the film is decreased, the monomers set a chain structure and remain at their preferred orientation.

While the chain is broken by C-OH bond, C=CH_2 bond and CH_2^{2-} bond are generated. In addition, C-CH_3 bond and CH_3^- bond are generated [20-22]. C-O-C bonds are generated by a substituting O atom for C atom in the chain, and polarized monomers lose its polarity by the substitution. In contrast, C=C and C-O bonds are broken by applying the electric field. While the film is liquefied by a temperature higher than melting point on an annealing process, ionized O atoms and O atom-contained molecules such as OH^- are possible to be desorbed by an electric field in the case of 1 MV/m. If an electric field was higher or lower than the critical point, O atoms and O atom-contained molecules is reacted unpredictably.

Variation in remanent polarization

Since O atoms are substituted for C atoms such as C-O-C bonds, the polarized monomer will be replaced and a remanent polarization will also be reduced. Table II shows P_r for a bias voltage of 100 V and 0.05 Hz. It is also found that P_r increases further when the electric field is applied. P_r of the film annealed with an electric field is almost twice P_r of the film annealed without an electric field. Pr as large as 44.3 is measured for the film formed by annealing with an electric field using the single solvent. This value is possibly comparable to that measured at 0.01 Hz for bulk-type P(VDF-TeFE) film [11].

As a result, the molecular structure of the polymer is affected by certain critical point of an electric field. In the case of an electric field lower than a critical point, the recombination probability is decreased by O atoms or O-related molecules, while more impurities intrude into the chain structure in the case of an electric field higher than a critical point. For this reason, the films are transformed to paraelectric material.

Table II. Variation of remanent polarization for applied electric fields

Applied electric field (MV/m)	Single solution (mC/m^2)	Mixed solution (mC/m^2)
0	30.9	18.2
0.5	Paraelectric break	
1	33.1	25.8
2	Paraelectric break	

CONCLUSIONS

We have studied about the deposition method for P(VDF-TeFE) film that can be applied to a micro generator using MEMS technology. Since an electric field which is applied during an annealing process affects properties of the film, we have investigated the critical point of an electric field. In order to block any contamination factor, the coating process was carried out in N_2 gas chamber and the annealing process was carried out in the vacuum chamber. On the electric field lower than the critical point, the electric field cannot affect the properties, but rather, disturb the recombination of monomers. Otherwise, the electric field allows impurities to intrude into chain structure of the polymer. In our result, we found that there is a suitable intensity of an electric field to have a decisive effect on the properties of the film.

ACKNOWLEDGMENTS

This study has been supported by the Global COE Program "Center for Electronics Devices Innovation" of the Japan Society for the Promotion of Science.

REFERENCES

1. C.B.Williams C. Shearwood, M. A. Harradine, P. H. Mellor, T. S. Birch, and R. B. Yates, IEEE proc. Circuits, Devices and Systems, vol. 148, no. 6, pp. 337-342 (2001).
2. R. Amirtharajah and A.P. Chandrakasan, IEEE J. Solid-State Circuits, vol. 33, no. 5, pp. 687–695 (1998).
3. S.J. Roundy, Doctoral dissertation, Dept. of Mechanical Eng., Univ. of California, Berkeley (2003).
4. N.S. Schenck and J.A. Paradiso, IEEE Micro, vol. 21, no. 3, pp. 30–42 (2001).
5. H. Su, A. Strachan, and W. A. Goddard, III: Phys. Rev. B 70, 064101 (2004).
6. H. Chen, X. Dong, T. Zeng, Z. Zhou, and H. Yang: Ceram. Int. vol. 33, pp. 1369 (2007).
7. S. Olcum et al., IEEE Transactions on Ultrasonic, Ferroelectric, and Frequency Control, vol. 52, no. 12 (2005).
8. S. M. Nakhmanson, M. Buongiorno Nardelli, and J. Bernholc: Phys. Rev. Lett. 92, 115504 (2004).
9. A. I. Baise, H. Lee, B. Oh, R. E. Salomon, and M. M. Labes: Appl. Phys. Lett. 26, pp. 428 (1975).
10. Z. Y. Wang, K. H. Su, H. Q. Fan, and Z. Y. Wen: Polymer, vol. 48, pp. 7145 (2007).
11. H. Kawai, Japanese Journal of Applied Physics, vol. 8, pp. 975-976 (1969).
12. J. B. Lando and W. W. Doll: J. Macromol. Sci. Phys. B 2, pp. 205 (1968).
13. J. C. Hicks, T. E. Jones, and J. C. Logan: J. Appl. Phys. 51, pp. 1135 (1980).
14. E. Fukada and S. Takashita: Jpn. J. Appl. Phys. 8, pp.960 (1969).
15. J. X. Wen, Polymer Journal, vol. 17, no. 2, pp. 399-407 (1985).
16. J. H. Jeong, C. Kimura, H. Aoki, M. Okuyama and T. Sugino: ECS Trans. 19, pp.59 (2009).
17. J. H. Jeong, C. Kimura, H. Aoki and T.Sugino: Jpn. J. Appl. Phys. 49, 04DK23 (2010).
18. A.P. Dementjev and K.I. Maslakov, Appl. Surf. Sci. 253, pp. 1095 (2006).
19. M. Schindler, F.C. Hawthorne, M.S. Freund, and P.C. Burns, Geochim. Cosmochim. Acta, vol. 73, pp. 2488 (2009).
20. S. Lanceros-Mendez, J. F. Mano, A. M. Costa, and V. H. Schmidt: J. Macromol. Sci., B 40, pp. 517 (2001).
21. J. Mihaly, S. Sterkel, H. M. Ortner, L. Kocsis, L. Hajba, E. Furdyga, and J. Mink: Croat. Chem. Acta, vol. 79, pp. 497 (2006).
22. J. Chlopek, A. Morawska-Chochol, and C. Paluszkiewicz: J. Mol. Struct., vol. 875, pp. 101 (2008).

Mater. Res. Soc. Symp. Proc. Vol. 1312 © 2011 Materials Research Society
DOI: 10.1557/opl.2011.904

Multi-layered PVDF Systems for Enhanced Energy Harvesting and Sensing

Jennifer Jones, Roberto S. Aga, Jr. and Richard Mu
Center for Physics and Chemistry of Materials, Fisk University
1000 17th Ave. N. Nashville, TN 37208-3051, U.S.A.

ABSTRACT

The peak output voltage (V_L) of polyvinylidene fluoride (PVDF) piezoelectric devices has been investigated as a function of load resistance (R_L). Two identical piezoelectric devices were mounted, back to back, on a vibrating cantilever giving almost similar V_L. Using R_L values in the range of 4.6 KΩ to 1 MΩ, it was observed that V_L was doubled when the devices were connected in parallel but it did not change when the devices were connected in series. For single, parallel and series configurations, V_L increases linearly with increasing R_L. These results are explained well by modeling the device as a discharging capacitor with internal source of charge generation.

INTRODUCTION

Scavenging energy from the environment to operate low power electronics or recharge portable battery system, has received considerable attention in recent years due to the rapid advancement in microelectronics [1]. Devices are getting smaller and their power consumption is also decreasing. It is now feasible to construct self-powered devices that can operate perpetually. Major applications of such devices include wireless sensors deployed in very remote locations [2]. If they operate on batteries, they will have finite lifespan and battery replacement could be a very expensive maintenance. For devices that run on batteries, incorporating a recharging system based on scavenging ambient energy can prolong their lifespan. One way to scavenge ambient energy is to use piezoelectric (PE) material. It can convert mechanical vibrations originating from sound, wind or motion into usable electricity.

Among the PE materials that have been investigated for energy harvesting application, polyvinylidene fluoride (PVDF) has received the most attention [3-6]. It is environmentally stable and can be fabricated in large areas. It possesses great PE properties with excellent mechanical flexibility. In general, a PE device has an output voltage (V_L) that is strongly dependent on the load resistance (R_L), which is an indication that it can be modeled as a current source [7]. However, results from this work indicate that it is not a rule of thumb. In fact, others consider it as a voltage source [8], which suggests that with the same R_L, the V_L of two PE devices connected in parallel should be the same as the V_L of a single PE device. But again, it is not a general observation from this work. In addition, V_L does not automatically add up when two PE devices are connected in series. Thus, understanding the role of R_L on the V_L of a PE device or a system of PE devices is crucial in optimizing their power conversion efficiency. It is also beneficial in designing integrated PE devices for specific applications such as high current or high voltage. For example, with the correct R_L value, stacking up N PVDF devices in a multi-layer structure may achieve V_L that is N times higher than a single layer. This work proposes a model that describes the dependence of V_L on R_L for a single PE device and uses it to understand the V_L when two PE devices are connected in parallel and in series.

EXPERIMENT

The PVDF PE devices studied in this work were purchased from Measurement Specialties and used as is. The device has a dimension of 6.1 x 1.2 cm^2. It comes with two wire leads attached to the top and bottom electrodes of the PVDF film. It has also a protective lamination that electrically insulates the electrodes from the environment while maintaining the mechanical flexibility. When the device is bent back and forth an AC voltage is generated across the two wire leads. The peak or amplitude of that voltage is the actual quantity that is referred in this paper as V_L. Two PE devices, labeled as D1 and D2, were mounted together, back-to-back, on a cantilever that could be driven by an electromagnet to vibrate at a resonant frequency of 10 Hz. The diagram of the experimental setup is illustrated in Fig. 1. The deflection amplitude of the cantilever was kept constant for all measurements. This was done by using a contact switch with an LED indicator. For each measurement, the cantilever deflection was slowly incremented by increasing the drive current of the electromagnet until the tip of the cantilever makes an electrical contact with the soft wire of the switch, which consequently activates the LED indicator. Resistors with different resistance values could be connected as a load to each PE device mounted on the cantilever. The two PE devices could also be connected in parallel or in series then connect the different load resistors for each configuration. Measurement of V_L was carried out at the resonant frequency of the cantilever using a FFT spectrum analyzer (SR760). Digital oscilloscope was also employed to observe the voltage waveform of the PE devices.

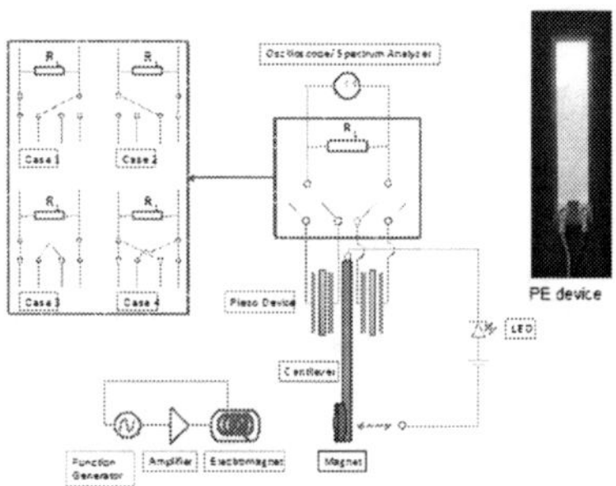

FIG. 1 Experimental setup.

RESULTS AND DISCUSSION

Fig. 2(a) shows the output voltage of D1 and D2 with R_L of 1 MΩ during the vibration of the cantilever. The similarity of the waveform suggests that the two PE devices are almost identical and they perform equally. In addition, when V_L was measured as a function of R_L, its behavior was also similar for D1 and D2 as depicted in Fig. 2(b). It increases linearly with increasing R_L values from 4.6 KΩ to 1 MΩ. It should be noted that the resistance values are the effective for a load resistor in parallel with the 1 MΩ input impedance of the FFT analyzer. The observed behavior of V_L with R_L can be explained well by modeling the PE device as a discharging capacitor with internal charge generation. Suppose G(t) is the generated charge in Coulomb at any time t due to the deflection of the device. The voltage across the device without any load is G(t)/C, where C is the capacitance. If there is a load R_L, there will be a continuous discharge process, which is defined here as Q(t). Thus, at any time t, the net charge in the model capacitor is G(t) – Q(t). The voltage across the capacitor, which is equal to the voltage across R_L, is then

$$\frac{G(t) - Q(t)}{C} = \frac{dQ(t)}{dt} R_L \qquad \text{Eqn. 1}$$

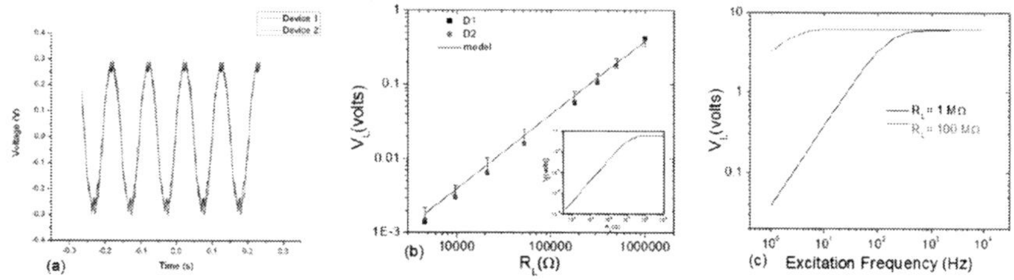

FIG. 2(a) Output voltage of D1 and D2 with load resistance 1MΩ. (b) Measured and modeled $V_L(R_L)$ behavior for D1 and D2. Inset: Saturation behavior of V_L. (c) Frequency dependence of V_L as predicted by the model.

It is worthwhile to mention that if there is no internal charge generation or G(t)=0, the above differential equation has the solution of $Q(t) = Q_0e^{-t/\tau}$, which describes a typical discharging capacitor with a time constant of $\tau = R_LC$. Based on the output waveform [Fig. 2(a)], it can be assumed that the charge generation of the PE devices mounted on the vibrating cantilever has a sinusoidal behavior. If there is no net charge at the start of vibration (t = 0), it can be assumed that G(t) = A Sin(ωt), where ω is the vibration frequency (10 Hz) and A is the maximum generated charge, which depends on several device parameters such as dimension, amplitude of deflection and piezoelectric properties of the material. Because it is very difficult to estimate A theoretically, it becomes a fitting parameter. For example, in Fig. 2(b) the solid curve is the predicted behavior of V_L versus R_L. It was calculated using several trial values of A to get the best agreement with the experiment. To get the analytic form of $V_L(R_L)$ used in Fig. 2(b), equation 1 has to be solved with G(t) = A Sin(ωt). This was done using Mathematica software package. After imposing the initial condition, which is Q = 0 at t = 0, the solution to Eq. 1 was determined to be

$$Q(t) = \frac{A}{(\omega R_L C)^2 + 1}[Sin(\omega t) - \omega R_L C\, Cos(\omega t)] + \frac{A\omega R_L C}{1 + (\omega R_L C)^2} e^{-t/R_L C}$$

The time varying output voltage of the device during vibration is

$$V(t) = \frac{dQ(t)}{dt} R_L = \frac{A R_L}{(\omega R_L C)^2 + 1}[\omega Cos(\omega t) + \omega^2 R_L C\, Sin(\omega t)] - \frac{A\omega R_L}{1 + (\omega R_L C)^2} e^{-t/R_L C}$$

V(t) has a transient and a steady state. The peak of the steady state of this waveform is the V_L. In Fig. 2(b), the predicted $V_L(R_L)$, which agrees fairly well with the experiment, was calculated using the measured capacitance value of the device (~ 1 nF), the angular frequency of the cantilever (20π rad/s) and a fitting parameter value of A= 6.2 x 10^{-9} Coulomb. With these values, the behavior of V_L can be predicted as R_L goes from zero to infinity. The inset of Fig. 2(b) is the calculated $V_L(R_L)$. It goes to zero as R_L goes to zero. When R_L is further increased, it saturates to a constant value (~6.2 V) and for this particular device, it occurs when $R_L > 10^8$ Ω. This predicted behavior is in fact expected experimentally for a PE device. The saturation voltage of ~6.2 V is equivalent to the open circuit voltage of the PE device for a particular stress. The model can also predict the frequency dependence of V_L. Fig. 2(c) plots V_L as a function of vibration frequency for two values of R_L (1 MΩ and 100 MΩ). V_L starts from zero at ω = 0, then increases linearly until it saturates around 6.2 V, which is the open circuit voltage of the PE device. The frequency in which this occurs depends on the value of R_L. The saturation occurs at lower frequency as R_L increases.

When D1 and D2 were connected in parallel, measured V_L was approximately doubled. Fig. 3(a) plots $V_L(R_L)$ for the parallel configuration. The scattered data points are from experiment and the red solid curve is the predicted $V_L(R_L)$ for a single PE device, which is also plotted in Fig. 2(b). This result is consistent with the model. For the parallel configuration, the differential equation that governs the current going into R_L is given by

$$\frac{G_1(t)-Q_1(t)}{C_1}=\frac{G_2(t)-Q_2(t)}{C_2}=\frac{d}{dt}[Q_1(t)+Q_2(t)]R_L=V(t)$$

Since D1 and D2 are identical, $Q_1(t)=Q_2(t)$, thus $V(t) = (dQ/dt)2R_L$. It can also be assumed that for D1 and D2, $G_1=G_2$, $C_1=C_2$. Thus, the differential equation for D1 (or D2) in parallel with a load of R_L is equivalent to the differential equation of D1 (or D2) single device with a load of $2R_L$. This is the reason why measured V_L increased by a factor of two. In this case, the PE device cannot be modeled as a voltage source. However, as R_L becomes very large, the factor two becomes insignificant in the parallel configuration and thus V_L saturates at the same value for a single and parallel-connected devices. This is illustrated in Fig. 3(a). The predicted $V_L(R_L)$ for the parallel configuration (black solid curve) starts out a factor of two higher than that of a single device but they coincide when $R_L > 10^8\ \Omega$. It should also be emphasized that it fits fairly well to the experimental data.

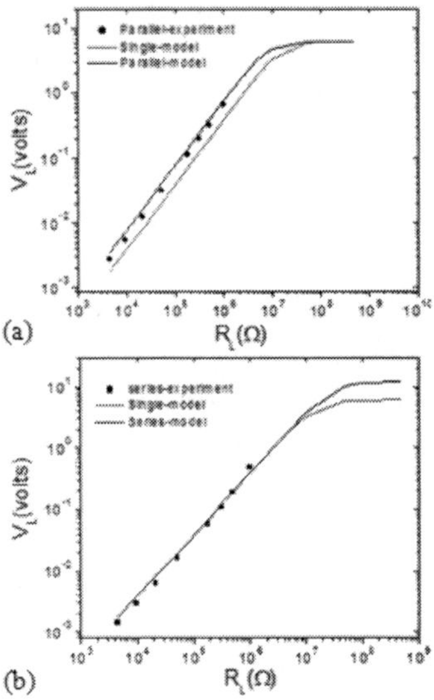

FIG. 3 (a) Measured and modeled $V_L(R_L)$ behavior for D1 and D2 connected in parallel. (b) Measured and modeled $V_L(R_L)$ behavior for D1 and D2 connected in series. Model for a single device is included for comparison.

When D1 and D2 were connected in series, measured V_L was approximately the same as the single device. Fig. 3(b) plots $V_L(R_L)$ for the series configuration. The scattered data points are from experiment and the red solid curve is the predicted $V_L(R_L)$ for a single PE device. This observed behavior is also in agreement with the model. For the series configuration, the differential equation that governs the current going into R_L is given by

$$\frac{G_1(t)-Q_1(t)}{C_1}+\frac{G_2(t)-Q_2(t)}{C_2}=\frac{d}{dt}Q_1(t)R_L=V(t)$$

Again, $G_1=G_2$, $C_1=C_2$ and $Q_1=Q_2$ for identical PE devices and thus, the above differential equation can be reduced to

$$\frac{G(t)-Q(t)}{C}=\frac{d}{dt}Q(t)\frac{R_L}{2}$$

It should be noted that this equation only represents the output voltage of one of the PE devices in series so to get the effective output voltage, the contribution from each should be added. This equation is equivalent to a single PE device with a load of $R_L/2$. Thus, V_L of each PE device in the series is half the V_L of a single device, and adding them yields the same values as the single device. This is the reason why the measured $V_L(R_L)$ for the series configuration is similar to that of a single device as evidenced in Fig. 3(b). However, as R_L becomes very large, V_L of each device in the series saturates at a value equal to the saturation value for a single device. In this case V_L of the series configuration becomes twice that of a single device. In Fig. 3(b), the predicted $V_L(R_L)$ for series configuration (black solid curve) and single device (red solid curve), coincide with each other at low R_L values. At around 10^7 Ω, V_L of series configuration starts to deviate to higher values. It is a factor of two higher than that of a single device when $R_L > 10^8$ Ω. It can be noticed in Fig. 3(b) that the measured V_L at R_L = 1 MΩ seems to have already deviated upwards. In fact, when R_L of 2 MΩ was tested for a single and series configuration, V_L of the series configuration (0.95 V) was significantly higher than the single device, with values of 0.95 V and 0.81V respectively.

CONCLUSIONS

In summary, the peak output voltage (V_L) of two identical PE devices (D1 and D2), mounted together on a vibrating cantilever was investigated as a function of load resistance (R_L). They exhibited similar performance with V_L increasing linearly as R_L increases from 4.6 KΩ to 1 MΩ. This behavior agrees with a model in which the PE device is considered a discharging capacitor with internal source of charge generation. The analytic form of the current delivered by the device to the load is obtained by solving the differential equation corresponding to the proposed capacitor. Modeled V_L is calculated by multiplying the peak current by R_L. When D1 and D2 were connected in parallel, V_L was doubled for the R_L values tested but when they were connected in series, V_L was the same as that of a single PE device. These observations are consistent with the model. Furthermore, the model predicts that as R_L becomes very large, V_L of parallel configuration would be equal to that of a single device. On the other hand, V_L of series configuration would be twice that of a single device. This work demonstrates that in general, PE devices can behave as a voltage source for large R_L values and a current source otherwise. However, the R_L values that determine whether a PE device will behave as a voltage or current source, also depend on other device parameters such as vibration frequency and device capacitance.

ACKNOWLEDGMENTS

This research is supported in part by NREL/DOE ACQ-4-33623-03 and NSF-CREST CA:HRD-0420516 and NSF-STC CliPS under grant #0423914.

REFERENCES

1. K A Cook-Chennault et al. *Smart Mater & Struc* **17** 043001 (2008)
2. H A Sodano et al. *Shock and Vibration Digest* 197 (2004)
3. A J Lovinger. Science **220** 1115 (1983)
4. J Granstrom et al. *Smart Mater & Struc* **16** 1810 (2007)
5. I Patel et al. Sensors and Actuators A. **159** 213 (2010)
6. R Guigon et al. *Smart Mater & Struc* **17** 015039 (2008)
7. C Chang et al. Nano Lett **10** 726 (2010)
8. H-S Yoon et al. J Intel Mat Sys & Struc **16** 877 (2005)

Mater. Res. Soc. Symp. Proc. Vol. 1312 © 2011 Materials Research Society
DOI: 10.1557/opl.2011.905

Polar-fluoropolymer Blends for High Energy Density Low Loss Capacitor Applications

Shan Wu[1, 2], Minren Lin[2], David S-G. Lu[2] and Qiming Zhang[1, 2, 3]
[1]Electrical Engineering Department, The Pennsylvania State University, University Park, Pennsylvania, USA.
[2]Materials Research Institute, The Pennsylvania State University, University Park, Pennsylvania, USA.
[3]Materials Science and Engineering Department, The Pennsylvania State University, University Park, Pennsylvania, USA.

ABSTRACT

Besides energy density, the electric loss at high electric fields is another major concern for many capacitor applications. This paper presents recent works in developing high energy density low loss polymer capacitors. In order to reduce the dielectric loss while maintaining high energy density in poly(vinylidene fluoride-hexafluoropropylene) P(VDF-HFP) and P(VDF-CTFE) (CTFE: Chlorotrifluoroethylene) based polymers, a polymer blend approach was investigated. We show that by blending P(VDF-CTFE) with a proper low loss polymer such as poly(ethylene-chlorotrifluoroethylene) (ECTFE) can lead to marked improvement in the loss of dielectric films. In this study, P(VDF-CTFE) blends films with different wt% of ECTFE have been examined to find a balance between dielectric constant and the loss. In addition, crosslink in the blends has been employed to further improve the dielectric performance of the blends. The results indicate that these blends exhibit an excellent performance: relatively high dielectric constant (~ 6~7) and low loss (~ 0.01) at 1 kHz. For the crosslink blend films, the high field loss is reduced to below 5% with a discharged energy density 4.3 J/cm^3 under a field of 300 MV/m.

INTRODUCTION

Polyfluoropolymers such as poly(vinylidene fluoride-hlorotrifluoroethylene) (P(VDF-CTFE)) has been developed as an outstanding dielectric material for energy storage applications recent years [1-6]. P(VDF-CTFE) exhibits a dielectric constant of 12 measured at 1 kHz and room temperature. Under an electric field > 600 MV/m, a charged energy density of more than 30 J/cm^3 can be reached, which represents a breakthrough in high energy density capacitor materials. On the other hand, the material shows a relatively high dielectric loss of 2.5%, while at high fields (>200 MV/m), the total electric loss can be 30%, which poses problems for several high energy density capacitor applications.

Blends, which consist of more than one dielectric material, have been exploited for improving the dielectric properties of polymers when the constituent materials are properly selected [7, 8]. PVDF blends have been investigated in the past and the blends films show enhanced polymer properties [9]. Here we investigate a polymer blend approach to reduce the dielectric loss while maintaining high energy density in the P(VDF-CTFE) based polymers. Furthermore, by employing crosslink in the two polymers of the blends, we show that the electric loss at high field can be further reduced [10-12]. In this study, Poly(ethylene-chlorotrifluoroethylene (ECTFE) is elected as the other polymer in the blends with P(VDF-

CTFE). There are several reasons for using ECTFE: (i) very lower dielectric loss (~ 0.0018 at 1 kHz) although the dielectric constant (~2.5) is also much lower than that of P(VDF-CTFE), (ii) common CTFE monomer in both P(VDF-CTFE) and ECTFE which makes it easier to form blends and for the subsequent crosslink, (iii) the melting temperature of 242 °C is within the temperature range for the melt process (extrusion) of the blend films.

EXPERIMENTAL

In this study, P(VDF-CTFE) 91/9 mol% copolymer and ECTFE were purchased from Solvay. Blends films with weight percent of ECTFE: 16 wt%, 30 wt%, 40 wt%, 50 wt%, are chosen to study how the blend composition influence the dielectric properties, energy density, and high field loss. Films were fabricated using extrusion process. For fabricating blend films of P(VDF-CTFE) and ECTFE, it was found that it is necessary to first blending the two polymers together at 260 °C and then extrude the blends pullets into films. This two-step process is needed due to large difference in the melting point (Mp) and melt flow index (MFI) between the two polymers. Table I lists Mp and MFI for P(VDF-CTFE) and ECTFE, and Table II is Extruder/Die Temperature for blends of ECTFE/P(VDF-CTFE) with ECTFE in the range of 16 wt% - 50 wt%.

Table I. Melting point and melt flow index of P (VDF-CTFE) and ECTFE

Polymer	Mp (°C)	MFI
P(VDF-CTFE)	169	15 (5 Kg/230°C)
ECTFE	242	18(2.16Kg/275°C)

Table II. Extruder/Die temperature for blends

wt% of ECTFE in the blends	Extruder/Die Temp
16% & 30%	260 ~270 °C
40%	265 ~275 °C
50%	275 ~285 °C

For the blend films, we also investigated the effect of crosslink on dielectric responses. The films were fabricated using a melt molding process for fabricating small quantity blend films. Here, 2,3-dimethyl-2,3-dipheybutan serves as the crosslink initiator. In this film fabrication process, powders of P(VDF-CTFE), ECTFE and initiator were mixed together. Freezer Mill was used for 2 hour to ensure that three components are mixed well. Then, the mixed powder was melt press into thin films under 3000 Lb pressure at 270°C for 2 hours followed by quenching into iced water to avoid any possible phase separation.

The thickness of the films for this study is in the range between 20 μm and 30 μm. They are electrode on both surfaces using Au sputtering. The polarization–electric field loop (P-E loop) is measured by a modified Sawyer–Tower circuit [13]. The stored (charged) energy density, released (discharged) energy density and efficiency are calculated from the P–E loop using the equation: $U_e = \int EdD$, the dielectric properties as function of temperature at different frequencies are characterized using a HP impedance analyzer (HP 4284) with a controlled temperature chamber (Delta 9023). The diameter of electrodes is 2 mm for the P-E loop measurement and 6 mm for the dielectric property measurement. In this study, the energy efficiency is introduced which is: efficiency=discharged energy density/total input energy density during the charge.

EXPERIMENTAL RESULTS

Fig. 1 presents P-E loops of pure P(VDF-CTFE) 91/9 mol% and ECTFE measured at room temperature and 10 Hz. P(VDF-CTFE) film shows a charged energy density of 36 J/cm^3, discharged energy density of 26 J/cm^3 under 600 MV/m. In general, there are two types of high field dielectric loss in polarfluoropolymers such as P(VDF-CTFE): ferroelectric loss and conduction loss. The loss increases non-linearly with the field and exhibits a peak of 35 % at about 250 MV/m, this type of loss is related with the ferroelectric switching by applied field so it is called ferroelectric switching loss; beyond 300 MV/m, the dielectric loss is mainly caused by another type of loss: conduction loss. There is a small reduce of dielectric loss with the field, and the dielectric loss is 29% at 500 MV/m. But the increasing conduction loss will lead to a larger loss again at field higher than 500 MV/m (31 % at 600 MV/m). In contrast, pure ECTFE film shows a P-E loop of a typical linear dielectric material, which has a much less dielectric loss (the observed loss is due to conduction which is caused by the dust particles and impurities introduced in the film fabrication process).

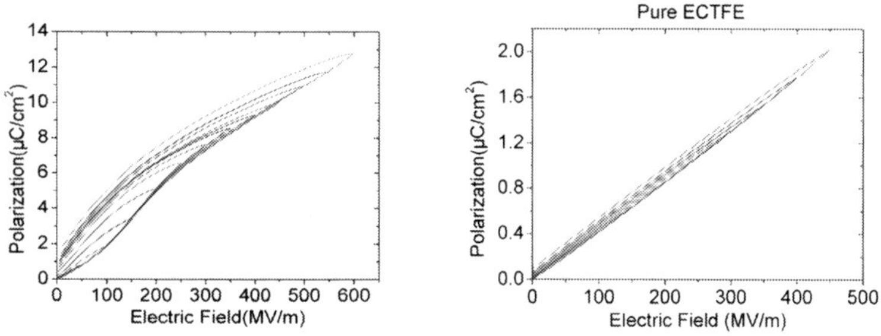

Figure 1. Polarization-Electric Field of Pure P(VDF-CTFE) and pure ECTFE

The Figure 2 shows the dielectric constant of pure P(VDF-CTFE) and blends with 30wt%, 40wt% and 50wt% ECTFE. As expected, the blends dielectric constant is reduced. Meanwhile, the dielectric loss is also reduced as shown in figure 2(b) for the blends with 30wt% ECTFE. Figure 3(a) is the comparison of P-E loops of the polymer blends under 300 MV/m field. The P-E loop of the blends now is much more close to a linear dielectric material compare to that of pure P(VDF-CTFE), which results in large increase in the efficiency while maintaining a relatively high energy density. From the P-E loop of pure ECTFE, there is still some conduction loss in the pure ECTFE films, which is caused by the impurities in the powder and introduced during the film process. The losses observed in these blends films at high fields are not very much higher than that from the pure ECTFE (see figure 4). In addition, the polymer blends films present relatively high breakdown fields that could reach 500 MV/m (figure 3(b)). This breakdown strength is a little lower than the original P(VDF-CTFE) (600 MV/m) because of some reasons such as interfacial issues and defects introduced during the process. These results indicate that the blends films with ECTFE can lead to significant improvement in the high field losses. For example, under 500 MV/m, films with 33wt% of ECTFE can reach a charged energy density of 12.7 J/cm^3 with more than 85% high field efficiency. For films with 40wt% ECTFE, the efficiency is higher than 90% and the charged energy density is higher than 10 J/cm^3.

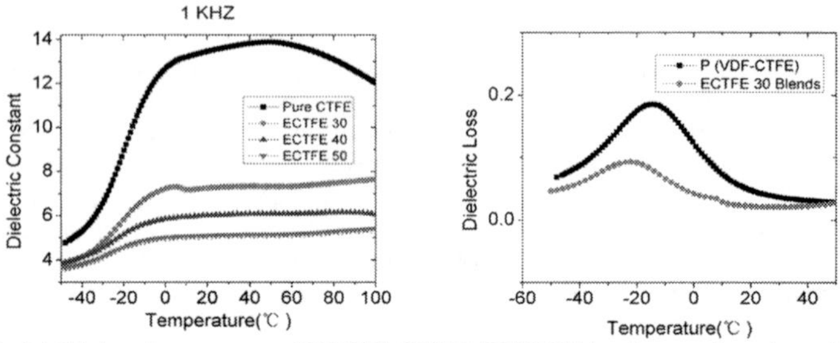

Figure 2. (a) Dielectric constant of P(VDF-CTFE)/ECTFE blends as a function of temperature measured at 1 kHz, (b) dielectric loss as a function of temperature for the blends with 0% and 30wt% ECTFE measured at 1 kHz.

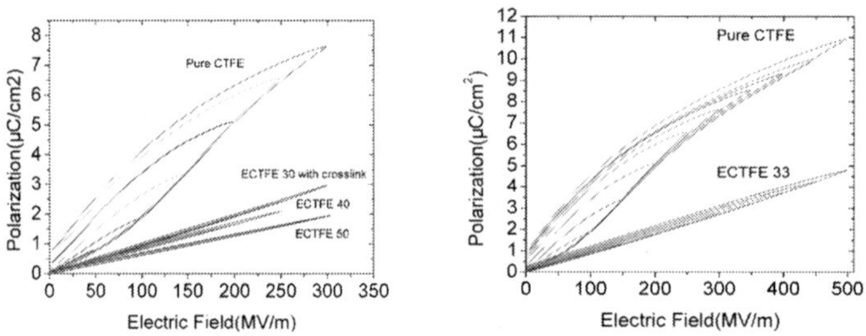

Figure 3. Polarization-Electric field loop of Pure P(VDF-CTFE), 30 wt% ECTFE, 40 wt% ECTFE, and 50 wt% ECTFE blends under 300MV/m (a) and 500 MV/m (b)

Crosslink in the PVDF based polymers is able to dramatically reduce the ferroelectric loss and conduction loss because the random defects caused by crosslink structures are much more effectively in limiting the formation of the large polarization domains under extremely high electric field than isolated CTFEs in P(VDF-CTFE) [14,15,16], which also results in an increased energy density. Figure 4 compares the P-E loop of blends films with or without crosslinking. The P-E loop of crosslinked blend films with 33wt% ECTFE much sharper than the films without crosslinking. Under 300 MV/m electric field the crosslinked blend film delivers a discharged energy density of 4.3 J/cm^3, which is higher than 3.9 J/cm^3 from the blend film without crosslinking. The most significant improvement here is that the crosslinked blend films exhibit efficiency over 95.2%, which is much higher than 85% efficiency from the non-crosslinked blend films.

Figure 5 summarizes the discharged energy density and efficiency for all the films studied, i.e., the pure P(VDF-CTFE) copolymer films and the blends films (no-crosslinked) as a function of applied electric field. Compared with the pure P(VDF-CTFE), all the blends films show much higher efficiency (> 85% under 500 MV/m electric field).

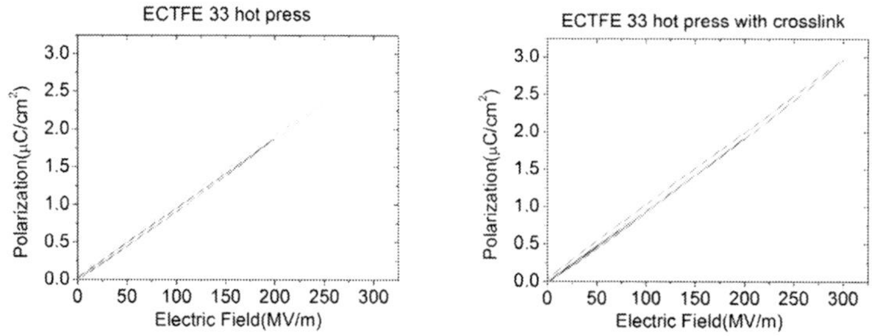

Figure 4. P-E response of the blend films with 33wt% (a) without crosslinking (b) with crosslinking.

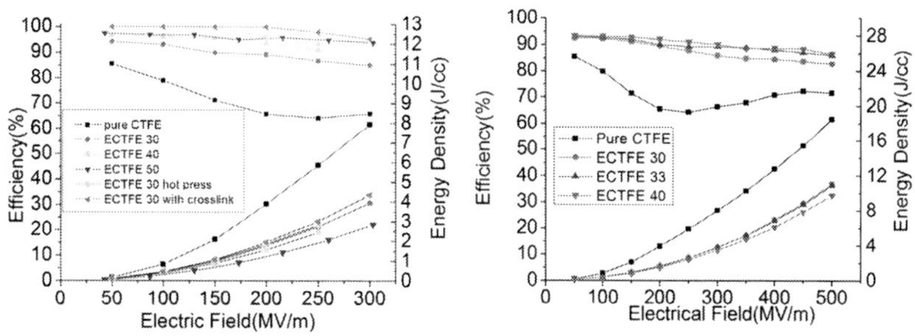

Figure 5. Discharged Energy density and efficiency of the pure copolymer and blends

Presented in figure 6 is dependence of the energy density and high field loss of the blends on the ECTFE wt% under an electric field of 300 MV/m. Because we cannot use the low field loss to predict the high field performance, here we define the high electric field loss that equals 1 – efficiency (%). Both energy density and high field loss decreases with the ECTFE content. What is remarkable is that the crosslinking in the blend films can increase the energy density and reduce the loss, compared with non-crosslinked films.

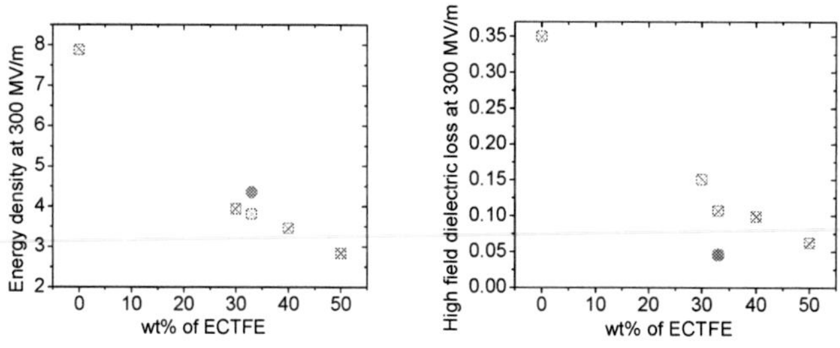

Figure 6. Discharged energy density (a) and efficiency (b) vs. the ECTFE wt% of the blends films under 300 MV/m. The dots are the data from crosslinked films, which exhibit higher energy density and efficiency than that of non-crosslinked films with the same blend composition.

CONCLUSIONS

The experimental results presented show that the approach of blend films can provide an effective avenue to developing dielectric materials which take advantages of the high energy density of one component and low dielectric loss of another component to realize both relative high energy density and low loss. For example, blend films with 40wt% ECTFE show the efficiency of higher than 90% and energy density higher than 10 J/cm^3 under 500 MV/m. Moreover, crosslink in the blend films provide further improvement the dielectric performance in terms of both energy density and efficiency, compared with non-crosslinked blend films.

ACKNOWLEDGMENTS

The authors would like to thank Dr. Xin Zhou for his assistance in the experiments. This work was supported by ONR under Grant No. N00014-08-1-0229 and N00014-05-1-0541 (MURI).

REFERENCES

1. B. Chu, X. Zhou, K. Ren, B. Neese, M. R. Lin, Q. Wang, F. Bauer, and Q. M. Zhang, Science 313, 334 (2006).
2. B. Chu, X. Zhou, B. Neese, Q. M. Zhang, and F. Bauer, IEEE trans. Dielectr. Electr. Insul. 13, 1162 (2006).
3. X. Zhou, B. Chu, B. Neese, M. R. Lin, and Q. M. Zhang, IEEE trans. Dielectr. Electr. Insul. 14, 1133 (2007).
4. Xin Zhou, Xuanhe Zhao, Zhigang Suo, Chen Zou, James Runt, Sheng Liu, Shihai Zhang, and Q. M. Zhang. Appl. Phys. Lett. 94, 162901(2009).
5. Z. Li, Y. Wang, and Z. –Y. Cheng, Appl. Phys. Lett. 88, 062904 (2006).
6. B. Chu, B. Neese, M. R. Lin, S. –G. Lu, and Q. M. Zhang, Appl. Phys. Lett. 93, 152903 (2008).
7. Z. Li, M. D. Arbatti, and X. Y. Cheng, Proc. SPIE 5385, 99(2004).
8. S. Zhang, B. Neese, K. Ren, B. Chu, and Q. M. Zhang, J. Appl. Phys. 100, 044113(2006).
9. H. S. Nalwa, Ed. Ferroelectric Polymers (Marcel Dekker, Inc. NY1995).
10. R. Casalini, C. M. Roland, Appl. Phys. Lett. 79, 16(2001).
11. R. Casalini, C. M. Roland, Journal of Polymer Science Part B: Polymer Physics, 40, 18(2002).
12. S. Liang, J. Claude, K. Xu, and Q. Wang, Macromolecules, 41, 6265-6268(2008).
13. J. K. Sinha, J. Sci. Instrum. 42, 696(1965).
14. Q. M. Zhang, V. Bharti, and X. Zhao, Science, 280, 2001(1998).
15. Q. M. Zhang, C. Huang, F. Xia, J. Su, and Y. Bar-cohen, SPIE Press, Bellingham, WA,2004.
16. F. Xia, Z. Cheng, H. Xu, H. Li, Q. M. Zhang, G. Kavarnos, R. Ting, G. Abdul-Sedat, and J. Belfield, Adv. Mater., 14, 1574(2002).

Mater. Res. Soc. Symp. Proc. Vol. 1312 © 2011 Materials Research Society
DOI: 10.1557/opl.2011.906

Influence of the pH on the Surface and Optical Properties of the Thin Film of Polyaniline / Polyethylene Terephthalate Composite. The AFM and Spectroscopies Studies.

Rafaella T. Paschoalin[1,2], Clarice Steffens[1,2], Alexandra Manzoli[1], Mhbuti R. Hlophe[3], Paulo S. P. Herrmann[1,2]

[1]National Nanotechnology Laboratory for Agribusiness (LNNA), Embrapa Agricultural Instrumentation, P. O. Box 741, 13560-970, São Carlos, SP, Brazil
[2]Biotechnology Graduate Program-UFSCar, São Paulo, Brazil
[3]North West University, Mafikeng Campus, South Africa

ABSTRACT

The effect of pH was investigated on the morphological and spectroscopic properties of PANI thin films on PET. PANI/PET strips were prepared by a pattern-printing technique and coated with a thin film of polyaniline *in situ* emeraldine oxidation state, doped with HCl (PANI-HCl), obtained by *in situ* polymerization. AFM, UV-VIS-NIR and FTIR/ATR spectroscopic results showed that pH has a great influence on the polymer layer morphology of the coating layers, while the intensity of vibrational bands decreases with rising pH, increase due to changes in the H-bonded structure in the polymer chain.

Keywords: Polyaniline, *in situ* polymerization, roughness, and conductivity.

1. INTRODUCTION

In the last few years, the interest in the development of new sensors for use in industry, agriculture, environment, medicine and other areas, has increased significantly. Among conducting polymers, polyaniline (PANI) has been studied more than any other, owing to its easy chemical synthesis and high environmental stability. PANI has electrical and optical properties suitable for use in optical sensors for monitoring pH [1], since its electronic absorption band sensitive to changes in pH is very broad, and optical sensors using PANI as the active layer are suitable for pH measurements in the range of 2 to 12 [2]. Given the advantages of low cost and short response time, pH optical sensors based on PANI thin film as the sensitive layer represent an interesting alternative to other sensors [3]. This paper reports the characterization of a low-cost optical sensor material assessing the influence of pH on a PANI film coated on PET. A printing technique was used to manufacture the coated pattern. PET (polyethylene terephthalate) and PANI thin film in the emeraldine oxidation state, deposited by *in situ* polymerization, were used as substrate and sensitive layer, respectively [4]. The morphology and roughness of the films were analyzed by atomic force microscopy (AFM). The degree of doping of the PANI was investigated by UV-VIS-NIR and FTIR/ATR spectrophotometry.

2. EXPERIMENTAL

2.1. Materials

Aniline (Aldrich) was purified by distillation. Ammonium persulfate ($(NH4)_2S_2O_8$), Sodium hydroxide (NaOH) and hydrochloric acid (HCl) were used as received (Merck). Water was purified to18 $M\Omega*cm^{-1}$ resistivity by the Milli-Q system (Millipore Inc.). The substrate PET (polyethylene terephthalate) was purchased from Nashua (XF-20, Nashua-USA).

2.2. Stripes

The stripes were developed by a pattern-printing technique [4,5] on a PET substrate. Briefly, a negative template was laser-printed, coated with PANI thin film by *in situ* polymerization and, thereafter, the toner was removed with toluene and M.E.K leaving the desired pattern.

2.3. In situ polymerization of PANI

All directions for formation of the PANI film *in situ* by polymerization in the emeraldine oxidation state on a surface have been published elsewhere [5,6].

2.4. UV-Visible-NIR Spectra

The UV-Visible-NIR spectra of PANI thin films were collected at various pHs in the range from 2 to 12 in a Shimadzu UV-1601PC model Spectrophotometer at wavelengths from 400 to 2000 nm.

2.5. Fourier transform infrared spectroscopy/ attenuated total reflectance (FTIR/ATR)

The FTIR / ATR spectra of the PANI films were collected at various pHs in the range from 2 to 12, in a Perkin-Elmer (Spectrum 1000) in the spectral region between 4000 and 650 cm^{-1}, with 16 scans per spectrum and 2 cm^{-1} resolution. A ZnSe crystal was used as internal reflection element, with an input face at 45 °. For each pH solution a single fresh strip of PET/PANI was used.

2.5. Atomic Force Microscope (AFM)

The morphology and roughness of PANI thin films on the PET substrate were analyzed with a Dimension V (Veeco) AFM, using a pyramidal silicon nitride tip attached to a cantilever with a spring constant of 0.03 N m^{-1}, in contact mode at a scan rate of 1 Hz. The images were processed with the aid of the data analysis package Gwydion© 2.1 data analysis software.

3. RESULTS AND DISCUSSION

3.1. Characterization of PANI thin films on PET in different pHs solutions.

3.1.a. UV-Visible-NIR Spectra

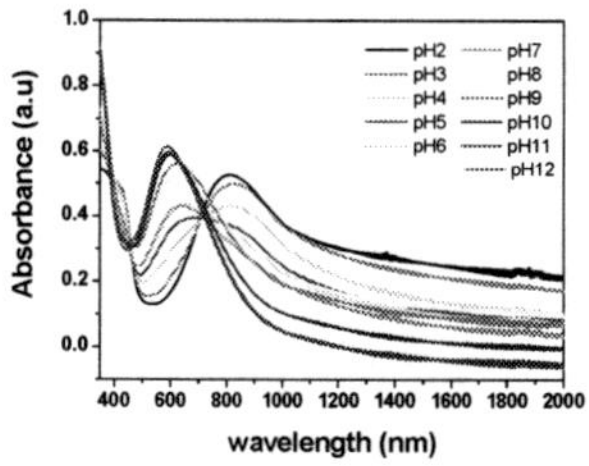

Figure 1. Spectra of UV-Vis-NIR of PANI thin films at various pHs.

In Figure 1 it is seen that of the band located at 600 nm increases in absorbance with pH increasing (basic solution). This band corresponds to an n-π * transition in the quinoid rings, characteristic of the dedoped form of PANI [7]. The other two bands located around 420 nm and 800 nm are characteristic of doped PANI and an these show an increasing interest with decreasing pH (acid pHs). With increasing pH, there was also a steady fall in the localized polaron peak and of the free carrier tail between 1000-2000 nm. An increase of the free carrier tail this is associated with higher conductivity [8].

3.1.b. Fourier transform infrared spectroscopy/ attenuated total reflectance (FTIR/ATR)

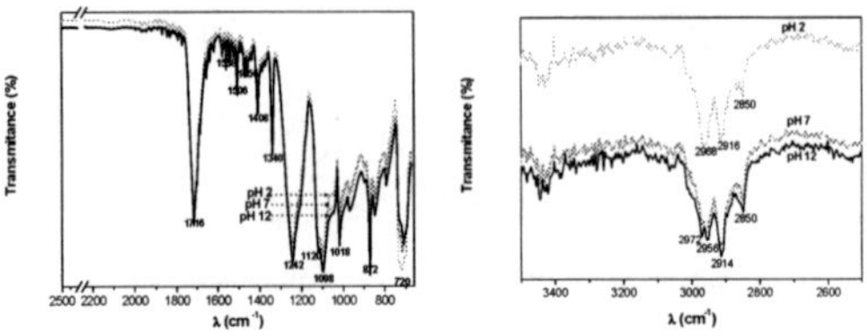

Figure 2. *FTIR / ATR of PANI thin film in different pHs*

The film of PET / PANI in acid conditions (pH = 2) in Figure 2 showed transmittance peaks at 2968, 2916, 2850, 1716, 1558, 1506, 1456, 1408, 1340, 1242, 1120, 1098, 1018, 872 and 720 cm^{-1}, which are consistent with values reported in the literature for PANI[9]. At pHs higher than 7 was observed two additional bands were seen at 2972 and 2956 cm^{-1}. The bands at 872 and 720 cm^{-1} are characteristic on p-substitution of the aromatic ring and they are assigned as C-H out-of- plane deformation vibration of the benzoic groups. The broad band at 720 cm^{-1} can be also attributed to N-H bond wagging of the secondary amine. The rings are mostly coupled at C4 and N positions, the corresponding C-N stretching modes are observed in the film of PET / PANI form at 1242 and 1098 cm^{-1}. The band around 1506 cm^{-1} results from the C-C ring-stretching mode and the band at about 1558 cm^{-1} (medium intensity) arises from a disorder-induced C-C ring stretching mode. The vibrational mode in the range of 1120 cm^{-1} is assigned as either C-H bending or ring amine bending vibrations. The bands at about 1408 and 1340 cm^{-1} have three contributions, coming from C-C ring deformation, C-H bending and C-N bending modes. The vibrational absorption at 1716 cm^{-1} is assigned as a C = N vibrational mode. These results are in agreement with an earlier report by Epstein and collaborators and Ping and collaborators [10,11]. Around 3446 cm^{-1}, two bands of different intensities appear (broad vibrational bands in the range of 3400 to 3350 cm^{-1} and 3300 to 3200 cm^{-1}), which are attributed to H bonding, due to OH stretching and N-H_2 asymmetric stretching vibrations. These bands can also be called overtones, which appear at half the intensity of that at 1716 cm^{-1}. The differences in the intensities of bands at 2968, 2916 (pH = 2), 2972, 2956 and 2914 cm^{-1} (pH = 12) are caused by the differences in the H-bonding between the doped and dedoped forms of PANI thin film on PET. Thus, the transmittance of vibrational bands decreases as pH rises owing to changing the H-bond pattern in the polymer chain. At basic pH, the bands also showed a broadening due to interaction of the OH group. Crosslinking of the chains is caused by hydrogen bonds formed by the amine and imine units. This generates a molecular conformation with is a twist between adjacent rings, causing a less planar structure between the rings and benzoid [12].

3.1.c. Roughness

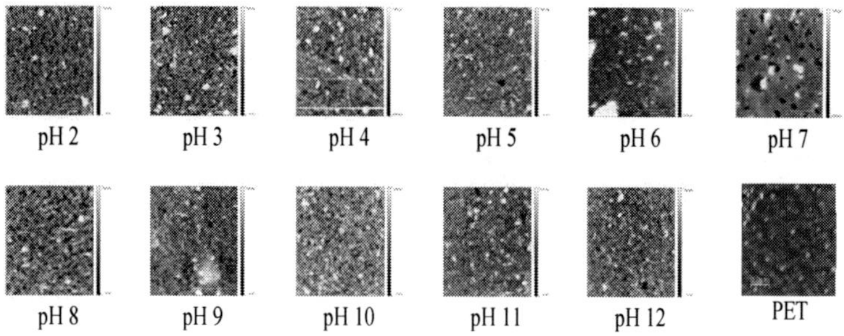

Figure 3. Topographic images of PANI thin films at different pHs. The area of each image is 100μm^2.

Table I. Average (ra) and root mean square (rms) roughness of PANI thin films at different pHs.

pH	**2**	**3**	**4**	**5**	**6**	**7**	**8**	**9**	**10**	**11**	**12**
ra (nm)	37.9	51.0	62.0	29.1	31.8	43.7	44.5	31.2	25.8	28.4	20.4
	±9.28	±9.46	±3.30	±0.82	±0.45	±0.72	±2.64	±0.54	±1.03	±0.57	±0.85
rms (nm)	52.6	74.0	90.0	38.2	45.7	66.8	58.1	42.0	33.6	39.2	27.0
	±24.0	±15.5	±4.30	±1.12	±1.00	±1.39	±3.99	±0.62	±2.77	±2.33	±0.42

An increase in roughness can be discerned from pH 2 to 4. In figure 3 and table I. At acid pH, the amine groups of PANI are more protonated, which favoring H bond interactions. Comparing pH 4 with pH 2, there are more interactions by H-bonding at pH 4, as there is a better balance equilibrium between protonated and deprotonated amines. As the folds and coils in polymer chains are formed by H-bonding, it is be expected that at pH 4 the polymer chains are more convoluted, that the PANI films are rougher. Thin films of PANI have a thickness around 140 nm, so the degree of roughness between pH 5 and 12 can be attributed to a solvent effect. At intermediate pH

(5 to 8), polymer-solvent contact is more favorable than polymer-polymer contact, making the folding more expanded. At more basic pH, the interaction of polymer with the solvent is weaker than the polymer-polymer interaction, making the folding more compact and thereby decreasing the roughness. Thus, the shrunk folded chain is smoother than the expanded folded chain and there was a decrease in the roughness of the films from pH 9 to 12.

CONCLUSION

The results showed that the polymer layer morphology and spectroscopic properties are influenced by the pH. The PANI doped film exhibited strong interactions of the type hydrogen bonding type between the amines and imines along the chain, that induces the formation of polymeric aggregates.

REFERENCES

1. D. Bloor, A. Monkman, *Synthetic Metals,* **21**, 1-3, 175-179 (1987).
2. U.W. Grummt, A. Pron, M. Zargorska, S. Lefrant, *Analytica Chimica Acta,* **98**, 2-3, 247-253 (2004).
3. P.T. Sotomayor, I.M. Raimundo, A.J.G. Zarbin, J.J.R. Rohwedder, G. Oliveira, O.L. Alves, *Sensors and Actuators B*, **74**, 157-162 (2001).
4. E.C. Venancio, L.H.C. Mattoso, P.S.D. Herrmann, A.G. MacDiarmid, *Sensors and Actuators B-Chemical,* **130**, 723-729 (2008).
5. C. Steffens, A. Manzoli, E. Francheschi, M. L. Corazza, F. C. Corazza, J. V. Oliveira, P.S.P. Herrmann, *Synthetic Metals,* **159**, 2329 (2009).
6. R.V. Gregory, W.C. Kimbrell, H.H. Kuhn, *Synthetic Metals*, **28**, 1-2, C823-C825 (1989).
7. A. E. Job, P. S. P. Herrmann, D. O. Vaz, L. H. C. Mattoso, *Journal of Applied Polymer Science*, **79**, 1220–1229 (2001).
8. W. Zheng, Y. Min, A. G. MacDiarmid, M. Angelopoulos, Y-H. Liao, A. J. Epstein, *Synthetic Metals*, **84**, 63-64 (1997).
9. A. F. Baldissera, D. B. Freitas and C. A. Ferreira, *Materials and Corrosion,* **61**, 790-801 (2010).
10. A.J. Epstein, R.P. Mccall, J.M. Grinder, A.G. MacDiarmid, *Spectroscopy of Advanced Materials*, ed. R.J.H. Clark and R.E. Hester (John Wiley, 1991), p. 355.

11. Z. Ping, G. Nauer, E.H. Neugebauer, J. Theiner, A. Neckel, *Electrochimica Acta,* **42**, 1693-1700 (1997).
12. M. Angelopoulos, R. Dipietro, W. Zheng, A.G. MacDiarmid, A.J., *Synthetic Metals,* **84**, 35-39 (1994).

ACKNOWLEDGEMENTS

The author would like to thank the CNPq and FAPESP with financial support of the following research projects: (IBAS-CNPq nº:490807/2007-0, INCT-NAMITEC process nº:573738/2008-4 (CNPq) and nº.2008/57862-6 (FAPESP)), FAPESP scholarship in the graduated level, project nº: 2010/04188-6 and project n° 2009/08244-0.

Mater. Res. Soc. Symp. Proc. Vol. 1312 © 2011 Materials Research Society
DOI: 10.1557/opl.2011.670

Ion Transport and Storage in Ionic Polymer Bending Actuators

Jun-Hong Lin[1], Yang Liu[2], Gokhan Hatipoglu[2], and Qiming Zhang[1,2]

[1]Department of Materials Science and Engineering, The Penn State University, PA16801, U.S.A.
[2]Department of Electrical Engineering, The Penn State University, PA16801, U.S.A.

ABSTRACT

The actuation of ionic polymer actuators is mainly caused by the ion transport and excess ions storage in the membrane and electrodes. To quantify the charge transport behavior, a time domain method based on Poisson-Nernst-Planck equations was applied. The time domain transient current in response to a step voltage can provide insights on the charge transport and storage behaviors in the membranes. In this study, we investigate the charge transport behavior of Aquivion ionomer with different uptakes of 1-ethyl-3-methylimidazolium trifluoromethanesulfonate (EMI-Tf). A critical uptake and voltage independent of charge transport behavior were observed. The results also show that bending actuations of the Aquivion membrane with 40wt% EMI-Tf is much larger than that of Nafion, indicating that the shorter flexible side chain ionomer possesses a better electromechanical coupling between the excess ions and the membrane backbones, while not affect the actuation speed.

INTRODUCTION

Ion transport and storage in ionomer membranes are of great interest for electroactive polymer (EAP) devices, such as actuators, sensors, energy harvesting devices, and supercapacitors.[1-3] Ionic liquids (ILs), which are a class of salt in liquid form that containing both ions and neutral molecules, possess many interesting properties that make them very attractive to be employed as electrolytes in electroactive polymer devices.[4-6] For example, the vapor pressure of ionic liquids is negligibly low and as a result they will not evaporate out of the EAP devices when operated in ambient condition. It has been demonstrated that comparing to water the use of ILs as solvent for EAP actuators can dramatically increase the lifetime of transducer.[12,13] Their high mobility leads to the potential of fast response of EAP devices while the wide electro-chemical window allows for higher applied voltages.[8-14]

This paper investigates charge dynamics of room temperature ionic liquid EMI-Tf in Aquivion® (Hyflon®) membranes swelled with different uptakes of ILs and the actuations of the membranes swelled with ILs above the critical uptake. EMI-Tf served in this study due to its comparably high conductivity (0.86 S/m), low viscosity (45 cP at 25°C) and larger electrochemical window(4.1V).[12] Hyflon® is known in the literature as short-side-chain ionomer (in comparison to Nafion® that is indicated as long-side-chain ionomer) and was originally developed by Dow Chemicals Company at the beginning of the '80 but now named Aquivion® by Solvay Solexis.[17-20]

When swelled with ILs the clusters expand. Above a certain IL uptake (critical uptake), these clusters will be connected with narrow channels forming percolation pathways for easy ion conduction, resulting in a high conductivity in the ionomer membrane. [15-20] Although in the literature, most ionic EAPs employ Nafion as the polymer matrix, [9-12] the short flexible side chains in Aquivion may provide better mechanical coupling between the ions and membrane backbones and, consequently, more efficient electromechanical transducers.

In this study, we investigate the charge transport and storage in Aquivion EAP membranes, where the Aquivion membrane is coated with Au electrodes on two surfaces and swelled with a given amount of EMI-Tf. Under applied voltage, anions and cations migrate towards the anode and cathode, and form electric double layers and diffusion layers. Moreover, the excess ions at the two electrodes cause strains in the membranes, and as a result, generate bending (actuation) of the membrane. Therefore, these ionomer membranes are attractive for ionic polymer actuators and transducers operated under low voltage (1-4V).

In general, charge transport is a result of drift and diffusion and can be described by Poisson-Nernst-Planck equations,

$$\varepsilon\varepsilon_0 \frac{\partial E}{\partial x} = \rho \quad (1), \qquad \psi_{\pm} = \pm \mu n_{\pm} \text{E-D} \frac{\partial n_{\pm}}{\partial x} \quad (2)$$

where ρ is the charge concentration, ε the dielectric constant of the medium, ε_0 the vacuum permittivity, ψ is the ion flux density (current density $J=q\psi$), μ is the ion mobility, n is the ion concentration, E electric field, D diffusion coefficient. For the ionic liquids studied here, we assume $n_+=n_-$, $\mu_+=\mu_-$, μ and D are related by the Einstein equation, $D = \mu kT / q$. [21,22]

For the ionomer membrane under a step voltage, the initial current density is $I_0=\sigma$ V S/d, where σ (=qnμ) is the conductivity, d is the membrane thickness, and S is the electrode area. When the applied voltage is in the order of 10kT, (where k is the Boltzmann's constant), the initial transient current follows the charging of an electric double layer capacitor C_D in series with a bulk resistor R, [23-26]

$$I(t) = I_0 \exp(-t/\tau_{DL}) \qquad (3)$$

where τ_{DL} = d λ_{DL}/2D = RC_D, describes the typical charging time for the electric double layer which has a thickness λ_{DL}, the Debye length,

$$\lambda_{DL} = (\varepsilon\varepsilon_0 kT / Z^2 q^2 n)^{1/2} \qquad (4)$$

where Z is the valence (=1 for EMI-Tf), and q=e, electron charge. It was further shown that at longer time, the charge diffusion from the bulk to the double layer region leads to a power law decay of the diffusion current (see figure 1(a), in which the initial current fits well by eq. (3), followed by a power law decay of the diffusion current, having a typical time constant $\tau\sim d^2/(4D)$). Therefore, by fitting experimental transient current I(t) with eq. (3), as illustrated in figure 1(a), σ, n, and μ can be obtained if ε of the ionomer membrane (with ILs) is known, [23-26] as shown in figure 2(b).

$$\sigma = \frac{I_0 d}{VS} \qquad \mu = \frac{qVS\varepsilon\varepsilon_0 d}{4kT\tau_{DL}{}^2 I_0} \qquad n = \frac{4kTI_0^2\tau_{DL}^2}{\varepsilon\varepsilon_0 q^2 V^2 S^2}$$

Impedance spectroscopy is employed to determine ε as shown in figure 1(b). Besides ε, σ (=d/(SR), where d is the thickness and S the surface area of the membrane), can also be determined from the Nyquist plot.

RESULTS

The current responses for Aquivion with 40wt% of EMI-Tf under various step voltages are measured. From these curves, conductivity σ under different applied voltages can be determined from I_0 (up to 4 volts in the experiment). On the other hand, the data shows that fitting to eq. (3) can be performed only for I(t) at $V \leq 1$ volts. Beyond that, there is no distinctive crossover region in the I(t) curve between the exponential decay of the drifting current and the power law decay of the diffusion current.

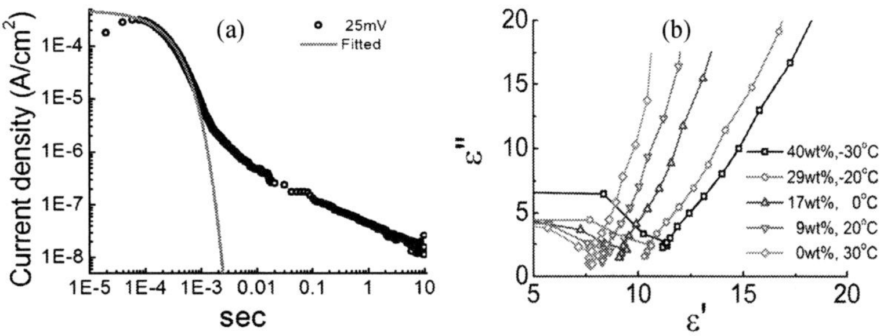

Figure 1(a) The transient current of Aquivion membrane with 40wt% uptake of EMI-Tf when a voltage step from 0V to 25 mV is applied.(b) The cole-cole plot of Aquivion membrane with various EMI-Tf uptakes under various temperatures

Results for the conductivity σ, free charge concentration n and mobility μ as a function of IL uptake are presented in figure 2. The data reveals that the mobile ion concentration increases almost linearly with the IL uptake. In contrast, the conductivity and mobility display abrupt changes with the increase of IL uptake, revealing a critical uptake of EMI-Tf in the Aquivion (EW790) membrane somewhere between 18 wt% and 29 wt%. Above it the conductivity and mobility dramatically increase. In the study of IL uptake in Nafion membranes, Leo et al. found that the critical uptake is closely related to the minimum amount of IL required to displace the counter ions away from the exchange sites. [12] Therefore, the critical uptake of EMI-Tf in Aquivion should be higher than that of EMI-Tf in Nafion membrane (23 wt%), due to that Aquivion (EW790) has a higher side chain density than Nafion(EW1100).Above the critical uptake, the ionomers form percolation path ways for charges to transport and also there are more mobile ions that do not strongly interact with exchange sites. Therefore, with the increase of IL content the mobile charge concentration increase and the mobility and conductivity are enhanced. With 40 wt% of EMI-Tf uptake, at which a substantial actuation is observed in both Aquivion and Nafion membranes, the conductivities of Aquivion and Nafion membranes are close to each other (Aquivion(0.11 mS/cm, Naifon 0.21 mS/cm) and their modulus are also similar.(Aquivion 61Mpa and Nafion 53Mpa).

In figure 2(a), σ determined from Nyquist plot under 0.1 V is also shown. The two methods yield nearly identical values of σ, indicating that both methods can be used to determine σ at low voltage. On the other hand, at high voltage, the ac electric impedance method to determine σ may become difficult due to non-linear electric response and even possible heating in the samples.

Figure 3(a) presents the conductivity σ of Aquivion membrane with different EMI-Tf uptake as a function of voltage, up to 4 volts. The data shows that within the experimental error ($\pm 10\%$), σ does not change with applied voltage for all EMI-Tf uptakes.

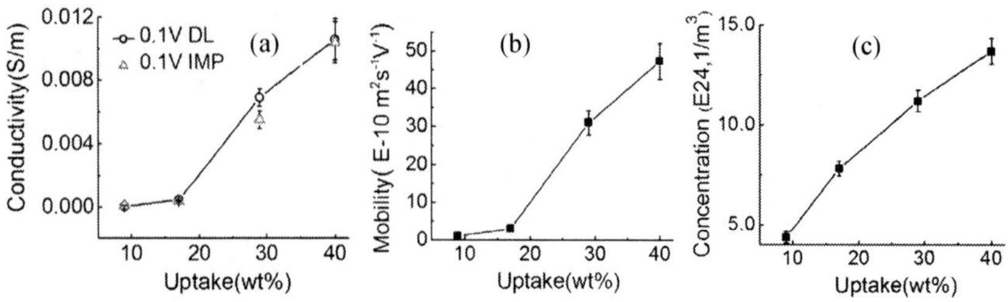

Figure 2 (a) conductivity (b) mobility and (c) mobile charge concentration as a function of EMI-Tf uptake in Aquivion membrane when a voltage step from 0V to 100mV is applied.

Figures 3(b) and (c) are the mobility and mobile ion concentration vs. applied voltage, measured up to 1 volt, and within the experimental error, no systematical variation with applied voltage was observed. These results indicate that the double layer charging time τ_{DL} as well as the overall charge dynamics do not change much with voltage, below 1 volt.

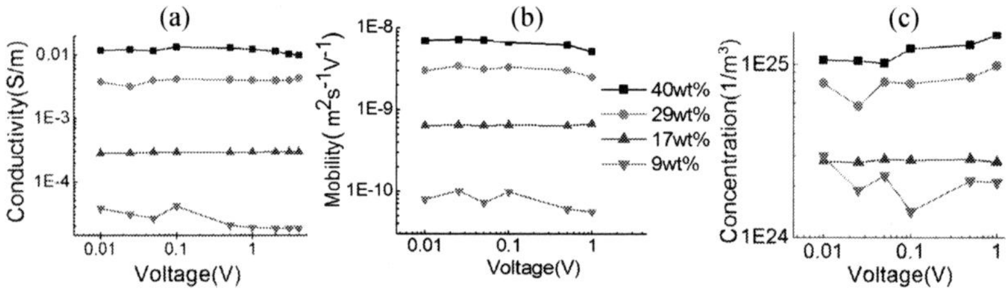

Figure 3 (a) Conductivity (b) mobility (c) mobile charge concentration of Aquivion membrane with different uptake under different applied voltage step.

Comparing the mobile ion concentration with the total ion concentration from the EMI-Tf, we found only about 0.5% to 0.55% of EMI-Tf is dissociated for all the ILs uptakes below 1 volt. This low dissociation ratio of ionic liquid may due to the fact that the Bjerrum length of EMI-Tf is larger than the ion pair distance and hence the Coulomb force dominates [31, 32].

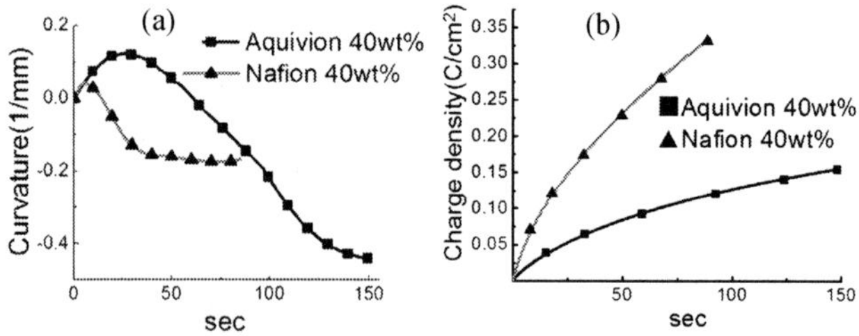

Figure 4 (a) The curvature change with time (b) the accumulated charge/area of a 62um thick of Aquivion and Nafion membrane when a voltage step from 0V to 4V is applied.

Presented in figure 4(a) is the curvature (1/R where R is the radius) of the bending membrane actuators vs. time after the application of a step voltage of 4 volts. The strain near the two electrodes (anode and cathode) is proportional to the curvature (1/R). [9] The data reveals several interesting features: (i) the initial slope of the curvatures of the two ionomers is nearly the same, indicating similar actuation speed of the two membranes; (ii) the maximum curvature of the Aquivion membrane is higher than that of the Nafion. As shown in earlier publications, the presence of both cations and anions in the membrane may cause cancellation in bending actuation. [9] The results in figure 4(a) indicate that the ions in Aquivion can generate much more bending strain, compared with that of Nafion membrane.

Figure 4(b) shows the charge stored in the two membranes under the same electric conditions as in figure 4(a). Combining the data in the two figures indicates that the bending strain generated per charge is much large in Aquivion than that in Nafion. The difference in the strain generation is likely caused by the weaker elastic coupling of the excess ions to the membrane backbones in Nafion, due to longer flexible side chains, compared with that of Aquivion. As shown in Kreuer's study [20], from the SAXS, the average width of hydrophilic channels of Aquivion is smaller and did not increase as much as that of Nafion with the increasing of water uptake. This implies that by reducing these soft side chain lengths, the excess ions in the ionomers will couple more effectively with the backbones to generate strain of the membranes.

CONCLUSION

In conclusion, a time domain approach is applied to study the charge dynamics in an ionomer membrane (Aquivion) swelled with ILs which, when combined with recently developed theoretical works, allows quantitative analysis of the membrane performance under real device working voltages (> 1 volts). A critical uptake of EMI-Tf in Aquivion membrane is observed, above which the charge mobility and mobile charge concentration increase markedly. It is also observed that the dissociation ratio of the swollen EMI-Tf remains at 0.5% for all IL uptakes. The experimental results reveal that the charge transport behavior does not change with applied voltage. Furthermore, it was found that the actuation of the ionic polymer membrane actuator is dominated by slow diffusion charges, which time constant scales with square of the membrane thickness. Therefore, by reducing the membrane thickness, the actuation speed can be increased. A comparison of the actuation strain shows that the short side chain Aquivion exhibits a better electromechanical coupling with ions than that of the long side chain Nafion, while the actuation speeds of the two membranes under same electrical stimulus are the same. Therefore, short side chain ionomers are preferred for ionic polymer actuator applications.

ACKNOWLEDGMENTS

This material is based upon work supported in part by the U.S. Army Research Office under Grant No. W911NF-07-1-0452 Ionic Liquids in Electro-Active Devices (ILEAD) MURI, by NSF under Grant No. CMMI 0709333, and by NIH under Grant No. R01-EY018387-02. The authors thank Ralph Colby and Sheng Liu for many stimulating discussions.

REFERENCES

1. Lu W, Fadeev AG, Qi BH, Smela E, Mattes BR, Ding J,and Forsyth M. Science 2002;297(5583):983-987.

2. McEwen AB, Ngo HL, LeCompte K, and Goldman JL. Journal of the Electrochemical Society 1999;146(5):1687-1695.
3. Ue M, Takeda M, Toriumi A, Kominato A, Hagiwara R, and Ito Y. Journal of the Electrochemical Society 2003;150(4):A499-A502
4. Galinski M, Lewandowski A, and Stepniak I. Electrochimica Acta 2006;51(26):5567-5580.
5. Ono S, Seki S, Hirahara R, Tominari Y, and Takeya J. Applied Physics Letters 2008;92(10)
6. Bar-Cohen Y and Zhang QM. Mrs Bulletin 2008;33(3):173-181.
7. Kim D, Kim KJ, and Tak Y. Applied Physics Letters 2007;90(18)
8. Bennett MD and Leo DJ. Sensors and Actuators a-Physical 2004;115(1):79-90.
9. Liu Y, Liu S, Lin JH, and Zhang QM. Applied Physics Letters 2010;96(22)
10. Liu S, Liu WJ, Liu Y, Lin JH, Zhou X, Janik MJ, Colby RH, and Zhang QM. Polymer International 2010;59(3):321-328.
11. Akle BJ, Leo DJ, Hickner MA, and McGrath JE. Journal of Materials Science 2005;40(14):3715-3724.
12. Bennett MD, Leo DJ, Wilkes GL, Beyer FL, and Pechar TW. Polymer 2006;47(19):6782-6796.
13. Watanabe M, Shirai H, and Hirai T. Journal of Applied Physics 2001;90(12):6316-6320.
14. Watanabe M, Shirai H, and Hirai T. Journal of Applied Physics 2002;92(8):4631-4637
15. Mauritz KA and Moore RB. Chemical Reviews 2004;104(10):4535-4585.
16. Hsu WY and Gierke TD. Journal of Membrane Science 1983;13(3):307-326.
17. Ghielmi A, Vaccarono P, and Arcella V. Journal of Power Sources 2005;145(2):108
18. Jalani NH and Datta R. Journal of Membrane Science 2005;264(1-2):167-175
19. Halim J, Buchi FN, Haas O, Stamm M, and Scherer GG.Electrochimica Acta 1994;39(8-9):1303
20. Kreuer KD, Schuster M, Paddison SJ, and Maier J. Journal of Power Sources 2008;178(2):499-509
21. Bazant MZ, Thornton K, and Ajdari A. Physical Review E 2004;70(2).
22. Kilic MS, Bazant MZ, and Ajdari A. Physical Review E 2007;75(2).
23. Beunis F, Strubbe F, Marescaux M, Beeckman J, Neyts K, and Verschueren ARM. Physical Review E 2008;78(1).
24. Strubbe F, Verschueren ARM, Schlangen LJM, Beunis F, and Neyts K. Journal of Colloid and Interface Science 2006;300(1):396-403.
25. Beunis F, Strubbe F, Neyts K, and Verschueren ARM. Applied Physics Letters 2007;90(18).
26. Marescaux M, Beunis F, Strubbe F, Verboven B, and Neyts K. Physical Review E 2009;79(1)
27. Wakai C, Oleinikova A, Ott M, and Weingartner H. Journal of Physical Chemistry B 2005;109(36):17028-17030.
28. Serghei A, Tress M, Sangoro JR, and Kremer F. Physical Review B 2009;80(18).
29. Krause C, Sangoro JR, Iacob C, and Kremer F. Journal of Physical Chemistry B 2010;114(1):382-386.
30. Lockett V, Sedev R, Ralston J, Horne M, and Rodopoulos T. Journal of Physical Chemistry C 2008;112(19):7486-7495.
31. Klein RJ, Zhang SH, Dou S, Jones BH, Colby RH, and Runt J. Journal of Chemical Physics 2006;124(14).
32. Fragiadakis D, Dou S, Colby RH, and Runt J. Journal of Chemical Physics 2009;130(6).

Mater. Res. Soc. Symp. Proc. Vol. 1312 © 2011 Materials Research Society
DOI: 10.1557/opl.2011.907

Investigation of Inherently Conductive Polymer as Structure Health Monitoring Sensor for Composite

Huaxiang Yang[1], Dongsik Kim[1], Abhishek K. Singh[1], Brady W. Pitts[1], Gregory J. Tregre[1], Patrick J. Kinlen[2]
[1] Crosslink, 46 Shelby Thames Drive, Hattiesburg MS 39402, U.S.A
[2] Crosslink, 950 Bolger Court, St. Louis, MO 63026, U.S.A

ABSTRACT

A soluble polyaniline was synthesized through emulsion polymerization and characterized using differential scanning calorimetry (DSC), thermogravimetric analysis (TGA), gel permeation chromatography (GPC), viscosity analysis, and coefficient of linear thermal expansion (CLTE) determination. The electrical conductivity is found to reach 1000 S/cm with specific post doping treatments. Multiple printing processes, such as inkjet printing, screen printing and aerosol jet printing etc, make it feasible to print a variety of sensor patterns. The electromechanical response of these sensors was used to measure strain/stress or damage of composite structures under various load conditions expected to be experienced by aircraft. These unique conductive polymer sensors provide a feasible, near real time monitoring system for composites without adding significant additional weight to the structure.

INTRODUCTION

Inherently conductive polymers (ICPs), which have a delocalized π-electron character, have attracted considerable attention because of their unique properties and potential applications. Among ICPs such as polyacetylene (PA), polypyrrole (PPy) and polythiophene (PT), polyaniline (PANI) is relatively popular because of its good conductivity, stability, ease of preparation, affordability and redox properties [1]. There are several methods for synthesis of PANI including heterophase polymerization [1,2], vapor deposition polymerization [3], and electro-chemical polymerization [4] etc. PANI produced by Emulsion polymerization with good solubility, processability and electrical conductivity [2] has been extensively investigated [1].

Structural health monitoring (SHM) is a process of implementing a damage sensing strategy which is expected to prevent catastrophic failure by, partially or completely, predicting the onset of failure. SHM systems may be classified as either passive or active sensory smart structures. A passive system contains only sensors which monitor structure's response excited by external signal. An active system has sensors and actuators in which actuators provide a well defined signal and sensors monitor the structure's response to the signal [5]. From the viewpoint of deformation/damage of the structure under load condition, there are two major sensors involved in the SHM systems: damage sensor and strain/stress sensor. To sense damage in a structure is helpful for structural health monitoring, timely repair and safety enhancement. To sense strain or stress in the structure is helpful for structural vibration control, load monitoring and recording load history. The ability to sense both strain and damage is more valuable than that of sensing either of them. For instance, acoustic emission can detect damage infliction but can not detect stress and strain in the absence of damage. [6] Ultrasonic inspection can detect damage such as cracking, but can not detect stress/ strain [7]. In contrast, electrical resistance measurement can provide simultaneous sensing of both damage and stress/strain because of their effects on resistance.

In regards to sensors in SHM applications, most of the research activities have focused on sensing materials such as carbon fibers (nanofibres) [8], carbon nanotubes [9], optical fibers [10], and piezoelectric materials [11]. In applications such as aircraft and spacecraft, the additional weight added to the structure by the sensing material is as critical as its sensitivity performance. From this point of view, inherently conductive polymers (ICP) are excellent candidate materials which show electrical resistivity changes under mechanical load. Recently, several investigations using poly(3,4-ethylenedioxylthiophene /polystyrene sulfonic acid (PEDOT/PSS) as SHM sensor were reported[12] However, overall there are few researchers investigating the use of ICP as SHM sensors. Since 2008, Crosslink has been working on solution processible PANI based SHM sensor for composites [13-15]. EMPAC™1003, developed by Crosslink, is a processable polyaniline solution with adjustable conductivity ranging from 10^{-3} to 10^{3} S/cm [16]. The advantage of this soluble PANI material is its ability to be applied using diverse and controllable techniques such as draw-down/brush coating, spray printing, screen printing, spin coating, inkjet printing, aerosol jet printing, etc. [14, 15]

This investigation will focus on characterization of PANI solution (PAC1003), through differential scanning calorimetry (DSC), thermogravimetric analysis (TGA), gel permeation chromatography (GPC), viscosity analysis and identification of the coefficient of linear thermal expansion (CLTE). With various printing techniques, PANI films were fabricated as strain gauge/damage sensors and their electromechanical response was evaluated.

EXPERIMENT

Materials: Aniline (99.5%) was purchased from Sigma-Aldrich. Dinonylnapthalene-sulfonic acid (DNNSA) was obtained from King Industries. Butyl cellosolve (BC), xylenes and N-methyl-2-pyrrolidone (NMP) were purchased from Fisher Scientific. Epoxy based silver ink AG 530 was obtained from Conducting Compounds, Inc. (Hudson, NH). SC-79 (epoxy resin) was donated by Applied Poleramics, Inc. (Benicia, CA). Unless otherwise noted, all other chemicals were reagent grade and used as received. **Synthesis of DNNSA doped polyaniline (PAC1003):** PAC1003 was synthesized using a modified procedure to that reported by Kinlen et al. **[2]**. Into a jacketed 1-L reactor were changed 0.1 Mole of DNNSA, 0.06 mole of aniline and 200 mL of water. The mixture formed a milky white emulsion. The emulsion was chilled to -2°C under N2 while mechanically stirred. To this mixture, 0.074 Mole of ammonium peroxydisulfate (APS) in 40 mL of water was added dropwise. The white color of the mixture changed to amber. The reaction was allowed to proceed for 23 hours, during which time the emulsion separated into a green organic phase and a colorless aqueous phase. The aqueous phase was discharged and the organic phase was washed with DI water. The concentrated organic phase was collected for use. **Characterization:** Molecular weight and molecular weight distribution were determined by GPC using a Polymer Laboratories PL-GPC 50 Integrated GPC System with a Reflect Index detector. The mobile phase was 0.01M NH_4HCO_2 in NMP. EasiCal ® polystyrene standards (Mw 580-337,400) (Polymer Laboratories) were used for calibration. DSC test was conducted on TA Instruments DSC Q1000. TGA of vacuum dried sample was performed on TA Instruments TGA Q500. The analysis was conducted from RT to 800°C at 10°C/min under N_2. Brookfield LVDV-II+Pro viscometer was used for viscosity analysis. PAC1003 was added into a disposable aluminum chamber and held for 15 minutes before running test. The CLTE test was run on the thermo-mechanical analyzer Q400. **Strain Gauge Fabrication:** PAC1003 was screen-printed on 0.002 in. thick Kapton film with an ATMA Model AT-600H/E machine. Epoxy based silver ink

(AG 530) was printed as electrodes for electrical measurements. **Mechanical Characterization of Strain Gauge:** Type-1 dog bone epoxy resin-based samples were fabricated per ASTM-D638 using silicone molds. The stress-strain-resistance characterization was performed on an Instron-3345 system. It is important to note that the goal over here is not to test the dog bone specimen but the strain gauge which is installed on the dog bone plaque. A contact extensometer with 1" nominal gauge length for strain measurements was placed opposite to the polymeric gauge to avoid interference with the measurement circuitry. The resistances of the strain gauges were measured using the Keithley2400® sourcemeter. **Humidity Effects Test:** Humidity tests were performed using an Espec SH-241 (Hudsonville, MI) environmental chamber which controls humidity from 30-95% RH. Multiple samples were measured concurrently using the Keithley 2700 digital multimeter in conjunction with a Keithley 7700, 20 channel data card with integrated cold-junction compensation for temperature measurements. For these tests, the strain gauges with an aspect ratio of 12.5 were screen printed on glass plates. Fifteen samples were screen-printed onto a single glass panel.

RESULTS AND DISCUSSION

Polymerization of aniline

Many researchers have been working on the aniline polymerization mechanism [17]. As was described in a previously published paper [2] and related references therein, the generally accepted mechanism begins with a slow oxidation of aniline by persulfate to form a radical cation $C_6H_5NH_2\bullet^+$, possibly followed by loss of a proton and second electron to give a nitrenium cation $C_6H_5NH^+$. The key step is the combination of two nitrenium cations or two radical cations to give a dimmer of aniline: the combination primarily occurs head-to-tail to give N-phenyl-1,4-phenylenediamine (NPPD). Once the dimer is formed, it is rapidly oxidized which can easily be nucleophilicly attacked by another aniline unit to form a trimer. With each aniline unit added, the oligomer is more easily oxidized and is more susceptible to nucleophilic attack by aniline monomer. Therefore, the polymer is eventually formed.

Characterization of PAC1003

Gel Permeation Chromatography (GPC). GPC chromatograms of deprotonated polyaniline are typically unimodal. The molecular weight and molecular weight distribution results are listed in **Table 1**. It shows good reproducibility using the previously described synthesis method.

Table 1 GPC data of PAC1003

Run	Mw	Mn	PDI
1	127871	76004	1.68
2	119563	73327	1.63
3	118729	74090	1.60
Average	**122054**	**74474**	**1.64**

Thermogravimetric Analysis (TGA). Thermal stability of PAC1003 was investigated on a TA Instruments TGA Q500. At about 200°C, the weight loss begins. This is primarily due to decomposition of sulfonic function group of the DNNSA molecule. When the temperature reaches ~250°C, degradation of the polymer backbone begins. This TGA result is consistent with literature [18], with the exception of differences due to dopant.

Differential Scanning Calorimetry (DSC): The DSC curve of PAC1003 shows a large endothermic peak around 70°C-100°C, which is similar with other doped PANI systems reported in the literature [18]. However, although the reported Tg of dodecyl benzenesulfonic acid (DBSA) doped PANI is approximately 130°C as characterized by conventional DSC method, we were unable to identify a clear Tg. The large endothermic peak can most likely be attributed to the removal of water from the polymer sample. This explanation is consistent with the results reported by R.V. Gregory et al [19]. A modulated DSC test on PANI/DNNSA carried out by Mississippi Polymer Institute (curve not included here) also showed no definitive glass transition peak. Further investigation is currently being done for clear understanding of such thermal behavior in this system.

Thermomechanical Analysis (TMA). The CLTE is an important factor for the ICP based strain gauge sensor, which may be affected by dimensional change due to temperature variation. The polyaniline sample was run through a heat-cool-heat cycle from 40°C to 200°C. The thickness of the preliminary sample was 117.1 microns, which was the maximum thickness possible due to brittleness of the polymer. The negative change in dimension observed in the first heating run is attributed to the removal of entrapped solvents/moisture. Data from subsequent runs are in question due to recent evidence that suggest degradation above 180°C. Further investigation of CLTE is undergoing.

Printability of PAC1003

Viscosity. The viscosity/rheology behavior of the ICP is a very important parameter for printing. A PAC1003 sample with a typical composition was selected as an example for this investigation. **Figure 1** shows the rheological behavior and viscosity change over time. PAC1003 showed a typical non-Newtonian behavior (**Figure 1**A). Additionally, the shelf life of PAC1003 was evaluated through viscosity changes over time. The PAC1003 with 30% solid content showed relatively stable viscosity (~83 cP) over a 6 month period after an initial decrease of ~10% from the initial viscosity value(**Figure 1**B).

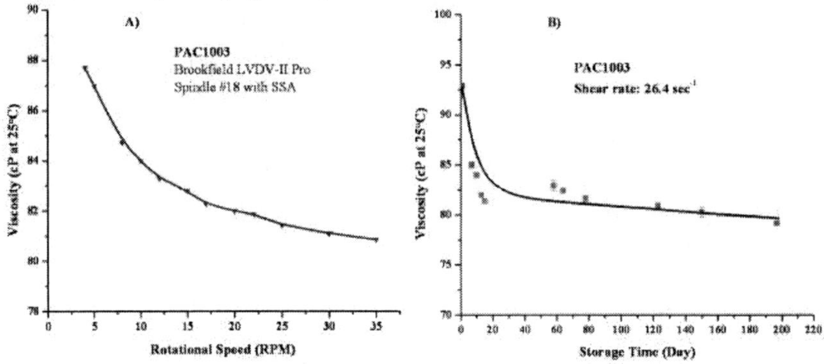

Figure 1 Viscosity test of PAC1003: A) Rheology behavior; B) Shelf live evaluation.

Printing methods. The advantage of the PAC1003 formulation is its ability to be easily modified to function in diverse printing techniques. Typical printing techniques used with this material are draw-down, spin coating, spray coating, inkjet printing, aerosol jet printing and screen printing. In addition, many materials have been explored as substrates for coating including glass, Teflon®, glass fabrics, polycarbonate, Kapton® film, aluminum, etc.

Mechanical Performance of Strain Gauge

This investigation mainly focused on the electro-mechanical response characteristic of ICP based strain gauge under different loading conditions. The specimen was loaded at a uniform crosshead displacement rate of 2 mm/min and released at rates of 1.5 mm/min and 2 mm/min within two steps, which was within the recommended limitations of ASTM. **Figure 2** shows that resistance change of strain gauge fits the loading profile very well, which means the sensor has the capability to monitor the substrate's (here, it is an epoxy "dog bone" structure) health status under load. In other words, the ICP based strain gauge can be used to monitor structure/substrate health in real time or near-real time.

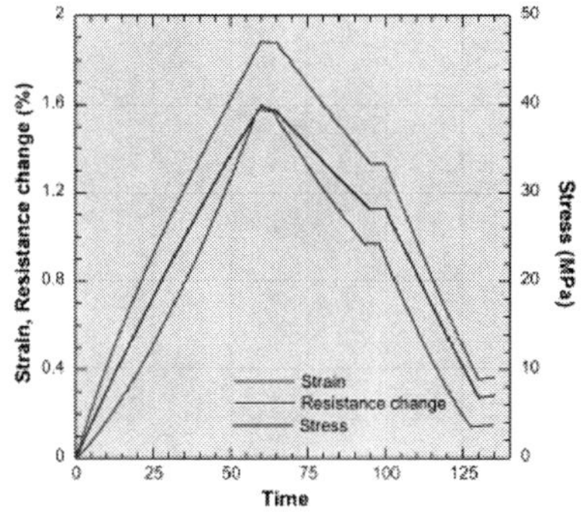

Figure 2 Electro-mechanical performance of PAC1003 based strain gauge under load condition.

Effect of Temperature

The effect of temperature on the ICP based strain gauge was investigated. This test was performed by measuring the resistance of a freestanding gauge which was exposed to the temperature change in the laboratory before a protective coating applied. **Figure 3** shows a typical performance curve (in terms of resistance change) of PAC1003 based strain gauge under varied temperature condition. As temperature increases, the resistance decreases and vice-versa.

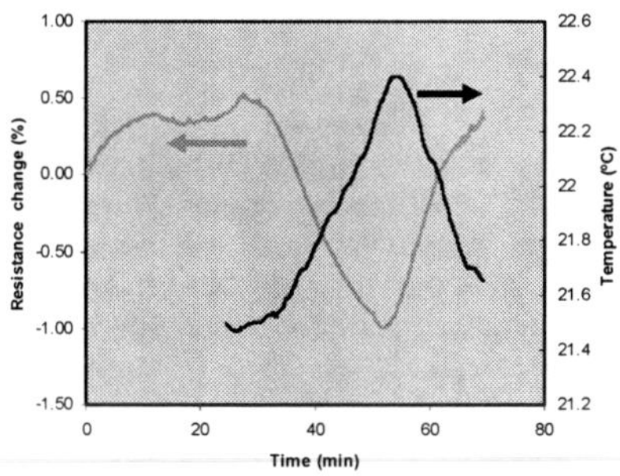

Figure 3 Effect of temperature on the ICP based strain gauge.

The resistance change with temperature variation was attributed to internal polymer chain movement and/or current transfer pathway changes. A more detailed temperature effect investigation is undergoing. In general, the ICP performs as a temperature sensor (thermometer) as well as a strain sensor.

Effect of Aspect Ratio (AR)

Comparative tests were performed on gauges with different aspect ratios. In this study the gauges had aspect ratios (AR=gauge length/width) as 12.5, 6, and 3. The PAC1003 based strain gauges were thermally treated at 150°C for 30 minutes. In **Figure 4**, PAC1003 based strain gauges showed no effects of varying aspect ratios.

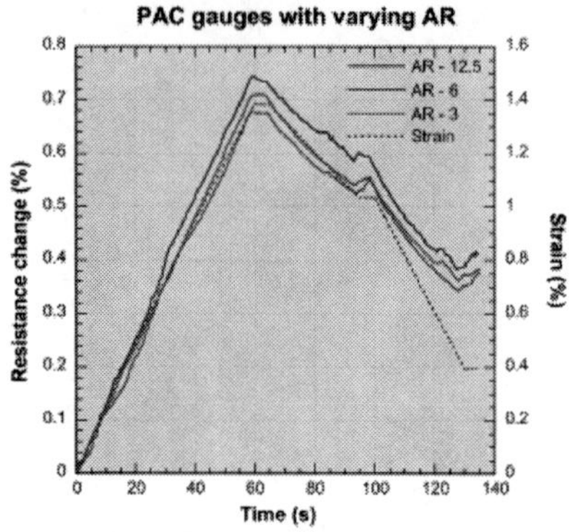

Figure 4 Effect of aspect ratio on PAC1003 based strain gauges.

Effect of Humidity at Room Temperature

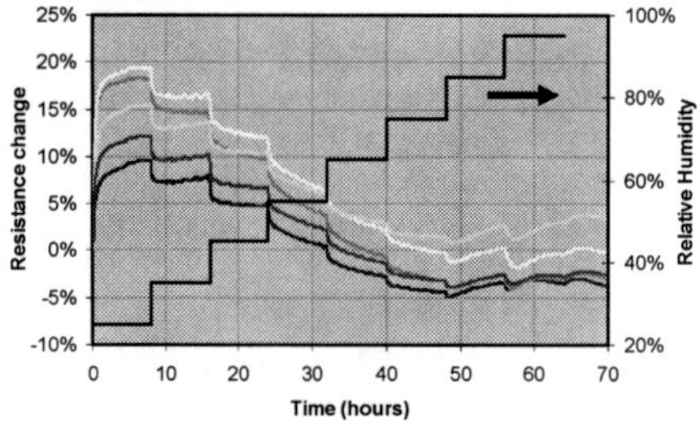

Figure 5 Resistance changes of PAC1003 vs. humidity variation at 25°C.

The effect of humidity on PAC1003 strain gauges is shown in **Figure 5**. The graph shows that the resistance of PAC1003 strain gauges decreases with increasing humidity within RH range of 20%-80%. At RH levels above 80%, the resistance changes showed an opposite effect. Further investigation into this phenomenon as well as evaluating the effects of humidity in combination with temperature is underway.

CONCLUSIONS

A soluble dinonylnaphthalene sulfonic acid (DNNSA) doped polyaniline (PANI), PAC1003, was synthesized through emulsion polymerization. The Mw is 122,000 with a PDI of 1.6. PAC1003 were stable to approximately 200°C (dopant degradation). No definitive glass transition was observed through convention DSC or modulated DSC characterization.

PAC1003 showed a non-Newtonian flow behavior and excellent printability. Various printing methods (screen printing, inkjet printing and aerosol jet printing etc.) have been applied. This unique property gives PAC1003 a variety of application possibilities. Investigation of electro-mechanical properties of Kapton film-supported PANI based strain gauge showed excellent structure health monitoring performance. No dependence on aspect ratio of the strain gauge was found. Additionally, PAC1003 also showed good sensing capabilities for temperature and humidity variations.

In summary, polyaniline based strain gauges showed good structure health monitoring capabilities. Further investigation and optimization of the strain gauges and protective coatings is underway at Crosslink.

ACKNOWLEDGMENTS

This research was supported by Wright-Patterson Air Force Base (Contract number: FA8650-09-C-5227). The authors are especially thankful to Siamack Mazdiyasni (WPAFB), Prof. James Rawlins, and Dr. Richard Cook (School of Polymers and High Performance Materials, The Univ. of Southern Miss.) for providing helpful insights. The authors would also like to thank Mr. Kevin Carroll (Crosslink, Fenton) for helping on screen printing of sensors.

REFERENCES

1. S. Palaniappan, A. John, *Prog. Polym. Sci.* **33**, 732-758 (2008).
2. P.J. Kinlen, J. Liu, Y. Ding, C. R. Graham, E. E. Remsen, *Macromolecules* **31**, 1735-1744 (1998).
3. J.Y. Kim, J.H. Lee, S.J. Kwon, *Synth. Met.* **157**, 336–342 (2007).
4. S. Bhadra, S. Chattopadhyay, N.K. Singha, D. Khastgir, *J. Appl. Polym. Sci.* **108**, 57–64 (2008).
5. M. Lin, F.-K. Chang, *Compos. Sci. Tech.* **62**, 919-939 (2002).
6. R.K.Verma, R.G. Kander, B.S. Hsiao, *J. Mater. Sci. Lett.* **13**, 438-442 (1994).
7. A. Sjogren, A. Krasnikovs, J. Varna. *Compos. Part A* **32**, 1237-1242 (2001).
8. Joung-Man Park, Dae-Sik Kim, Sung-Ju Kim, Pyung-Gee Kim, Dong-Jin Yoon, and K. Lawrence DeVries, *Compos. Part B* **38**, 847-861 (2007); **37**, 612-626 (2006).
9. E. Broitman, P. Alonso, R. Aragon, and R. Zimmerman, *Adv. Mat. Sci. & Tech.* **1**, 13-16 (1996).
10. K. S. C. Kuang, W. J. Cantwell, *J. Thermoplast. Compos. Mater.* **16**, 213-229 (2003).
11. I. Chopra, *AIAA Journal* **20**, 2145-2187 (2002).
12. W.A. Daoud, J. H. Xin, Y.S. Szeto, *Sens. Actuators B* **109**, 329-333 (2005).
13. P. J. Kinlen, H. Yang, D. Kim, A. K. Singh, B. Pitts, *Patent No.* **US 2010095781** (2010).
14. A.K. Singh, D. Kim, H. Yang, B. W. Pitts, G.J. Tregre, P.J. Kinlen, *SAMPE, 2009,* Wichita, KS, Oct 19-22, 2009.
15. D. Kim, A.K. Singh, H. Yang, B. W. Pitts, G.J. Tregre, P.J. Kinlen, *ASME 2009,* Oxnard, CA, Sept 21-23, 2009.
16. Y.-G. Kim, S.Viswanathan, J. Mbugua, J.-H. Jung, P.J. Kinlen, *Proc. of 18th Int. Seminar on Double Layer Capacitors & Hybrid Energy Storage Devices* 218-223 (2008).
17. F. Lux, *Polymer* **35** (14), 2915-2936 (1994).
18. M.J.R. Cardoso, M.F.S. Lima, D.M. Lenz, *Materials Research* **10** (4), 425-429 (2007).
19. L. Ding, X. Wang, R.V. Gregory, *Synth. Met.* **104**, 73-78 (1999).

Mater. Res. Soc. Symp. Proc. Vol. 1312 © 2011 Materials Research Society
DOI: 10.1557/opl.2011.908

Conductive Polymer Bilayers – A Spectroelectrochemical Look at Their Doping Reactions

Henrik Gustafsson[1], Carita Kvarnström[2] and Ari Ivaska[1]
[1]Process Chemistry Centre/Laboratory of Analytical Chemistry, Åbo Akademi University, Biskopsgatan 8, FI-20500 Åbo-Turku, Finland
[2]Turku University Centre for Materials and Surfaces/Laboratory of Materials Chemistry and Chemical Analysis/Department of Chemistry, University of Turku, Vatselankatu 2, FI-20014 Turku, Finland

ABSTRACT

The doping behavior of bilayers of electropolymerized poly(3,4-ethylenedioxythiophene) doped with hexafluorophosphate and drop-cast poly(benzobisimidazobenzophenanthroline) was studied with cyclic voltammetry, and with *in situ* ultraviolet-visible and attenuated total reflectance Fourier transform infrared spectroscopy. The spectroelectrochemical characterization of the polymer bilayers provided us with fundamental information of the properties of the materials, information being essential in order to be able to use the conducting polymers in future real-life applications, *e.g.*, as organic semiconducting components.

INTRODUCTION

In future electronic applications, conjugated conducting polymers may play an important role as organic semiconductors. Poly(3,4-ethylenedioxythiophene), PEDOT (see Figure 1), is one of the few conducting polymers that are both p- and n-dopable. We have previously [1,2] studied the doping behavior of electrochemically polymerized PEDOT doped with hexafluorophosphate, PEDOT(PF_6), on aluminum.

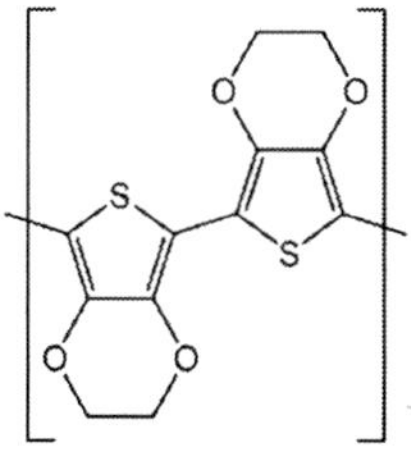

Figure 1. PEDOT structure.

Poly(benzobisimidazobenzophenanthroline), BBL (see Figure 2), is an n-dopable ladder polymer. Few applications have been found for pristine BBL, since the unmodified polymer has low solubility even in strong acidic solutions. Derivatization of the BBL end chains with poly(ethylene oxide), PEO, gives a water dispersible polymer which is easier to process.

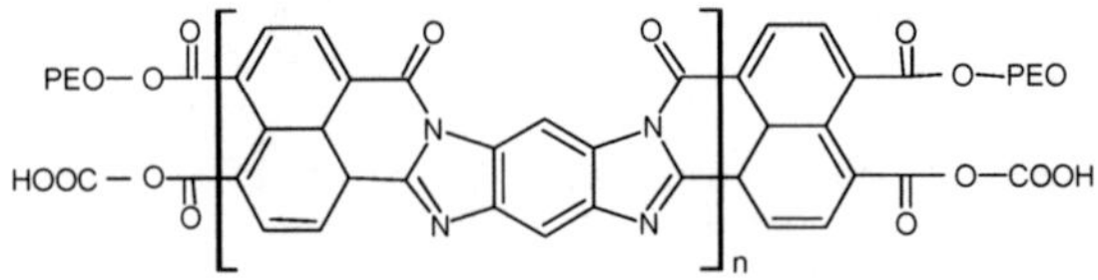

Figure 2. BBL-PEO structure; n≈20, Mw_{PEO}≈550.

Attenuated total reflection Fourier transform infrared (ATR-FTIR) spectroscopy has been shown to be a useful technique for *in situ* studies of the doping behavior of conducting polymer films. Neutral, p-doped and n-doped films all show different FTIR spectra with doping-induced vibration bands arising during oxidation and reduction [3].

In this work, we study bilayers of electropolymerized PEDOT(PF_6) and drop-cast BBL-PEO. In such a system, PEDOT acts as electron donor while BBL acts as acceptor. The doping behavior of the bilayer structures was studied with cyclic voltammetry (CV), and with *in situ* ultraviolet-visible (UV-vis) and ATR-FTIR spectroscopy.

EXPERIMENT

Acetonitrile, ACN (anhydrous, 99.8%), and the monomer 3,4-ethylenedioxythiophene, EDOT, were used as obtained from Sigma-Aldrich. BBL-PEO was synthesized by Sami-Pekka Hirvonen and Prof. Heikki Tenhu at the Laboratory of Polymer Chemistry, University of Helsinki, Finland. Tetrabutylammonium hexafluorophosphate, $TBAPF_6$ (Fluka), was dried under vacuum before use. Great caution was taken in order to protect the electrolyte from atmospheric oxygen and water throughout the experiments, all solutions were deoxygenated with nitrogen before use.

A conventional three-electrode one-compartment electrochemical cell was used in all electrochemical experiments, and a silver wire coated with silver chloride was used as the reference electrode (RE). All potentials are referred to this pseudo RE, calibrated vs. the ferrocene/ferrocinium redox couple in 0.1 M $TBAPF_6$–ACN ($E_{½}$≈0.4 V vs. Ag/AgCl). In CV experiments, a glassy carbon (GC) rod was used as the counter electrode (CE), and a polyvinyl chloride-sealed 7.1 mm^2 GC disk as the working electrode (WE). In UV-vis experiments, a Pt wire was used as the CE, and a SnO_2 glass (with a gold plate contact) as the WE. The SnO_2 glass was sonicated in acetone, methanol, isopropyl alcohol, and ultrapure deionized water, respectively, and finally rinsed with acetone, before use. A 1 cm path length quartz cuvette was used for UV-vis measurements. In ATR-FTIR experiments, a Pt wire was used as the CE and a ZnSe reflection element with a 30 nm Pt layer as the WE.

A PerkinElmer Lambda 25 UV/VIS Spectrometer was used to record the UV-vis spectra and a Bruker IFS 66/S spectrometer with an MCT detector to record the ATR-FTIR spectra. A BAS CV-50W Voltammetric Analyzer was used for the *in situ* UV-vis experiments, while all other electrochemical experiments were performed using an Autolab PGSTAT 100 potentiostat with general-purpose electrochemical system software.

PEDOT(PF_6) films were electropolymerized on the WEs by potential scanning between -0.9 and +1.25 V in a monomer solution containing 0.1 M EDOT and 0.1 M $TBAPF_6$ in ACN. Aqueous BBL-PEO solution was drop-cast on the WEs, and left to dry overnight. All polymer films were rinsed with ACN before any further experiments were made. The polymer films were p- and n-doped either stepwise, keeping the potential constant while acquiring the UV-vis spectra, or with CV, cycling the films in the appropriate potential range (all other experiments except UV-vis). Polymer bilayers were made either by electropolymerization of PEDOT(PF_6) on WE/BBL-PEO, or by drop-casting of BBL-PEO on WE/PEDOT(PF_6).

DISCUSSION

The doping process of both single layers of BBL-PEO and PEDOT(PF_6) and of their bilayers was studied by spectroelectrochemistry, correlating spectroscopic changes with the doping level, which is controllably varied by changing the potential of the WE. Here, we present only some of the results.

Cyclic voltammetry

Cyclic voltammograms of GC/BBL-PEO/PEDOT(PF_6) being both n- and p-doped, as well as of single films of GC/BBL-PEO and GC/PEDOT(PF_6) being n-doped, are shown in Figure 3. Distinct reduction and oxidation current peaks are seen in the bilayer voltammogram: BBL-PEO has two n-doping peaks at -1.04 and -1.23 V, while the n- and p-doping peaks of PEDOT(PF_6) are obtained at -1.79 V and +0.04 V, respectively. Compared with voltammograms of the single polymer films, the BBL-PEO reduction current peaks are smaller and the n-doping of both polymers takes place at more negative potentials (n-doping peaks of single BBL-PEO films at -0.97 and -1.15 V and of single PEDOT(PF_6) at -1.69 V). Furthermore, a weak BBL-PEO reduction shoulder seen around -0.6 V is covered by a PEDOT(PF_6) rereduction peak. The PEDOT(PF_6) film, covering the BBL-PEO film, is in a neutral state in the potential region where BBL-PEO gets n-doped, hence hindering the electron transfer.

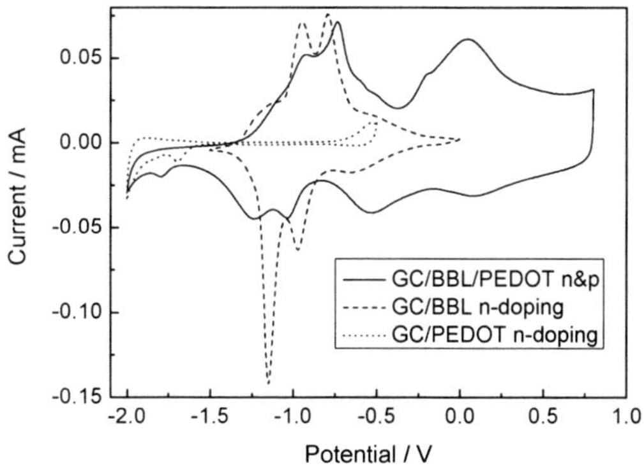

Figure 3. Cyclic voltammograms of n- and p-doping of a GC/BBL-PEO/PEDOT(PF_6) electrode, and of n-doping of GC/BBL-PEO and GC/PEDOT(PF_6) electrodes. Electrolyte solution: 0.1 M $TBAPF_6$–ACN. Potential range: -2.0 V to +0.8 V (BBL/PEDOT), -1.5 V to 0 V (BBL), -2.0 V to -0.5 V (PEDOT). Scan rate=20 mV/s, only the third scans are shown.

Ultraviolet-visible spectroscopy

In situ UV-vis spectroscopy is a technique that gives information on electrochromic properties of conductive polymers. Absorption spectra of SnO_2/BBL-PEO/PEDOT(PF_6) being n-doped are shown in Figure 4. In its neutral and n-doped state, PEDOT(PF_6) has a very strong absorption maximum around 600 nm. More changes are seen in spectra of single BBL-PEO films being n-doped (not shown): absorption bands appear around 900, 740, 580, 940, and 525 nm as the potential turns more negative. With BBL-PEO covered by PEDOT(PF_6), similar changes are seen at the higher wavelengths, whereas the changes in absorption at lower wavelengths are hidden by the strong PEDOT(PF_6) absorption band.

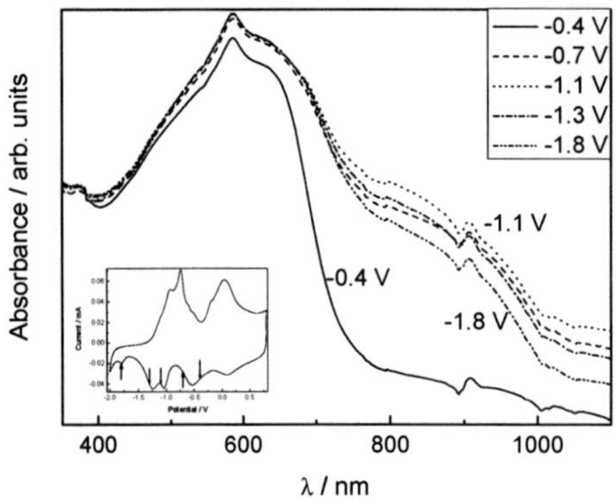

Figure 4. *In situ* UV-vis absorption spectra of n-doping of SnO_2/BBL-PEO/PEDOT(PF_6). Wavelength scan: 1100 nm to 350 nm. Inset: BBL/PEDOT (figure 3) cyclic voltammogram with added markers showing the potentials at which the spectra are recorded.

Fourier transform infrared spectroscopy

With *in situ* ATR-FTIR spectroscopy, structural changes and modifications in the electronic band structure of the polymer chain can be detected during electrochemical reactions such as doping. ATR-FTIR spectra of ZnSe/Pt/BBL-PEO/PEDOT(PF_6) being n-doped are shown in Figure 5. During the recording of each spectrum, the potential is scanned ~82.5 mV in the negative direction. All the spectra are divided by a reference spectrum, recorded at -665 mV–-747.5 mV. During polymer film doping, charge carriers are formed. This can be seen as electronic absorption bands in the spectra at wavenumbers above 2000 cm^{-1}. Doping-induced infrared active vibration bands appear at lower wavenumbers.

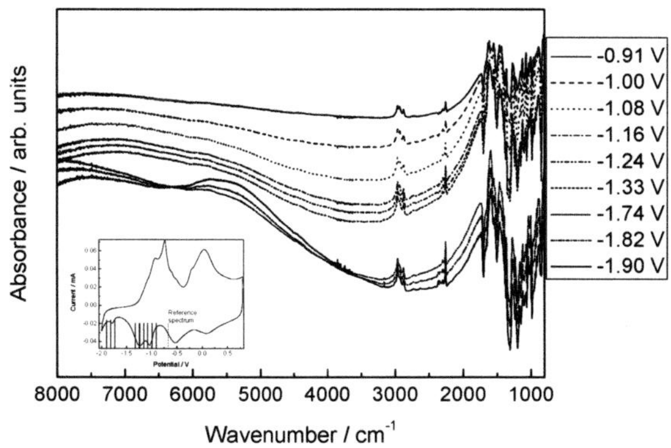

Figure 5. *In situ* ATR-FTIR difference spectra of n-doping of ZnSe/Pt/BBL-PEO/PEDOT(PF_6). Wavenumber region: 8000-800 cm^{-1}. Inset: BBL/PEDOT (figure 3) cyclic voltammogram with added markers showing the potentials at which the recording of the spectra starts.

CONCLUSIONS

Conducting polymers are considered interesting as organic semiconductors for future electronic devices. This study deals with bilayers of n-dopable BBL-PEO and both p- and n-dopable PEDOT(PF_6), showing that the polymers retain their dopability also in bilayer structures, even at potentials where one of the polymers is in a neutral, virtually nonconducting form. In the case of a WE/BBL-PEO/PEDOT(PF_6) structure, the n-doping of the polymers requires more energy than with single polymer films. Bilayer structures, in which the charge carriers are able to cross the interface between the polymers, might become of use in "sandwich" type or bulk heterojunction organic electronics applications, *e.g.* in future photovoltaic devices.

ACKNOWLEDGMENTS

The grants and scholarships provided by the Fortum Foundation and the Research Institute of Åbo Akademi Foundation are acknowledged. This work is part of the activities of the Åbo Akademi Process Chemistry Centre of Excellence Program (Academy of Finland, 2000-2011).

REFERENCES

1. H. Gustafsson, C. Kvarnström and A. Ivaska, *Thin Solid Films* **517**, 474-478 (2008).
2. F. Sundfors, H. Gustafsson, A. Ivaska, and C. Kvarnström, *J Solid State Electrochem* **14**,

1185-1195 (2010).

3. C. Kvarnström, H. Neugebauer, A. Ivaska, and N. S. Sariciftci, *J Mol Struct* **521**, 271-277 (2000).

Mater. Res. Soc. Symp. Proc. Vol. 1312 © 2011 Materials Research Society
DOI: 10.1557/opl.2011.909

Stability Improvement of Polymer Based Solar Cells by Thermally Evaporated Cr_2O_3 Interfacial Layer

Mingdong Wang [1], Fangyan Xie [2], Shizhao Zheng [3], Jianbin. Xu [1*]
[1] Department of Electronic Engineering, and Materials Science and Technology Research Center, The Chinese University of Hong Kong, Hong Kong
[2] Instrumental Analysis & Research Center, Sun Yat-Sen (Zhongshan) University, Guangzhou, P. R. China
[3] Department of Physics, The Chinese University of Hong Kong, Hong Kong, China.

ABSTRACT

We report on Cr_2O_3 layer by thermal evaporation of Cr_2O_3 powder as cathode interfacial layer to improve the stability in air for the bulk heterojunction solar cells of poly(3-hexylthiophene) (P3HT) and [6,6]-phenyl C_{61} butyric acid methyl ester (PCBM). Devices with Cr_2O_3 interfacial layers show higher power conversion efficiency (PCE) and stability than those without interfacial layer. Devices with Cr_2O_3 show improved stability approximately 100 times than that of devices without interfacial layer or with LiF interfacial layer.

INTRODUCTION

Polymeric photovoltaic devices promise low-cost production of lightweight, flexible solar cells. Recent progress in polymeric solar cells, has led to power conversion efficiencies (PCEs) of 6-7% [1-2]. In organic/polymeric solar cells, the active layer usually has a thickness of around 100 nm, due to its poor electric conductivity. Electrode interface engineering is a key strategy to improve the PCE through optimization of the short circuit current (J_{sc}), open circuit voltage (V_{oc}), fill factor (FF), and the stability [3-7].

It is preferable for the cathode interface to have a low work function contact for efficient electron extraction. Al is the most commonly used electrode material in solar cells, which is generally deposited by thermal evaporation, although this process frequently alters the quality of the metal/organic interface, because the hot metal atoms may react with oxygen species remaining in the vacuum chamber and diffuse into the organic layers. Several cathode interfacial layers such as LiF [8,9], Ca [10], Ba [11], TiO_x [12,13], ZnO [14], and Cs_2CO_3 [15] are commonly employed to protect the metal/organic interface in organic devices. Here we report on an investigation on thermally evaporated Cr_2O_3 thin film as a cathode interfacial layer for polymer-fullerene solar cells to improve performance and stability. Compared to LiF, Cr_2O_3 is more stable and economic.

EXPERIMENTS

The device structure is shown in figure 1 and the fabrication consists of the following steps. Firstly, ITO glass was treated with oxygen plasma for ~2 min.

Secondly, the PEDOT:PSS was spun cast on ITO glass with a thickness of ca. 40 nm, and then annealed at 145 °C for 10 min in air. A mixed solution composed of P3HT and PCBM in dichlorobenzene was spun cast on the top of PEDOT:PSS layer at a slow speed of 500 rpm for 5 seconds and a high speed of 1200 rpm for 50 seconds, the thickness of active layer is 90-110 nm, and then annealed at 120 °C for 10 min in nitrogen atmosphere. The mixed solution had a P3HT:PCBM weight ratio of 1:1 with a concentration of 20 mg/ml. The thin Cr_2O_3 layer (ca.5 nm) was deposited by thermal evaporation of Cr_2O_3 powder under background vacuum of 1×10^{-6}~ 2×10^{-6} mbar. Finally, a 90 nm-thick film of Al was thermally evaporated onto the top of the device. The active area of device was 2 mm×1 mm. Electrical measurements were performed by a semiconductor characterization system (Keithley 236) at room temperature in air under the spectral output from solar simulator (Newport) using an AM 1.5G filter with a light power of 100 mW/cm^2. The light intensity was precisely calibrated by a standard solar cell. The devices with LiF as an interfacial layer (ITO/PEDOT:PSS/P3HT:PCBM/LiF/Al, with 1 nm-thick LiF), and without the interfacial layer (ITO/PEDOT:PSS/P3HT:PCBM/Al) were also fabricated and tested under the same conditions. For all devices, no external package or encapsulation was applied after device fabrication. The thicknesses of evaporated layers were measured with a thickness monitor (Sigma SQM-160), and also determined by AFM.

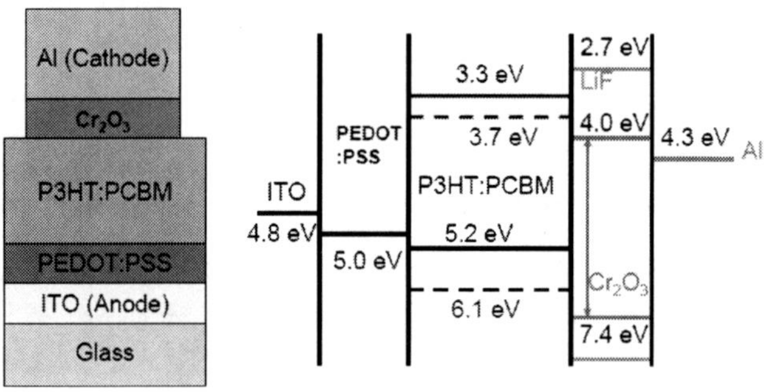

(a) (b)

Figure 1 (a) photovoltaic device structure; (b) energy-band diagram composed of different organic layers and interfacial layers.

RESULTS AND DISCUSSION

Figure 2 shows the current density-voltage (*J-V*) characteristics under illumination. The device without interfacial layer (only Al) has a poor efficiency of 1.5%, with V_{oc}=0.4 V, J_{sc}=8.55 mA/cm^2, and FF=0.43. By comparison, it is found that device performance is obviously improved by including the interfacial layer of either LiF or Cr_2O_3. The device with Cr_2O_3 (ITO/PEDOT:PSS/P3HT:PCBM/Cr_2O_3/Al) exhibits a PCE as high as 2.8%, with V_{oc}=0.58V, J_{sc}=10.0 mA/cm^2, and FF=0.50. The performance is improved mainly due to the increase of short circuit current and open circuit voltage. Cr_2O_3 can improve the organic/cathode contact such that the V_{oc} loss due to the energy level offset can be eliminated. In addition to the comparable efficiency improvement, devices with Cr_2O_3 show a dramatic improvement of stability in comparison with those with LiF interfacial layer. The detailed efficiencies, J_{sc}, V_{oc} and FF are listed in table 1.

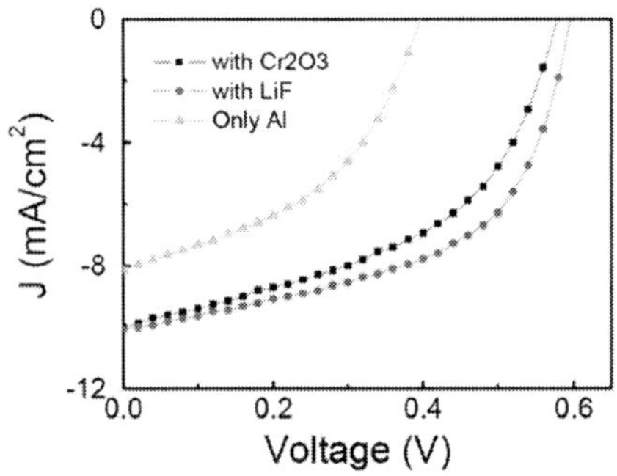

Figure 2 *J-V* characteristics of polymer-fullerene bulk heterojunction devices under illumination with an incident optical intensity of 100 mW/cm^2.

Table 1 Solar cell parameters extracted from Figure 2

Sample	Efficiency	V_{oc}(V)	J_{sc} (mA/cm^2)	FF
Only Al	1.5%	0.40	8.55	0.43
With LiF	3.2%	0.59	10.08	0.54
With Cr_2O_3	2.8 %	0.58	10.0	0.50

To collect the compositional information of chromium oxide films, XPS analysis was performed. In Figure 3 (a), the peak at 577.6 eV relates to Cr $2p_{3/2}$ of chromium oxide (Cr_2O_3) [16,17] while the peak at 587.4 to Cr $2p_{1/2}$ of chromium oxide. Cr_2O_3 is a stable semiconductor [18], with its conduction band level of 4.0 eV [19], just between the work function of Al and *LUMO* of PCBM (shown in figure 1(b)), which renders it an efficient electron transport material. The large band gap of 3.4 eV is also capable of blocking the holes and excitons efficiently. As expected, a low work function material, small *LUMO* or small conduction band minimum of interfacial layers in cathode can improve V_{oc}. Cr_2O_3 has a slightly lower conduction band minimum (shown in Figure 1) than that of Al. As previous reported CrO_x [20], the Cr_2O_3 film by thermal evaporation of Cr_2O_3 powder has much lower Cr metal content than the reported CrO_x by thermal evaporation of Cr metal, which will affect carrier collection. But inserting Cr_2O_3 layer can result in a higher V_{oc}.

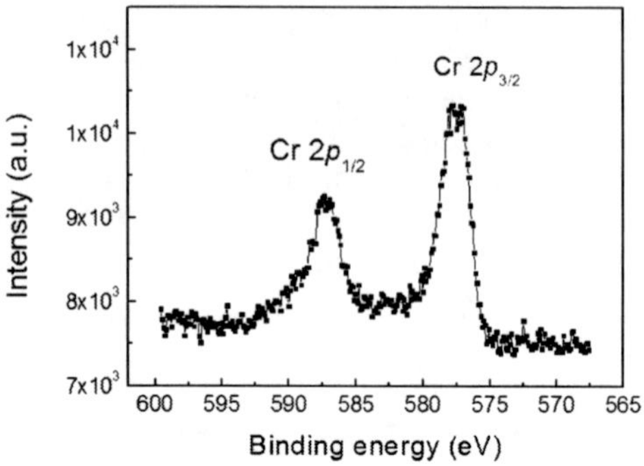

Figure 3 Fitted core level XPS spectra of Cr 2*p* peaks of the as-deposited Cr_2O_3 film.

Stability is another important issue for polymeric solar cells [20-22]. It is reported that a 30 nm-thick TiO_x buffer layer between the active layer and Al cathode and the inverted structure with deposition of Al before spinning active layer obviously improved the stability [13,23]. Here our experiments have shown that the thermally evaporated Cr_2O_3 layer indeed improved the lifetime of polymeric solar cells in air. Figure 4 shows the comparison of efficiency versus storage time of the solar cell with Cr_2O_3 layer, LiF layer, and without the interfacial layer (only Al). For all devices, no extra package or encapsulation was used. For the devices with LiF interfacial layer and without the interfacial layer stored in ambient air, dramatic decreases of PCE were observed as the storage time increased, and they lost 90% of the initial efficiencies after ~5 hours, mainly because of the loss of photocurrent. However, for the device with Cr_2O_3 interfacial layer, the degradation in the first several hours was much slower. It still remained nearly 85% of the initial power conversion efficiency after 5 hours, and 35% after 5 days.

Based on the experimental results above, it can be deduced that the Cr_2O_3 layer plays a key role in retarding degradation. During the thermal evaporation process, the Al atoms with high kinetic energy flux can diffuse into polymer to form a thick interfacial layer. Around 5 nm thick interfacial layer between the P3HT:PCBM blend and the Al contact was found [24]. Theoretically, Lögdlund et. al. [25] proposed a direct reaction for the formation of Al-C bound with high reductive power due to Al. An alternative to Al-C was single-electron transfer with the creation of anion radicals on the polymer [26]. Both organo-aluminum compounds and anion radicals are highly reactive species that can react with any proton donors present, such as water, or oxygen. Besides, Norrman et al. [27] reported that oxygen and water, which can diffuse through the aluminum electrode via microscopic pinholes and metal grains to organic-Al interface and the active layer. It indicates that the thick organic-aluminum interface layer between polymer P3HT:PCBM and the Al caused by the evaporated Al is easily oxidized resulting an barrier for carrier extraction when it is exposure to air. The thermally evaporated Cr_2O_3 buffer layer can improve the device stability because it can effectively avoid or minimize the formation of organic-aluminum interface (interdiffusion area) caused by the evaporation of Al. This interdiffusion area between organic active layer and Al is easily oxidized under oxygen and water molecules, to form a charge blocking layer at cathode, thus finally reduce the J_{sc}. Cr_2O_3 buffer layer leads to a more air-resistive cathode/organic interface and eventually improved device stability. Although the detailed mechanisms of the stability and degradation issues are rather complicated and certainly not yet fully understood for the time being, a thin Cr_2O_3 buffer layer is found to be sufficient to enhance the stability of polymeric solar cells.

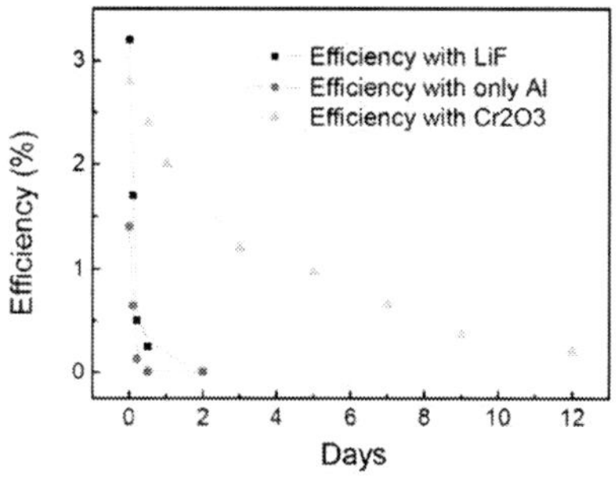

Figure 4. Comparison of the power conversion efficiencies as a function of storage time for polymeric solar cells with the interfacial layers of Cr_2O_3, LiF, and without the interfacial layer (only Al). Note that the device characteristics are monitored with increasing storage time for the same devices.

CONCLUSION

We have found that a Cr_2O_3 interfacial layer, which is formed by thermal evaporation Cr_2O_3 powder, has an ideal energy-band match to improve the power conversion efficiency of polymer-fullerene BHJ organic solar cells. Moreover, Cr_2O_3 layer also function as a protective layer to significantly enhance the device stability assumably by minimizing the organic-Al interfacial caused by the thermal evaporation of Al.

ACKNOWLEDGMENTS

This work is in part supported by Hong Kong Research Grants Council, particularly, under Grant No. CUHK02/CRF/08 and CUHK41821/09E. We would like to thank the National Science Foundation of China for the support, particularly, via Grant Nos. 60990314 and 60928009.

REFERENCES

[1] S. H. Park, A. Roy, S. Beaupre, S. Cho, N. Coates, J. S. Moon, D. Moses, M. Leclerc, K. Lee, A. J. Heeger, Nat. Photonics 3, 297 (2009).

[2] H.Y.Chen, J. Hou, S.Zhang, Y. Liang, G.Yang, Y. Yang, L.Yu, Y. Wu, G. Li, Nat. Photonics 3, 649 (2009).

[3] W. Cai, X. Gong, Y. Cao, Sol. Energy Mater. Sol. Cells 94, 114 (2010)
[4] J. Xue, P. Rand Barry, S. Uchida; S. R. Forrest. J. Appl. Phys. 98, 124903 (2005).
[5] H.Y Wei, J. H. Huang, K. C. Ho, Chih-Wei Chu, ACS Appl. Mater. Interfaces 5, 1281 (2010).
[6] T. Kuwabara, Y.Kawahara, T.Yamaguchi, K.Takahashi, ACS Appl. Mater. Interfaces, 10, 2107 (2009).
[7] T. Tromholt, S. A.Gevorgyan, M. Jørgensen, F. C. Krebs, O. S. Kristian, ACS Appl. Mater. Interfaces, 1, 2768 (2009).
[8] S. E. Shaheen, G. E. Jabbour, M. M. Morrell, Y. Kawabe, B. Kippelen, N. Peyghambarian, M. F. Nabor, R. Schlaf, E. A. Mash, N. R.Armstrong, J. Appl. Phys.84, 2324 (1998).
[9] L. S.Hung, R. Q. Zhang, P. He, M. G. Mason, J. Phys. D: Appl. Phys. 35, 103 (2002).
[10] J. M. Bharathan, Y.Yang, J. Appl. Phys. 84, 3207 (1998).
[11] M. O. Reese, M. S.White, G. Rumbles, D. S.Ginley, S. E. Shaheen, Appl. Phys. Lett. 92, 053307 (2008).
[12] S. J. Yoon, J. H. Park, Lee, H.K. Park, O. Ok, Appl. Phys. Lett. 92, 143504 (2008).
[13] J.Y. Kim, S. H. Kim, H. H.Lee, K.Lee, W. Ma, X. Gong, A. J. Heeger, Adv. Mater. 18, 572 (2006).
[14] A. K. Kyaw, X. W.Sun, C. Y.Jiang, G. Q. Lo, D. W.Zhao, D. L. Kwong, Appl. Phys. Lett. 93, 221107 (2008).
[15] F.C.Chen, J.L.Wu, S. S.Yang, K.-H. Hsieh, W. C. Chen, J. Appl. Phys. 103, 103721 (2008).
[16] J. F.Weaver, H. A.E. Hagelin-Weaver, G. B. Hoflund, G. N.Salaita, Appl.Surf. Sci. 252, 7895 (2006).
[17] H. C. Barshilia, N. Selvakumar, K. S. Rajam, A. Biswas, J. Appl. Phys. 103, 023507 (2008).
[18] S. Pokhrel, C.E. Simion, V. Quemenera, N. Barsan, U. Weimar, Sens. Actuators B 133,78 (2008).
[19] Y. Xu, M. A.A. Schoonen, American Mineralogist 85, 543 (2000).
[20] M.D.Wang, Q. Tang, J. An, F. Y. Xie, J. Chen, S.Z. Zheng, K. Y. Wong, Q. Miao, J. B. Xu, ACS Appl. Mater. Inter. 2, 2699 (2010).
[21] F. C.Krebs, T.Tromholt, M. Jørgensen, Nanoscale, 2, 873 (2010).
[22] F. C. Krebs, T. D. Nielsen, J. Fyenbo, M.Wadstrøm, M. S. Pedersen, Energy Environ. Sci. 3, 512 (2010).
[23] B. Zimmermann, U.Wurfel, M. Niggemann, Solar Energy Materials & Solar Cells 93, 491 (2009).
[24] C.Y. Nam, D.Su, C. T. Black, Adv. Funct. Mater. 19, 1 (2009).
[25] M.Lögdlund, J. L.Bredas, J. Chem. Phys. 101, 4357 (1994).
[26] M. Jørgensen, K. Norrman, F. C. Krebs, Sol. Energy Mater. Sol. Cells 92,686 (2008).
[27] K. Norrman, F. C.Krebs, Sol. Energy Mater. Sol. Cells, 90,213 (2006).

Mater. Res. Soc. Symp. Proc. Vol. 1312 © 2011 Materials Research Society
DOI: 10.1557/opl.2011.910

A Hexagonal Pillar Array of Thermo-responsive Soft Actuators Prepared by Nanoimprinting

Ching-Mao Wu[1*], Szu-Yin Lin[1], and Kuo-Tung Huang[1]

[1]Material and Chemical Research Laboratories, Industrial Technology Research Institute, Chutung, Hsinchu, Taiwan 31040

ABSTRACT

Thermo-responsive actuation (thermomechanical effects) based on nematic liquid crystal elastomers (LCEs) have become a research priority in the preparation of soft actuators. Nematic LCEs combine the anisotropic features of liquid crystal phases with the rubber elasticity of polymer network. When heated at nematic to isotropic phase transition temperature (*N-to-I* temp.), a uniaxial thermomechanical deformation of LCEs will undergo at nearly constant volume due to a change of LC director order. Recently, an array of the micro-sized LCE pillars related to such thermomechanical effects have been successfully constructed through a soft lithography technology (i.e., replica molding). The prepared LCE pillars are mono-dispersive and micro-sized. They also possess *N-to-I* temp. higher than 100℃, largely limiting the available application. By contrast, the present study will report a hexagonal array of nano-sized thermo-responsive pillar actuators that are able to contract and expand in response to temperature changes around a lower *N-to-I* temp. is manufactured *via* using reactive rod-like liquid crystal and ultraviolet nanoimprinting technology. According to atomic force microscope (AFM) observation, a hexagonal array of pillars can be easily constructed by nanoimprinting and a responsive surface with a thermo-stimuli-driven roughness change is achieved. The room-temperature AFM scans quantitatively represent the single pillar shows a diameter of ca. 270 nm and 140 nm in depth, and the pitch meaning the averaged inter-pillar distance is measured as ca. 425 nm, thus lying in a nano-sized range. Furthermore, temperature-variable AFM is also utilized to demonstrate the pillar behaves as a thermally-stimulated nano-sized actuator. In our case, when heated above *N-to-I* phase transition temperature (ca. 65℃), it is clearly observed that the pillar diameter is expanded in the order of over 12-15 % and then reversibly contracted in response to temperature drop.

INTRODUCTION

In the field of smart materials, the materials bearing thermo-stimuli-driven deformation in shape or size have received a lot of attention due to the recent scientific interests in preparing artificial muscles or actuators.[1] Also, thermo-responsive actuation (thermomechanical effects) based on nematic liquid crystal elastomers (LCEs) have gradually become a research priority in the preparation of soft actuators.[1-4] Nematic LCEs combine the anisotropic features of liquid crystal phases with the rubber elasticity of polymer network. When heated at temperature close to nematic to isotropic phase transition temperature (*N-to-I* temp.) of LCEs, a uniaxial thermomechanical deformation of LCEs will undergo at nearly constant volume due to a change of LC director order.[3-5] Recently, Keller et al.'s pioneering study reported an array of the micro-sized side-on LCE pillars concerning such thermomechanical effects had been successfully constructed through a soft lithography technology (i.e., replica molding). The prepared LCE pillars are mono-dispersive and micrometer size (diameter~20μm). Side-on LCEs within single pillar were applied magnetic field to align the nematic director of the molecules along the long axis of the pillars. The contraction of the one pillar was observed around 35-40%.[5] In turn, Keller et al. also presented a report of micron-sized main-chain LCE actuators with ultra-large amplitude contractions of around 300-400%.[6] Nevertheless, the used LCE haves *N-to-I* temp. higher than 100℃, largely limiting the available application.[1,5]

For the sake of lowing *N-to-I* temp. of LCEs and promoting nano-sized patterning process of mass production in industry, the present study will report a hexagonal array of nano-sized thermo-responsive pillar actuators that are able to contract and expand in response to temperature changes around a lower *N-to-I* temp. is manufactured *via* using reactive rod-like liquid crystal and ultraviolet nanoimprinting technology. It has been reported that nanoimprinting lithography (NIL) is a versatile and powerful technique for patterning nanostructures at low cost and high throughput.[7,8] With atomic force microscope (AFM) observation, a hexagonal array of pillars can be easily constructed by nanoimprinting and a responsive surface with a thermo-stimuli-driven roughness change is achieved. The room-temperature AFM scans quantitatively represent the single pillar shows a diameter of ca. 270 nm and 140 nm in depth, and the pitch meaning the averaged inter-pillar distance is measured as ca. 425 nm, thus lying in a nano-sized range. Furthermore, temperature-variable AFM is also utilized to demonstrate the pillar behaves as a thermally-stimulated nano-sized actuator. In our case, when heated above *N-to-I* phase transition temperature (ca. 65℃), it is clearly observed that the pillar diameter is expanded in the order of 12-15 % and then reversibly contracted in response to temperature drop.

EXPERIMENTAL

Materials and Nanoimprinted LC Film Preparation.

The preparations of the nanoimprinted LC film formed on a glass substrate having alignment film thereon are described as below. The aim of using alignment film was to align LC molecules along one direction.

Alignment Film Preparation. The alignment layer material was polyamide (PI) prepared from poly(amic acid) (abbreviated as PAA). The PI precursor PAA solution was AL-12G (commercially available from Daily Polymer, Taiwan) having 7 wt % solid content. The PAA solution was filtered and then directly used without additional purification. The filtered PAA solution was two-step spin coated, 500 rpm for 10 seconds and then 1000 rpm for 30 seconds, on a dried clean non-alkali glass substrate, pre-baked at 110 ℃ for 10 minutes, and baked at 230 ℃ for 1 hour to obtain a clean, transparent, and slightly yellow film formed on the glass substrate. After rubbed several times with a rubbing machine with a roller covered a velvet fabric, the alignment film was obtained. The thickness of the PI alignment film was about 0.5 μm.[9]

Nanoimprinted LC Film Preparation. The coating for forming the LC film of the present study was prepared by directly using a rod-like photopolymerizable liquid crystal monomer solution (RMS03-011, commercially available from Merck). The coating was filtered and then used without additional purification. The filtered coating was single step spin coated, 1000 rpm for 30 seconds, on the PI alignment film. Subsequently, the coating was dried at room temperature for about 1 minute to partly remove organic solvent. The coating was then heated to a temperature lower than the clear point of the monomers, and higher than the nematic phase point of the monomers. Herein, the heating temperature was 50 ℃, and the clear point of the monomers was about 67℃. After maintaining the heating temperature for 1~3 minutes, the liquid crystal molecules of the coating were aligned along the rubbed direction of the PI alignment film. In turn, the LC film was imprinted at ca. 55℃ for 5 minutes under 1 bar of imprinted pressure and vacuum through a PDMS soft mold in a homemade imprint system. After imprinting, the sample was pre-UV-cured at 55℃ for 30 minutes with a UV lamp in the imprint system. After taking from the system and cooled down to room temperature, the sample was further exposed to UV light (wavelength of 365±10 nm and intensity of 9 mW/cm^2) for about 1 minute for photopolymerization. Finally, the mold was removed, thereby forming a nanoimprinted LC coating of high hardness and high transparency on the alignment film.

Characterization of Surface Morphology.

The surface morphology of nanoimprinted LC film was examined by a commercial atomic force microscopy SPA 300HV (Seiko Instruments Industry, Co., Ltd., Japan) operated in the tapping model. The machine was also equipped with a heating stage so that a variable-temperature AFM could be executed. The used Si_3N_4 tips (SI-DF20) contained a force constant of 11 N/m and a resonance frequency of 125 KHz. The resultant AFM images were further treated by tilt collection and flat *via* manufacturer-provided software.

RESULTS AND DISCUSSION

Before proceeding to imprint, we need to examine the quality of PDMS soft mold. Figure 1 displays the AFM images of the used PDMS mold. It can be seed that a well-defined hexagonal array of cavity pits marked by dark color is clearly observed (Figure 1(a)). Based on the analysis of surface profile, the diameter of cavity pit is ca. 192±13 nm and inter-cavity-pit distance or pitch is about 395±10 nm. Lower magnification topographic image showing a larger area of AFM scan also shows some defects, as shown in Figure 1(b), which seem unavoidable in the replicating from original Ni mold to PDMS mold. However, the defects probably result from the signal mismatching during tip scanning. A phase image which is concurrently collected during scanning displays the better image quality, thus signaling the PDMS mold used herein still possesses the well-accepted quality of surface relief patterns (Figure 1(c).

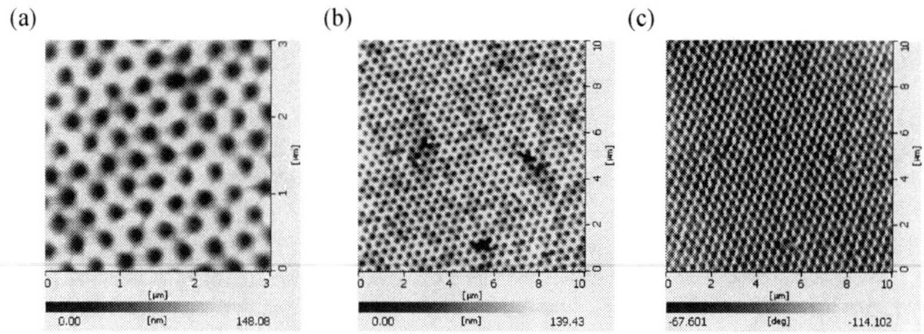

Figure 1. AFM images of the PDMS mold used here: (a) higher magnification topographic image; (b) lower magnification topographic image and its corresponding phase image (c).

Figure 2 shows the AFM images of the imprinted LC film at room temperature (~20 ℃). The images show the high-quality surface relief patterns. A well-defined hexagonal array of pillars is clearly prepared. The prepared LC pillars are mono-dispersive and nano-sized dimensions. The room-temperature AFM scans quantitatively represent the single pillar shows a diameter of ca. 270 nm and 140 nm in depth, and the pitch meaning the averaged inter-pillar distance is measured as ca. 425 nm, thus lying in a nano-sized range. It is noted that the structural parameters (pitch and diameter) of the pillars on the imprinted LC film do not match with original measured size of the pattern on PDMS. It is well-known that PDMS is thermally-sensitive polymer. Consequently, the mismatching in structural sizes is possibly attributed that the extent of thermal expansion of PDMS is much larger than the used LC materials during heating. In our imprinting process, the imprinting temperature is set as 55℃, and thus when heated to 55℃ from room temperature, the size of surface patterns on PDMS will gradually deform (expand or construct), therefore leading to a larger structural sizes been replicated on the LC surface. For soft PDMS mold, such a mismatching of the structural size seems unavoidable.

(a)

(b)

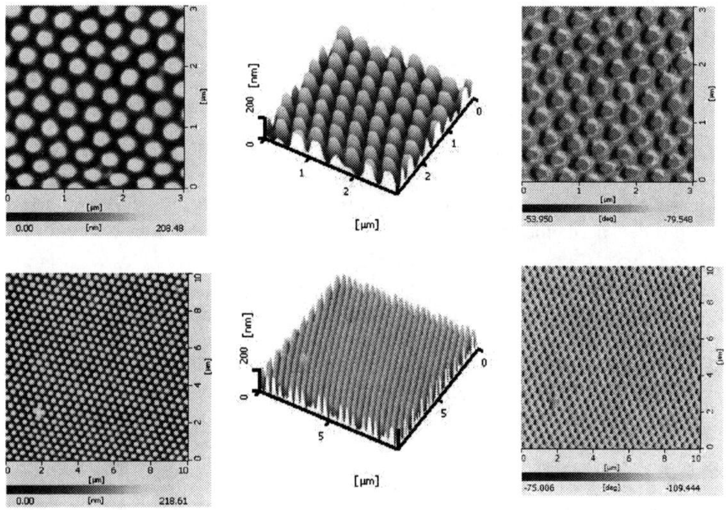

Figure 2. AFM images of the nanoimprinted LC film scanned at room temperature: (a) higher magnification with 3x3 μm^2 of scanning area; (b) lower magnification with 10x10 μm^2 of scanning area. Left, middle, and right images depict the topography, semi-3D, and phase model, respectively.

To clarify the thermo-responsive actuation (thermomechanical effect) based on nematic liquid crystal elastomers within one pillar, we use a variable-temperature AFM to execute *in-situ* observation. For each AFM image scanned at specific temperature, a retention period of ca. 30 minutes is necessary so that a thermal equilibrium could be achieved. On the other hand and in this case, due to thermal sensitivity stemming from PDMS mold, we prepare another nanoimprinted LC film with imprinted area of 3x3 cm^2 and imprinted temperature of ~40 ℃.

The room-temperature AFM scans quantitatively represent the single pillar shows a diameter of ca. 207±5 nm and the pitch meaning the averaged inter-pillar distance is measured as ca. 309±7 nm. Starting with room temperature (~20℃), the sample is heated and AFM images are collected *in situ*. Figure 3 representatively displays the AFM topographic images measured at 60 ℃ and 80 ℃. Obviously, the image collected at 60℃ is similar to that at room temperature. The pillar diameter and pitch are 213±3 nm and 314±5 nm, respectively, which are close to that at room temperature. Nevertheless, when heat to 80 ℃, the diameter is expanded into 246±6 nm but no significant difference in pitch, as shown in Figure 3(b). The diameter of pillar expanding along surface direction implies the LC molecules are orientated along the direction normal to surface. During *N-to-I* phase transition, the soft chain segments between the mesogens would gradually soften and relax, leading to mesogens with ordered orientation become disordered and consequently the diameter of pillar become larger and height of pillar is shorten simultaneously.

To further clarify the *N-to-I* phase transition temperature, we execute a series of *in-situ* variable-temperature AFM over a temperature range of 20 ℃ to 150 ℃. The results show a constant pitch ~ 310 nm is measured but a sudden temperature raising occurs at around 65℃, as plotted in Figure 4, thus indicating the *N-to-I* phase transition temperature is near 65℃. Also, a sharp temperature raising occurs between 60℃ and 70℃, signifying a thermo-responsive actuation should be fast. Furthermore, when the LC film is cooled down to room temperature from 150 ℃, it is found that the diameter of pillar will return to original room-temperature status (Figure 4), indicating the thermo-responsive actuation belongs to a reversible process. In contrast to the AFM image in Figure 2(a), we also note that the diameter of pillar shown in Figure 4 become smaller, implying an annealing effect undergoes during thermal history. The LC molecules will be re-packed more closely, leading to a smaller diameter measured. In summary, temperature-variable AFM demonstrates the pillar behaves as a thermally-stimulated nano-sized actuator. In this case, when heated above *N-to-I* phase transition temperature (ca. 65℃), it is clearly observed that the pillar diameter is expanded in the order of 12-15 % and then reversibly contracted in response to temperature drop.

For the actuation examination, we attempt to conduct an optical microscopy (OM) equipped with a CCD video system and a hotplate machine to directly execute a real-time observation of thermo-stimuli-driven actuation. Figure 5 displays the OM micrograph of the nanoimprinted LC film under 1,000 magnification and 25℃. The hexagonally-arrayed pattern on nanoimprinted film is clearly visible, but the image sharpness is poor. When heated at a heating rate of 2℃/min from 25 ℃ to 80℃, the OM micrograph does not show the significant difference. Consequently, it is very difficult to directly observe the actuation of these nano-sized actuators through optical microscopy. Nowadays, we have re-examined the actuation *via* grazing-incidence small-angle X-ray scattering (GISAXS) with a synchrotron source. The related experimental works are currently undergoing.

CONCLUSIONS

In the present study, a hexagonal array of LC-based pillars can be easily constructed by nanoimprinting and a responsive surface with a thermo-stimuli-driven roughness change is achieved. The room-temperature AFM scans quantitatively represent the single pillar shows a diameter of ca. 270 nm and 140 nm in depth, and the pitch meaning the averaged inter-pillar distance is measured as ca. 425 nm, thus lying in a nano-sized range. Furthermore, temperature-variable AFM is also utilized to demonstrate the pillar behaves as a thermally-stimulated nano-sized actuator. In our case, when heated above *N-to-I* phase transition temperature (ca. 65℃), it is clearly observed that the pillar diameter is expanded in the order of 12-15 % and then reversibly contracted in response to temperature drop.

ACKNOWLEDGMENTS

This work was supported by the Industrial Technology Research Institute of Taiwan under Grant 9354DA1250.

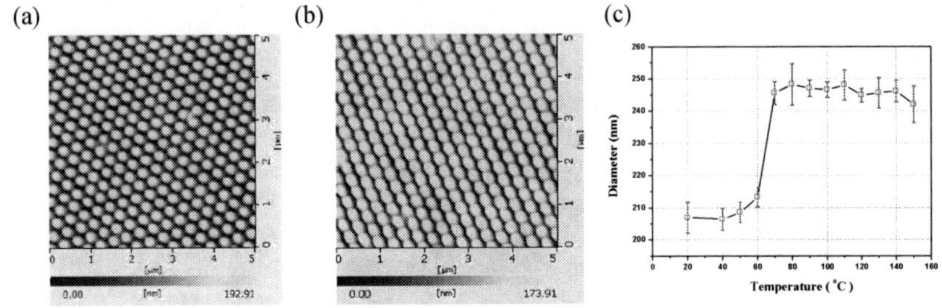

Figure 3. Variable-temperature AFM topographic images of the nanoimprinted LC films scanned at (a) 60 ℃ and (b) 80 ℃. The pillar diameters of the nanoimprinted LC film as a function of temperatures (c).

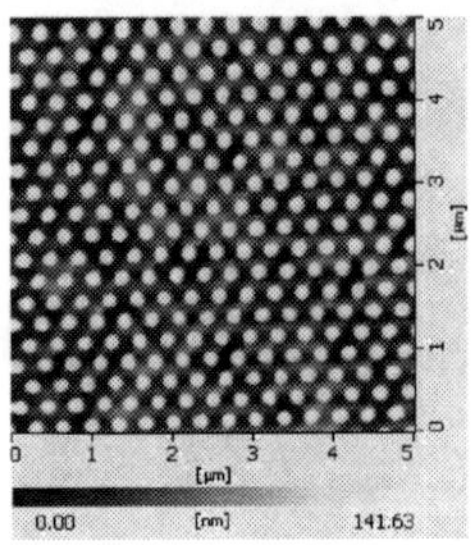

Figure 4. AFM topographic image collected at room temperature when cooled down to room temperature from 150 ℃.

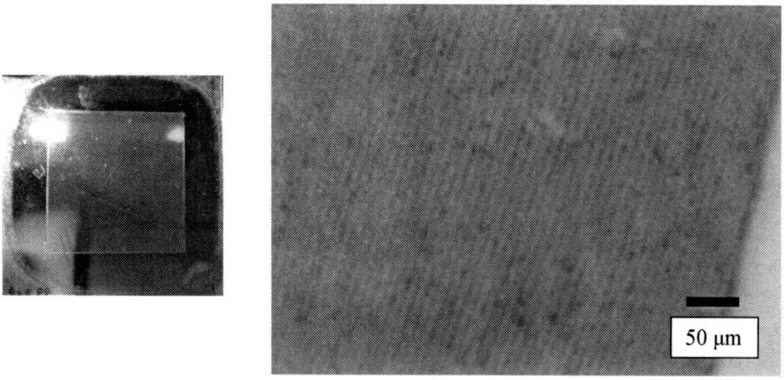

Figure 5. The nanoimprinted film used to execute the observation of thermo-stimuli-driven actuation (left) and its observed OM micrograph with a magnification of 1,000 and 25 ℃ (right).

REFERENCES

1. M. H. Li and P. Keller, *Philos. Trans. R. Soc. Lond. A* **364**, 2763-2777 (2006)
2. C. Ohm, C. Serra, and R Zentel, *Adv. Mater.* **21**, 4859-4862 (2009)
3. Y. Yu and T. Ikeda, *Angew. Chem. Int. ed.* **45**, 5416-5418 (2006)
4. T. Ikeda, J. Mamiya, and Y. Yu, *Angew. Chem. Int. ed.* **46**, 506-528 (2007)
5. C. Ohm, M. Brehmer, and R Zentel, *Adv. Mater.* **22**, 3366-3387 (2010)
5. A. Buguin, M. H. Li, P. Silberzan, B. Ladoux, and P. Keller, *J. Am. Chem. Soc.* **128**, 1088-1089 (2006)
6. H. Yang, A. Buguin, J. M. Taulemesse, K. Kaneko, S. Mery, A. Bergeret, and P. Keller, *J. Am. Chem. Soc.* **131**, 15000-15004 (2009)
7. S. Y. Chou, P. R. Krauss, and P. J. Renstorm, *Appl. Phys. Lett.* **67**, 3114 (1995)
8. L. J. Guo, *Adv. Mater.* **19**, 495 (2007)
9. C. M. Wu, M. T. Wu, and K. T. Huang, *US Patent* 20100079720 (2010)

Mater. Res. Soc. Symp. Proc. Vol. 1312 © 2011 Materials Research Society
DOI: 10.1557/opl.2011.911

A Patterned Conducting Polyaniline Layer on a Non-Conducting Polymer Matrix

Edward Song[1] and Jin-Woo Choi[1, 2]
[1]Department of Electrical and Computer Engineering, Louisiana State University, Baton Rouge, LA 70803, USA
[2]Center for Advanced Microstructures and Devices, Louisiana State University, Baton Rouge, LA 70803, USA

ABSTRACT

Polyaniline is one of the most studied conducting polymers and its properties have been used in many sensor applications including gas sensors and pH sensors. In this work, we present a method that is able to embed a conducting polyaniline film in a desired pattern onto a non-conducting polymer matrix. We have developed an embedded polyaniline film on a polydimethylsiloxane (PDMS) matrix using a cast molding technique. The polyaniline film was grown by electrochemically polymerizing polyaniline from an electrolyte containing sulfuric acid and aniline monomer. A three-electrode cell system was used with a standard Ag/AgCl reference electrode, a gold auxiliary electrode, and a patterned gold working electrode. After the polyaniline film was formed on the patterned working electrode, PDMS was poured atop and cured. Cured PDMS was then debonded along with the polyaniline film embedded on the PDMS layer or block. The gold layer of the working electrode can be patterned using a photolithography steps so that the designed pattern can be transferred by the polyaniline growth on the working electrode. The greenish color of the polyaniline film indicated that the embedded polyaniline was in the conducting state (emeraldine salt form) and the conductivity of the polyaniline has also been verified by applying a DC voltage across the film and measuring the current through the polyaniline layer. The measured conductivity of the PANi layer at room temperature was in the range of 4.5 ~ 4.6 S/m. The advantage of this technique is that the PDMS matrix holds the polyaniline in place while allowing the polyaniline film to undergo reversible doping/dedoping chemistry when electrolyte solution comes into contact with the surface of the film. The developed technology can be used for various chemical sensor applications, for example, a pH sensor or gas sensors where a flexible and all-polymer apparatus is needed.

INTRODUCTION

Conducting polymer (CP) has a unique property among polymers that is able to conduct electricity along its conjugated backbone structure[1]. Various forms of CPs have found applications in many areas including organic electronics[2], supercapacitors[3], and chemical sensors[4]. In particular, polyaniline (PANi) has attracted much attention due to its environmental stability, good conductivity, ease of processing, and simple doping/dedoping chemistry[5]. These properties make PANi a good candidate to be used as electrodes for microfluidic chips[6]. Polymer electrodes for microfluidic devices also have the advantage of being flexible, simple to fabricate, and more cost effective compared to the metallic electrodes on glass substrates. In addition, unlike metallic electrodes, embedded polymer electrodes in a PDMS matrix could avoid leakage and sealing issue in the polymer-based chip fabrication. Here, we present a method for fabricating a patterned layer of PANI embedded in PDMS block and the electrical properties of the material. Under certain growth conditions, PANi grows in a one-dimensional nanowire shape with the diameters of the nanowires on the order of tens of nanometers. This morphology of PANi film has increased surface area which allows PANi to accept or donate protons more readily. We have adopted the electrochemical growth method of a PANi layer of nanowire networks[7-9] because this method of growth is site-specific[10] meaning the PANi grows on the surface of lithographically patterned gold surface. Once the PANi layer is successfully grown, a patterned and conducting PANi layer can be formed in an electrically insulating PDMS block by casting a PDMS elastomer on top of the grown PANi layer. Furthermore, the electrochemical growth method circumvents the post-processing steps which are necessary for other growth techniques such as template-based growth or chemical synthesis.

EXPERIMENTAL DETAILS AND RESULTS

PANi film was grown using three-electrode electrochemical cell configuration, where a Ag/AgCl electrode (Weiss Research) was used as a reference electrode, a gold deposited glass slide was used as a counter electrode, and a lithographically patterned Cr/Au electrode (10 nm Cr and 50 nm Au) on glass slide was used as a working electrode. The electrolyte solution contained 0.05 M aniline monomer (Wako Chemicals) and 1 M H_2SO_4 (EMD Chemicals). All chemicals including the aniline monomer was used as received without further treatment. A simple potentiostat circuit was built and connected to the electrochemical cell in order to maintain the potential between the working and the reference electrodes. A potential of 0.795 V with respect to the reference was applied at the working electrode. PANi was grown for approximately 10 minutes until a thick layer of dark green PANi was deposited on the gold surface. After the growth, the working electrode was gently rinsed with DI water and dried in the convection oven at 40°C. The dried electrode was then placed in a petridish and PDMS polymer (Sylgard 184, Dow Corning) mixed with a curing agent at 10:1 ratio was poured on top of the electrode and cured. Once the cured PDMS was debonded from the electrode substrate, the patterned PANi film was successfully transferred to the PDMS block.

SEM images of the grown PANi nanowires on gold surface

The scanning electron micrograph (SEM, Hitachi S-4500II field emission) images in Figure 1 are taken from a sample of a 1 cm × 1 cm area of gold surface on which a nanowire framework of PANi is grown electrochemically by the method detailed in the previous section.

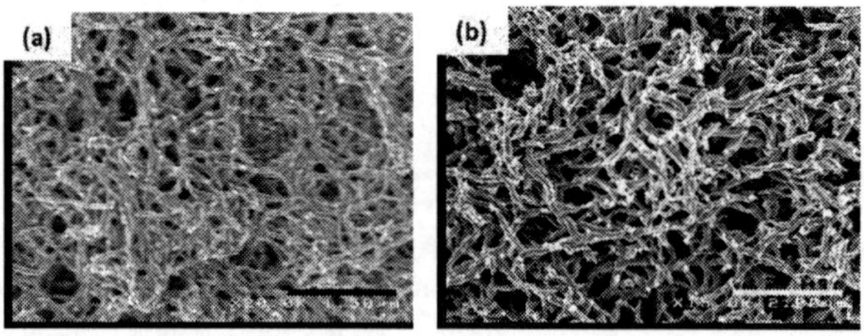

Figure 1. SEM images of PANi nanowires in emeraldine salt form grown on gold surface: (a) scale bar: 1.5 μm and (b) scale bar: 2 μm.

PANi layer embedded in a PDMS block

Figure 2 shows the growth of the PANi film grown on the gold surface which replicated the pattern of the gold electrode. By casting a PDMS on top of this electrode the pattern of the conducting polymer is transferred to the PDMS polymer matrix. The resolution of the PANi pattern can mainly be determined by the gold electrode width that can be defined by photolithography. We have demonstrated a grid pattern of PANi with line width and the spacing between them of 1 mm in Figure 2 (a). However, our recent experimental results show that a feature size in the order of tens of micrometers is achievable through this technique.

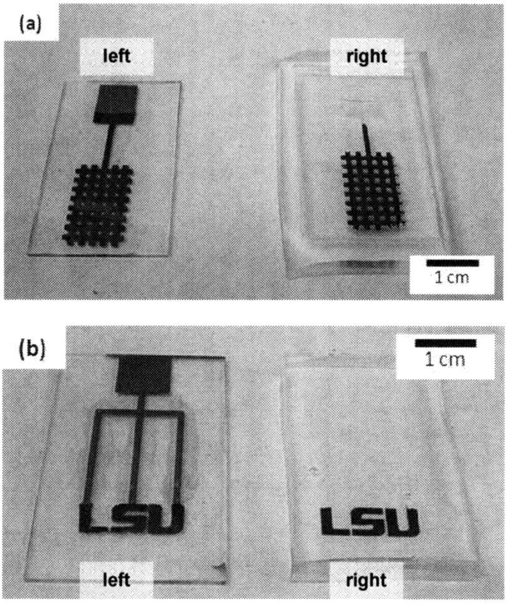

Figure 2. Lithographically patterned gold electrode (left) and the transferred PANi film on PDMS block (right): (a) grid pattern and (b) LSU logo pattern.

DISCUSSION

The range of conductivity that we were able to obtain at room temperature was between 4.5 S/m to 4.6 S/m, which demonstrates its ability to be used as a conductive electrode for the fully polymeric device. A strip of PANi layer with a width of 2 mm and a length of 10 mm embedded in PDMS block was fabricated and the conductivity of this PANi film was measured by applying a DC voltage and measuring the current. Figure 3 shows the conductivity of the PANi electrode with varying voltages. At a given voltage, the current tended to increase over time. This could be due to Joule heating during the measurement, which needs to be confirmed with further investigation. However, the increasing current eventually settles to a constant value.

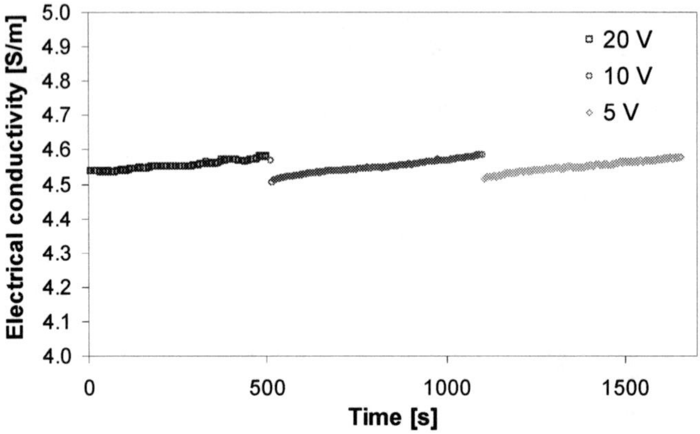

Figure 3. Measured conductivity of the embedded PANi film in PDMS with varying DC voltages.

CONCLUSIONS

In this paper, we have demonstrated the patterning ability of the PANi conducting polymer layer in an electrically insulating PDMS block and that the embedded PANi layer maintained its conductive nature as illustrated in the resistance measurement. This material may have interesting applications in fully polymeric device where the patterned PANi layer could act as an electrode for the device. The conductivity of the PANi material is significantly influenced by its redox state which is mainly determined by the proton concentration of the environment. Hence maintaining the pH level of the PANi layer is important in regulating the electrical conductivity of the material. PANi is also known to respond to various chemical species such as HCl and NH_3, hence the development of a polymer based chemical or gas sensor could also be achieved. As an extension to this work, possible future plans include, further investigation on the cause for the rising current under steady state environment, AC voltage analysis of the PANi layer, and fabrication and characterization of the all polymer pH sensor.

REFERENCES

1. A. G. MacDiarmid, Angewandte Chemie International Edition **40**, 2581-2590 (2001).
2. G. B. Blanchet, Y.-L. Loo, J. A. Rogers, F. Gao and C. R. Fincher, *Applied Physics Letters* **82**, 463-465 (2003).
3. C. Meng, C. Liu, L. Chen, C. Hu, and S. Fan, *Nano Letters* **10**, 4025-4031 (2010).
4. J. Janata, and M. Josowicz, *Nature Materials* **2**, 19-24 (2003).
5. D. Li, J. Huang, and R. B. Kaner, *Accounts of Chemical Research* **42**, 135-145 (2009).

6. R. D. Henderson, R. M. Guijt, P. R. Haddad, E. F. Hilder, T. W. Lewis, and M. C. Breadmore, *Lab on a Chip* **10**, 1869-1872 (2010).
7. V. Gupta and N. Miura, *Electrochemistry Communications* **7**, 995-999 (2005).
8. J. Wang, S. Chan, R. R. Carlson, Y. Luo, G. Ge, R. S. Ries, J. R. Heath, and H.-R. Tseng, Nano Letters **4**, 1693-1697 (2004).
9. L. Liang, J. Liu, C. F. Windisch, G. J. Exarhos, and Y. Lin, *Angewandte Chemie International Edition* **41**, 3665-3668 (2002).
10. Y. Wang, K. K. Coti, J. Wang, M. M. Alam, J.-J. Shyue, W. Lu, N. P. Padture, and H.-R. Tseng, *Nanotechnology* **18**, 424021 (2007).

Mater. Res. Soc. Symp. Proc. Vol. 1312 © 2011 Materials Research Society
DOI: 10.1557/opl.2011.912

Electrochromic performance of WO_3 thin films with solvent-free viscous electrolytes based on polyethylene glycol-titanium oxide nanocomposites

Narcizo Mendoza[1,2], Liliana Hechavarría[1], Francisco Paraguay-Delgado[2] and Hailin Hu[1]
[1]Centro de Investigación en Energía, UNAM, Priv. Xochicalco S/N, Temixco, Morelos, 62580, México.
[2] Centro de Investigación en Materiales Avanzados, Miguel de Cervantes 120, Chihuahua, Chih, 31109, México.

ABSTRACT

Polymeric nanocomposites of polyethylene glycol (PEG) with titanium oxide compound, PEG-Ti, have been prepared by sol-gel method from liquid PEG and titanium isopropoxide in acidic medium. Lithium salt (LiX) has been added into PEG-Ti to form PEG-Ti-LiX polymeric electrolytes. Electrochromic devices based on tungsten oxide thin films and PEG-Ti-LiX electrolyte may show excellent optical transmittance transient, however it depends on the type of lithium salt used during the sol-gel process. With LiI, the color change speed of the devices is very fast but they show a yellow color at bleaching state. The use of $LiClO_4$ makes the devices totally transparent in visible region but the optical contrast is small. Possible molecular structure model of these polymeric electrolytes have been analyzed to explain the relation between electrochromic performance of tungsten oxide and electrolyte chemical composition.

INTRODUCTION

Electrolytes work as ionic conductor -reservoir and as a separator of electrodes in electrochemical devices such as lithium batteries, dye-sensitized solar cells, electrochromic windows and sensors [1,2]. They can be classified as liquid, viscous or solid according to their viscosity. An electrolyte with volatile solvents shows fast ionic conduction, but the leak or evaporation of solvents is inevitable. The use of this type of electrolytes obligates a rigorous device sealing to reduce or prevent solvent loss. On the other hand, electrolytes without volatile solvents, or simply solvent-free electrolytes, are chemically stable and easy to be handled, which is desirable for device applications.

A polymer electrolyte is usually prepared by dissolving a salt in a polymeric matrix. PEO is a commercially available polyether: $HO-(CH_2-CH_2-O)_nH$. As the lithium salt is added into it, the ether oxygen atoms of the repetitive unit are responsible to dissociate such salt, interact with lithium ions and transport them from one to another. In comparison with those with solvents, the disadvantage of solvent-free electrolytes in comparison with those with solvents is their relatively low ionic conductivity, which can prejudice charge transfer process at electrode/electrolyte interfaces of electrochemical devices. Several approaches have been reported to improve the ionic conductivity of polymeric electrolytes, and one of them is the use of composite materials; those of PEO composites with ceramic nanoparticles (TiO_2, SiO_2, etc.) show better ionic conductivity than that of pure salt-in-polymer electrolytes [3-6]. In a recent work we reported a PEG-Ti nanocomposite made by sol-gel route. The electrochromic performance of tungsten oxide with PEG-Ti-LiI as electrolyte was improved [7,8]. In this work we will show the effect of type of lithium salts as well as source of acid used in sol-gel process on the electrochromic performance of tungsten oxide thin films. Moreover we will analyze such effect along with a major change in titanium molar concentration in the polymeric composites.

EXPERIMENTAL

Tungsten peroxy-ester solution was prepared by sol-gel method according to Cronin et. al. [9]. This solution was spin-coated onto clean conductive glass (ITO) by 2000 rpm for 60 s, followed by the annealing in air at 250 °C for 1 h. The obtained thin films by this process were amorphous tungsten oxide of about 350 nm thick ready for electrochromic device use. PEG –Ti composite materials were prepared by sol-gel method starting from dissolving titanium isopropoxide in an acidic solution with PEG of molecular weight equal to 600 and 2-propanol under stirring, as reported previously [7]. Either LiI or $LiClO_4$ was added in viscous PEG-Ti composites to form PEG-Ti-LiX electrolytes. The molar ration between an average PEG molecule and titanium atom was chosen as PEG:Ti = 24:1, 8:1 and 4:1. PEG-Ti-LiI and PEG-Ti-$LiClO_4$ electrolytes were prepared by dissolving LiI or $LiClO_4$ in viscous PEG-Ti complexes with a molar ratio of one atom of Li into 8 oxygen atoms in PEG. All of these electrolytes were vacuum-dried at 80 °C for 3 h before use.

Tungsten oxide electrochromic devices were prepared by collocating polymer electrolyte (PE) between ITO glass and WO_3 coating on ITO to form a device configuration of ITO/WO_3/PE/ITO. Transient optical transmittance curves of tungsten oxide based devices with different PEG-Ti-LiX electrolytes were determined under positive and negative potential bias. Electrochemical impedance spectroscopy (EIS) study was realized at room temperature (25 °C) on tungsten oxide based electrochromic devices with different polymeric electrolytes by using a dynamic-EIS system (VoltaLab PGZ301). The DC potential was -1.5 or -2.0 V and the amplitude of the AC signal was 50 mV. The frequency varied from 0.1 Hz to 100 kHz.

RESULTS AND DISCUSSION

The incorporation of titanium oxide compound in PEG changes its physical appearance. The increasing viscosity of PEG-Ti composite materials with the increasing titanium precursor amount in original sol-gel solution is a clear evidence of the formation of longer or crosslinked molecular chains of PEG by titanium compounds in polymeric composites. The formation of titanium oxide compounds in viscous PEG-Ti (8:1) complexes was observed by Transmission Electron Microscope (not shown here). The image shows a kind of dark pots that are related to interferences of crystalline planes (Moire bands), which indicate that those particles or clusters are crystalline. The average diameter of those dark areas was about 1.8 to 7 nm.

As reported before, the oxidation reversibility of tungsten oxide has improved considerably by introducing titanium oxide compound inside the PEG-LiI electrolytes with a PEG:Ti molar ratio equal to 24:1, but the reduction one was still poor [7]. Now the titanium percentage in the polymer composites has been increased up to four molecules of PEG vs. one atom of titanium. Figure 1 shows the transient optical transmittance curves at 650 nm of ITO/WO_3/PE/ITO devices under oxidation (+1.0 V) and reduction (-1.5 V) bias, in which PE were PEG-Ti-LiI with PEG:Ti molar ratio as 24:1 and 4:1. This curve indicates that the more the titanium percentage, the faster the color change speed of the devices.

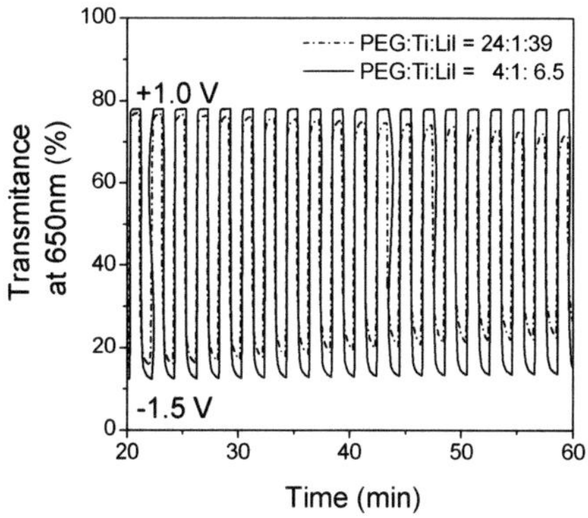

Figure 1. Optical transmittance transients at 650 nm of ITO/WO_3/PE/ITO devices with polymer electrolyte (PE) as PEG-Ti-LiI with PEG:Ti molar ratio of 4:1 and 24:1.

The fast oxidation speed should come from a fast lithium ion insertion into the electrochromic element as it was positively polarized. A faster cation insertion should be related to better lithium iodide dissociation inside the polymer matrix. The formation of iodine and tri-iodide inside the PEG-Ti composite materials after addition of LiI indicates that iodide ions have been oxidized. This is chemically possible considering that titanium oxide has been known to be a weak Lewis acid that can accept unshared electron pair of base. Since the titanium oxide compounds are associated with PEG macromolecules to form crosslinked network or larger molecular chains, the average size of negative charges in the composite electrolytes becomes much larger than that of iodides in PEG. Under external polarization, the separation of lithium ions from negative charges inside the composite polymer should be easier and, as a result, the cation insertion process should be faster. In the case of reduction, the improvement of colouring speed is also notable but not so large, because of different mechanisms of charge transfers in each process [7].

The formation of iodine and tri-iodides inside the PEG-Ti composites after the dissolution of LiI makes the suggestion of substitution this by some other lithium salt that is free of iodides and may avoid the yellow color of tungsten oxide during the bleaching process. $LiClO_4$ was chosen to be added into PEG-Ti composites to form PEG-Ti-$LiClO_4$ electrolytes due to its colorless appearance and its good solubility in most of the solvents. Effectively the substitution of LiI by $LiClO_4$ leads to more transparent PEG-Ti-$LiClO_4$ composite electrolytes. However, the WO_3 electrochromic devices with PEG-$LiClO_4$ electrolyte (without titanium oxide compound) show very small electrochromic capacity under +1 V and -2 V bias (Figure 2); with -1.5 V the tungsten oxide could not be colored in this case. When titanium oxide compounds are incorporated in the electrolytes, the electrochromic performance of the corresponding WO_3 devices is improved, although the coloring process is not quite completed under -2V reduction polarization, and color change speeds are slow. Although Figure 2 only exhibits the electrochromic performance of one

PEG-Ti-$LiClO_4$ device (PEG:Ti:$LiClO_4$ = 8:1:13), the variation in titanium concentration did not improved the observed electrochromic properties in these devices.

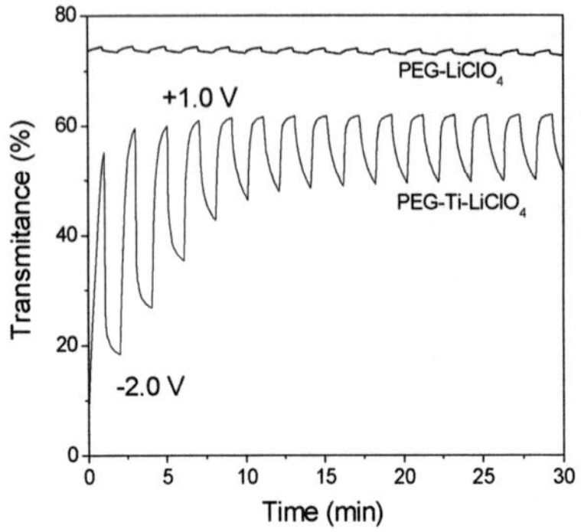

Figure 2. Optical transmittance transients at 650 nm of ITO/WO_3/PE/ITO devices with polymer electrolyte (PE) as PEG-$LiClO_4$ or PEG-Ti-$LiClO_4$ with PEG:Ti molar ratio of 8:1.

The poor electrochromic performance of WO_3 with lithium perchlorate comes first from the lower dissociation of $LiClO_4$ in PEG; it took longer time to dissolve this than LiI in the polymer as well as in the PEG-Ti composites. Secondly, there must be some kind of interactions between titanium oxide compounds with ClO_4^- ions that results in a better dissociation of $LiClO_4$ in comparison with that in PEG only, provided the electrochromic performance of the devices has been improved again with the presence of titanium oxide compounds. However, these interactions are not sufficiently strong to lead charge transfer between perchlorate ions and titanium complexes, and as a result, the electrochromic performance improvement of tungsten oxide devices is limited.

The effect of different lithium salts in electrolytes on electrochromic behavior of tungsten oxide thin films can be compared by studying the charge transfer phenomena at electrolyte/electrode interfaces observed with electrochemical impedance spectroscopy (EIS). Figure 3a shows Nyquist plots of tungsten oxide based electrochromic devices (ECDs) with PEG-Ti-LiI as electrolytes at -1.5 V polarization on WO_3 electrode and its equivalent circuit, and Figure 3b the same plots for ECDs with PEG-Ti-$LiClO_4$ as electrolytes.

In all cases semicircle form of the plots at middle-low frequency range is observed, indicating a charge transfer process occurred at interface of electrolyte with tungsten oxide electrode under reduction process by negative bias. It is observed that ionic resistance in electrolyte (high frequency resistance $\mathbf{R_b}$) is lower with LiI than with $LiClO_4$, as well as the charge transfer resistance at the interface of electrolyte/tungsten oxide electrode ($\mathbf{R_1}$). The charge storage capacity at the same interface (**CPE1**) is larger with LiI than with $LiClO_4$. It is observed also from Figure 3a that at -1.5 V the impedance resistances of those ECDS with LiI, reduce by more than one order of magnitude after the introduction of titanium oxide compounds, and the more the titanium percentage, the larger the reduction of impedance. In the case of polyelectrolytes with $LiClO_4$, this reduction is only about 40-70 percentage at – 2V polarization. The less the values of impedance, the faster the charge transfer processes occurred in tungsten oxide electrodes under negative bias, just as observed through the color change in this electrochromic material. Similar EIS results were obtained for the same ECDs under positive polarization (+1V).

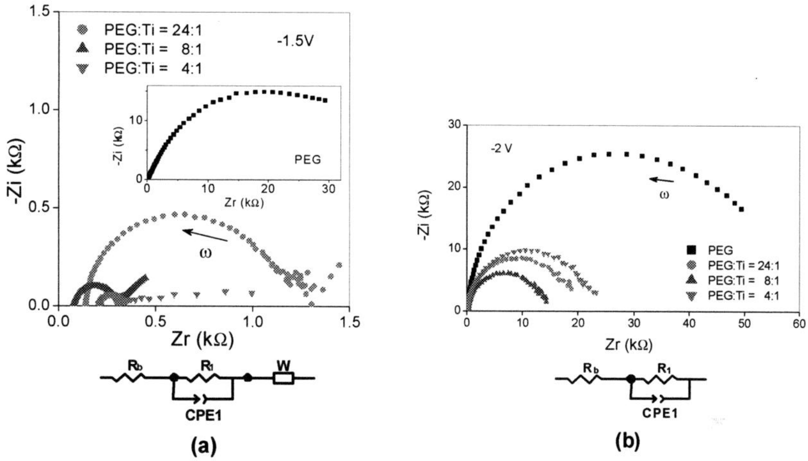

Figure 3. Nyquist plots and equivalent circuits of ITO/WO_3/PE/ITO with PE as: (a) PEG-Ti-LiI at -1.5 V and (b) PEG-Ti-$LiClO_4$ at – 2V for PEG:Ti molar ratio of 24:1, 8:1, 4:1 and 1:0.

CONCLUSIONS

Titanium oxide compounds of maximum size of 1.8 to 7 nm have been formed inside the polyethylene glycol (PEG) by sol-gel method as the molar ratio between PEG and titanium precursor (PEG:Ti) varied from 24:1 to 4:1. For molar ratio of PEG:Ti from 24:1 to 4:1, the larger the concentration of titanium, the better the WO_3 color change speed as well as its optical contrast and stability. They are associated with PEG molecules through hydroxyl-oxygen interactions and form longer or crosslinked molecular chains. As the lithium iodide salt was added into the PEG-Ti composite polymer, part of the iodide ions were oxidized into iodine and tri-iodides. At the same time, titanium oxide compounds inside the PEG-Ti complexes should be negatively charged, leading to a faster lithium ion insertion in positively polarized tungsten oxide thin films. When LiI was substituted by $LiClO_4$, both optical contrast and color change speeds were reduced. Electrochemical impedance spectroscopy study confirms the fast charge transfer process occurred in those polymeric composite electrolytes with LiI salt.

REFERENCES

[1] J. W. Schultze and H. Karabulut, *Electrochim. Acta* **50,** 1739 (2005).
[2] M. Grätzel, J. *Photochem. & Photobiology C: Photochemistry Reviews* **4** 145 (2003).

[3] L. M. Bronstein, J. C. Karlinsey, A. Ryder, and J. M. Zwanziger, *Chem. Materials* **13,** 3678 (2001).
[4] G. Katsaros, T. Stergiopoulos, I. M. Arabatzis, K. G. Papadokostaki, and P. Falaras, *J. Photochemistry and Photobiology A. Chemistry* **149,** 191 (2002).
[5] V. Di Noto, R. Gliubizzi, E. Negro, and G. Pace, *J. Phys. Chem. B* **110,** 24972 (2006).
[6] M. S. Akhtar, J. M. Chun, and O. B. Yang, *Electrochem. Commun.* **9,** 2933 (2007).
[7] L. Hechavarria, N. Mendoza, P. Altuzar, and H. Hu *J. Solid State Electrochem.* **14,** 323 (2010).
[8] N. Mendoza, F. Paraguay-Delgado, L. Hechavarría, M. E. Nicho, and H. Hu, submitted.
[9] J. P. Cronin, D. J. Tarico, A. Agrawal, and R. L. Zhang *USA Patent* 5,277,986 (1994).

Mater. Res. Soc. Symp. Proc. Vol. 1312 © 2011 Materials Research Society
DOI: 10.1557/opl.2011.913

Thermo-mechanical Behavior of (Meth)Acrylate Shape-Memory Polymer Networks

Carl P. Frick[1], Nishant Lakhera[1] and Christopher M. Yakacki[2,3]

[1]Mechanical Engineering, University of Wyoming, Laramie, Wyoming, USA
[2]MedShape Solutions, Inc., Atlanta, Georgia, USA
[3]School of Materials Science & Engineering, Georgia Institute of Technology, Atlanta, USA

ABSTRACT

Our overall approach is based on developing a photocrosslinkable polymer network with a favorable shape-memory response, using polymer chemistry and crosslinking density to control thermo-mechanical properties. Three polymer networks were created and thermo-mechanically tested, each from tert-Butyl acrylate linear builder co-polymerized with a poly(ethylene glycol) dimethacrylate cross-linker. By systematically altering the molecular weight and the weight fraction of the cross-linker, it was possible to create three polymers that exhibited the same glass transition temperature, but varied by almost an order of magnitude in rubbery modulus. Therefore, the mechanical stiffness could be tailored to suit a given application. Recovery behavior of the polymers was characterized over a range of deformation temperatures. It has been implicitly assumed a linear relationship between Free-Strain (i.e. no actuation force) and Fixed-Stress (i.e. maximum actuation force), however, this has never been confirmed experimentally. The energy per unit volume performed by the shape-memory polymer was quantified, and observed to be a function of strain recovered. The maximum recoverable work was shown to increase with cross-linking density, although the overall efficiency is similar for all materials tested.

INTRODUCTION

Shape-memory polymers (SMPs) have attracted increased attention over the last several years due to their ability to temporarily store a deformed shape, and subsequently recover the deformation upon exposure to heat. This capability has led to several proposed applications, including self-deployable structures for biomedical and space applications, among others [1-4]. Inherent advantages include high recovery strain, low density, and low cost. For example, in comparison with NiTiNOL shape-memory alloys that can recover approximately 4% to 8% of applied displacement, whereas SMPs can recover strains on the order of 50% to 500% [5, 6].

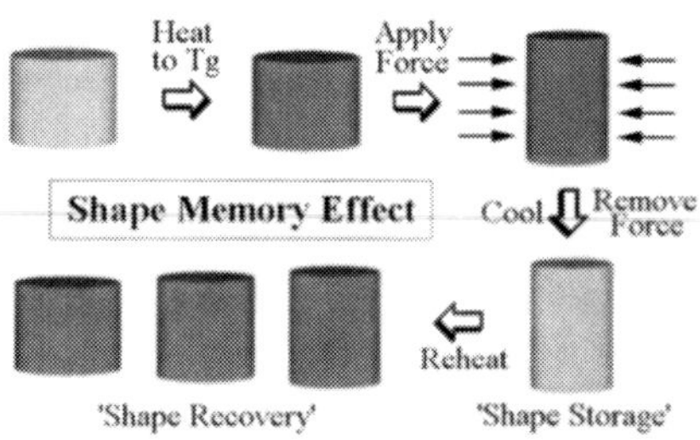

Figure 1: Schematic illustration of the Shape-Memory Effect. The material is mechanically deformed, cooled to store the deformed shape, and then heated to elicit shape recovery.

The shape-memory effect in polymers is schematically outlined in Figure 1. The polymer is first synthesized into a permanent shape by standard polymer processing techniques. Subsequently, the polymer is heated above a critical temperature, such as the glass transition temperature, T_g, and thermo-mechanically deformed into a temporary shape, a process known as shape storage. The polymer will remain in the stored shape until it is reheated in the vicinity of its glass transition temperature, upon which it will experience controlled shape recovery.

Thermoplastic and thermoset polymers with various chemistries are capable of demonstrating shape-memory if the polymer structure can be adequately controlled and the transition temperatures exist in a useful range [7]. From a macroscopic viewpoint, a polymer will demonstrate a useful shape-memory effect if it possesses a distinct and significant glass transition, a modulus temperature plateau in the rubbery state, and a large difference between the maximum achievable strain during deformation and permanent plastic strain after recovery [8]. Although numerous polymer systems will exhibit the shape-memory effect to some degree, the subset of appropriate polymers is significantly reduced with the requirements of an adjustable thermal transition in an appropriate temperature range and large recoverable strain and stress.

The purpose of this research is to better characterize partial recovery behavior. The majority of research studies demonstrate the shape-memory effect by observing deformation recovery with no applied force (Free-Strain) or by fixing deformation and measuring the applied force (Fixed-Stress). These two tests represent the extremes in the SMP behavior, demonstrating either the maximum recoverable strain or the maximum actuation stress. However for the majority of SMP applications, neither condition is likely to be realized. It is far more probable that in actual service, the SMP will be partially constrained, meaning that the material will be allowed to recover under some applied load below the maximum actuation stress. For example, SMPs have been investigated for use as self-deployable stents [8-10]. Stents are expandable scaffolds shown to have a clinical benefit after balloon angioplasty. For this application the SMP stent would be deformed to a minimal profile, and when placed in position, allowed to expand against the blood vessel sidewall. It is critical that the SMP not reach full displacement recovery because the frictional force must be large enough to keep the stent in place, as well as prevent vasospasms and restenosis. However, the SMP stent must recover enough displacement such that it allows for improved blood flow. In this example, the SMP has partially recovered the maximum possible displacement, and partially realized the maximum possible actuation force. It has been implicitly assumed a linear relationship between these two maximums; however, this has not been explicitly demonstrated.

The focus of this research is to investigate the thermo-mechanical and shape recovery behavior of photopolymerizable (meth)acrylate shape-memory polymers. Photopolymerizable networks refer to polymer solutions that can be solidified upon exposure to ultra-violet light. Such polymers have several inherent advantages including the possibility of *in situ* fabrication, relatively quick synthesis rates, and simple one-step processing into complex geometries [11]. Furthermore, (meth)acrylate-based networks are advantageous because their material properties can be tuned by control of polymer chemistry [11, 12]. Tailoring of mechanical properties is critical to match the needs of any specific application. Therefore, three previously investigated (meth)acrylate SMP networks were chosen [8]. Each SMP was fabricated from *tert*-Butyl acrylate linear builder co-polymerized with a poly(ethylene glycol)$_n$ dimethacrylate cross-linker. The relative weight fractions of the chemical constituents were tailored such that all three materials had the same glass transition temperature resulting in shape-memory behavior at similar temperatures, but varied significantly in rubbery modulus allowing for a range of Fixed-

Stress actuation forces. The recoverable energy per unit volume exhibited by the three SMP networks was quantified as a function of strain recovered. Results show that the maximum work output scaled with the magnitude of the rubbery modulus, however, the efficiency was similar for all materials tested.

EXPERIMENTAL METHODS

Materials

tert-Butyl acrylate (*t*BA), poly(ethylene glycol)$_n$ dimethacrylate (PEGDMA) with typical molecular weight of M_n = 550, di(ethylene glycol) dimethacrylate (DEGDMA), and photoinitiator 2,2-dimethoxy-2-phenylacetophenone (DMPA) were all ordered from Aldrich and used in their as-received conditions.

Synthesis of Polymer

A crosslinker solution (XLS) was created by mixing 30 wt.% DEGDMA with 70 wt.% PEGDMA. The *t*BA-*co*-XLS networks were then synthesized by free-radical polymerization using 0.1 wt% of DMPA photoinitiator. The polymer solution was injected between two glass slides separated by a spacer, or into a cylindrical polyethylene mold, depending on subsequent testing. For the slides, polymerization was performed in a UV crosslinker oven (UVP, CL1000) for 15 minutes. For the cylindrical molds, specimens were polymerized under a UV lamp (Blak Ray B100 A/R) for 15 minutes. Following UV curing, all polymer samples were placed in an oven at 90° C for one hour to ensure complete polymerization.

Dynamic Mechanical Analysis (DMA)

The rectangular samples cured from the slides then where cut into smaller rectangles and sanded to remove surface imperfections. These samples were tested with a Dynamic Mechanical Analyzer (DMA - TA Instruments Q800) to characterize the thermo-mechanical properties of the polymer networks. Samples were tested under tensile loading with the sample dimensions measuring 27 x 5.3 x 1.57 mm^3. Samples were thermally equilibrated at -50 °C and heated to 150°C at a rate of 5°C/min while subjected to a 0.1 % dynamic strain at 1 Hz. The glass transition temperature (T_g) was defined at the peak of the tan delta curve. The onset of glass transition (T_{onset}) was calculated using an intersecting line method with the starting point of the left line being 50°C less than T_g and the starting point of the right line being T_g. The rubbery modulus was measured at the lowest point of the rubbery plateau.

Load Controlled Strain Recovery Experiments (Compressive)

Load controlled tests were conducted on a MTS hydraulic load frame utilizing cylindrical samples of each network cut to length to diameter ratio of 1.5 (the diameter used was close to 10 mm, although it varied slightly due to the machining process). Load controlled strain recovery experiments involved a three-step process that included isothermal compression, cooling while displacement is maintained, and thermally induced shape-memory recovery under a constraining load. During deformation, samples were equilibrated in a temperature-controlled chamber at 70°C and held for 10 minutes to ensure constant temperature throughout the sample. Samples were then deformed to approximately 30% strain. Loading rate used was 0.01 mm/sec. The samples were then held for 20 minutes to allow relaxation. Following stress relaxation, samples were cooled to 0°C at a rate of 5 °C/min and held until the stress value fell to zero. This indicates the sample contracting away from the platens. Following cooling, the load was increased to fraction of the total applied stress during isothermal compression. Samples were then heated up to 120°C and recovered strain was measured. Temperature was ramped at a rate of 2°C/min.

Work and Efficiency Calculations

In order to better characterize the partial recovery behavior, the work performed by the SMP during heating (i.e. the energy recovered) was calculated. The work is defined by:

$$W_R = \sigma_A \varepsilon_R \tag{1}$$

where σ_A is the constant applied constraining stress during recovery heating, and ε_R is the recovered strain at 100°C. This temperature was chosen because it is well above the glass transition region. Strictly speaking, the work calculated in equation (1) is the energy recovered per unit volume.

The work efficiency of the SMP was also calculated. To accomplish this, it was required to first calculate the work necessary to deform the SMP. Based on the observed linear elastic stress-strain behavior (see following Results & Discussion section, Figure 5) the energy required to initially deform the SMP was defined as the area underneath the stress-strain curve during initial compression. The energy per unit volume required to deform the SMP was calculated by:

$$U = \frac{1}{2}\sigma\varepsilon \tag{2}$$

Where σ and ε are the stress and strain required to initially deform the sample, respectively. Efficiency is therefore defined as the ratio of the work recovered during heating relative to the work required to deform, given by:

$$\eta = \frac{W_R}{U} \tag{3}$$

RESULTS & DISCUSSION

The thermo-mechanical behavior of the three polymers used in this study is shown in Figure 2. Each polymer was created by photopolymerizing *tert*-Butyl acrylate linear builder with poly(ethylene glycol)$_n$ dimethacrylate cross-linker. Respective weight fractions and molecular weight of the constituents were varied such that all three materials had the same glass transition temperature (T_g = 55°C) but varied in rubbery modulus from 2 MPa to 13 MPa as a function of cross-linking density. These polymers were originally developed for use as cardiovascular devices [8]. They exhibit shape storage at room temperature when the SMPs are in the glassy state, and shape-memory recovery at body temperature or above, as the temperature

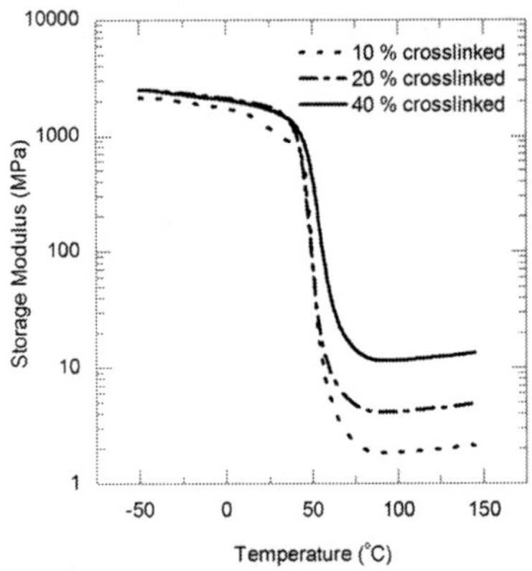

Figure 2: Dynamic mechanical analysis measurements of copolymer comprised of tert-Butyl acrylate linear builder polymerized with poly(ethylene glycol)$_n$ dimethacrylate cross-linker. Polymer networks have been tailored to vary rubbery modulus with a constant glass transition temperature.

approaches the glass transition of the SMP. Rubbery modulus values were designed to vary over a wide range in order to demonstrate the feasible design space. By systematically varying the rubbery modulus it is possible to tailor the stiffness to better match with a potential application. The addition of cross-linker beyond 40 wt.% was shown to widen the glass transition, and consequently made shape storage at room temperature impractical.

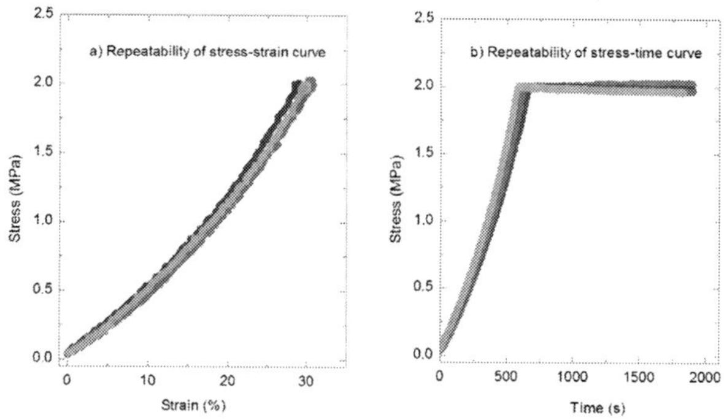

Figure 3: (a) Stress-strain and (b) stress-time response of shape-memory polymers tested at 70°C, comprised of 20 wt.% crosslinker. Samples manufactured using 10 wt.% and 40 wt.% crosslinker exhibited analogous elastic behavior. The magnitude of the stress scaled cross-linking density.

Partial shape-memory behavior of the polymers was investigated by compressing each of the three networks at 70°C, cooling to 0°C while maintaining deformation, and re-heating to 120°C under a prescribed load. Figure 3 shows compressive stress-strain and stress-time behavior of the 20 wt.% cross-linked SMP specimens tested at 70°C. All samples demonstrate elastic loading up to the maximum applied strain (approximately 30% shown in Figure 3a). The compressive specimens were then isothermally maintained under the applied strain for 20 min. (Figure 3b) to ensure that no time-dependent relaxation occurred. The observed linear elastic behavior is expected for (meth)acrylate networks above the glass transition temperature. Compression of the 10 wt.% and 40 wt.% cross-linked samples demonstrated qualitatively similar behavior. Maximum force was found to increase with increasing cross-linking density, with values of approximately 0.9 MPa, 2.0 MPa, and 5.8 MPa measured for the 10 wt.%, 20 wt.%, and 40 wt.% cross-linked samples, respectively.

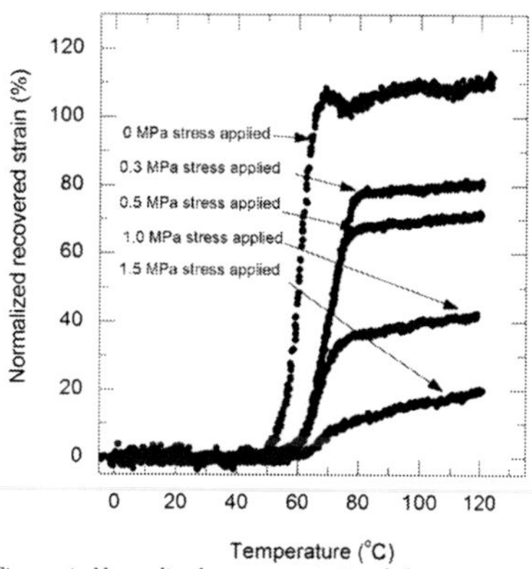

Figure 4: Normalized recovery strain of shape-memory polymers comprised of 20 wt.% crosslinker shown in Figure 2.

Following compression, all specimens were held in place and cooled to 0°C, allowing for shape storage (Figure 1). As the temperature approaches 0°C, the stress value drops to zero. This is both a reflection of the material entering into the glassy regime, and thermal contraction away from the compression platens. Once the material is stabilized at 0°C, constant force value is then applied. The magnitude of the constraining stress is lower than the maximum value applied during initial compression. Samples are then heated while this force is maintained, and allowed to recover. The partial recovery behavior of the samples shown in Figure 3 is displayed in

Figure 4. As temperature increases, the SMP remains stable until values approach the glass transition region. As the specimens undergo the phase transformation, the SMPs exhibit relatively rapid displacement recovery over a narrow temperature range (approximately 55°C to 75°C). As the maximum recoverable displacement is approached, the recovered strain plateaus with increasing temperature. The maximum recovered strain values match expected linear elastic behavior according to storage modulus measurements (Figure 2). This behavior is quantified in Figure 5, which shows the maximum normalized recovered strain as a function of applied stress during recovery for all materials tested. As expected, applied stress values are higher for specimens with larger wt.% cross-linker. The relationship between recovered strain and applied stress is observed to be near linear. These results explicitly confirm the implicit assumption held by previous SMP researchers that a linear relationship exists between the actuation force and the recoverable strain.

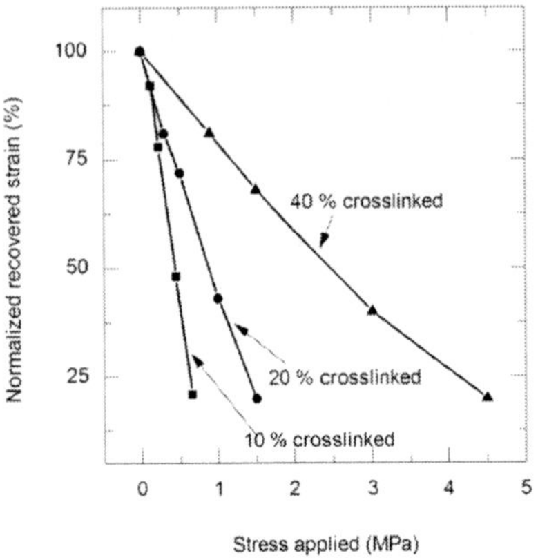

Figure 5: Normalized recovery strain as a function of applied stress for all polymers tested. Stress applied scales with increasing cross-linking density. Results indicate a linear relationship.

Using values shown in Figure 5, energy recovered by the SMPs per unit volume were calculated. This recovered energy, referred to as work performed by the SMP, is shown in Figure 6 as a function of normalized recovered strain. The performed work is not defined at either end of the curve, as either no actuation force or strain recovery is possible for either Fixed-Strain or Fixed-Stress testing, respectively. Therefore the work is fixed at zero for normalized recovered strain values of either 0% or 100%, by definition. In between these two extremes, the work performed by the SMP is shown to be symmetric as a function of recovered strain, exhibiting a maximum at approximately 50% recovery. This result has important implications for engineering design using SMPs. Figure 6 indicates that for a maximum recoverable work performance the system should be designed such that half the applied strain is recovered. Because the rubbery modulus increases with increasing cross-linking density, the magnitude of the recoverable work increases with cross-linking density. However, it is also important to consider the energy required for deformation. Materials with higher cross-linking density will also require more energy for initial deformation. Figure 7 shows work efficiency as a function of normalized strain recovered. Efficiency is defined as the recovered work divided by work required to deform the sample. Figure 7 clearly shows that the efficiency of the SMPs tested in this study are independent of cross-linking

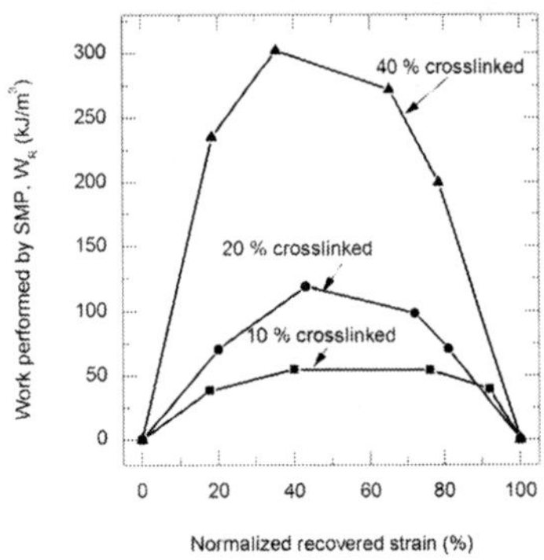

Figure 6: Work per unit volume performed by the SMP during recovery, as a function of normalized strain recovery.

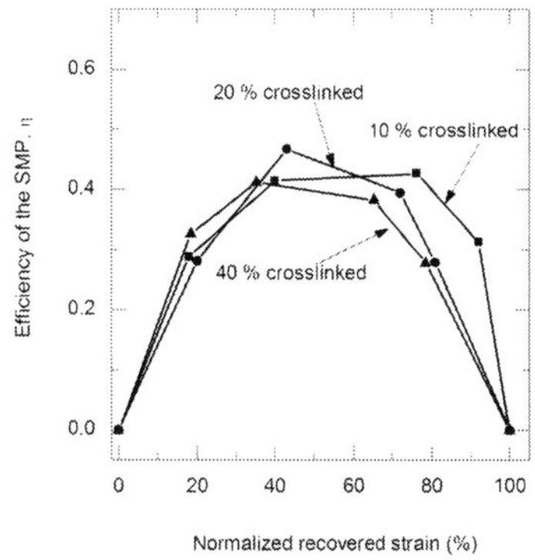

Figure 7: Efficiency of the SMP, defined as the work per unit volume observed during recovery divided by the work required for deformation.

density. Nominal values reach a maximum between 35% to 45%.

CONCLUSIONS

1. A linear relationship between applied stress and strain recovered is observed during partial recovery. This is extremely relevant from a design perspective, as most SMP applications will not experience either Free-Strain or Fixed-Stress recovery.
2. Work performed by the shape-memory effect during partial recovery scales positively with cross-linking density. This result is implicitly a function of the rubbery modulus. SMPs tested with higher rubbery modulus values would yield more recovery strain for the same constraining force values.
3. Efficiency, defined as the recovered work during heating divided by the work required to initially deform, is independent of cross-linking density for SMPs deformed above T_g. Even though SMPs with higher rubbery modulus values will give a higher work output, they require more energy for initial deformation. The maximum efficiency of the shape memory effect is achieved in a partially constrained environment.

REFERENCES

[1] B. Dietsch, T. Tong, J. Adv. Mater. 39 (2007) 3.
[2] W.M. Sokolowski, S.C. Tan, J. Spacecr. Rockets 44 (2007) 750.
[3] X. Lan, Y.J. Liu, H.B. Lv, X.H. Wang, J.S. Leng, S.Y. Du, Smart Mater. Struct. 18 (2009) 6.
[4] C.M. Yakacki, K. Gall. Shape-Memory Polymers for Biomedical Applications. Shape-Memory Polymers, vol. 226. Berlin: Springer-Verlag Berlin, 2010. p.147.
[5] J.A. Shaw, S. Kyriakides, J. Mech. Phys. Solids 43 (1995) 1243.
[6] W. Voit, T. Ware, R.R. Dasari, P. Smith, L. Danz, D. Simon, S. Barlow, S.R. Marder, K. Gall, Adv. Funct. Mater. 20 (2010) 162.
[7] C. Liu, H. Qin, P.T. Mather, J. Mater. Chem. 17 (2007) 1543.
[8] C.M. Yakacki, R. Shandas, C. Lanning, B. Rech, A. Eckstein, K. Gall, Biomaterials 28 (2007) 2255.
[9] G.M. Baer, W. Small, T.S. Wilson, W.J. Benett, D.L. Matthews, J. Hartman, D.J. Maitland, Biomed. Eng. Online 6 (2007) 8.
[10] H.M. Wache, D.J. Tartakowska, A. Hentrich, M.H. Wagner, J. Mater. Sci.-Mater. Med. 14 (2003) 109.
[11] D.M. Brey, J.L. Ifkovits, R.I. Mozia, J.S. Katz, J.A. Burdick, Acta Biomater. 4 (2008) 207.
[12] K.S. Anseth, C.N. Bowman, L. Brannonpeppas, Biomaterials 17 (1996) 1647.

Mater. Res. Soc. Symp. Proc. Vol. 1312 © 2011 Materials Research Society
DOI: 10.1557/opl.2011.671

Thermo-switchable Polymer Dielectrics

Ross S. Johnson, Fenil M. Kholwadwala, Shawn M. Dirk

Organic Materials Department
Sandia National Laboratories
P.O. Box 5800, Albuquerque, New Mexico 87185

ABSTRACT

We are interested in utilizing the thermo-switchable properties of precursor poly(*p*-phenylene vinylene) (PPV) polymers to develop capacitor dielectrics that will fail at specific temperatures due to the material irreversibly switching from an insulator to a conducting polymer. By utilizing different leaving groups on the polymer main chain, the temperature at which the polymer transforms into a conductor can be varied over a range of temperatures. Electrical characterization of thin-film capacitors prepared from several precursor PPV polymers indicates that these materials have good dielectric properties until they reach elevated temperatures, at which point conjugation of the polymer backbone effectively disables the device. Here, we present the synthesis, dielectric processing, and electrical characterization of a new thermo-switchable polymer dielectric.

INTRODUCTION

Capacitors that have a built in safety mechanism may be useful for a variety of applications including hybrid and all electric vehicles. In the event of a crash of such a vehicle, a charged capacitor may represent a significant electrical hazard to first responders as a typical capacitor bank uses 1000 mF of capacitance at a voltage of up to 2000 V. We have previously reported that precursor polymers to poly[(2,3-diphenyl-*p*-phenylene)vinylene] (DP-PPV) can be utilized as capacitor dielectrics [1-3]. At high temperatures (> 200 °C), conjugation of the polymer backbone greatly increases electron flow through the capacitor causing a large (~90%) reduction in capacitance and a 75-fold increase in dissipation factor, effectively disabling the device from storing charge [1]. To gain a better understanding of how the chemical structure affects the dielectric properties of these types of polymers, a new chlorinated variant was synthesized, characterized, and its electrical properties were evaluated.

EXPERIMENT

Our synthetic route (scheme 1) was based upon previous literature reports [4-6] and full details will be published elsewhere. Commercially available benzimidazole was reacted with two equivalents of octyl bromide to produce the benzimidazolium salt **2** in 46% yield, which was used as a condensation catalyst in the following reaction. Coupling 4-chlorobenzaldehyde **3** in the presence of the condensation catalyst **2** under basic aqueous conditions followed by oxidation with iron trichloride produced 4,4'-dichlorobenzil **4** in 69% yield. The benzil derivative was then condensed with diethyl 1,3-acetonedicarboxylate to afford the cyclopentanone derivative **5** in 64% yield. A Diels-Alder reaction with norbornadiene produced the terephthalate derivative **6** in 89% yield. Reduction of hindered diesters is typically carried out using $LiAlH_4$; however, we obtained a complex mixture of products as a result of the partial reduction of the phenyl chlorines. Better results were obtained with $LiBH_4$, which produced the desired diol in 93% yield, but required long (36 h) reaction times. The monomer **8** was obtained in 95% yield after stirring the diol **7** in thionyl chloride for 16 h. The precursor polymer **9** was afforded in 37% yield by reaction of the monomer **8** with 1 equivalent of *t*-BuOK. The fully conjugated polymer **10** could be prepared by heating the precursor polymer to high temperatures.

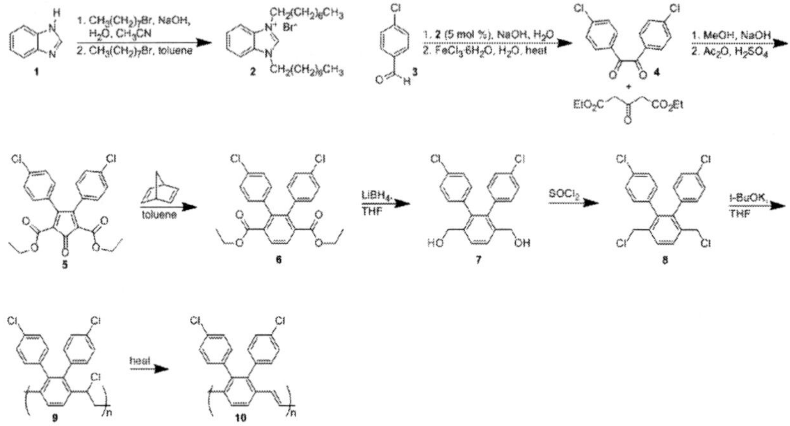

Scheme 1. Synthesis of the Cl-DP-PPV polymer.

Thermogravimetric analysis (TGA) was performed to determine the temperature at which the precursor polymer conjugated by eliminating HCl. TGA showed a 10.6% weight loss (10.2% theoretically) from 188-289 °C, indicating the thermo-conversion temperature profile (Figure 1). UV-Vis spectroscopy was performed to further investigate the thermo-conversion process. A solution of the precursor polymer (1.0 mg/mL in $CHCl_3$) was drop-cast onto a quartz slide, the sample was air dried, and the spectrum was recorded (Figure 1). Subsequent measurements were performed after the sample was heated in an oven (measurements were taken at ~25 °C

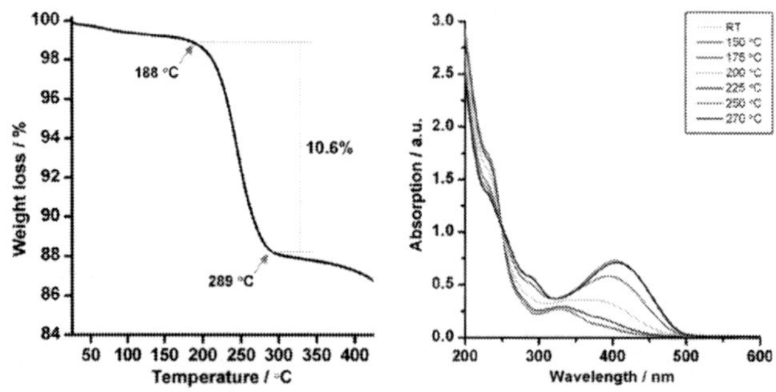

Figure 1. TGA (left) and UV-Vis (right) of the Cl-DP-Cl precursor polymer.

increments from 125-270 °C; the sample was subjected to 30 min at each temperature). The UV-Vis profile began displaying minor changes after the sample was subjected to 175 °C and 200 °C. A new, distinct band appeared after treatment at 225 °C, which continued to grow in intensity and red shift as the temperature was increased. The λ_{max} of the band was 406 nm, and is associated with $\pi \rightarrow \pi^*$ transitions along the conjugated polymer backbone. The UV-Vis experiments support the TGA results and demonstrate the polymer undergoes an irreversible conversion from a non-conjugated to a conjugated polymer at high temperatures.

Electrical characterization was performed on the precursor polymer to determine the material's performance as a dielectric. A viscous solution of the polymer (305 mg in 4.50 mL $CHCl_3$) was filtered through a cotton plug, and a 22 μm thick film was prepared using a drawdown machine. Using a sputter coater and a mask, circular gold electrodes (6.30 mm in diameter, ~50 mm thick) were deposited on one side of the film; the other side was blanket coated with gold (~50 nm thick). Capacitance and dissipation factor (Figure 2) were recorded on 17 electrodes at frequencies of 500 Hz, 1 KHz, 10 KHz, 100 KHz, and 1 MHz using an Agilent LCR meter. The material displayed a low dissipation factor averaging 0.004 over the entire frequency range, indicating that the polymer stores charge very efficiently. The average dielectric constant over the entire frequency range was 4.7, which is notably higher than our previously reported DP-PPV polymer (4.2) [1]. The increase in dielectric constant is attributed to the additional carbon-chlorine dipoles (1.25 D) [7] as well as the induced dipole orientation polarization [8] from the electronegative chlorine atoms, which are anticipated to create permanent dipoles in both phenyl rings.

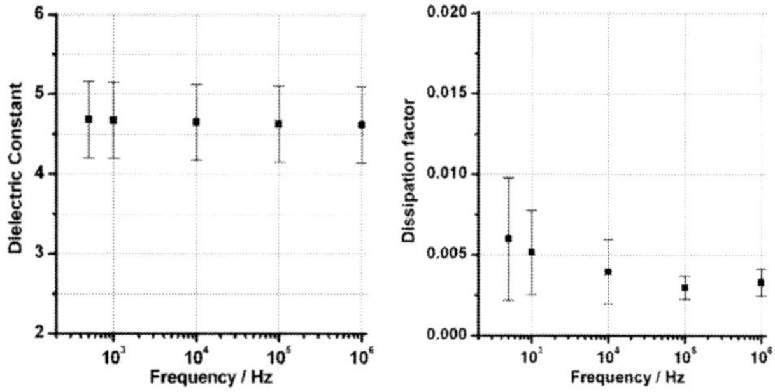

Figure 2. Dielectric constant (left) and dissipation factor (right) as a function of frequency for thin-film capacitors prepared from the Cl-DP-PPV precursor polymer.

Variable temperature electrical characterization was performed to determine whether the thin-film capacitors would fail as the polymer conjugated. Single electrodes were heated on a Signatone DC hot chuck and capacitance and dissipation factor were recorded as a function of temperature (Figure 3). To prevent the polymer from curling as the temperature was increased, the sample was taped down with Kapton® adhesive tape at the edges of the film. It was found that neither capacitance nor dissipation factor appreciably changed until the temperature reached ~200 °C (the same temperature at which the polymer was observed to conjugate by TGA and

UV-Vis). As the temperature continued to increase, the capacitance fell to 9% of the initial value and the dissipation factor spiked, increasing approximately 23 times. This data clearly demonstrates a massive reduction in the ability of the capacitor to store charge after the thermo-conversion process.

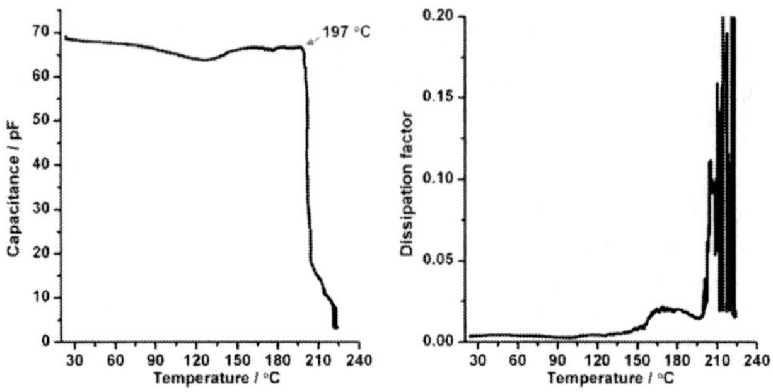

Figure 3. Capacitance (left) and dissipation factor (right) as a function of temperature for thin-film capacitors prepared from the Cl-DP-PPV precursor polymer.

CONCLUSIONS

The synthesis of a new chloro-phenyl substituted PPV polymer was presented. The incorporation of chlorine in the phenyl rings was found to increase the dielectric constant by ~0.5 compared to DP-PPV while maintaining a low dissipation factor. The precursor polymer was demonstrated to function as a good dielectric until conjugation of the polymer backbone caused a massive reduction in the ability of the capacitor to store charge. If incorporated into electronic devices, these types of advanced dielectrics would effectively shut-down the capacitors in the event of a fire or overheating, fundamentally improving the safety of the device.

ACKNOWLEDGMENTS

This work was supported by the Laboratory Directed Research and Development program at Sandia National Laboratories. Sandia is a multiprogram laboratory operated by Sandia Corporation, a Lockheed Martin Company, for the United States Department of Energy's National Nuclear Security Administration under Contract DE-AC04-94AL85000.

REFERENCES

1. R. S. Johnson, K. N. Cicotte, P. J. Mahoney, B. A. Tuttle, S. M. Dirk, *Adv. Mater.* **22**,1750-1753 (2010).
2. R. S. Johnson, K. N. Cicotte, S. M. Dirk, *Polym. Mater. Sci. Eng.* **101**, 1776-1777 (2009).
3. R. S. Johnson, Fenil M. Kholwadwala, Shawn M. Dirk, *Mater. Res. Soc. Symp. Proc.* Vol. 1197E, 1197-D07-43 (2010).
4. H. V. Huynh, L. R. Wong, P. S. Ng, *Organometallics,* **27**, 2231-2237.

5. X. Jing, X. Pan, Z. Li, Y. Shi, C. Yan *Synth. Commun.* **39**, 492-496, 2009.
6. W. C. Wan, H. Antoniadis, V. E. Choong, H. Razafitrimo, Y. Gao, W. A. Feld, B. R. Hsieh, *Macromolecules* **30**, 6567-6574 (1997).
7. B. A. Hess, R. J. Buenker, *Chem. Phys.* **101**, 211-218 (1986).
8. G. Hougham, G. Tesoro, A. Viehbeck, J. D. Chapple-Soko, *Macromolecules* **27**, 5964-5971 (1994).

Mater. Res. Soc. Symp. Proc. Vol. 1312 © 2011 Materials Research Society
DOI: 10.1557/opl.2011.914

Poly(vinylidene fluoride-trifluoroethylene) (72/28) interconnected porous membranes obtained by crystallization from solution

Armando Ferreira[1], Jaime Silva[1,2], Vitor Sencadas[1], José Luís Gómez-Ribelles[3,4,5] and Senentxu Lanceros-Méndez[1,*]

[1]Centro/Departamento de Física, Universidade do Minho, 4710-057 Braga, Portugal.

[2] IPC-Institute for Polymers and Composites, University of Minho, 4800-058, Guimarães, Portugal.

[3]Centro de Biomateriales y Ingineria Tisular, Universidad Politécnica de Valencia, 46022, Valencia, Spain.

[4]Regenerative Medicine Unit, Centro de Investigación Príncipe Felipe, Autopista del Saler 16, 46013 Valencia, Spain.

[5]CIBER en Bioingeniería, Biomateriales y Nanomedicina, Valencia, Spain.

ABSTRACT

Electroactive macroporous poly[(vinylidene fluoride)-co-trifluoroethylene] membranes have been processed by solvent evaporation at room temperature with different polymer/solvent concentrations. The pore architecture consists on interconnected spherical pores and this morphology is independent of the membrane thickness. The porosity of the produced membranes increases from 72% for the higher polymer concentration in the polymer/solvent solution (15/85), up to 80% for the lowest polymer concentration in the polymer/solvent solution.

Fourier transform infrared spectroscopy and differential scanning calorimetry measurements reveal that the polymer crystallizes in the ferroelectric phase and the polymer/solvent ratio does not influences the Curie transition and the melting temperature of the polymer.

INTRODUCTION

Poly(vinylidene fluoride) (PVDF) and its copolymers with trifluoroethylene (P(VDF-TrFE)) have attracted technological and scientific interest due to their outstanding electroactive properties [1-2] among polymer materials and as they can be implemented from industrial [3] to biomedical applications [4]. Concerning applications as a biomaterial, PVDF and VDF (vinylidene fluoride) copolymers can be unique materials due to their piezoelectric properties that can be used in cell culture "ex-vivo" under mechanical and electrical excitation in tissue engineering techniques [5].

Laroche *et al.* reported that PVDF can be applied in the form of non absorbable syntetic sutures for vascular surgery [6]. It was observed some surface oxidation and water imbibitions of the polymer with no significantly modification of the tensile strength of the PVDF suture, which retained 92.5% of its original value.

PVDF porous structures can also be taylored for applications as smart scaffolds with controled fiber distribution and electroactive β-phase content [7-8]. A number of experimental procedures are available to produce porous PVDF with controlled pore morphology. One of the most important requirements for cell culture is an interconnected pore structure that allows cell seeding and diffusion of nutrients and waste products of the cell metabolism.

One of the successful procedures to prepare porous PVDF is thermally induced phase separation, TIPS [9]. In this procedure, PVDF is dissolved at high temperature in a mixture of a high boiling point good solvent and a certain amount of a bad solvent. The decrease of temperature in controlled conditions produces phase separation when entering into a two phase coexisting zone of the phase diagram [5]. Depending of the solvents and the initial composition of the solution, liquid-liquid spinodal phase separation can take place or liquid-solid phase separation can occur when PVDF crystallizes from the solution. Nucleation and growth of the coexisting phases will determine later the pore morphology when, after cooling to low temperatures, the solvents are extracted in order to leave the porous PVDF membrane [9].

Following a previous work [5], this paper reports on the pore size and architecture, polymer phase and thermal behavior of electroactive macroporous poly(vinylidene fluoride-trifuoroethylene), P(VDF-TrFE) (72/28), membranes with a well organized porous structure and thickness from 10 to 250 μm produced by solvent evaporation at room temperature starting with a diluted solution of the co-polymer in dimethylformamide (DMF).

EXPERIMENTAL

Samples with thickness between 10 and 250 μm were prepared by spreading a solution of poly(vinylidene fluoride-trifluorethylene) P(VDF-TrFE) (72/28) (*Solvay*, Brussels, Belgium) in N,N dimethylformamide (DMF, *Merck*) on a glass slide. The initial concentration of the solution and the viscosity (measurement performed in a *Fungilab* ViscoStar apparatus) are shown in Table 1. Solvent evaporation and crystallization of P(VDF-TrFE) took place at room temperature. The P(VDF-TrFE) (72/28) co-polymer obtained by this procedure is in the ferroelectric phase, being therefore also piezoelectric [1-2].

Scanning electron microscopy (SEM) measurements were performed in a *FEI Nova NanoSEM 200* with an accelerating voltage of 20 kV, in previously gold coated samples (*Polaron, model SC502 sputter coater*). The samples analysed in cross-section were previously immersed in liquid nitrogen for 180 s in order to ensure that the sample temperature decreases below glass transition (*Tg*), and then were mechanically broken.

Infrared measurements were performed at room temperature in a *Perkin-Elmer Spectrum 100* apparatus in ATR mode from 4000 to 650 cm^{-1}. FTIR spectra were collected with 64 scans and a resolution of 4 cm^{-1}. The thermal behavior of the samples was analyzed by differential scanning calorimetry measurements (DSC) with a *Perkin Elmer Diamond* DSC apparatus. The samples were cut into small pieces, placed into 40 μl aluminum pans and heated between 30 and 200 °C at a heating rate of 10 $°C.min^{-1}$. All experiments were performed under a nitrogen purge.

Table 1. Viscosity and porosity of the films of P(VDF-TrFE) for the different polymer concentrations.

Polymer/Solvent Fraction	**Polymer Concentration**	**Viscosity**	**Porosity**
	g.cm^{-3}	**cP**	**%**
1g P(VDF-TrFE) / 5 mL DMF	0,20 (15/85)	629,3	72
1g P(VDF-TrFE) / 6.7 mL DMF	0,15 (10/90)	267,6	76
1g P(VDF-TrFE) / 10 mL DMF	0,10 (5/95)	81,2	80

A suitable method to estimate the porosity of the samples is the measurement with a pycnometer. The porosity of the samples was measured by an improved weight-method. The weight of the pycnometer filled with ethanol, was weighted and labeled as W1; the sample, whose weight was Ws, was immersed in ethanol. After the sample was saturated by ethanol, additional ethanol was added to complete the volume of the pycnometer. Then, the pycnometer was weighted and labeled as W2; the sample filled with ethanol was taken out of the pycnometer. The residual weight of the ethanol and the pycnometer was labeled W3. The porosity of the scaffold was calculated according to [10]:

$$\varepsilon = \frac{W_2 - W_3 - W_s}{W_1 - W_3} \qquad \text{Eq. 1}$$

The mean porosity of each membrane was obtained as the average of the values determined in three samples.

RESULTS AND DISCUSSION

Membranes of P(VDF-TrFE) with thickness between 10 to 250 μm were produced by solvent evaporation at room temperature from a homogeneous solution in DMF. The microstructure of the polymer membrane is shown in figure 1. It can be observed that all samples present a spherical porous structure with diameter around 15 μm. In figure 1b smaller pores are observed in the walls of the bigger porous structure with 4 μm diameter . This fact demonstrates the interconnectivity of the porosity that results from the spaces between the polymer microspheres that form the pores walls, while larger secondary pores may appear due to some defects in the structure. In a previous work, it was found that the thickness of the membrane does not have influence in the porous structure, as this microstructural effect is due to solvent evaporation [5]. In this way, samples prepared by spin-coating or solvent casting present the same porous morphology.

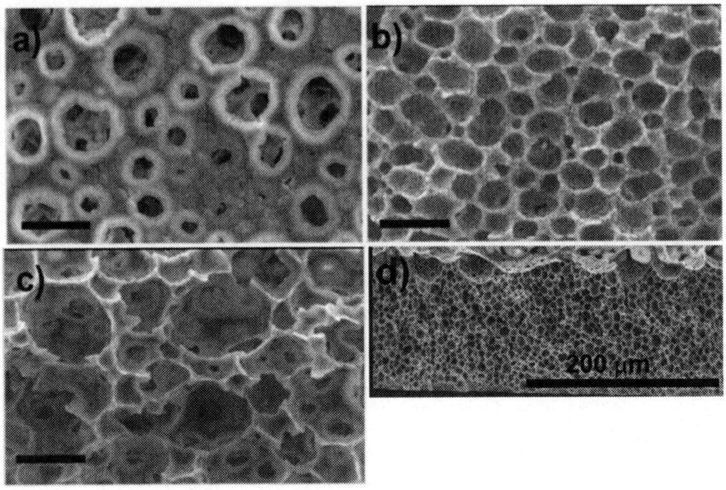

Figure 1 – SEM micrographs of the P(VDF-TrFE) samples crystallized at room temperature: a) topography of the sample with 72% porosity; b) Cross-section for the sample with 76% porosity; c) Cross-section for the sample with 72% and d) Cross-section for the sample with 80% porosity. The scale bar corresponds to 20 microns.

The nature of the crystalline phase present in the polymer can be identified by differential scanning calorimetry (DSC) and Fourier Transformed infrared spectroscopy (FTIR). FTIR spectra for the samples crystallized at room temperature show that the crystallization of the polymer occurs in the ferroelectric phase (figure 2a) due to the presence of the absorption band at 840 cm^{-1}, characteristic of the ferroelectric phase of the polymer [11].

In the DSC thermograms (figure 2b) two peaks are observed: the one that occurs at the lower temperature corresponds to the ferroelectric-paraelectric transition (FE-PE, Curie transition) with maxima at ca. 117 °C. The higher DSC endotherm corresponds to the melting of the paraelectric phase and it is located at ca.145 °C. It can be observed that the crystallization process does not affect the FE-PE transition temperature and the melting temperature of the polymer, what leads to conclude that the solvent evaporation does not play an important role in the final polymer crystalline properties, but in the crystallization dynamics, which affects the porosity and consequently the microstructure of the P(VDF-TrFE) membranes depending on the specific place of the P(VDF-TrFE)/DMF solution phase diagram in which the isothermal crystallization begins [5].

In this way, the final polymer porous microstructure taking place in the P(VDF-TrFE)/DMF system undergoing solvent evaporation is explained by a liquid/liquid spinodal decomposition followed by polymer crystallization [5].

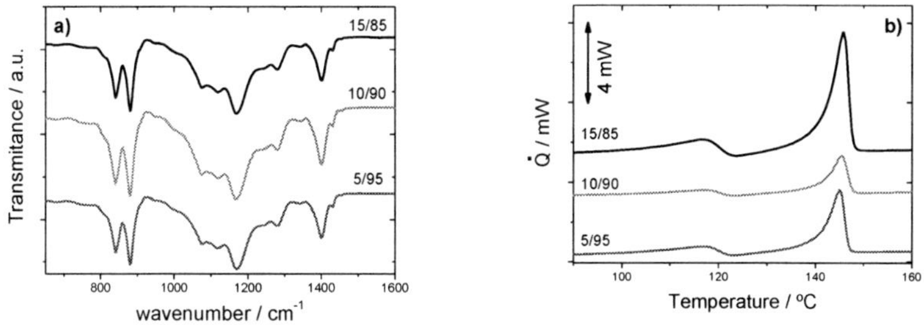

Figure 2 – *a)* Infrared measurements for the PVDF-TrFE samples obtained by room temperature crystallization from P(VDF-TrFE)/DMF solutions with different amounts of DMF solvent (15/85, 10/90 and 5/95, corresponding to 72, 76 and 80% of porosity, respectively); *b)* DSC results for the same samples.

CONCLUSIONS

Room temperature isothermal solvent evaporation from an homogeneous P(VDF-TrFE)/DMF solution allows to obtain a well organized tridimensional porous structure in membranes from 10 up to 250 μm thick. The pore architecture consists on spherical pores of around 15 μm with small porous of 4 μm in the walls of the first ones.

The polymer/solvent ratio determines the final porosity of the membrane, the mechanism being a liquid/liquid spinodal decomposition followed by polymer crystallization. The porosity ranges from 72% up to 80%, being the highest degree of porosity obtained for the P(VDF-TrFE)/DMF solution with the highest initial amount of DMF. The degree of crystallinity and ferroelectric phase of the obtained samples are not affected by the initial P(VDF-TrFE)/DMF ratio.

ACKNOWLEDGMENTS

This work is funded by FEDER funds through the "Programa Operacional Factores de Competitividade – COMPETE" and by national funds by FCT- Fundação para a Ciência e a Tecnologia, project references NANO/NMed-SD/0156/2007, PTDC/CTM/73030/2006 and PTDC/CTM/69316/2006. V.S. thanks the support of the FCT under grant SFRH/BPD/63148/2009 (V.S.) Work supported also by the Spanish Ministry of Education and Ministry of Science and Innovation (Project No. MAT2007-66759-C03-01).

REFERENCES

1. Lovinger AJ. Developments in Crystalline Polymers. London: Elsevier Applied Science, 1982.

2. Nalwa HS. Ferroelectric Polymers: Chemistry, Physics, and Applications. New York: Marcel Decker, Inc., 1995.
3. Qiming Zhang, Huang C, Xia F, and Su J. Electroactive polymer (EAP) actuators as artificial muscles: reality, potencial and Challenges New York, 2004.
4. Klinge U, Klosterhalfen B, Öttinger AP, Junge K, and Schumpelick V. Biomaterials 2002;23(16):3487-3493.
5. Ferreira A, Silva J, Sencadas V, Ribelles JLG, and Lanceros-Méndez S. Macromolecular Materials and Engineering 2010;295(6):523-528.
6. Laroche G, Marois Y, Schwarz E, Guidoin R, King MW, Pâris E, and Douville Y. Artificial Organs 1995;19(11):1190-1199.
7. Sencadas V, Hernández JCRH, Ribeiro C, Goméz-Ribelles JL, and Lanceros-Mendez S. 2009, 1140. Materials Research Society Symposium Procceedings, vol. 1140. Boston: MRS, 2009.
8. Ribeiro C, Sencadas V, Ribelles JLG, and Lanceros-Méndez S. Soft Materials 2010;8(3):274 - 287.
9. Ji G-L, Zhu B-K, Cui Z-Y, Zhang C-F, and Xu Y-Y. Polymer 2007;48(21):6415-6425.
10. Ribeiro C, Sencadas V, Caparros C, Gómez Ribelles JL, and Lanceros-Méndez S. Journal of Nanoscience and Nanotechnology 2010;To be published.
11. Simoes RD, Rodriguez-Perez MA, De Saja JA, and Constantino CJL. Polymer Engineering & Science 2009;49(11):2150-2157.

Mater. Res. Soc. Symp. Proc. Vol. 1312 © 2011 Materials Research Society
DOI: 10.1557/opl.2011.672

Modified Polyaniline Nanofibres for Ascorbic Acid Detection

Larisa Florea[1], Emer Lahiff[1] and Dermot Diamond[1]
[1] CLARITY: The Centre for Sensor Web Technologies, Dublin City University, Dublin 9, Ireland.

ABSTRACT

Polyaniline nanofibres (PAni) can be surface modified to improve electroactivity over a broader pH range. The technique we describe here can be used to attach carboxylic acid terminated substituents. Modified nanofibres maintain their high surface area, and ability to switch between different redox states. These properties make the material suitable for sensing applications. Unlike unmodified PAni, the functionalised material is self-doping and hence more stable in higher pH solutions. Here we demonstrate how modified PAni fibres can be used for the detection of ascorbic acid.

INTRODUCTION

Ascorbic acid (AA) is an effective reducing agent, and a powerful antioxidant in food. Various methods have been reported for the determination of ascorbic acid, such as spectrophotometry [1], fluorimetry [2], chromatography [3] and electrochemical detection [4]. Modified electrodes have been applied for the determination of AA electrochemically [5,6], and using conducting polymer electrodes specifically has advantages such as three-dimensional mediator distribution, broad potential window and good electrode stability [7-9].

Polyaniline (PAni) is an example of a stable conducting polymer whose properties (optical, electrical and electrochemical) change in response to the immediate molecular environment of the material. PAni can be cheaply and conveniently synthesized, and thus has huge potential for sensor applications. The nanofibre form of PAni can be synthesised directly by chemical means, such as interfacial polymerisation. Synthesis by this method is scalable, whereby increasing the amount of reactant results in a greater yield of the nanofibre product. By using PAni nanofibres (versus bulk) the surface area exposed to the target molecules can be dramatically increased, resulting in a sensor material with enhanced sensitivity and improved response times [10-14]. Our focus in this paper is on the covalent modification of these nanostructures post-polymerisation [15]. We functionalise the nanofibres with carboxylic acid terminated side-chains, resulting in self-doped PAni. Using the technique described, control over the extent of functionalisation can be achieved [16] and, the modified fibres can be used for the successful detection of ascorbic acid.

EXPERIMENT

Aniline (BDH), HCl (Fisher Scientific), ammonium peroxydisulfate (Aldrich), ascorbic acid (Aldrich), mercaptopropanoic acid (Aldrich), mercaptohexadecanoic acid (Aldrich) and

Phosphate Buffered Saline (Aldrich) were used. The aniline monomer was purified by vacuum distillation before use. Other chemicals were used as received.

Polyaniline nanofibres were synthesised by interfacial polymerisation between an aqueous and an organic layer, as reported previously [10]. The product was purified by centrifugation (4000 rpm/5 min/3 cycles) and suspended as a colloid in deionised water. Fibres were then modified with an acid-terminated thiol by refluxing at 100 °C for 2 hours in an aqueous buffer at pH 4 [17]. The product was purified by centrifugation (4000 rpm/5 min/3 cycles) and re-suspended as a colloid in deionised water.

Electrochemistry of the fibres was investigated using a CH Instruments Electrochemical Analyser. Cyclic voltammetry results were obtained using phosphate buffered saline solution (PBS, pH = 7.4) as the electrolyte at a scan rate of 10 mV/s. The test solutions were deaerated by passing a stream of pure nitrogen for at least 20 min prior to experiments, and all experiments were carried out at room temperature around 20 °C. A glassy carbon working electrode, a Pt wire counter-electrode and a Ag/AgCl reference electrode were used. Samples were drop cast onto a polished glassy carbon electrode, to form insoluble coatings after drying in air for several hours.

Modified nanofibres were imaged using field emission scanning electron microscopy (FESEM) at an accelerating voltage of 20 kV on a S-4300 Hitachi system. UV-vis spectroscopy was carried out on a Perkin Elmer UV-vis NIR Lambda 900 Spectrometer at 1 nm resolution.

DISCUSSION

Polyaniline nanofibres were modified by a simple reflux process to attach carboxylic acid terminated thiol side-chains (PAni-COOH). The covalent attachment was verified using FTIR and NMR as previously published by the authors [16]. Post-modification nanofibres maintain their nanostructure, and hence high surface area (Figure 1).

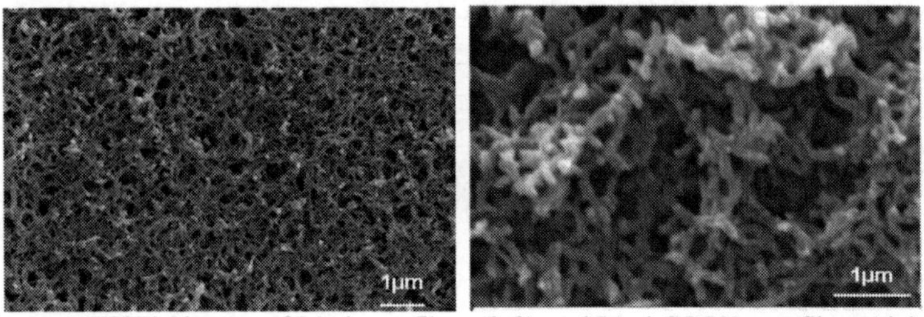

Figure 1: FESEM images of PAni nanofibres (left), and PAni-COOH nanofibres (right). Images confirm that the PAni maintains its nanostructured form post-functionalisation.

The modified polyaniline nanofibres maintain their ability to switch between different redox states, making them suitable for sensing applications. Cyclic voltammetry of PAni-COOH nanofibres is similar to that of PAni, under acidic conditions. Redox cycling for both PAni and PAni-COOH is reversible, and current flow for both oxidation and reduction is approximately equal (Figure 2).

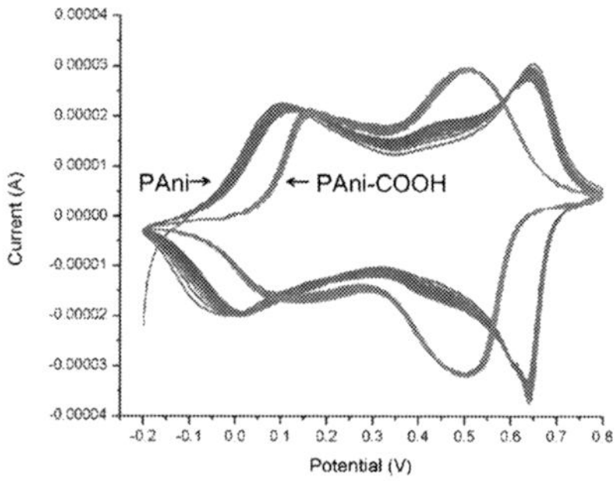

Figure 2: Cyclicvoltammetry of PAni nanofibres was compared with that for the functionalised material (pH=1). Modification causes a slight change in the electrochemical profile, however PAni-COOH remains clearly redox active.

Polyaniline nanofibres are of interest for sensing applications as they adapt their redox state, and chemical structure, depending on the local environment. The UV-Vis spectra of doped polyaniline (known as the emeraldine salt form), shows three characteristic absorption bands at 350nm, 420nm and 770 nm. These bands correspond to π-π* band-gap absorption, low wavelength polaron-π* band and high wavelength π-polaron band, respectively. The undoped form of polyaniline (known as the emeraldine base form), shows a low wavelength π-π* band at 320 nm and a strong absorption band at 575 nm. These are attributed to local charge transfer between a quinoid ring and the adjacent imine-phenyl-amine unit [18].

PAni at pH > 4 has the emeraldine base form, which does not have the polaron transition (due to low concentration of acid, H^+). In the UV-Vis absorption spectra of carboxylated polyaniline (PAni-COOH), the generation of the polaron was observed, even with this system at pH 7.4 (Figure 3). Therefore, UV-vis spectra demonstrate that the PAni-COOH is self-doped in media of pH > 4. The intensity of the polaron transition band is greater when the mole ratio between the monomer and carboxylic acid is increased. This self-doping polyaniline shows better dispersability and stability (in PBS), over time, when compared with unmodified polyaniline.

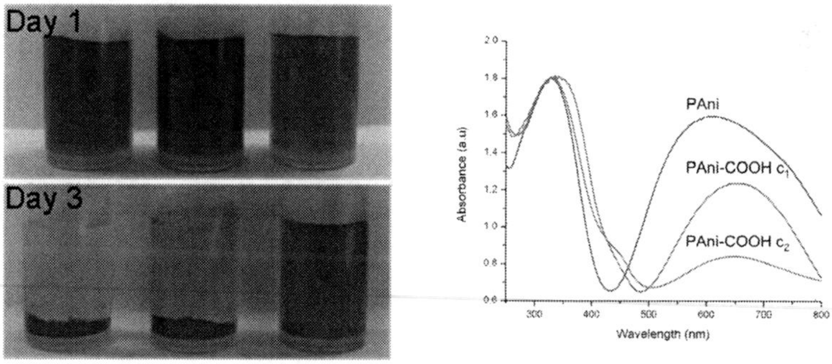

Figure 3: Pictures of PAni dispersion (at pH 7.4), on day 1 and day 3. Images left-to-right show unmodified PAni, PAni-COOH c_1 (concentration 1), and PAni-COOH c_2 (concentration 2), where c_1<c_2. UV-Vis Spectra of PAni and PAni-COOH (in PBS, pH 7.4), are shown on the far right.

Ascorbic acid detection using the PAni-COOH electrode

As PAni-COOH is more stable than PAni in neutral pH environments, the material was examined as a simple biosensor. Here we demonstrate the material for use in electrochemical ascorbic acid (AA) detection. AA has an oxidation peak at ~ 0.5V. Using the PAni-COOH modified electrode however, we can detect the AA at a much lower potential of 0.05 V. This peak corresponds to the oxidation peak of PAni-COOH, and it was observed to increase when AA was added (Figure 4). The magnitude of this peak continues to increase for greater concentrations of AA.

The reaction process can be described as: AA in solution diffused toward the polymer modified electrode surface and reacted with the oxidized state polyaniline which led to the increase of the reduced state polyaniline. This increasing of reduced state polyaniline resulted in the increase of oxidation currents [9]. CV curves by continuous cyclic sweep in AA solution did not bring about any change, which indicated that this polymer-modified electrode was not polluted by the adsorption of AA.

For higher concentrations of AA (> 3mmol/l), a second redox peak appears at ~0.3V. This is not a PAni redox peak and we suggest that it reflects a shifted AA peak. As the amount of AA increases, the position of this second redox peak shifts further towards the 0.5V of AA. For example, at 10mmol/l AA the peak is at 0.4V (data not shown).

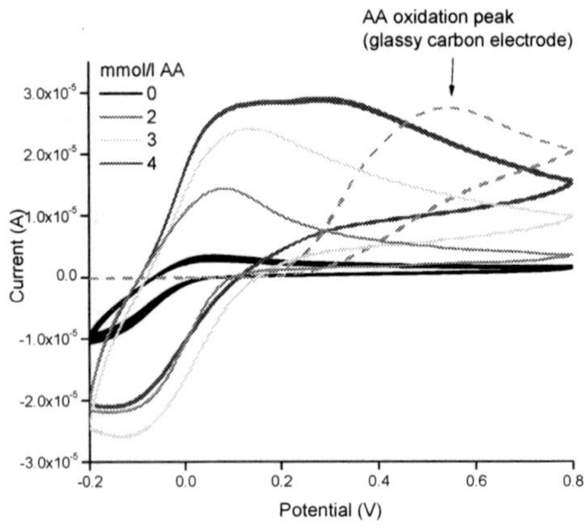

Figure 4: The CVs of a PAni-COOH modified electrode in the presence of 0, 2, 3, and 4 mmol/l ascorbic acid. The CV obtained for a bare glassy carbon electrode in the presence of ascorbic acid is also shown for comparison.

CONCLUSIONS

One of the major limitations of PAni is that it loses the electroactivity at pH >4. By functionalising PAni with carboxylic acid terminated side-chains we create a self-doping nanomaterial that presents good electroactivity over a wide pH range. Here we demonstrate that this new material exhibits good electrocatalytic activity for the oxidation of ascorbic acid. The carboxyl group-functionalised polyaniline offers new opportunities in the area of self-assembly, and chemical and biological sensors. As the carboxyl dopant group is a structural part of the main polymer chain it will not leach away, and can be further functionalised by other guest molecules via esterification. This polymer could be further applied for the construction of an enzyme biosensor, which would greatly widen the use of conductive polyaniline in biological electrochemistry.

ACKNOWLEDGMENTS

The authors acknowledge funding from Science Foundation Ireland under SFI 07/CE/I1147 - "CLARITY: Centre for Sensor Web Technologies" and IRCSET – The Embark Initiative.

REFERENCES

1. O. W. Lau, S. F. Luk and K. S. Wong, Analyst **112** (7), 1023-1025 (1987).
2. N. Moll and J. P. Joly, Journal of Chromatography **405**, 347-356 (1987).
3. H. P. Huang, R. X. Cai, Y. M. Du and Y. N. Zeng, Anal. Chim. Acta **309** (1-3), 271-275 (1995).
4. K. Matsumoto, J. J. B. Baeza and H. A. Mottola, Anal. Chem. **65** (13), 1658-1661 (1993).
5. J. Facci and R. W. Murray, Anal. Chem. **54** (4), 772-777 (1982).
6. A. P. Doherty, M. A. Stanley and J. G. Vos, Analyst **120** (9), 2371-2376 (1995).
7. A. M. Yu, D. M. Sun, H. Y. Gu and H. Y. Chen, Analytical Letters **29** (15), 2633-2643 (1996).
8. E. W. Kristensen, W. G. Kuhr and R. M. Wightman, Anal. Chem. **59** (14), 1752-1757 (1987).
9. J. J. Xu, D. M. Zhou and H. Y. Chen, Fresenius Journal of Analytical Chemistry **362** (2), 234-238 (1998).
10. J. X. Huang, S. Virji, B. H. Weiller and R. B. Kaner, Journal of the American Chemical Society **125** (2), 314-315 (2003).
11. S. Virji, R. B. Kaner and B. H. Weiller, Journal of Physical Chemistry B **110** (44), 22266-22270 (2006).
12. S. Virji, R. B. Kaner and B. H. Weiller, Chemistry of Materials **17** (5), 1256-1260 (2005).
13. S. Virji, J. X. Huang, R. B. Kaner and B. H. Weiller, Nano Letters **4** (3), 491-496 (2004).
14. S. Virji, J. D. Fowler, C. O. Baker, J. X. Huang, R. B. Kaner and B. H. Weiller, Small **1** (6), 624-627 (2005).
15. E. Lahiff, Bell, S, Diamond, D., Mater. Res. Soc. Symp. Proc. **1054**, 1054-FF1005-1005. (2008).
16. E. Lahiff, T. Woods, W. Blau, G. G. Wallace and D. Diamond, Synthetic Metals **159** (7-8), 741-748 (2009).
17. E. Lahiff, Scarmagnani, S, Schazmann, B, Cafolla, A, Diamond, D., International Journal of Nanomanufacturing. **5** (1/2), 88-99 (2010).
18. S. Atkinson, H. S. O. Chan, A. J. Neuendorf, S. C. Ng, T. T. Ong and D. J. Young, Chemistry Letters (3), 276-277 (2000).

Mater. Res. Soc. Symp. Proc. Vol. 1312 © 2011 Materials Research Society
DOI: 10.1557/opl.2011.915

Temperature-adaptive Insulation Based on Multicomponent Fibers of Various Cross-sections

Barry S. DeCristofano, Stephen A. Fossey, Elizabeth A. Welsh, Jeffrey Perry and Deana Archambault

U. S. Army Natick Soldier R, D & E Center, 15 Kansas Street, Natick, MA 01760, U.S.A.

ABSTRACT

We have developed a thermal insulation based on bi-component fibers that adapts to its thermal environment, providing greater insulation at low temperatures than at warmer temperatures. The analysis of bi-metallic strips done by Timoshenko [1] for strips of rectangular cross-section concluded that "... curvature is proportional to the difference in elongation of the two metals and inversely proportional to the thickness of the strip." We have extended Timoshenko's formulation and applied it to bi-component fibers of circular and triangular cross-sections. In each case, the curvature resulting from the balance of the axial forces and bending moments has been brought into a standard form inversely proportional to (A+Bn+C/n) where n is the ratio of the moduli and A, B, and C are functions of the geometry of the two components. An important consequence of this result is that for any "n" there is a maximum curvature where (A+Bn+C/n) is a minimum. We have used the process of melt-spinning to produce fibers with circular and triangular cross-sections, varying the proportion of the two components. The polymers used have widely different coefficients of thermal expansion. These fibers spontaneously form mats at room temperatures. The experimentally measured thickness changes are in good agreement with the analytical results for fiber bending. The most effective samples to date change thickness by more than 1.5% per degree C (30% over a temperature range from approximately 20 °C to 0 °C).

INTRODUCTION

Materials that undergo a dimensional change as a result of some change in their environment have been combined in devices in various ways. The ubiquitous bi-metallic strip in a thermostat leverages differing coefficients of thermal expansion (CTE) to control the flow of current to provide heating or cooling as the temperature changes. Voltage applied to tri-layered polymers cause a redox process to differentially swell the layers as ions are exchanged between them and the bending that results can be used for micro manipulation tasks [2]. In some instances, these effects can be integrated into structures to counter the thermal expansion of other components and maintain a constant dimensional aspect [3].

Previous work with circular bi-component polymer fibers demonstrated that a change in the thickness of a mat of these fibers with temperature could be achieved [4]. One application of such a mat would be for providing an amount of insulation in clothing that is tailored to the thermal environment in which it is worn [5].

As an extension of this work, we investigated bi-component polymer fibers of circular and triangular cross-sections in an attempt to optimize the temperature-induced bending and therefore, the change in insulation thickness. After spinning and forming the mats, the change in thickness of the mats as a function of temperature was measured.

THEORY

The analysis by Timoshenko for a bi-layer strip of rectangular cross-section resulted in the following (Figure 1):

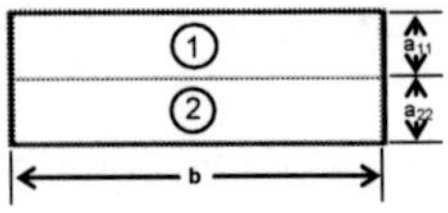

$$I_1 = \frac{{a_{11}}^3 b}{12} \qquad I_2 = \frac{{a_{22}}^3 b}{12} \tag{1}$$

Figure 1. Cross-section of a rectangular bi-component bar and area moment of inertia equations.

Balancing the forces and bending moments on a cross-section of the strip resulting from a difference in the CTE values, α_1 and α_2, as the temperature, t, changes from some baseline temperature, t_0, leads to an equation for the curvature of the strip, $1/\rho$:

$$\frac{1}{\rho_{rect}} = \frac{6(\alpha_2 - \alpha_1)(t - t_0)(1+m)^2}{h\left[3(1+m)^2 + (1+nm)\left(m^2 + \frac{1}{nm}\right)\right]} \qquad m = \frac{a_{11}}{a_{22}} \quad and \quad n = \frac{E_1}{E_2} \tag{2}$$

E_1 and E_2 are the elastic moduli of the materials and h is the total height. For the fibers under consideration here, the corresponding analyses are not as straightforward because of the geometries involved. In the case of the triangle, the layer cross-sections are a triangle and a trapezoid. For the circle, except for the case where the layers are of equal height (then both halves are semi-circles), there are two unequal sections of the circle as shown in Figure 2.

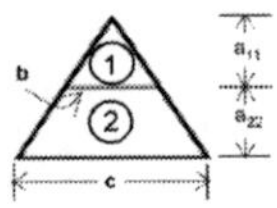

$$I_1 = \frac{{a_{11}}^3 b}{36} \qquad I_2 = \frac{{a_{22}}^3 (b^2 + 4bc + c^2)}{36(b+c)} \tag{3}$$

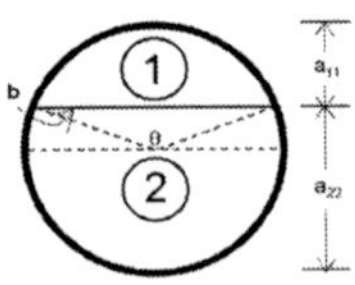

$$I_1 = r^4\left[\frac{1}{8}\left(\theta - \sin\theta + 2\sin\theta\sin^2\frac{\theta}{2}\right) - \frac{8}{9}\frac{\sin^6\frac{\theta}{2}}{(\theta - \sin\theta)}\right]$$

$$I_2 = r^4\left\{\frac{1}{8}\left(\theta - \sin\theta + 2\sin\theta\sin^2\frac{\theta}{2}\right) - \frac{4}{9}\frac{\sin^6\frac{\theta}{2}}{\left[\pi - \frac{1}{2}(\theta - \sin\theta)\right]}\right\} \tag{4}$$

Figure 2. Cross-sections and moment equations for triangular and circular cross-sections.

The curvature equations for these configurations are found to be:

$$\frac{1}{\rho_{tri}} = \frac{18(\alpha_2 - \alpha_1)(t - t_0)(1+m)^2}{h\left\{9(1+m)^2 + 2(1+m)\left[(m^2+1)(1-a_1) + \frac{2(a_1 - a_1^2)}{(a_1+1)^2} + nm^3\frac{(a_1 - {a_1}^2)}{(a_1+1)} + \frac{1}{n}\frac{({a_1}^2 + 4a_1 + 1)(1 - a_1)}{m({a_1}^2 + a_1)}\right]\right\}} \tag{5}$$

$$\frac{1}{\rho_{circ}} = \frac{2EF(\alpha_2 - \alpha_1)(t - t_0)(1 + m)^2}{h\left\{EF + [n(BE - C) + k(BF - C)]\left(\frac{1}{nk} + 1\right)\right\}} \quad E = \frac{(\theta - \sin\theta)}{2} \quad F = \pi - \frac{(\theta - \sin\theta)}{2} \quad k = \frac{A_1}{A_2} \tag{6}$$

Though it is possibly hard to see from equations 2, 5 and 6, each of these results can be written in the form $1/\rho \propto 1/(A+Bn+C/n)$, where n is the ratio of the moduli and A, B, and C are functions of the geometry of the two components and the fraction of the full cross-section that each occupies. This is more easily seen when m=1 (Table 1).

cross-section	curvature equation when m=1
rectangle	$\frac{1}{\rho_{rect}} = \frac{24(\alpha_2 - \alpha_1)(t - t_0)}{h\left[14 + n + \frac{1}{n}\right]}$
triangle	$\frac{1}{\rho_{tri}} = \frac{72(\alpha_2 - \alpha_1)(t - t_0)}{h\left[41 + \frac{2}{3}n + \frac{26}{3}\frac{1}{n}\right]}$
circle	$\frac{1}{\rho_{circ}} = \frac{\frac{\pi}{2}(\alpha_2 - \alpha_1)(t - t_0)}{h\left[\left(\frac{\pi^2}{8} - \frac{8}{9}\right) + \left(\frac{\pi^2}{16} - \frac{4}{9}\right)n + \left(\frac{\pi^2}{16} - \frac{4}{9}\right)\frac{1}{n}\right]}$

Table 1. Curvature equations when the cross-section is equally divided between the components.

An important consequence of this result is that for any "n" there is a maximum curvature where (A+Bn+C/n) is a minimum. Plotting these functions allows visualization of the design envelopes for the fibers as shown in Figures 3 (a), (b) and (c).

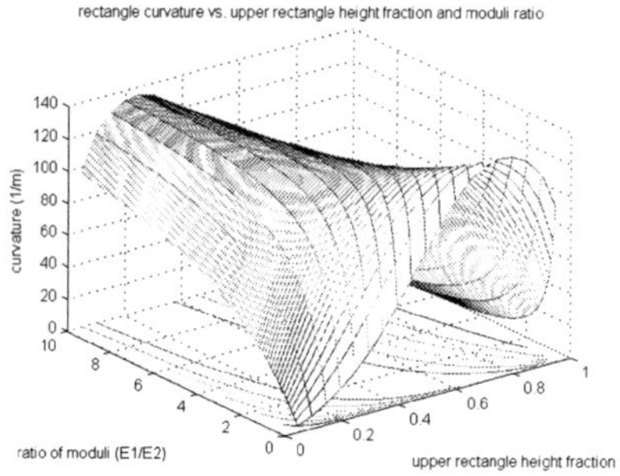

(a)

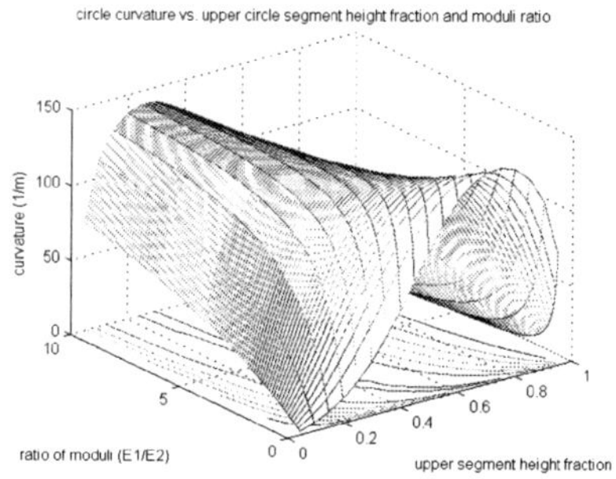

(b)

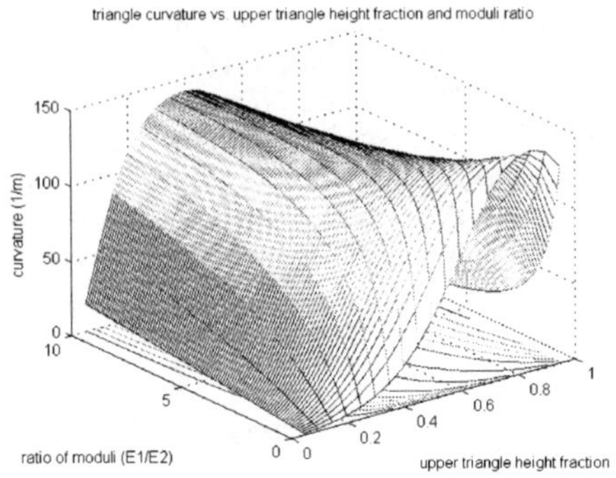

(c)

Figure 3. Surfaces and contours of curvature vs. moduli ratio and fraction of component 1 for (a) rectangular, (b) circular and (c) triangular cross-sections.

APPROACH

Component fractions of the cross-sections

One thing noticeable here is the effect of the area fractions at the higher moduli ratios. For polymers, the elastic modulus is inversely proportional to the coefficient of thermal expansion. So, selecting polymers with a wide difference in CTE results in a modulus ratio greater or less than 1. It is apparent from the plots (especially from the projections of the contour lines of constant curvature); the curvature is only weakly dependent on the modulus ratio above a value of 5 or so. From these plots, the optimum height fraction for each geometry can be determined. At modulus ratio between 5 and 10, the optimum height fractions of component 1 are between 0.24 and 0.31 for rectangles, between 0.27 and 0.33 for circles and between 0.40 and 0.47 for triangles. The optimum moves towards a greater fraction of component 1 as the moduli ratio decreases.

Fiber spinning

We chose commercial grades of isotactic and syndiotactic polypropylene (i-PP, s-PP) as the components. The isotactic polymer is expected to be of relatively high crystallinity and therefore have a relatively high modulus and low coefficient of thermal expansion. In contrast, the syndiotactic polymer crystallizes slowly and is expected to have a high CTE and a low modulus. Uehara [6] found well drawn isotactic polypropylene to have a fiber modulus of 20 GPa while well drawn syndiotactic polypropylene had a fiber modulus of 3 GPa – a ratio of 6.67-to-1. A typical CTE for an amorphous polymer is around 10^{-4} m/m · °C. Crystalline and highly oriented polymeric fibers can have a negative CTE along the length of the fiber. The chain axis CTE for isotactic PP crystals is negative, -1×10^{-5} m/m · °C [7].

As reported elsewhere [8], we spun from commercial grades of polypropylene. Fibers had an edge length of approximately 50 microns. Fibers were spun with a draw ratio of 2.5-to-1 and collected onto rolls of 15 cm diameter. Figure 4 shows a typical fiber cross-section with a small amount of dye added to the middle section. The bottom section (C) was i-PP, the top section (A) s-PP and the center section (B) a random ethylene propylene copolymer (co-EP) with a low modulus to limit interfacial stress between the other two components but with a coefficient of thermal expansion similar to s-PP.

(a) (b)

Figure 4. (a) Schematic drawing of a fiber; (b) Micrograph of actual fibers.

In these trials, the high modulus polymer was the bottom component giving us a modulus ratio of 1/6.67 or 0.15. From the equations, for a system with this modulus ratio, the maximum curvature is predicted to occur when the fraction of the upper component is between 0.8 and 0.85 (lower component fraction between 0.2 and 0.15). The fiber closest to having this component fraction was Fiber 1, as seen in Table 2.

Fiber	polymer / area fraction of each polymer		
	A (s-PP)	B (co-EP)	C (i-PP)
1	0.5	0.2	0.3
2	0.45	0.2	0.35
3	0.40	0.20	0.40
4	0.35	0.20	0.45
5	0.30	0.20	0.50

Table 2. Proportions of the A, B and C components of the spun fibers.

When the fibers were cut from the rolls they spontaneously curled to form batting. The curling is the result of relaxation of stresses from the spinning process and the change in length as the fibers are cooled from the spinning temperature.

The battings were then die cut into 20 in^2 (129 cm^2) circular samples for testing. The thickness of the samples was then measured in a temperature controlled chamber by compressing the samples with a pressure of 0.02 psi. The thickness of the samples was first measured at room temperature then the environmental chamber was cooled to zero degrees Celsius.

Figure 5 shows a plot of the temperature response for a series where the area fraction of the base of the triangle is systematically varied. The results are consistent with the analytical model. The best sample shows a change in thickness of 1.8% per degree C while the lowest response is 0.3% or less. An area fraction of 0.3 is the approximate lower limit for the fiber spinning apparatus. Plotted with the data are the curvatures as calculated from the analytical model.

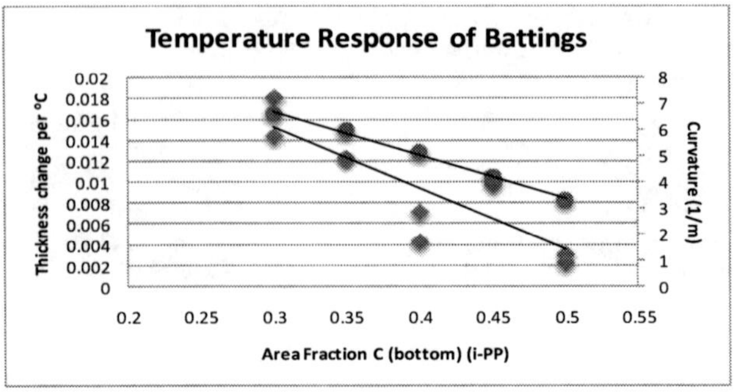

Figure 5. Thickness change (blue) and calculated curvature (red) as a function of component C.

CONCLUSIONS

We have analyzed thermally-induced curvature of bi-component fibers of various cross-sections with respect to the ratio of the moduli of the components and the fraction of the cross-section that each component occupies. Graphical analysis of these functions showed qualitatively that to achieve maximum curvature of the fibers, the design space of interest resides along the 'spine' of the surfaces. Quantitatively, the functions allow the proportions of the components to be selected to maximize the fiber curvature for a given pair of polymers. Experimental fibers of varying component fractions were spun, formed into mats and their thickness measured at various temperatures. The experimentally measured thickness changes are in good agreement with the analytical results for fiber bending. Based on a one dimensional heat transfer model the change in thickness appears to be sufficient for a practical adaptive thermal insulation.

REFERENCES

1. **Timoshenko, S.** Analysis of Bi-metal Thermostats. *J. Opt. Soc. Am.* 1925, **11** (3), pp.233-255.
2. **Alici, Gursel and Huynh, Nam N.** Predicting force output of trilayer polymer actuators. *Sensors and Actuators A.* 2006, Vol. 132, pp. 616-625.
3. **Slepyan, L., Ryvkin, M. and Fuchs, M.B.** Temperature Insensitive Curved Bimetal Elements. *Acta Astronautica.* 1995, Vol. 37, pp. 95-103.
4. *Variable Loft Thermal Insulation for Temperature Adaptive Clothing.* **Fossey, S.A. and du Plessis, A.J.** Roseville, MN USA : Industrial Fabrics International Association. Saftey & Protective Fabrics Proceedings 2004.
5. *Variable Loft Thermal Insulation.* **Fossey, S.A, Gilbert, D. and Hunt, V.** Atlanta, GA USA : s.n., 2006. AATCC 2006 International Conference.
6. **Uehara, H., Yamazaki, Y. and Kanamoto, T.** Tensile Properties of Highly Syndiotactic Polypropylene. 1996, Vol. 37, 1, pp. 57-64.
7. **Lacks, D.J. and Rutledge, G.C.**. Mechanisms for Axial Thermal Contraction in Polymer Crystals: Polyethylene vs Isotactic Polypropylene. *Chemical Engineering Science.* 1994, Vol. 49, pp. 2881-2888.
8. *Temperature Adaptive "Smart" Thermal Insulation Based on Multicomponent Fiber Spinning.* **Fossey, S.A, DeCristofano, B.S., Welsh, E.A., Perry, J., Archambault, D.** Orlando, FL : s.n., 2010. 2010 Army Science Conference.

Mater. Res. Soc. Symp. Proc. Vol. 1312 © 2011 Materials Research Society
DOI: 10.1557/opl.2011.916

Development of a dual growth factor loaded biodegradable hydrogel and its evaluation on osteoblast differentiation *in vitro*

Deepti Dyondi [1,2], Thomas J Webster [2] and Rinti Banerjee [1]

[1]Department of Biosciences and Bioengineering, Indian Institute of Technology Bombay, Mumbai 400076, Maharashtra, India.

[2] School of Engineering and Department of Orthopedics, Brown University, Providence 02906, RI, United States.

ABSTRACT

Hydrogels with their tunable properties are attractive candidates for developing tissue engineering scaffolds for various applications (including bone and cartilage). The current work involved studying the synergistic effect of basic fibroblast growth factor (bFGF) and platelet derived growth factor BB (PDGF-BB) entrapped within injectable porous gels for bone regeneration applications. An *in situ* gelling system was developed using bacterial polysaccharides gellan and xanthan gum by temperature and ionic gelation with Ca^{+2}. After the initial characterization of the hydrogels, a dual growth factor release system was developed wherein growth factors were encapsulated within chitosan nanoparticles embedded in the gels as well as directly within the gel. The hydrogel structure was characterized by SEM and TEM and *in vitro* growth factor release studies showed a slow release profile in PBS. Further, human fetal osteoblasts were entrapped within the hydrogel and a 21 day osteoblast differentiation study was conducted. An improvement in osteoblast total protein synthesis and collagen content was observed by day 21 compared to control gels without growth factors. Although further evaluation regarding mechanical properties and expression of osteogenic differentiation marker genes will be necessary, the present study suggests that injectable scaffolds can be used for the delivery of multiple growth promoting agents to support osteoblast differentiation.

INTRODUCTION

With its ability to deliver osteoconductive as well as osteoinductive (such as growth factors and cytokines) materials to the defect site [1], bone tissue engineering has emerged as a promising alternative to the currently available bone restoration methods (autografts, allografts etc) that offer several complications of undesirable tissue response and increased risk of infection at the implant site.

The use of biodegradable polymers (such as hydrogels) allows for a great deal of flexibility with respect to their tunable mechanical properties and natural degradation times within the body. The polymer system could be tuned to gel within the body thus entrapping molecules for delivery to the bone defect site including stem cells, growth factors, drug molecules and recombinant proteins [2, 3].

The current study explores natural polymers namely gellan and xanthan for developing a nanoparticulate *in situ* gelling system which along with cells and growth factors may be delivered at the defect site. Gellan gum, an anionic exocellular polysaccharide derived from the bacteria *Sphingomonas elodea*, in combination with hyaluronic acid, in the presence of calcium, has been shown to gel at 37 °C which can, thus, be used as a temperature- and cation- triggered injectable *in situ* gelling system. [4]

Gellan, in combination with xanthan gum, an anionic exocellular polysaccharide produced from bacteria *Xanthmonas campestris,* was evaluated as an injectable scaffold supplemented with growth factors and with osteoblasts entrapped within the hydrogel matrix. These growth factors were entrapped either within the gel or nanoparticles for sustained release purposes. While, chitosan nanoparticles were developed and characterized for the delivery of the growth factor bFGF, PDGF BB was directly entrapped within hydrogel matrix. The objective of the current study was to fabricate and evaluate the combinatorial effect of ascorbic acid, PDGF-BB and bFGF entrapped within chitosan nanoparticles and a hydrogel matrix on osteoblast differentiation (including mineralization).

EXPERIMENTAL DETAILS

Hydrogels were prepared by combining the two polymers: gellan and xanthan in various ratios with and without calcium chloride ($CaCl_2$)) and were characterized for gelling times, feasibility of injection and viscosity profiles (Contraves low shear 30 viscometer, Zurich, Switzerland). Viscoelasticity measurements were measured using Gemini 200 - Rotational Rheometer from Malvern UK at a frequency range of 0.1 Pa-100 Pa at a strain rate of 0.06 using a 20 mm parallel plate geometry at a temperature of 37 °C. Scanning Electron Microscope (SEM) (Hitachi S-3400N) of gellan xanthan gels was done by lyophilizing the gels and placing a small piece carefully over 0.5 cm x 0.5 cm glass slides to prepare them for gold sputtering. The samples were then visualized under different SEM magnifications at an accelerating voltage of 10 kV.

Chitosan nanoparticles were synthesized using the ionic gelation technique and were used for encapsulation of growth factor bFGF while PDGF BB was entrapped within hydrogel matrix. *In vitro* release profile was studied for all the growth factors in PBS (pH 7.4) at 37° C in a water bath shaken at 150 rpm. 1 ml of the sample was withdrawn at prefixed time intervals and stored at -80°C for further quantification and replaced by the same volume of PBS. The release was carried out in triplicate and quantification was completed using human PDGF-BB and bFGF ELISA kits (RayBio®). The absorbance was measured by a UV spectrophotometer (Perkin Elmer Lambda 25 UV/VIS spectrometer, Shelton, USA). The concentration corresponding to the absorbance was determined from the concentration vs. absorbance calibration curve.

The *in vitro* toxicity of the polymers was measured using L929 mouse fibroblast cells (National Centre for Cell Sciences, Pune, India) as well as human fetal osteoblasts (bone-forming cells; CRL-11372 American Type Culture Collection) and a MTT (3-(4,4-dimethylthiazol-2-yl)-2,5-diphenyl tetrazolium bromide) assay. To determine osteoblast differentiation on the materials of interest, human fetal osteoblasts (CRL 11372, ATCC; bone forming cells) were cultured in Dulbecco's modified Eagle medium (DMEM, Hyclone) supplemented with 10% fetal bovine serum (FBS; Hyclone) and 1% penicillin/streptomycin (Pen/Strep; Hyclone), 50 μg/mL L-ascorbic acid (Sigma), and 10 mM β-glycerophosphate

(Sigma) for 14 days under standard cell culture conditions, a 37 °C, humidified, 5% CO_2, 95% air environment. After 14 days of culture, hydrogel constructs were papain digested (125 μg/ml Papain, 5 mM cysteine, 100 mM Na_2HPO_4, 5 mM EDTA pH 7.5) and the lysate was further used to determine total protein content, total collagen content and alkaline phosphatase activity via well-established commercially available kits (as described below). Lastly, calcium deposited by osteoblasts were dissolved in 0.6 mol/L hydrochloric acid and tested for calcium concentration using a commercially available kit [5].

DISCUSSION

Gellan: xanthan hydrogels formed were tested for gelling times by an inverted tube test for various ratios of the two polymers (gellan and xanthan) to find the most appropriate concentration for a gelling time of 5-10 minutes for developing a tissue engineered scaffold that can entrap growth factors and cells. The viscosity for 0.9% gellan: xanthan at room temperature was found to be around 400 mPa.s which increased to a value of 40,000 mPa.s upon the addition of heat and a cation trigger.

Mechanical spectra of GX3 indicated a higher storage modulus (G') than the loss modulus (G") at all frequencies (Fig 1b). SEM micrographs of the hydrogels showed a uniform crosslinked structure, a property important for the mechanical strength as well as the water holding capacity of the gels (Fig. 1a).

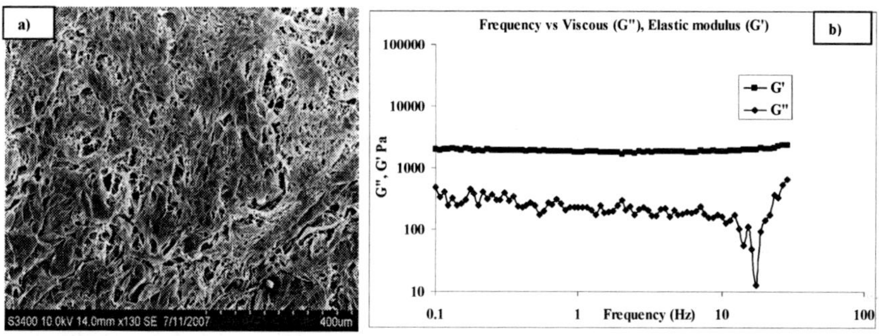

Figure 1. a) SEM image of 0.9 % 9:1 gellan xanthan gels without Ca^{2+}, original magnification x 130, scale bar 400 μm, b) Storage and loss shear moduli plotted against frequency for *in situ* 0.9 % 9:1 gellan: xanthan gels with 3 mM Ca^{2+}.

Chitosan nanoparticles: particle size, zeta potential and *in vitro* release

Particle size of the chitosan nanoparticles as determined by photon correlation spectroscopy (PCS) measurements was found to be 296.6 ± 60.64 nm with a polydispersity index of 0.34. A zeta potential of +31.9 was observed. The particles had a smooth morphology with no aggregation as observed by SEM.

The release profiles indicated a slow and sustained delivery of growth factor molecules by entrapping these factors within gels and nanoparticles. bFGF within chitosan nanoparticles showed a cumulative release percentage of 80% by day 20 (Fig. 2). bFGF encapsulated within the hydrogel-entrapped chitosan nanoparticles had a further slower release with a cumulative percentage release of around 49% by day 21.

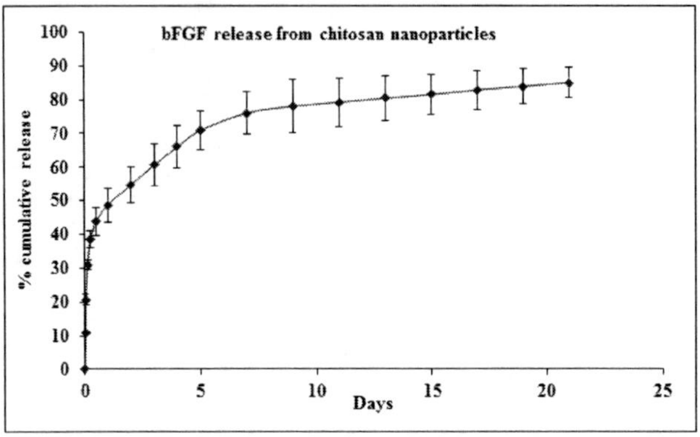

Figure 2. Cumulative release of bFGF from chitosan nanoparticles *in vitro* at 37°C in PBS

In vitro toxicity

>95% of L929 cells were viable on the polymers. Microscopy images showed that L929 cells grew and maintained their morphology over the hydrogels, which is an important property as the hydrogels should support cell growth and not cause any adverse cell responses when injected into the tissue defect site. >80% viability was observed for osteoblasts after 48 h further confirming the low toxicity of the novel biomaterials developed here.

Total protein and calcium

Total protein synthesis from osteoblasts as determined by the BCA assay was higher when both growth factors were loaded into the hydrogels, specifically with ascorbic acid and PDGF entrapped within the hydrogels and bFGF entrapped within the chitosan nanoparticles in the hydrogels. Results further indicated increased osteoblast total protein synthesis in single and dual growth factor groups as opposed to the control groups (i.e., cells on tissue culture polystyrene (TCPS) and cells in gellan: xanthan gels with no growth factors) (Fig. 3). Although calcium was deposited by osteoblasts during this study, no significant differences in calcium deposition by osteoblasts were found between the samples tested here.

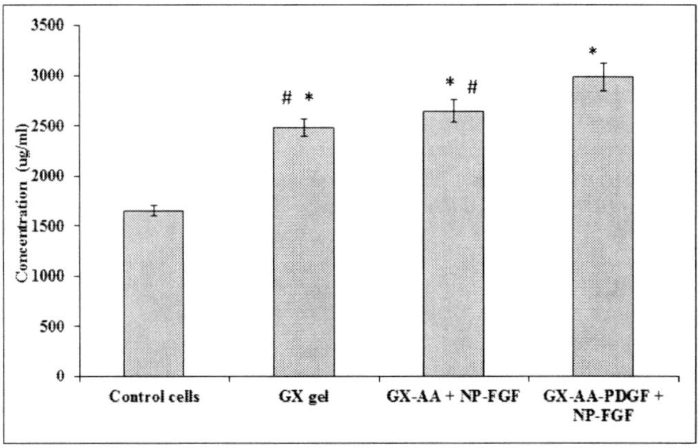

Figure 3. Total protein synthesis in hydrogels by osteoblasts by day 14. Data = mean ± SEM; N=3. Control cells: cells on TCPS (control), GX: Gellan : xanthan, GX-AA+NP-FGF: Gellan : xanthan with ascorbic acid and nanoparticles loaded with bFGF, GX-AA-PDGF-BB+NP-FGF: PDGF-BB and ascorbic acid entrapped within hydrogels and bFGF containing chitosan nanoparticles within the gel matrix. *two-tailed p value <0.05 as compared to TCPS control, # two-tailed p value <0.05 as compared to the dual growth factor release system

Total collagen synthesis

Total collagen synthesis by osteoblasts was found to be higher for bFGF containing hydrogel constructs as compared to those without any growth factors (gel controls) as well as tissue culture polystyrene controls. Hydrogels with a combination of PDGF and bFGF showed slightly higher amounts of collagen synthesized as hydrogels with bFGF alone (p values <0.05 for all the sample groups).

CONCLUSIONS

The proposed hydrogel system has the advantage of being a less invasive surgical form for bone repair which can be administered using injection under local anaesthesia, as was shown by the gelling time and viscosity studies. The use of a dual release system allows the feasibility of delivering two different growth factors in a desirable temporal manner maintaining the roles they play in different stages of bone formation. Importantly, molecules like ascorbic acid help in the regeneration of bone in a less expensive manner as compared to growth factors.

An important result from the present study was that increased total protein and collagen synthesis by osteoblasts with the present dual growth factor release was observed as compared to single growth factor release.

In summary, an *in situ* gelling hydrogel system was developed and characterized to study the combinatorial effect of two of the growth factors (specifically, PDGF-BB and bFGF) known in bone development. The gellan: xanthan hydrogel system, thus, can be used as a platform technology for the delivery of several drugs and growth factor molecules. At present, sustained release of three agents namely ascorbic acid, PDGF and bFGF were achieved. This implies that small molecular weight moieties as well as labile peptide moieties can be successfully encapsulated and released from the injectable scaffold. Further, the scaffold allowed proliferation of cells in a three dimensional matrix. Further studies evaluating osteoblast differentiation marker genes are ongoing. Addition of nanocrystals of calcium phosphate or hydroxyapatite may help to improve osteoblast differentiation in the injectable scaffold.

ACKNOWLEDGEMENTS

The authors would like to acknowledge the Indo-U.S. Joint Centre on Biomaterials for Healthcare funded by the Indo-U.S. Science and Technology Forum (IUUSTF), New Delhi, India, for providing funds for travel and work; Malvern Aimil Instruments Private Limited, Navi Mumbai, India for viscoelastic studies and Dr. Yupeng Chen and Dr. Bahar Bilgen from Brown University, Providence, Rhode Island, USA for providing their valuable input in the research. The authors would also like to acknowledge Centre for Research in Nanotechnology & Science (CRNTS), IIT Bombay for providing instruments facility for material characterization.

REFERENCES

1. J. A. Burdick, M. N. Mason, A. D. Hinman, K. Thorne, K. S. Anseth. *Journal of Controlled Release* **83**, 53 (2002).

2. G. Balasundaram, T. J. Webster. *Macromol. Biosci.* **7**, 635 (2007).

3. M. A. Bokhari, G. Akay, S. Zhang, M. A. Birch. *Biomaterials* **26**, 5198 (2005).

4. S. Suri, R. Banerjee. *Journal of Biomedical Materials Research* **79A(3)**, 650 (2006).

5. Y. Chen, T. J. Webster. *J. Biomed Mater Res A* **91A**, 296 (2009).

Polymer-Based Nanocomposites

Mater. Res. Soc. Symp. Proc. Vol. 1312 © 2011 Materials Research Society
DOI: 10.1557/opl.2011.201

Microstructural and Mechanical Properties of Polyester/Nanoclay Nanocomposites: Microstructure-Mixing Strategy Correlation

Hamid Dalir [1], Rouhollah D. Farahani[1], Vireya Nhim[2], Benjamin Samson[3], Martin Lévesque[1] and Daniel Therriault[1]
[1] Center for Applied Research on Polymers and Composites (CREPEC), Department of Mechanical Engineering, École Polytechnique de Montréal, C.P. 6079, Succ. Centre-ville, Montréal, QC H3C 3A7, CANADA.
[2] School of Mechanical Engineering, Ecole Nationale Supérieure d'Arts et Métiers (ENSAM), 151 Boulevard Hôpital, Paris 75013, FRANCE.
[3] Department of Mechanics, École Polytechnique de Paris, 32 Boulevard Victor, Paris 75015, FRANCE.

ABSTRACT

Different nanoclay mixing strategies using a three-roll mill and ultrasonication is proposed to obtain the desired polyester/nanoclay dispersion, intercalation, and exfoliation. The dispersion states of the modified nanoclay in polymer with 2, 4 and 6 wt% loading were characterized with X-ray diffraction, scanning electron microscopy (SEM), and low and high magnification transmission electron microscopy (TEM). The mechanical properties of the clay-reinforced polyester nanocomposites were a function of the nature and the content of the clay in the matrix. The nanocomposite containing 4 wt% modified Cloisite® 15A exhibits excellent improvement in modulus (by ~51%) and tensile strength (by ~12%) with a decrease in fracture strain (by ~26%) and fracture energy (by ~17%). These mechanical characteristic changes can be attributed to the dispersion, intercalation, and exfoliation of the nanoclays inside the polyester matrix.

INTRODUCTION

Polymer/clay nanocomposite (PCN) is a class of material consisting of either thermoset, thermoplastics, rubber or co-polymers as matrix and nanoclay as the reinforcement fillers [1,2]. A combination of organic polymeric matrix and inorganic nano-layered clay platelets have lately evoked a great deal of academic and industrial research activities due to the formation of nanocomposite materials with significantly enhanced physical properties such as thermal properties (e.g., thermal stability [3], flame retardant [4], thermal conductivity [5]), mechanical properties (e.g., mechanical strength [6], hardness [7], abrasion resistance [8]), electro-rheological properties [9], permeability properties (e.g., gas barrier [10], pervaporation [11]) and corrosion protection properties [12] of polymers.

The thermomechanical responses of polymers are favorably altered by the addition of a trace amount of nanofillers. To obtain optimal nanocomposite formation, the following four structural parameters should be maximized: (1) the particle aspect ratio, (2) particle dispersion, (3) particle packing (or alignment), and (4) polymer to particle interfacial stress transfer [13–17].

Until now, many published research works have focused on the investigation of the mechanical properties of polyester resin-clay nanocomposites based on comparative studies [18-25]. The addition of clay in the polyester usually leads to three different microstructures depending on the nature of the material and the method of preparation. When the polyester is unable to intercalate between the silicate sheets, a phase separated composite is obtained in which the materials stay in the same range as traditional microcomposites. Intercalated structure, in which a single (or more) polymer chain intercalated between the silicate layers results in a well ordered multilayer morphology built up with alternating polymeric and inorganic layers. When the silicate layers are completely and uniformly dispersed in a continuous polymer matrix, an exfoliated or delaminated structure is obtained. The intercalated and exfoliated structures are formed only when the clay is organically treated with surfactants, otherwise it will lead to the formation of phase separated microscale conventional composites [18,19].

In this article, we first prepared polyester polymer/nanoclay hybrid in the styrene ($C_6H_5CH{=}CH_2$) solution by the use of predispersed nanoclay. Then, the hybrid resins were cured by NOROX® MEKP-9 to obtain polyester-based nanoclay hybrid sample. Three different kinds of modified nanoclays were chosen for investigating the effect of the nature and loading of clay on the properties of the nanocomposites. Obviously, the clay in the matrix influences the microstructure and performance of the polyester. An understanding of the extent of improvement and the reasons are important, since polyester-based nanocomposites have been widely used in automobiles, space, defense, adhesives, coatings, and so on. This is part of our ongoing studies on structure-property relationship of various special purpose elastomeric nanocomposites.

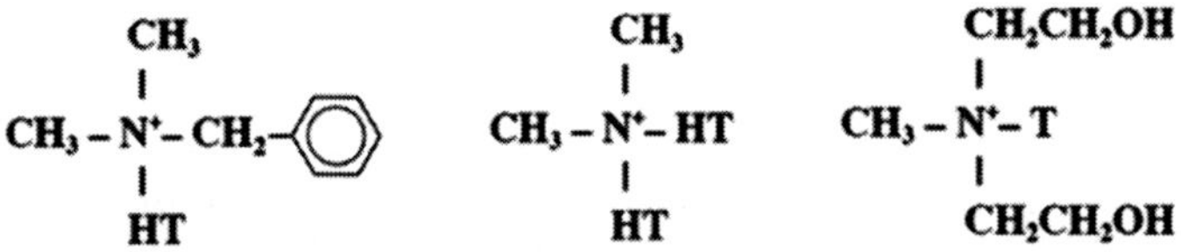

Fig. 1. Structure of the surface modified (a) Cloisite® 11B, (b) Cloisite® 15A, and (c) Cloisite® 30B nanoclays.

Table 1. Nanoclay specifications and types of the surface modifiers.

Nanoclay	Surface modifier	Concentration	Moisture (%)	Interlayer spacing (Å)
Cloisite® 11B	2MBHT	95 meq/100g	< 4%	18.4
Cloisite® 15A	2M2HT	125 meq/100g	< 2%	30.8
Cloisite® 30B	MT2EtOH	90 meq/100g	< 2%	18.5

EXPERIMENTAL DETAILS

Nanocomposite preparation

The nanocomposite material was a mixture of one of the three different types of surface modified nanoclays (Cloisite® 11B, Cloisite® 15A, and Cloisite® 30B, Southern Clay Products, Louisville, US), shown in Fig. 1 with different surface modifiers as listed in Table 1. A special catalyst cured thermosetting polyester (RL7520, Progress Plastiques Cie, Drummondville, CA) having a viscosity of 200 mPa.s, gel time of 75 min, modulus of 3.6 GPa, and strength of 80 MPa was used with MEKP-9 (NORAC Inc., California, US) as the catalyst. The desired amounts of nanoclays (2, 4, 6 wt%) were first dispersed in dichloro-methane 99.9% (DCM, Sigma-Aldrich Ltd., Ontario, CA) solution. After ultrasonication (ultrasonic cleaner 8891, Cole-Parmer) at 42 kHz for 1 h, polyester was slowly added to the solution over a stirring hot plate (model SP131825, Barnstead international). The samples were further mixed by five different mixing strategies listed in Fig. 2.

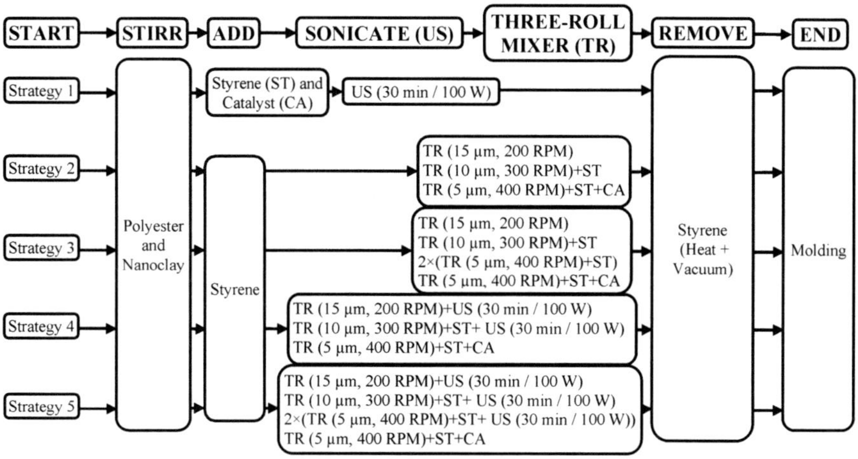

Fig. 2. Schematic description of various processing techniques.

In strategy 1, Styrene (ST, Sigma-Aldrich Ltd., Ontario, CA) and catalyst were added to the mixture and then ultrasonicated (US) for 30 min. In strategies 2 and 3, the nanocomposite mixtures were passed several times through a three-roll mixing mill (TR, Exakt 80E, Exakt Technologies Inc.) where the gap between the rolls and the speed of the apron roll were adjusted at 15 μm-200 RPM, 10 μm-300 RPM and either one or three times at 5 μm-400 RPM, for three or five passes, respectively. For strategies 4 and 5, the same procedure and parameters as those of strategies 2 and 3 were chosen, respectively, however US was performed after each TR process, except the last one.

To investigate the effect of the type and the amount of the clay on the mechanical properties and the microstructure of the polyster/clay nanocomposite, three concentrations (2, 4, and 6 wt%) of the modified clays were used. The obtained nanoclay-polyester mixtures were then immediately degassed and poured into tension test sample molds (ASTM D638) and placed under vacuum for 5 to 8 h at room temperature, until all solvent was evaporated.

Characterization and properties

The degree of intercalation and exfoliation was analyzed based on scanning electron microscope (SEM, JEOL JSM-7600TFE), transmission electron microscope (TEM JEOL JEM-2100F) images and X-ray diffraction (XRD, X'Pert PRO Diffractometer). The interlayer separation measurements were carried out with Pro X-ray diffractometer with Cu target and Ni filter at a scanning rate of 2°/min, with CuKα radiation (λ_w = 1.541 Å) operating at 30 kV and 15 mA. Tensile tests were carried out using tensile test machine (MTS 810) with a 25 kN load cell at the crosshead speed of 1 mm/min (ASTM D638).

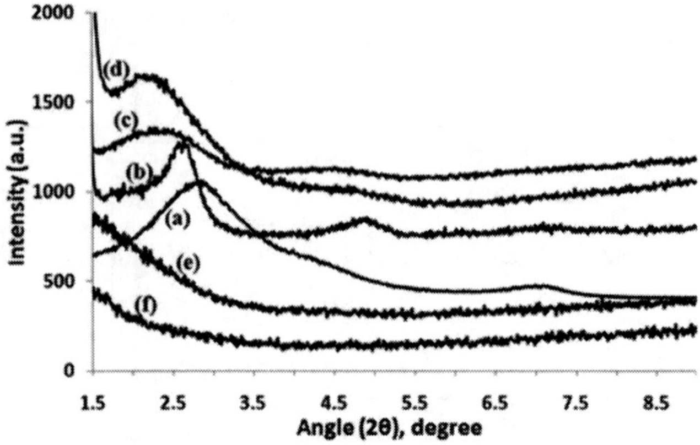

Fig. 3. XRD patterns of 4 wt% Cloisite® 15A-polyester samples prepared by (a) stirring hotplate (initial mixing), (b) strategy 1, (c) strategy 2, (d) strategy 3, (e) strategy 4, and (f) strategy 5.

Table 2. Interlayer spacing-mixing strategy correlation of different kinds of nanoclays.

Nanoclay	Stirring hotplate	Strategy 1	Strategy 2	Strategy 3	Strategy 4	Strategy 5
Cloisite® 11B	18.4	19.4 (+5%)	22.7 (+23%)	23.9 (+30%)	27.3 (+48%)	29.8 (+62%)
Cloisite® 15A	30.8	33.4 (+8%)	40.0 (+30%)	41.1 (+33%)	---	---
Cloisite® 30B	18.5	21.3 (+15%)	22.5 (+22%)	23.6 (+28%)	25.2 (+36%)	25.7 (+39%)

RESULTS AND DISCUSSION

To compare the degree of intercalation and exfoliation of different types of nanoclays inside the polyester resin, XRD patterns of 4 wt% Cloisite® 11B-, Cloisite® 15A-, and Cloisite® 30B-polyester nanocomposites prepared by stirring hotplate and five different mixing strategies were obtained. As it can seen from Fig. 3, the Cloisite® 15A-polyester nanocomposite prepared by stirring hotplate shows a peak at 2.87° derived from the basal spacing in the range between 0 and 10°. On the other hand, the nanocomposites obtained by mixing strategies 1, 2, and 3 shifted to lower peaks at 2.64°, 2.21°, and 2.15°, respectively. For the nanocomposites mixed with strategies 4 and 5, no peaks were observed. The interlayer distance (*d*-spacing) of clays inside nanocomposites prepared by different mixing strategies is obtained from Bragg's law:

$$2d\sin\theta = n\lambda_w \quad (1)$$

where d is the interlayer distance of clay, λ_w is the wavelength used, n is the order which is equal to one for the first order, and θ is the measured angle. Therefore, the nanoclay interlayer distance calculated from the above equation is 30.8 Å, 33.4 Å, 40.0 Å, and 41.1 Å for nanocomposites prepared by stirring hotplate and mixing strategies 1, 2, and 3, respectively. Detecting no peaks in XRD graphs for nanocomposites prepared according to mixing strategies 4 and 5 confirms the formation of well intercalated and exfoliated nanoclay structure. As it can be seen in Table 2, for the two other types of nanoclays, namely Cloisite® 11B and Cloisite® 30B, none of the above mentioned strategies lead to a fully exfoliated structure.

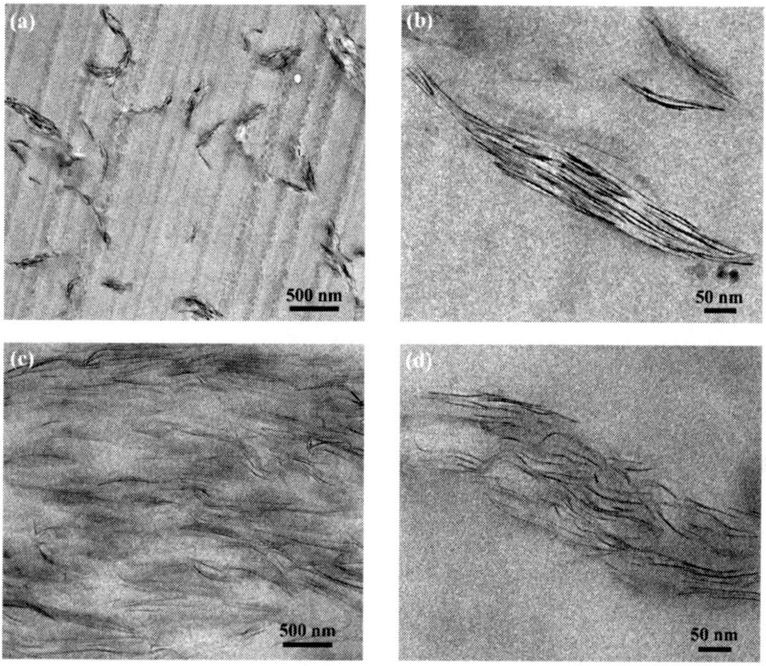

Fig. 4. Low and high magnification TEM micrographs of 4 wt% Cloisite® 15A-polyester samples prepared according to (a,b) strategy 2, and (c,d) strategy 4.

Figure 4 shows the TEM images of 4 wt% Cloisite® 15A-polyester samples prepared by strategy 2, and strategy 4. The bright region represents the matrix portion and the dark lines represent the clay nanolayers. The lower magnification of TEM micrographs (Figs. 4a,c)) helps us to study the dispersion of clay particles in the matrix and the higher magnification ones (Figs. 4b,d)) are used to observe the clay microstructures (intercalation and/or exfoliation). The nanoclay aggregates in Fig. 4a show that the dispersion of clay nanolayers has not been uniformly done throughout the polyester. The corresponding higher magnification image (Fig. 4b) shows the existence of intercalated nanoclay structure inside the matrix. However, the nanoclay-polyester samples prepared according to mixing strategy 4, show a more uniform dispersion (Fig. 4c) of intercalated and exfoliated (Fig. 4d) nanoclays inside the polyester matrix which is in agreement with the results obtained from XRD patterns.

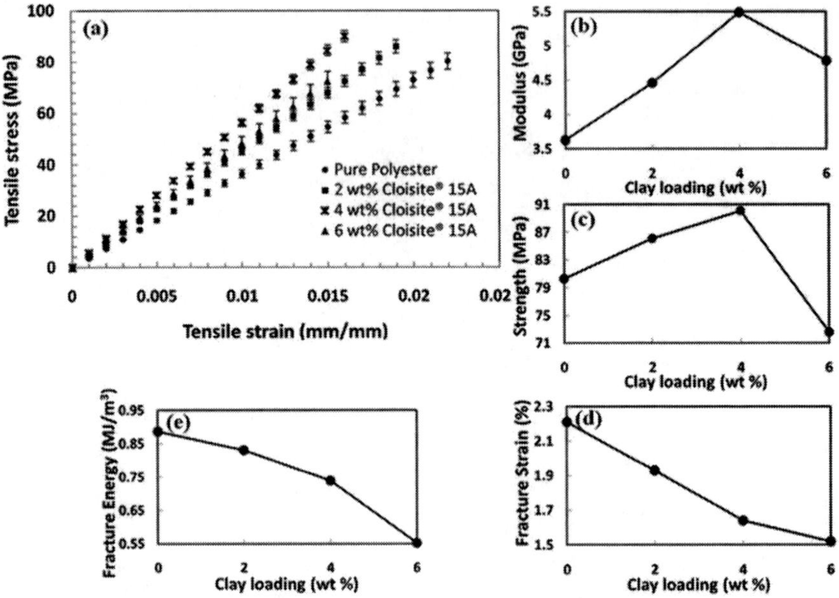

Fig. 5. Averaged mechanical properties of pure polyester and 2, 4, 6 wt% Cloisite® 15A-polyester samples. (a) Stress-strain curves measured from tensile tests, (b) modulus, (c) strength, (d) fracture strain, and (e) fracture energy.

Table 3. Summary of tensile test results and variation of the mechanical properties.

Samples	Modulus (GPa)	Modulus var. (%)	Strength (MPa)	Strength var. (%)	Fracture strain (%)	Fracture strain var. (%)
Pure polyester	3.63±0.18	---	80.3±4.2	---	2.21±0.01	---
2 wt% clay 15A	4.46±0.09	+23	86.1±3.5	+7	1.93±0.04	-13
4 wt% clay 15A	5.49±0.07	+51	90.1±2.7	+12	1.64±0.03	-26
6 wt% clay 15A	4.78±0.17	+32	72.6±5.1	-10	1.52±0.05	-31

Fig. 5a shows the averaged stress-strain curves obtained from tensile tests of minimum three specimens for each of the four types of samples made of pure polyester, and 2, 4, 6 wt% Cloisite® 15A. As it can be seen all specimens failed immediately after the stress reached the maximum value. The stress-strain curves showed considerable linearity before reaching the maximum stress. Table 3 shows the averaged tensile test results and their deviations over three repetitions for each experiment. The first column represents the different Cloisite® 15A weight percentages at which the experiments were done. The other columns list the measured mechanical properties (e.g., modulus, strength, and fracture strain) and their variations. Tensile results for Cloisite® 15A-polyester samples containing 2 wt% nanoclay show an increase of 23% in modulus and 7% in strength, and in reverse a decrease of 13% in fracture strain and 7% in fracture energy. Optimal loading of clay was found at 4 wt% nanoclay with an improvement of about 51% in modulus and 12% in strength and a reduction in fracture strain by 26% which leaded to 17% decrease in fracture energy (see Fig. 5b-e). Samples having 6 wt% nanoclay show a decrease of 13% in modulus, 20% in strength, 8% in fracture strain, and 25% in fracture energy in comparison with those having 4 wt% nanoclay. Therefore, according to the results obtained in our research, increasing the nanoclay amount beyond 4 wt% (e.g., 6 wt%) with even the best mixing strategies (strategies 4 and 5) presented here, will lead to nanocomposites having weaker mechanical properties than those of 4 wt%, which can be attributed to the existence of nanoclay aggregates in nanocomposite structure acting as stress concentrators.

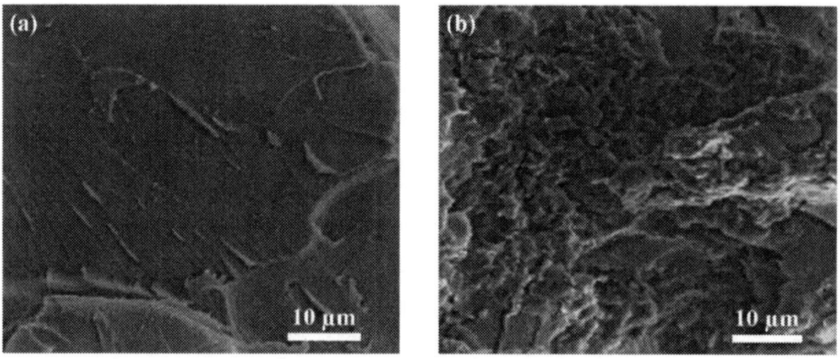

Fig. 6. SEM images showing tensile fracture surface morphologies for the 4 wt% Cloisite® 15A-polyester nanocomposites prepared according to (a) strategy 2, and (b) strategy 4.

The roughness of the fracture surface has generally been associated to fracture properties and fracture energy release rates [26,27]. A smooth fracture surface is attributed to brittle failures and rougher fracture surfaces are attributed to tougher nanocomposites [27]. Fracture surfaces were analyzed to qualitatively determine the degree of dispersion and the fracture properties of nanocomposites prepared by different processing methods. Fig. 6a shows the relatively smooth tensile fracture surface of the 4 wt% Cloisite® 15A-polyester nanocomposites prepared according to strategy 2. As discussed for the TEM observations, samples from strategy 4 had the best dispersion of clay particles. This is consistent with significantly rougher fracture surfaces as can be observed in Fig. 6b.

CONCLUSIONS

This study was aimed at finding an efficient processing technique for the synthesis of nanoclay reinforced polyester resin systems by assessing and comparing five different mixing strategies. The effects of sonication energy, three-roll mixing mill gap between the rolls, the speed of the apron roll, number of passes, and the associated processing issues for each of the techniques were assessed by evaluating the nanoclay morphology and the tensile properties of the resulting polyester nanocomposites. It was found that the addition of 4 wt% Cloisite® 15A, according to strategy 4, will lead to a uniform dispersion of intercalated and exfoliated nanoclays inside the polyester matrix which results in having nanocomposites with significant improvement in their mechanical properties.

Direct sonication and use of styrene as a solvent resolved the drawbacks of foreign solvent residue and reduced the time required for solvent removal. Strategy 4 was found as the most efficient mixing process that enables incorporation of relatively large amount of polyester and nanoclay content with minimal processing problems, lower processing time and desirable tensile properties. Strategy 4 consists of using three passes three-roll mixing mill with direct sonication of nanoclays after each mixing process except the last one. Obtaining such mechanically improved polyester based nanocomposites and understanding the reasons behind, are crucially important for their applications in many industries such as automobiles, space, defense, adhesives, coatings, and so on.

ACKNOWLEDGMENTS

The authors acknowledge the financial support from NSERC (the Natural Sciences and Engineering Research Council of Canada). They also acknowledge the support of NRC-IMI (the National Research Council-Industrial Materials Institute) for helping us with TEM and SEM imaging and members of CREPEC (Centre de Recherche En Plasturgie Et Composites), Prof. François Trochu and Dr. Nadir Kchit, for their technical assistance in performing some of the mechanical tests.

REFERENCES

1. T. J. Pinnavaia. *Science* **4595** (1983), pp. 365.
2. Y. Fukushima *et al. Clay Minerals* **23** (1988), pp. 27.
3. T. Lan, P. D. Kaviratna, and T. J. Pinnavaia, *Chem Mater* **6** (1994), pp. 573.
4. J. W. Gilman, C. L. Jackson, and A. B. Morgan. *Chem Mater* **12** (2000), pp. 1866.
5. K. J.Yao, M. Song, D. J. Hourston, and D. Z. Luo. *Polymer* **43** (2002), pp. 1017.
6. H. L. Tyan, Y. C. Liu, and K. H. Wei. *Chem Mater* **11** (1999), pp. 1942.
7. J. M. Yeh, C. L. Chen, C. C. Huang, and F. C. Chang. *J Appl Polym Sci* **99** (2006), pp. 1576.
8. C. Li and G. L. Wilkes. *Chem Mater* **13** (2001), pp. 3663.
9. M. S. Cho, H. J. Choi, and W. S. Ahn. *Langmuir* **20** (2004), pp. 202.
10. J. M. Yeh, C. L. Chen, Y. C. Chen, and C. Y. Ma. *Polymer* **43** (2002), pp. 2729.
11. J. M. Yeh, M. Y. Yu, and S. J. Liou. *J Appl Polym Sci* **89** (2003), pp. 3632.
12. J. M. Yeh, S. J. Liou, C. Y. Lai, and P. C. Wu. *Chem Mater* **13** (2001), pp. 1131.
13. L. N. Sridhar, R. K. Gupta, and M. Bhardwaj. *Ind Eng Chem Res* **45** (2006), pp. 8282.
14. N. K. Bose and M. R. Kamal. *Polym Eng Sci* **49** (2009), pp. 641.
15. C. H. Lee, H. B. Kim, S. T. Lim, and H. J. Choi. *J Mater Sci* **40** (2005), pp.3981.
16. M. Maiti, M. Bhattacharya, and A. K. Bhowmick. *Rubber Chem Technol* **81** (2008): pp. 384.
17. M. Bhattacharya and A. K. Bhowmick. Polymer **49** (2008), pp. 4808.
18. L. L. Pluart, J. Duchet, and H. Sautereau. *Polymer* **46** (2005), pp. 12267.
19. G. Akorna and I. Ugi. *Angew Chem Int Ed Engl* **16** (1977), pp. 259.
20. C. P. R. Nair, G. Glouet, and Y. Guilbert. *Polym Degrad Stsb* **26** (1989), pp. 305.
21. Y. H. Yu, C. C. Jen, H. Y. Huang, and P. C. Wu. *J Appl Polym Sci* **91** (2004), pp. 3438.
22. J. M. Yeh and C. P. Chin. *J Appl Polym Sci* **88** (2003), pp. 1072.
23. X. Kornmann, H. Lindberg, and L. A. Berglund. *Polymer* **42** (2001), pp. 1303.
24. X. Kornmann, H. Lindberg, and L. A. Berglund. *Polymer* **42** (2001), pp. 4493.
25. P. B. Messersmith and E. P. Giannelis. *Chemistry of Materials* **6** (1994), pp. 1719.

26. H. Miyagawa, A. Mohanty, and M. Misra. *Ind Eng Chem Res* **43** (2004), pp. 7001.
27. L. Wang, K. Wang, L. Chen, Y. Zhang, and C. He. *Composites A* **37** (2006), pp. 1890.

Mater. Res. Soc. Symp. Proc. Vol. 1312 © 2011 Materials Research Society
DOI: 10.1557/opl.2011.263

Multiscale Modeling of Reinforced Epoxy Resins by Carbon Nanotubes and Graphene

Kelvin Suggs,[1,2] Vernecia Person,[1] Chantel Nicolas,[1] and Xiao-Qian Wang[2]

[1]Department of Chemistry, [2]Department of Physics, Center for Functional Nanoscale Materials,

Clark Atlanta University, Atlanta, Georgia 30314, U.S.A.

ABSTRACT

Nanocomposites are of increasing interest due to their unique structural, electronic, and thermal properties. Simultaneously, multiscale molecular modeling is becoming more robust. Therefore computational models are able to be examined with increased accuracy, complexity, and dimension. Graphene based molecules are lauded for their conductive properties as well as their architecture-like geometry which may allow bottom up nanoscale fabrication of nanoscopic structures. Furthermore, these macrocycled molecules allow high interactivity with other molecules including highly tensiled polymers that yield other novel supramolecular structures when interacted. These supramolecular structures are being investigated in lieu of a variety of potential applications. Nanocomposites of cured epoxy resin reinforced by single-walled carbon nanotubes exhibit a plethora of interesting behavior at the molecular level. A fundamental issue is how the self-organized dynamic structure of functional molecular systems affects the interactions of the nano-reinforced composites. A combination of force-field based molecular dynamics and local density-functional calculations shows that the stacking between the aromatic macrocycle and the surface of the SWNTs manifests itself via increased interfacial binding. First-principles calculations on the electronic structures further reveal that there exists distinct level hybridization behavior for metallic and semiconducting nanotubes. In addition there is a monatomic increase in binding energy with an increase in the nanotube diameter. The simulation studies suggest that graphene nanoplatelets are potentially the best fillers of epoxy matrices. The implications of these results for understanding dispersion mechanism and future nanocomposite developments are discussed.

INTRODUCTION

Lightweight and high strength composite materials provide an appealing combination that propels composites to new perspective in advanced aerospace vehicles and propulsion systems.[1,2] Epoxy resins are leading candidates for these types of applications due to their excellent combination of physical and mechanical properties.[3] Nanocomposites of cured epoxy resin reinforced by single-walled carbon nanotubes exhibit a plethora of interesting behaviors at the molecular level.[4,5] We have employed a combination of force-field-based molecular mechanics and first-principles calculations to study the corresponding binding and charge-transfer behavior.

The simulation study of various nanotube species and curing agent configurations provides insight into the optimal structures in lieu of interfacial stability.[6,7] An analysis of charge distributions of the epoxy functionalized semiconducting and metallic tubes reveals distinct level hybridizations. These results are important for understanding dispersion mechanism and future nano reinforced composite developments.

Epoxy resins serve as key elements in the field of high-performance materials, coatings, and composites. These materials have demonstrated high cured glass transition temperature and a satisfactory combination of low and stable melt viscosity, thermal and mechanical performance, cost, as well as free of toxic monomers.[8,9] The availability of a wide variety of epoxy systems, along with different physical performances associated with various cross-linking degrees, makes these thermosetting polymers extremely suitable for applications such as structural, medical, or aerospatial.[10] Although epoxy resins are inherently brittle, curing processing in the manufacture of high-performance composites offers high modulus, stiffness and thermal stability. In this regard, remarkable progress has been made in the curing of epoxy resins, and several studies have been performed to investigate the curing mechanism and the influence of various parameters on processing and material properties.[11,12] On the other hand, carbon nanotubes have been employed as reinforcers, which are connected to their unique properties and outstanding performance. While recent demonstration of the potential of these types of resins offers an unprecedented combination of high temperature laminate performance[13,14] with superior processing characteristics, there is a striking lack of theoretical prescriptions for developing nano-reinforced epoxy resins. To gain the understanding necessary to develop such prescriptions, it is desirable to consider realistic models for nano-reinforced composites.

Epoxy-based polymer materials have diverse applications such as metal coatings, automotive primer, printed circuit boards, semiconductor encapsulants, adhesives, and aerospace composite materials.[15,16] The current trial-and-error formulation approach faces many challenges to satisfy an ever increasing demand for a "designer" range of properties of epoxy resin end-products, a greater economic pressure for efficiency, and a larger focus on the environmental impact requirements. Lightweight multifunctional composites with enhanced properties can potentially be produced by effectively incorporating individual carbon nanotubes (CNTs) into polymer matrices. Composites produced by conventional methods, such as solution mixing and melt blending, usually fail to yield significant improvements of bulk mechanical and electrical properties, when compared to individual nanotubes.[10-12] A controlled nanostructure of CNT/epoxy thin sheet can be achieved through impregnating CNT networks with epoxy solution, and the simulation study will be crucial for understanding the effects of composition, aspect ratio, diameter, chirality and functionalization. By changing various parameters, such as concentration of epoxy solution, oxidation state of CNT material and impregnation protocol, it is possible to tune the mechanical/ electrical properties of the prepared composites. The enhancement of Young's modulus and a tensile strength will be beneficial for a variety of applications such as conductive composite films for electromagnetic shielding, as well as for components in the lightweight multifunctional composites.[10-17]

The goal of our multiscale modeling work is to provide fundamental understanding of the mechanisms at the molecular level of the curing reaction and the effects of catalysts and accelerators using multiscale modeling methods. Such information will significantly aid the formulation process to meet its demands. The multiscale modeling approach is based on a combination of force-field-based molecular dynamics and local density-functional calculations. Force-field-based molecular dynamics can be used to pre-select molecular geometries, and first-principles calculations can be employed to determine the electronic structure of the nano-reinforced composites. A fundamental issue is how the self-organized dynamic structure of functional molecular systems affects the interactions of the nano-reinforced composites. The proposed multiscale approach will provide a systematic investigation of the π–π stacking between the aromatic macrocycle and the surface of the SWNTs that manifests itself *via* increased interfacial binding. First-principles calculations on the electronic structures can further characterize level hybridization behavior for metallic and semiconducting nanotubes.

COMPUTATIONAL METHODS

In this section, we summarize the simulation methods that will be used in this study of nanoreinforced composites. All of them are state-of-the-art multiscale modeling methods. We will use the density functional theory[23] to study the system's energetics in order to address the size-dependent and shape-dependent stability and atomic arrangements of these nanocomposites. The atomic motions are investigated using molecular dynamics with forces obtained from the density functional theory.

Density functional theory

To properly handle a many-electron system so that one can derive its various properties from fundamental quantum mechanics is a constant challenge in theoretical physics and chemistry. Although the interaction between electrons is well known, the facts that electrons, with a spin quantum number of ½, have to obey specific statistical rules and that one normally has to deal with quite a few of them at the same time make this problem immensely formidable. One approach that has become the standard one for large-scale electronic simulations is the density functional theory[23] in the so-called Kohn-Sham framework.[24] It is based on a theorem stating that the ground-state energy of a many-electron system can be represented as a functional of the electron density only. Thus one can hope to obtain the electronic energy without dealing with the many-body wave functions that are highly multidimensional with the notorious property of being antisymmetric with respect to particle exchange. Being a scalar in the real space, the electron density is a much simpler quantity to manage, making it possible to investigate more complex systems. By minimizing the energy functional with respect to possible density distributions one can then determine the ground-state electronic energy for a given atomic arrangement. The energy minimization procedure is most conveniently carried out by a mapping of the truly interacting system to an auxiliary system of noninteracting particles with the same density distribution.[24]

The fact that the effective potential is a simple local function makes a tremendous difference in practical calculations. Other quantum-chemistry schemes such as the Hartree-Fock method commonly involves nonlocal operators which require much more computational resources. It is not surprising that the density-functional theory has become the prevailing approach in modern electronic-structure calculations with wide applications in quantum chemistry and materials physics. In practical calculations, approximations to the energy functional are required. Commonly used ones include the local-density approximation (LDA)[23], in which the density is assumed to be locally uniform and the result for a homogeneous electron gas is used point by point based on the local density, and the generalized gradient approximation

(GGA), in which the gradient correction to the LDA is added. These are also the approximations we will use in our studies.

In order to study systems of hundreds of atoms, we will focus on the properties of the valence electrons and employ norm-conserving pseudopotentials to model the effects of core electrons. The one-particle orbital will be expanded in terms of plane waves to eliminate any bias in the basis functions. Evaluating relevant quantities in either the real space or momentum space will carry out the self-consistent solution of the corresponding Kohn-Sham orbitals.

Molecular dynamics with ab initio forces

In first-principles (*ab initio*) molecular-dynamics simulations (FPMD), the forces underlying the atomic motion (Helmann-Feynman forces) originate from the electronic distributions that are evaluated in the course of the simulations, concurrent with the instantaneous spatial configuration of the atoms. This is the source of the name *ab initio*, namely, free of *a priori* assumptions. The FPMD simulation method that have gained a primary position in theoretical investigation if materials are those based on the Kohn-Sham density functional with the LDA[20,23], LSDA, or GGA approximations for exchange and correlation functional. Since the forces are calculated completely based on density-functional theory, they provide a more accurate description of interatomic interactions than simpler methods such as model interatomic potentials, We plan to use FPMD in a large part of our planned investigations, employing mostly the Bohn Oppenheimer method, formulated and implemented for simulations of dynamics at finite temperature, in addition to electronic structure and geometrical optimization. It is worth noting that our FPMD codes have been rewritten and optimized by us for massive parallel computations, allowing treatment of rather complex and large systems.

Force-field-based molecular-dynamics simulations

Proteins undergo large conformational changes in connection with their function. The conformational changes are associated with ligand binding thereby allowing proper positioning of the substrate for catalysis, or allowing function within a receptor-signaling pathway. Molecular dynamics simulations are very useful in characterizing the energetics and dynamics of the relationships of the various conformations.[18] The time scales of these conformational changes are typically too long for conventional molecular dynamics simulations. This presents a challenge both to computational resources and theoretical methods. It is important to study energetic relationships between two known conformations and the prediction of how the ligand-enzyme structure changes as a result of mutations. We have been employing the force-field-based molecular dynamics package, Nanoscale Molecular Dynamics (NAMD), to study a selected number of biomolecule systems. NAMD[25] is a parallel, object-oriented molecular dynamics code designed for high-performance simulation of large biomolecular systems. Based on CHARMM++ parallel objects, NAMD scales to hundreds of processors on high-end parallel platforms.

Single-walled carbon nanotubes

SWNTs are rolled graphene sheets along a certain chiral vector. The structure of a SWNT is uniquely characterized by (n, m) chiral indices. Among chiral vectors, there are two distinctive high-symmetry directions, corresponding to $(n, 0)$ "zig-zag" and (n, n) "armchair." From the rolling graphene model, the chiral angle θ and diameter d_0 of a (n, m) SWNT ($0 \le m \le n$) are[19-21]

$$d = \frac{d_0}{\pi}\sqrt{n^2 + m^2 + nm} \qquad (1)$$

$$\theta = tan^{-1}\frac{\sqrt{3}m}{2n+m} \qquad (2)$$

respectively, where d_0 is the bond length of graphene. A SWNT is considered to be metallic if $n - m$ is divisible by 3, and semiconducting otherwise. However, the electronic structure of metallic SWNTs is very sensitive to radial deformations because of the presence of degenerate low energy electronic states in these systems. The effect of breaking symmetry due to non-covalent functionalization remains a paucity of classification from first principles calculations.

The SWNTs involved in the present study were constructed based on a sp^2 hybridization model. The initial value of $b = 1.42$ Å was used. The geometric structures of the SWNTs were fully relaxed in the molecular dynamics through intensive simulated annealing. A systematic evaluation of the available empirical force fields showed that the MM+ and CHARMM force fields provide consistent results for both the carbon nanotubes and epoxy resins.

Epoxy resins

The molecular and chemical structures of the EPON resin 862 are shown in Fig. 1, along with two prototype curing agents, diethyltoluenediamine (DETDA) and diethyltoluenamine (DETA). The consensus is that DETDA represents characteristic features of the curing agent W.

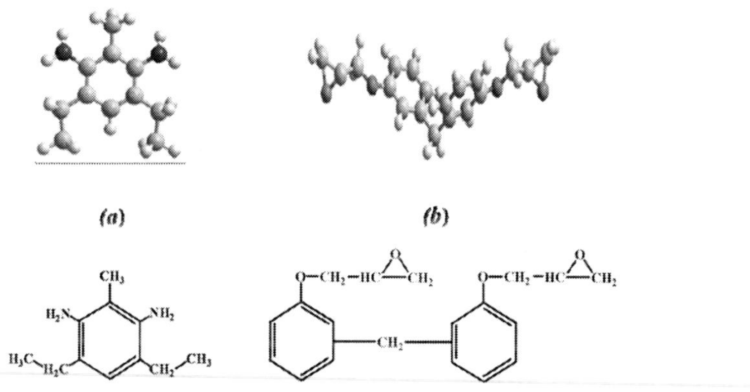

Figure 1. Molecular and chemical structures of (a) diethyltoluenediamine (DETDA), (b) EPON resin 862, respectively. Colored with blue, red, cyan, and white are nitrogen, oxygen, carbon, and hydrogen atoms respectively.

The terminal groups of the EPON resin 862 can open up and connect to curing agents. The formation of a chain-like or crosslinked structure15 is reminiscent of the cured state of the nano-reinforced composites. In order to explore the nature at the molecular level of nanotube composites, we have performed intensive geometry optimization studies of conformations via force-field-based molecular dynamics. Whereas our molecular dynamics simulation confirms that the EPON resin 862 self-assembles onto the surface of the SWNTs, the cured state of the EPON resin 862 adopts energetically favorable configurations by wrapping around the tube helically. Illustrated in Figure 2 is the optimized structure of a SWNT wrapped with chain-like cured EPON 862. A careful examination of the dynamical structural and energetic changes reveals that electrostatic interactions within the EPON backbone are primarily responsible for the wrapping into a helical structure. The aromatic rings in the EPON resins 862 are planar to the surface of the SWNT, and the structural changes proceed through a rearrangement of the torsional angles.

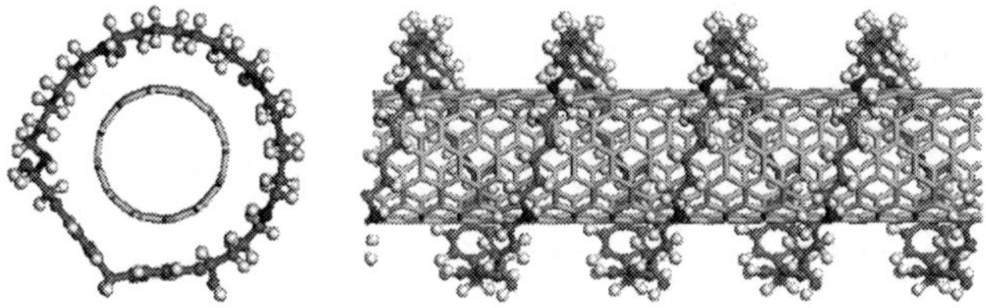

Figure 2. Top and side views of armchair (6, 6) helically wrapped with EPON resin 862. The carbon, oxygen, nitrogen, and hydrogen atoms are colored with grey, red, blue, and white, respectively.

RESULTS AND DISCUSSION

The binding of the epoxy resin to the SWNT is a combination of electrostatic and van der Waals (vdW) interactions. As a consequence, the electrostatic interactions are stronger than vdW binding for non-conjugated molecules, which may explain why the nanotubes are generally dispersed better with conjugated molecules. In general, the adsorption energy can be estimated from the difference between the potential energy of the composite system and the potential energies of the epoxy resin and corresponding SWNTs as follows:

$$E = E_{\mathrm{SWNT}} + E_{\mathrm{Epon}} - E_{\mathrm{total}}, \tag{3}$$

where E_{total} is the total potential energy of the nanohybrid, E_{SWNT} is the energy of the nanotube without the porphyrin and E_{Ppon} is the energy of the epoxy resin without the nanotube.

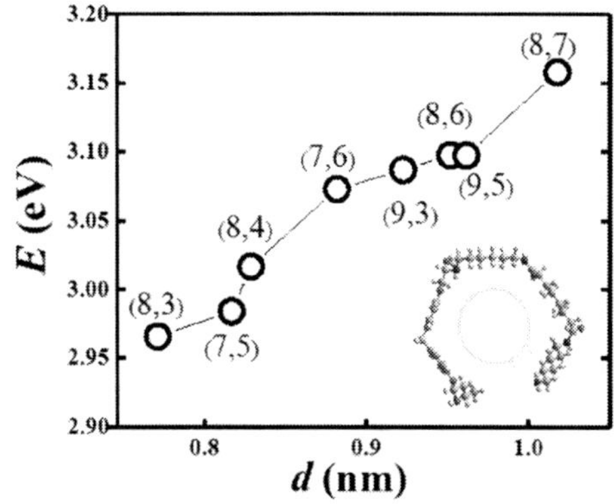

Figure 3. Calculated binding energy of nano-reinforced epoxy resin of carious diameters on the basis of fore-field-based molecular dynamics.

We illustrate in Figure 3 the calculated adsorption energies using force field-based MD. As seen from Figure 3, the adsorption energy increases with increasing diameter of the tube. This is attributed to an improved geometric match between the epoxy resin conformation and the carbon nanotube, as well as a reduction in π-orbital misalignment.

Another important aspect of the nanotube reinforced epoxy resin is the charge transfer behavior. It is well known that classical molecular dynamics cannot properly describe charge transfers. To this end, we plan to carry out intensive first-principles calculations based on density functional theory. For large systems, a semi-empirical density-functional molecular dynamics will be employed, which is capable of handling a few thousands of atoms. The proposed project will provide insight to the charge transfer behavior of carbon nanotube and graphene-reinforced epoxy resins. In lieu of the rapid progress in this field, the combination of computational simulation and experimental characterization will be very fruitful in the future development of nano-reinforced composite materials.

In addition to the helical wrapping conformations, the cured epoxy network can be formed through cross-linking of the EPON 862 resin with curing agents. During the curing reaction, the amine groups of curing agents react with the epoxide groups of epoxy resins. While for the chain-like conformation the curing reaction is limited at the $-NH_2$ site of the curing agents, the cross-linking activity can happen at the NH site as well. As a result, the cross-link expands in all directions and forms a network of macromolecules. It is worth noting that DETA-cured EPON resin 862 can assume either chain-like or cross-linked conformations. Although the helically wrapped chain configuration is energetically preferred, the availability of additional reaction sites at NH for the cross-linked structure may lead to increased density for the nano-reinforced composite. As the curing reaction is a dynamic process, the interplay between the energy and molecular density can be studied on the basis of quantum calculations.

Figure 4. Molecular dynamics simulation of EPON 862 and curing agent W chain interacts noncovalently with a graphene nanoplatelet. Electronic structure characteristic of nanohybrid is investigated.

CONCLUSIONS

In summary, we have shown in this paper that multiscale modeling is complementing experimental studies in providing useful information of nano-reinforcement of composites. Our results show that the stacking between the aromatic macrocycle and the surface of the SWNTs manifests itself via increased interfacial binding. First-principles calculations on the electronic structures further reveal that there exists distinct level hybridization behavior for metallic and semiconducting nanotubes.[22] In addition, there is a monotonic increase in binding energy with an increase in the nanotube diameter. Our simulation studies suggest that graphene nanoplatelets are potentially the best fillers of epoxy matrices.

ACKNOWLEDGMENTS

The authors thank E. Mintz, I. M. Khan, and M. W. Shute for stimulating discussions. This work was supported in part by the National Science Foundation (Grant Nos. DMR-0934142 and HRD-0630456), NASA (Grant No. NCC3-1044), and Air Force Minority Leaders Program (Grant No. CL-ATL-10-5567-0016-02-C2)

REFERENCES

1. E. Bekyarova, E. T. Thostenson, A. Yu, M. E. Itkis, D. Fakhrutdinov, T. W. Chou and R. C. Haddon, Functionalized single-walled carbon nanotubes for carbon fiber-epoxy composites, *Langmuir*, 2007, **23**, 3970.
2. S. J. V. Frankland, A. Caglar, D. W. Brener and M. Griebel, Molecular simulation of the influence of chemical crosslinks on the shear strength of carbon nanotube–polymer interfaces, *J. Phys. Chem. B*, 2002, **106**, 3046.
3. J. Gao, G. Hou, Y. Wang, H. Li and Y. Liu, Curing and thermal properties of PEPEB liquid crystalline diepoxide/aromatic diamine, *Polym.-Plast. Technol. Eng.*, 2006, **45**, 94.
4. E. W. Wong, P. E. Sheehan and C. M. Lieber, Nanobeam mechanics: elasticity, strength, and toughness of nanorods and nanotubes, *Science*, 1997, **277**, 1971.
5. P. M. Ajayan, L. S. Schadler, S. C. Giannaris and A. Rubio, Single-walled carbon nanotube-polymer composites: strength and weakness, *Adv. Mater.*, 2000, **12**, 750.
6. D. W. Steuerman, A. Star, R. Narizzano, H. Choi, R. S. Ries, C. Nicolini, J. F. Stoddart and J. R. Heath, Interactions between conjugated polymers and single-walled carbon nanotubes, *J. Phys. Chem. B*, 2002, **106**, 3124.
7. M. Kanungo, H. Lu, G. G. Malliaras and G. B. Blanchet, Suppression of metallic conductivity of single-walled carbon nanotubes by cycloaddition reactions, *Science*, 2009, **323**, 234.
8. D. A. Britz and A. N. Khlobystov, Noncovalent interactions of molecules with single walled carbon nanotubes, *Chem. Soc. Rev.*, 2006, **35**, 637.
9. M. Zheng, A. Jagota, E. D. Semke, B. A. Diner, R. S. Mclean, S. R. Lustig, R. E. Richardson and N. G. Tassi, DNA-assisted dispersion and separation of carbon nanotubes, *Nat. Mater.*, 2003, **2**, 338.
10. R. J. Chen, Y. Zhang, D. Wang and H. Dai, Noncovalent sidewall functionalization of single-walled carbon nanotubes for protein immobilization, *J. Am. Chem. Soc.*, 2001, **123**, 3838.
11. A. Nish, J.-Y. Hwang, J. Doig and R. J. Nicholas, Highly selective dispersion of single-walled carbon nanotubes, *Nat. Nanotechnol.*, 2007, **2**, 640.
12. O. O. Ogunro and X. Q. Wang, Quantum electronic stability in selective enrichment of carbon nanotubes, *Nano Lett.*, 2009, **9**, 1034.
13. M. M. J. Treacy, T. W. Ebbesen and J. M. Gibson, Exceptionally high Young's modulus observed for individual carbon nanotubes, *Nature*, 1996, **381**, 678.
14. J. Gou, B. Fan, G. Song and A. Khan, Study of affinities between single-walled nanotube and epoxy resin using molecular dynamics simulation, *Int. J. Nanosci.*, 2006, **5**, 131.
15. H. B. Fan and M. M. F. Yuen, Material properties of the cross-linked epoxy resin compound predicted by molecular dynamics simulation, *Polymer*, 2007, **48**, 2174.
16. J. Gou, B. Minaie, B. Wang, Z. Liang and C. Zhang, Computational and experimental study of interfacial bonding of single-walled nanotube reinforced composites, *Comp. Mater. Sci.*, 2004, **31**, 225.
17. R. R. Johnson, A. T. C. Johnson and M. L. Klein, Probing the structure of DNA-carbon nanotube hybrids with molecular dynamics, *Nano Lett.*, 2008, **8**, 69.
18. F. Jensen, *Introduction to Computational Chemistry*, 2nd ed, John Wiley & Sons, Hoboken, 2006.
19. M. Al-Haik, M. Y. Hussaini and H. Garmestani, Adhesion energy in carbon nanotube-polyethylene composite: effect of chirality, *J. Appl. Phys.*, 2005, **97**, 074306.
20. *DMol3*, Accelrys Software Inc., San Diego, 2009.

21. A. Nduwimana and X.-Q. Wang, Charge carrier separation in modulation doped coaxial semiconductor nanowires, *Nano Lett.*, 2009, **9**, 283.
22. Kelvin Suggs, X.-Q., Wang, Structural and Electronic Properties of Carbon Nanotube-Reinforced Epoxy Resins", *Nanoscale,* **2010**, *2, 385.*
23. P. Hohenberg and W. Kohn, "Inhomogeneous Electron Gas", Physical Review, 136,B864-871(1964); M. C. Payne, M. P. Teter, D. C. Allan, T. A. Arias, and T. D. Joannopoulos, "Iterative minimization techniques for *ab initio* total energy calculations: molecular dynamics and conjugate gradients", *Rev. Mod. Phys.* **64**, 1045 (1992)
24. W. Kohn and L. J. Sham, "Self-Consistent Equations Including Exchange and Correlation Effects", Physical Review, 140, A1133-1138 (1965); W. Kohn, "Density-functional and density-matrix method scaling linearly with the number of atoms", *Phys. Rev. Lett.* **76**, 3168 (1996).
25. James C. Phillips, Rosemary Braun, Wei Wang, James Gumbart, Emad Tajkhorshid, Elizabeth Villa, Christophe Chipot, Robert D. Skeel, Laxmikant Kale, and Klaus Schulten, "Scalable molecular dynamics with NAMD", *Journal of Computational Chemistry***,** 26:1781-1802, (2005).

Mater. Res. Soc. Symp. Proc. Vol. 1312 © 2011 Materials Research Society
DOI: 10.1557/opl.2011.109

Length scales of interactions in magnetic, dielectric, and mechanical nanocomposites

R. Skomski, B. Balamurugan, E. Schubert,* A. Enders, and D. J. Sellmyer

Department of Physics and Astronomy and Center for Materials Research and Analysis, University of Nebraska, Lincoln, NE 68588

**Department of Electrical Engineering, University of Nebraska, Lincoln, Nebraska*

ABSTRACT

It is investigated how figures of merits of nanocomposites are affected by structural and interaction length scales. Aside from macroscopic effects without characteristic lengths scales and atomic-scale quantum-mechanical interactions there are nanoscale interactions that reflect a competition between different energy contributions. We consider three systems, namely dielectric media, carbon-black reinforced rubbers and magnetic composites. In all cases, it is relatively easy to determine effective materials constants, which do not involve specific length scales. Nucleation and breakdown phenomena tend to occur on a nanoscale and yield a logarithmic dependence of figures of merit on the macroscopic system size. Essential system-specific differences arise because figures of merits are generally nonlinear energy integrals. Furthermore, different physical interactions yield different length scales. For example, the interaction in magnetic hard-soft composites reflects the competition between relativistic anisotropy and nonrelativistic exchange interactions, but such hierarchies of interactions are more difficult to establish in mechanical polymer composites and dielectrics.

Keywords: Maxwell-Garnett equation, Bruggeman composites; Dielectric Energy Density; Breakdown; Fracture; Polymers; Rubber; Nanocomposites; Coercivity; Energy Product

INTRODUCTION

Nanocomposites are widely used in technology, because they combine the advantages of single-phase materials and range from naturally occurring biological structures and traditional materials to artificial materials used in transport, space, microelectronic, and other high-tech applications. Examples are naturally occurring skeletal materials, such as bones and wood, whose nanostructure ensures stiffness without brittleness, and in artificial mechanical materials, such as concrete, fiber composites, reinforced polymers for car tires to [1, 2, 3]. In magnetism, aligned two-phase permanent magnets have been predicted to yield energy products beyond those of single-phase rare-earth permanent magnets [4]. Many multiferroic and multifunctional materials are also structured as two-phase composites. Some of these structures are straighforward mixtures, but many exhibit macroscopic or nanoscale interactions between the phases.

Examples are magnet-polymer composites, which have been investigated in the light of future applications such as materials with negative index of refraction [5, 6, 7].

A traditional approach to composite materials is the description in terms of effective materials constants. Some mixtures of materials obey mixing rules of the type

$$A_{\mathrm{eff}} = (1 - f)\,A_{\mathrm{m}} + f A_{\mathrm{i}} \tag{1}$$

where A_i and A_m are the materials constants of the matrix and inclusion phases, respectively, and f is the volume fraction of the inclusion phase. However, Eq. (1) is limited to noninteracting mixtures and to interacting phases where $A_i \approx A_m$. Consider, for example, the resistivity of insulating oxide particles ($\rho_i = \infty$) in a metallic matrix of finite resistivity ρ_m. For arbitrarily small but nonzero volume fractions f, Eq. (1) then leads to the unphysical prediction of an infinite resistivity $\rho_{\mathrm{eff}} = \infty$ ($A_{\mathrm{eff}} = \infty$). Similar violations of the mixing rule exist in other systems. For instance, dielectric and magnetic susceptibilities rarely obey Eq. (1).

For small volume fractions, the unphysical divergences are removed by using the generalized Maxwell-Garnett equation

$$\mathrm{A}_{\mathrm{eff}} = \mathrm{A}_{\mathrm{m}} \left(1 + f \frac{A_{\mathrm{i}} - A_{\mathrm{m}}}{g\,A_{\mathrm{i}} + (1 - \mathrm{g})\,A_{\mathrm{m}}}\right) \tag{2}$$

This equation contains a shape parameter g, which can derived for specific systems by explicit single-inclusion calculations. Examples are dielectrics [8], magnets [9, 10], conductors [10], and mechanical composites [2, 11, 12]. In many but not all cases (see below), the shape factor is equal to the inclusion's depolarizing factor, $g = N$. In particular, $N = 1/3$ for spheres, $N = 0$ for strongly elongated ellipsoids (rods or needles) and $N = 1$ for oblate or flat ellipsoids (platelets). Figure 1 shows the geometry of typical composites. Most of these structures can be prepared by traditional methods, but some need new approaches, such as glancing-angle deposition (GLAD) onto rotating substrates in the case of magnetic nanospirals [13]. The extension to arbitrary volume fractions is generally nontrivial, but there exists a powerful effective-medium approach known as the *Bruggeman theory* [1], which covers arbitrary volume fractions and percolation phenomena on a mean-field level (see next section).

The question arises how nanoscale interactions affect the performance of composite materials and whether the underlying physics exhibits universal tendencies across systems and phenomena. This question has three important ramifications. First, figures of merit such as the electrostatic energy density stored in dielectrics, the energy product of permanent magnets and the energy dissipated in viscoelastic composites, are integrals over nonlinear functions, and different figures of merit may be differently affected by nanoscale inhomogeneities. Second, effective materials constants are only one aspect of the performance of composites. Phenomena such as dielectric breakdown, coercivity, and mechanical fracture usually involve nanoscale materials imperfections and are poorly described by effective medium theories [4, 9, 14]. Third, nanoscale interactions directly modify the materials constants when the structural features size becomes comparable to the nanoscale interaction lengths. The purpose of this paper is to analyze how these mechanisms operate in nanostructures, comparing various systems and elaborating similarities and differences.

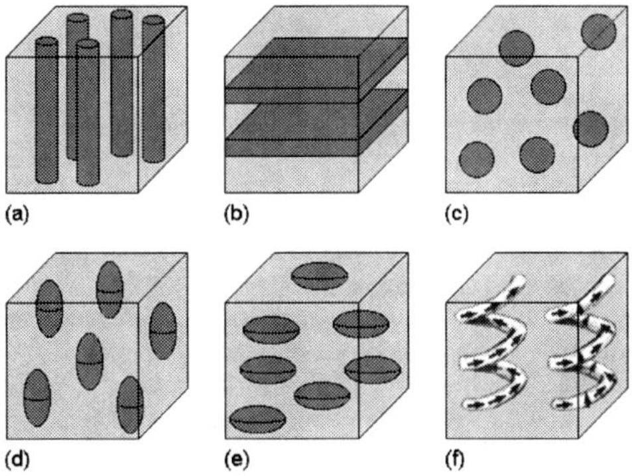

Fig. 1. Geometry and depolarization factors for some two-phase nanostructures: (a) rods ($N = 0$), (b) platelets ($N = 1$), (c) spheres ($N = 1/3$), (d) prolate ellipsoids ($N < 1/3$), (e) oblate ellipsoids ($N > 1/3$), and (f) nanospirals.

SCIENTIFIC BACKGROUND: BRUGGEMAN MODEL

The Bruggeman approach is a length-scale-free and selfconsistent mean-field method to calculate materials properties of composites [1, 15]. The idea is to start from the Maxwell-Garnett equation (1) and to treat arbitrary volume fractions f by selfconsistently embedding the matrix and inclusion phases in an effective medium. It is also known as the Polder–van Santen formula, self-consistent field theory (SCF) or simply effective medium theory [16, 17]. In solid state physics, the Bruggeman theory is closely related to the coherent-potential approximation (CPA) for binary alloys [18, 19, 20].

The theory was originally derived for dielectric materials but can be used for a broad variety of static and dynamic phenomena. Examples are effective dielectric susceptibilities (static and dynamic dielectric response of inhomogeneous media, materials with negative index of refraction) [10, 21], effective magnetic permeability [4, 10, 22, 23], optical properties [8, 21], mechanical composites (elasticity of reinforced construction materials and filled polymers, such as car tires) [12, 20, 23, 24], geoscience [17], rheology (viscosity and viscoelasticity of colloidal suspensions, such as blood, food, gels) [25], electrical conduction (insulating inclusions in metals and metal-superconductor composites) [10, 26, 27], thermal conduction (heat insulation using composite construction materials) [10], and diffusivity (hydrogen transport) [10]. In general, both isotropic and anisotropic composites can be treated [26, 23, 28, 29].

The starting point is a generalized potential $\Phi(\mathbf{r})$ whose negative gradient is the (generalized) field, $\mathbf{f} = - \nabla\Phi$. For example, the electric field in a static dielectric can be written as $\mathbf{E} = - \nabla\Phi$. The field creates a generalized flux density $\mathbf{b}$, and this flux density is described by the materials equation $\mathbf{b} = A\, \mathbf{f}$, where A is a generalized materials constant (compliance). In the dielectric analogy, this is the electric permeability. The flux is usually fed into the system by applying an external force or field but has no sources inside the material, so that $\nabla \mathbf{b} = 0$. This yields the flux-conservation relation

$$\nabla \cdot (A(\mathbf{r}) \nabla \Phi) = 0 \qquad (3)$$

In any single-phase region of the composite, this reduces to $\nabla^2 \Phi = 0$, accompanied by appropriate boundary conditions. This equation does not contain any characteristic length, so that is can be used for macroscopic, micron-size and nanoscale inclusions. The respective parallel and perpendicular components of the force **f** and of the flux density **b**, are continuous at interfaces, and far away from the inclusion, the field $\mathbf{f} = -\nabla \Phi(\mathbf{r})$ is homogeneous. The solution of Eq. (3) yields the Maxwell-Garnett equation [2, 10, 12, 15, 23, 24].

In contrast to Eq. (2), which is limited to small volume fractions f, the Bruggeman approach describes arbitrary f on a mean-field level. This is achieved by selfconsistently embedding both phases in an effective medium averaging over the total volume. The procedure yields

$$f \frac{A_i - A_{eff}}{g A_i + (1 - g) A_{eff}} + (1 - f) \frac{A_m - A_{eff}}{g A_m + (1 - g) A_{eff}} = 0 \qquad (4)$$

which is basically a quadratic equation for A_{eff}. Aside from the input parameters A_m, A_i, and f, Eq. (3) contains a single parameter, namely the shape constant g. This parameter depends on the considered physical property and on the geometry of the composites. Many electromagnetic and transport systems belong to the first group, where $g = N$. In particular, $g = 0$ for long needles, $g = 1/3$ for spherical inclusions, and $g = 1$ for multilayers of embedded plates [23, 30]. In mechanical composites, $g \neq N$ depends on Poisson's ratio and on the considered elastic modulus, because Eq. (3) must be modified [2, 23].

Equation (4) yields a *percolation* transition with a dimensionality-independent mean-field exponent $\gamma = 1$ [19]. For example, metallic regions in an insulator do not affect the resistivity until the metallic volume fraction reaches the percolation threshold $f_c = 1/3$. By putting $A_i = 0$ and investigating the transition $A_{eff} \to 0$ we obtain a threshold $f_c = 1 - g$, whereas $A_i = \infty$ and $A_{eff} \to \infty$ yields $f = g$. The metal-insulator percolation threshold $f_c = g$ is obtained by considering the effect of insulating inclusions ($\sigma = \infty$) in a metallic matrix ($\sigma = \sigma_m$), which yield volume fractions of $1 - g = 2/3$ for the insulating phase and $g = 1/3$ for the metallic phase. The conductivity of superconducting spheres in a normal conductor is of the type $A_i = \infty$ and yields percolation at $f = 1/3$. More generally, the switching between inverse quantities, such as resistivity ρ and conductivity $\sigma = 1/\rho$, is realized by the transformation $g \to 1 - g$ in Eqs. (2) and (4). The mean-field character of Bruggeman percolation means that fluctuations are treated in a very crude way, on the level of Ornstein-Zernike correlation functions with the critical exponent $\nu = 1/2$ [23, 31]. This leads to inaccuracies in the predictions for A_{eff} near the percolation threshold.

The applicability of the Bruggeman theory to stationary transport processes, such as electrical conduction, thermal conduction, and diffusion, may be shown in two alternative ways. The first derivation exploits that conductivity and diffusion (A) involve generalized forces $\nabla \Phi$ that have the character of concentration, potential, or temperature gradients. The flux density **j** is conserved, $\nabla \cdot \mathbf{j} = 0$, and obeys Fick's law, for example $\mathbf{j} = - D \nabla c$ for diffusion. This yields Eq. (3) in the form $\nabla\cdot(D(\mathbf{r})\nabla c) = 0$ and leads to $g = N$. A second derivation is based on Maxwell's equation $\nabla\times\mathbf{H} = \mathbf{j} + \partial\mathbf{D}/\partial t$, which shows that the time derivative of the displacement is a current density, in analogy to the current necessary to charge a capacitor. Application of $\mathbf{E}(t) = \mathbf{E}_o \exp(i\omega t)$ in combination with $\mathbf{D} = \varepsilon\,\mathbf{E}$ and $\mathbf{j} = \sigma\,\mathbf{E}$ yields the complex permittivity $\varepsilon^* = \varepsilon - i\sigma/\omega$. This makes it possible to treat static and stationary phenomena on a common footing, by considering complex quantities such as $\varepsilon'' = \varepsilon' + i\varepsilon''$.

A somewhat less well-known approach is the use of the Bruggeman theory for *mechanical* composites. There is a rich literature on small volume fractions [2, 32, 33], and true Bruggeman equations have later been obtained for various elastic, viscous, and viscoelastic materials [20]. It is instructive to extract generalized shape constants g for mechanical composites. For *spherical* inclusions, the shear modulus G and the bulk modulus K are described by

$$g = \frac{2\,(4 - 5\,\nu_0)}{15\,(1 - \nu_0)} \qquad (5a) \qquad \text{and} \qquad g = \frac{1 + \nu_o}{3\,(1 - \nu_o)} \qquad (5b)$$

respectively. In both equations, is Poisson's ratio of the matrix. Taking typical solids with $\nu_o \approx 0.35$ yields $g = 0.46$ for G and $g = 0.69$ for K. Incompressible materials (rubbers) have $\nu_o = 1/2$, that is, $g = 2/5$ for G and $g = 1$ for K.

The shear stress in a viscoelastic material contains an elastic component $\sigma = G\,\varepsilon$ and a viscous component $\sigma = \eta\,d\varepsilon/dt$, and the complex shear modulus $G^* = G + i\omega\eta$ [2, 27]. When the material as incompressible, then both the viscosity and shear modulus exhibit $g = 2/5$. The corresponding Maxwell-Garnett relation, $\eta = \eta_o(1 + 2.5 f)$, was first derived by Einstein [2, 32, 33], but $g = 2/5$ is easily derived by putting $\nu_o \approx 1/2$ in Eq. 5(a). Incompressibility is realized in ordinary liquids, which are largely imcompressible by nature, and in elastic (and viscoelastic) rubbers, where $K \gg G$ and $\nu_o \approx 1/2$ [11]. In a very good approximation, this limit is realized for carbon-reinforced rubber composites, as used in car tires [11, 34]. Since $g = 2/5$, percolation is predicted when the volume fraction of rigid spheres reaches 40%. This is close to the experimental value of about 49% [2].

NONLINEAR AND LOCAL ENERGY AND ENTROPY EFFECTS

A counterintuitive feature of many composites is that an improvement of materials constants may actually lead to a *deterioration* of figures of merit. The first example comes from permanent magnetism, where a increased fraction of rare-earth atoms may be used to improve the coercivity (magnetic hardness) H_c of the material but deteriorates the energy product by reducing the magnetization. In more detail, H_c scales as $2K_1/M_s$ and can be made arbitrarily large by choosing materials with a small M_s. However, the energy product of a permanent magnet does not exceed $M_s^2/4$, so that this coercivity increase is not an option. The second example is metallic inclusions in dielectrics, which enhance the permittivity but tend to reduce the breakdown field and the electrostatic-energy density in a capacitor made from the composite material [17, 21, 35, 36, 37]. Elastic moduli are important for construction materials, but very often the main considerations are mechanical toughness and fraction behavior [38].

This deterioration of figures of merit of composites violates effective-medium theory, which predicts materials constant A_{eff} intermediate between A_m and A_i. There are two reasons for this effect, namely nonlinear and local mechanisms. First, the underlying materials equations are nonlinear, characterized by field-dependent materials constants. Even in materials with linear equations of state, figures of merits often have the character of energy densities rather materials constants, which introduces nonlinearity. Compared to the determination of effective materials constants, whose physics largely reduces to the determination of a single parameter (g), the nonlinearity yields an essential differentiation between different systems, not only between different classes of materials but also between different figures of merit for a single class of materials. The reason is that there are usually many ways of constructing energy functions from materials constants. For example, losses in soft-magnetic materials are described by the area of the *M-H* hysteresis loop, whereas the energy product of permanent magnets is equal to the largest rectangle fitting under the *B-H* loop in the 2nd quadrant of the magnetization curve, where $B = M + H$ [22]. In dielectric materials, the electrostatic energy $W = \int \eta \, dV$, where η is the electrostatic energy density. In the linear regime, $\eta = \int \mathbf{E} \, d\mathbf{D}$ yields $\eta = \frac{1}{2} \varepsilon_o \varepsilon_r E_{max}^2$, but D is in general a nonlinear function of the field, which affects η [26, 33, 39]. In practice, the saturation polarization puts an upper limit to η, and E is limited by the breakdown field. For example, an external electric field may rotate or align nearly free dipoles, but the energy stored by this mechanism is very low.

Second, the Bruggeman theory is a mean-field approach, considering inclusions in an average environment and unable to account for long-range *fluctuations*. For example, electric breakdown may be realized via a percolating backbone, where the creation of a single percolation bond initiates the collapse of the electric field. The Bruggeman theory is poorly equipped to handle this scenario and yields, for example, incorrect critical exponents. Fluctuations often determine the main figures of merit of materials. For example, mechanical fracture tends to start at cracks [14], the dielectric storage capacity is limited by the maximum local electric field [35] (as contrasted to the average field), and the coercivity of permanent magnets is determined by the local rather than average anisotropy [9, 22]. In fact, the dielectric contrast caused by the embedding of a single high-ε particle in a low-ε matrix is known to shift field strength and energy density from the particles into the matrix and to substantially reduce the breakdown field. The effect depends on geometrical factors such as surface curvature of the particle surface [35].

Materials-specific breakdown, fracture, and coercivity mechanisms often yield a *logarithmic* dependence of materials constants and figures of merit on the system size L. Figures of merit describing the failure of dielectric and mechanical materials are often a power of $1/\ln(L)$, and a similar size dependence is encountered in permanent magnets [14, 22]. In the former case, the cracks responsible for mechanical failure may be modeled as ellipsoidal inclusions, which assimilates the mechanical problem to the dielectric problem [14]. The latter case may be understood by considering defect-containing models hard-magnetic particles in bonded or sintered magnets [22].

NANOSCALE EFFECTS

Breakdown phenomena are often realized on a nanoscale or even atomic scale [14], as epitomized by the mean free path of the electrons that realize the breakdown (scattering or capturing of electrons). A different question is the effective *interaction length* $1/\kappa$ in dielectrics. Aside from affecting the break-down field, interactions may also yield a direct modification of materials constants if the inclusion size become comparable to the interaction range. The Bruggeman theory is unable to distinguish between macroscopic and nanoscale effects, because the partial differential equation $\nabla^2\Phi = 0$ does not involve a characteristic length. Nanoscale effects would require

$$\nabla^2\Phi - \kappa\,\Phi = 0 \tag{6}$$

where $1/\kappa$ can be interpreted as a screening length [4, 25, 40]. This screening length reflect the competition between different local and non-local energy contributions and is sometimes of the order of a few nanometers. It has far-reaching consequences for the understanding of advanced mechanical and electromagnetic composites, because the inclusion size tends to interfere with the length scale of the interaction.

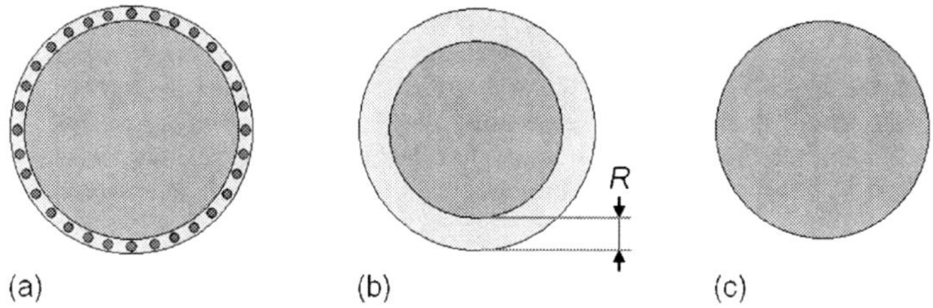

Fig. 2. Range of interaction effects: (a) atomic, (b) nanoscale, and (c) macroscopic.

Magnetic Composites

The magnetic permeability of some soft-magnetic composites is described by Eq. (4), but in general the finite saturation magnetization puts a limit to the applicability of the Bruggeman theory. Hard-soft nanocomposites magnets for permanent-magnet applications, which have been investigated quite intensively [4, 9, 22, 41], exhibit nontrivial nanoscale interactions. In these materials, the energy product is determined by the coercivity, which is, in a crude nucleation-field approximation [4, 9], obtained as the lowest-lying H eigenvalue of

$$-A\,\nabla^2 m + (K_1(\mathbf{r}) - \tfrac{1}{2}\mu_o M_s H)\,m = 0 \tag{7}$$

Here A is the exchange stiffness and K_1 is the first uniaxial anisotropy constant. Equation (7) has the structure of Eq. (6), and both coercivity and energy product depend on the size of the hard and soft regions, as described by the local anisotropy $K_1(\mathbf{r})$. The corresponding interaction length is

$$\frac{1}{\kappa} = \sqrt{\frac{A}{K_1(\mathbf{r}) - \tfrac{1}{2}\mu_o M_s H}} \tag{8}$$

This means that g is a function of the radius of the hard inclusions rather than a constant. Physically, the hard-magnetic spheres create an 'interaction cloud' of a exchange-coupled soft-magnetic material.

The length $1/\kappa$ is essentially equal to the domain-wall width of the hard phase. In practice, it reaches several nanometers, in spite of the atomic origin of the two parameters A and $K_1(\mathbf{r})$. The reason is the relativistic smallness of K_1 [9, 42], meaning that $1/\kappa$ describes a competition between Heisenberg exchange, which is characterized by Curie temperatures of several 100 K, and magnetocrystalline anisotropy, which corresponds to energy equivalents of 1 K or less [23]. Consequently, magnetic domain walls are quite broad, from about 20 interatomic distances (5 nm) for very hard materials to more than 1000 interatomic distances in soft magnets. This length scale also determines the spin structure in nanospirals, Fig. 1(f). Note that the ferromagnetic (and ferroelectric) domain-wall width must not be confused with the actual domain size, which may be much larger.

The nanoscale range of the coupling is limited to the coercivity. Equations such as (7) cannot be used, for example, to determine effective Curie temperatures T_c. As analyzed elsewhere [23, 43], Curie temperatures differing by more than a few K mean that the phase with the lower T_c slightly polarizes the phase with the higher T_c, but the corresponding decay length is only a few interatomic distances. In a strict sense, the residual coupling is a thermodynamically equivalent to a ferromagnetic phase transition with a relatively high T_c, but for any practical purposes the Curie-temperature behavior is two-phase like.

Mechanical Composites

There is a wide range of mechanical composites, but our focus is on a specific case, namely reinforced rubbers, as used for car tires. The reinforcement is caused by small silica or carbon-black particles, and it is well-known that the use of nanoparticles drastically improves mechanical properties such as modulus and toughness. For 'ideal' or 'phantom' polymer networks, the shear modulus Is $G_o \approx nk_BT$, where n is the cross-link density, and Young's modulus is three times as large. For large particles, one can use the Einstein formula, $G = G_o (1 + 2.5 f)$, which corresponds to Eq. (5) with $g = 2/5$.

High-performance fillers have particle radii of about 10 nm, which is comparable to the average end-to-end distance $R_N \sim N^{1/2}$ of the polymer chains [11, 34]. In other words, the average cross-link distance, which determines n and G, becomes comparable to the radius of the carbon or silica nanoparticles and leads to a complicated chain behavior in filled rubber [44, 45]. Figure 3 illustrates some of these chain configurations. Since the surface of the particle creates additional crosslinks, there is an additional 'Bueche' contribution proportional to the number of chains that touch a particle, which enhances G [34] and means that g becomes a function of R_N/R. However, $G_o \approx nk_BT$ derives from the entropy reduction of the chains as the end-to-end distance increases under strain. Some chains, such as C in Fig. 3(a), do not exhibit a change in the end-to-end distance, so that their zeroth-order contribution to G_o vanishes. This effect is also governed by the ratio R_N/R but qualitatively different from an enhanced cross-link density.

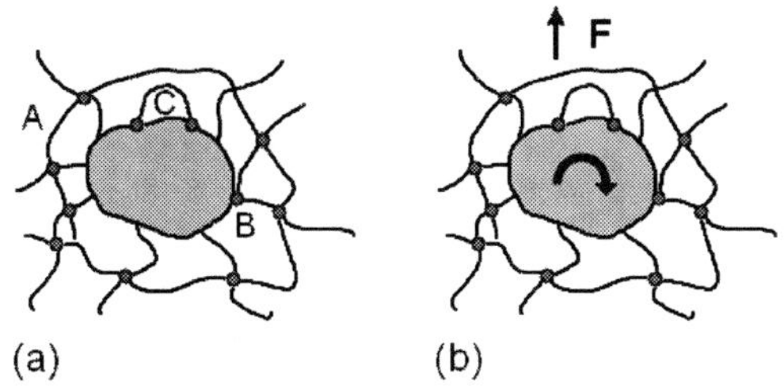

Fig. 3. Rubber reinforcement due to nanoparticles: (a) types of chains (A: phantom chain, B: extra crosslink, C: dead chain) and (b) behavior of the chain under strain. The toughness enhancement probably reflects the release of excessive strain due to particle motion and rotation (b).

Dielectric Composites

A key materials constant for dielectrics is the permittivity ε, as it appears in the electric energy density $w = \frac{1}{2}\varepsilon E^2$. Since composites have materials constants intermediate between the materials constants of the individual phases, it is tempting to use composites where ε is enhanced by embedding inclusions that have a high ε, such as TiO_2 [25, 46]. In fact, metallic [35, 36] particles have an infinite permittivity und are ideally suited for ε-enhancement. The dielectric constant can be enhanced by embedding highly permittive particles in a matrix or using structures such as metallic grains separated by non-metallic grain boundaries [37]. In former years, it was difficult to create nanoparticles of very small sizes, which has lead to problems such as eddy currents in metallic inclusions, but a recent development is that nanoparticles can now be produced on length scales of 5 nm or less [47], as compared to the submicron particles used in earlier research. However, as discussed above, the positive effect on ε is typically overcompensated by the reduction in break-down field.

An interesting question is how nanoscale phenomena affect dielectric composites and whether nanostructuring can be used improve the dielectric energy density. Gradient expressions similar to the exchange-stiffness term of Eq. (7) can be added to a Landau-type potential energies for dielectric polymers, but the question is the range of the exchange interaction. An exchange length of a few nanometers has recently been used to discuss core-shell particles in a polymeric matrix [25, 48], but the interaction range $1/\kappa$ is treated as an adjustable phenomenological parameter. Mindlin [49] considered the polarization gradient in the stored energy function of elastic dielectrics, but the interaction range is usually rather small, only a few Å in typical dielectric materials. In fact, Askar *et al.* [50] used a long-wavelength approximation to calculate an interaction length of about $1/\kappa \sim 0.2$ nm for some ionic compounds, without indication that other materials might exhibit robust interactions on a much larger length scale.

In magnetism, the exchange length reflects the relativistic competition between anisotropy energy E_a per atom and the Heisenberg exchange J, and this leads to a physically well-based hierarchy in terms of Sommerfeld's fine structure constant, $E_a/J \sim (1/137)^2$ [42]. In dielectric materials, there is no comparable hierarchy. The interface acts as a perturbation and locally changes the polarization of the adjacent phases, but the extra polarization costs energy and decreases quite rapidly with increasing distance from the interface. In fact, cohesive and dielectric energies are of comparable orders of magnitude, as epitomized by typical ferroelectric Curie and melting temperatures, respectively.

Figure 4 illustrates how nanoscale interactions may be realized in dielectrics. The model assumes dipoles separated by a segment length a and having a polarization $P_i(x) = P_o \cos(\theta_i)$, subjected to an electric field that points in the z direction. The dipoles can freely rotate in the y-z plane, but there is a coupling $- K \cos(\theta_i - \theta_j)$ between neighboring dipoles. The free chain (a) gets easily polarized by the electric field (high ε but low energy density $w \sim 1/\varepsilon$), whereas the coupled chain (b) can store a relatively large energy associated with the torque interaction constant K. The permittivity and energy density depend on the interaction length $L \sim a \sqrt{K/P_o E}$ and on the particle spacing, and by adjusting the latter, one can optimize the performance of the structure.

In a few cases, electric interactions may be long range, with a $1/r^m$ power-law distance dependence of the interaction strength. One example is the switching of the magnetization in insulating multiferroic composites that contain dielectric and rare-earth magnet regions [23, 51]. Here the electric dipoles create a crystal-field contribution that scales as $1/r^4$ and interacts with the electric quadrupole moment of the tripositive rare-earth ions. Due to the rigid spin-orbit coupling of the rare earths, this can be used to switch the magnetization direction.

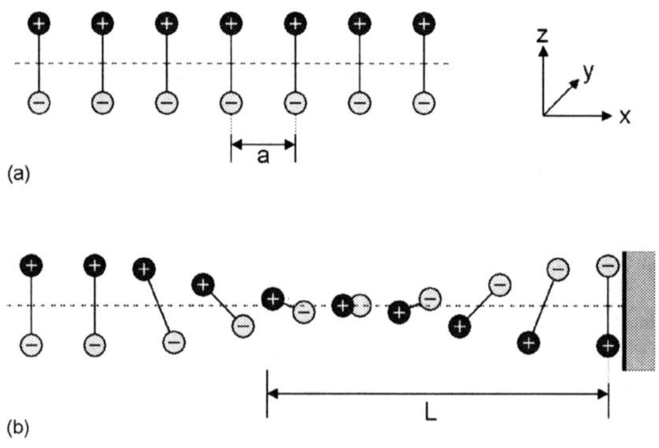

Fig. 4. Nanoscale interaction effects in a dielectric polymer model: (a) free chain and (b) chain adsorbed to a nanoparticle (right).

DISCUSSION AND CONCLUSIONS

In summary, we have investigated how different magnetic, mechanical and dielectric properties are affected by nanostructuring. Length-scale independent effective-field theories, whose physics is contained in a single shape parameter g, yield reasonable volume-averaged materials parameters and even provide a qualitative description of percolation phenomena. However, physical figures of merit tend to have the character of energy densities, which introduces a considerable complexity and diversity in the description of composite materials. Examples are the dependence of the dielectric energy on the permittivity of the oxide or metal inclusions and the energy-product enhancement in magnetic two-phase nanocomposites. A further complication is due to nanoscale phenomena, which affects coercivity, break-down, and fracture, and often lead to a logarithmic dependence of figures of merit on the macroscopic size of the considered body. Nanoscale interaction lengths are materials- and property-specific, as contrasted to the length-scale independent character of effective-field theories.

ACKNOWLEDGEMENT

This work is supported by NSF-MRSEC, ONR, and NCMN. The authors are grateful to S. Ducharme and K. Kraemer for discussing various aspects of dielectric composites.

REFERENCES

[1] D. A. G. Bruggeman, Ann. Phys. (5) **24** (1935) 637.
[2] R. M. Christensen, *Mechanics of Composite Materials*, Wiley, New York 1979.
[3] C. Verdier, J. Theor. Med. **5** (2003) 67.
[4] R. Skomski, J. Phys.: Condens. Matter **15** (2003) R841.
[5] A. Chipara, D. Hul, J. Sankar, D. Leslie-Pelecky, A. Bender, L. Yue, R. Skomski, and D. J. Sellmyer, Composites B: Engineering **35** (2004) 235.
[6] M. Chipara, R. Skomski, D. J. Sellmyer, Mater. Lett. **61** (2007) 2412.
[7] N. Ali, M. Chipara, S. Balascuta, R. Skomski, and D. J. Sellmyer, J. Nanosci. Nanotechnol. **9** (2009) 4437.
[8] J. C. M. Garnett, Philos. Trans. R. Soc. (London) A **203** (1904) 385.
[9] R. Skomski and J. M. D. Coey, Phys. Rev. B **48** (1993) 15812.
[10] Z. Hashin and S. Shtrikman, J. Appl. Phys. **33** (1962) 3125.
[11] B. Erman and J. E. Mark, *Structures and Properties of Rubberlike Networks*, Oxford University Press, Oxford 1997.
[12] J. M. Dewey, J. Appl. Phys. **18** (1947) 578.
[13] D. Schmidt, A. C. Kjerstad, T. Hofmann, R. Skomski, E. Schubert, and M. Schubert, J. Appl. Phys. **105** (2009) 113508.
[14] B. K. Chakrabarti and L. G. Benguigui, *Statistical Physics of Fracture and Breakdown in Disordered Systems*, University Press, Oxford 1997.
[15] R. Skomski, J.-Y. Li, J. Zhou, and D. J. Sellmyer, in: *Materials for Space Applications*, Eds. M. Chipara, D. L. Edwards, R. S. Benson, S. Phillips, Mater. Res. Soc. Symp. Proc. **851** (2005) NN1.7.
[16] D. Polder and J. H. van Santen, Physica **12** (1946) 257.
[17] K. K. Kärkkäinen, A. H. Sihvola, and K. I. Nikoskinen, IEEE Trans. Geosci. and Remote Sensing **38** (2000) 1303.
[18] B. Velický, S. Kirkpatrick, and H. Ehrenreich, Phys. Rev. **175** (1968) 747 .
[19] S. Kirkpatrick, Phys. Rev. Lett. **27** (1971) 1722.
[20] T. C. Choy, *Effective Medium Theory*, University Press, Oxford 1999.

[21] W. T. Doyle, J. Appl. Phys. **85** (1999) 2323.
[22] R. Skomski and J. M. D. Coey, *Permanent Magnetism*, Institute of Physics, Bristol 1999.
[23] R. Skomski, *Simple Models of Magnetism*, University Press, Oxford (2008).
[24] Z. Hashin, J. Appl. Mech. **29** (1962) 143.
[25] J. Y. Li, L. Zhang, and S. Ducharme, Appl. Phys. Lett. **90** (2007) 132901.
[26] O. Levy and D. Stroud, Phys. Rev. B **56** (1997) 8035.
[27] I. M. Ward and D. W. Hadley, *Mechanical Properties of Solid Polymers*, Wiley, New York 1993.
[28] P. Banerjee, I. Perez, L. Henn-Lecordier, S. B. Lee, and G. W. Rubloff, Nature Nanotechnology **4** (2009) 292.
[29] A. Lakhtakia, B. Michel, and W. S. Weiglhofer, J. Phys. D: Appl. Phys. **30** (1997) 230.
[30] J. A. Osborn, Phys. Rev. **67** (1945) 351.
[31] J. M. Yeomans, *Statistical Mechanics of Phase Transitions*, University Press, Oxford 1992.
[32] T. S. Chow, *Mesoscopic Physics of Complex Materials*, Springer, Berlin 2000.
[33] A. Einstein, Ann. Phys. **19**, 289 (1906); Erratum: **34** (1911) 591.
[34] R. Skomski, *Theory of Elasticity of Filled Polymer Networks*, THLM Leuna-Merseburg (Diplomarbeit, unpublished, 1986).
[35] P. M. Duxbury, P. D. Beale, H. Bak and P. A. Schroedert, J. Phys. D: **23** (1990) 1546.
[36] Zh.-M. Dang, Y.-H. Lin, and C.-W. Nan, Adv. Mater. **15** (2003) 1625.
[37] J.-H. Kim, Y.-W. Lee, M. G. Kim, A. Souchkov, J. S. Lee, H. D. Drew, S.-J. Oh, C. W. Nan, and E. J. Choi, Phys. Rev. B **70** (2004) 172106.
[38] G. Murphy, *Advanced Mechanics of Materials*, McGraw-Hill, New York 1946.
[39] H. E. Alper and R. M. Levy, J. Phys. Chem. **94** (1990) 8401.
[40] R. Skomski, J. Magn. Magn. Mater. **272-276** (2004) 1476.
[41] R. Coehoorn, D.B. de Mooij, and C. de Waard, J. Magn. Magn. Mater. **80** (1989) 101.
[42] R. Skomski, H.-P. Oepen, and J. Kirschner, Phys. Rev. B **58** (1998) 3223.
[43] R. Skomski and D. J. Sellmyer, J. Appl. Phys. **87** (2000) 4756.
[44] A. Botti, W. Pyckhout-Hintzena, D. Richter, V. Urban, and E. Straube J. Chem. Phys. **124**, 174908 (2006).
[45] E. Straube, V. Urban, W. Pyckhout-Hintzen, D. Richter, and C. J. Glinka, Phys. Rev. Lett. **74**, 4464 (1995).
[46] G. F. Dionne, J. F. Fitzgerald, and R. C. Aucoin, J. Appl. Phys. (1976) 1708.
[47] B. Balamurugan, K. L. Kraemer, N. A. Reding, R. Skomski, S. Ducharme, and D. J. Sellmyer, ACS Nano **4** (2010) 1893.
[48] J. Y. Li, Phys. Rev. Lett. **90** (2003) 217601.
[49] R. D. Mindlin, International Journal of Solids and Structures **4** (1968) 637.
[50] A. Askar, P. C. Lee, and A. S. Cakmak, Phys. Rev. B **1**-3537 (1970) 3525.
[51] R. Skomski, A. Kashyap, and A. Enders, J. Appl. Phys., in press (2011).

Mater. Res. Soc. Symp. Proc. Vol. 1312 © 2011 Materials Research Society
DOI: 10.1557/opl.2011.217

Study of ITO@PMMA Composites by Transmission Electron Microscopy

Elen P. S. Arlindo[1], Marcelo O. Orlandi[1]
[1] Department of Physical Chemistry, *São Paulo State University, CEP 14801-970, P.O. Box 355, Araraquara, SP, Brazil.*

ABSTRACT

ITO nanowires were synthesized by carbothermal reduction process, using a co-evaporation method, and have controlled size, shape, and chemical composition. The electrical measurements of nanowires showed they have a resistance of about 10^2 Ω. In order to produce nanocomposites films, nanowires were dispersed in toluene using an ultrasonic cleaner, so the PMMA polymer was added, and the system was kept under agitation up to obtain a clear suspension. The PMMA polymer was filled with 1, 2, 5 and 10 % in weight of nanowires, and the films were done by tape casting. The results showed that the electrical resistance of nanocomposites changed by over 7 orders of magnitude by increasing the amount of filler, and using 5 wt% of filler the composite resistance decreased from 10^{10} Ω to about 10^4 Ω, which means that percolation threshold of wires occurred at this concentration. This is an interesting result once for nanocomposites filled with ITO nanoparticles is necessary about 18% in weight to obtain percolation. The addition of filler up to 10 wt% decreased the resistance of the composite to 10^3 Ω, which is a value close to the resistance of wires. The composites were also analyzed by transmission electron microscopy (TEM), and the TEM results are in agreement with the electrical ones about percolation of nanowires. These results are promising once indicates that is possible to produce conductive and transparent in the visible range films by the addition of ITO nanowires in a polymeric matrix using a simple route.

INTRODUCTION

Nanocomposites are materials that have two or more solid phases, and at least one of them should present one dimension below 100 nm. They are known as multifunctional materials because they have, usually, more than one property increased for some special application [1-4]. Up to now, a great effort has been done by researches to achieve multifunctional nanocomposites of polymeric matrices filled with metallic, ceramic or inorganic phase, and in this work we studied the influence of conductive ITO (Indium Tin Oxide) nanowires on the electrical and the optical properties of Polymethyl methacrylate (PMMA) polymer [5-8]. PMMA is a polar polymer which is extensively commercially used because of its excellent transparence in the visible range of electromagnetic radiation, good mechanical properties and light weight. However, the extremely low conductivity of the PMMA can be a drawback depending of the application. By other hand ITO nanowires can have good conductivity and transparence in the visible range depending of the In:Sn ratio in the structure [9-11].

The main goal of this work is to develop a method of introducing different amounts of high aspect ratio ITO nanostructures (nanowires) inside a polymeric matrix (PMMA) followed by a study of the electrical properties and transmission electron microscopy (TEM).

EXPERIMENTAL

The ITO nanowires were synthesized by a carbothermal reduction method using the co-evaporation of oxides [12]. In this work, the starting materials used were SnO_2 (Sigma-Aldrich, 99.9% of purity) and In_2O_3 (Sigma-Aldrich, 99.99% of purity) mixed in a molar proportion of 1:1 with the reducing agent, which in the present study was the carbon black (Union Carbide, > 99% of purity). In a typical procedure both mixtures were placed in separate alumina boats which were then inserted side by side in the hot zone of a tube furnace (EDG, model FT-HI 40). The synthesis was performed at 1150 °C for 60 minutes with the extremities of the tube sealed with aluminum cover and PTHF o-rings. A N_2 gas flux of 80 sccm was kept throughout the synthesis. Details of the ITO nanostructures synthesis are given in references [12, 13].

To prepare the nanocomposites, the material collected after the synthesis was separated using an ultrasonic cleaner in an alcohol medium. Then, the PMMA (Sigma-Aldrich, average molecular weight of 12000 g/mol) was dissolved in Tetrahydrofuran (THF) and a known amount of ITO nanowires was inserted into the solution at 1, 2, 5 and 10 wt% in relation to the amount of PMMA. The suspension was kept under stirring for 15 minutes and then deposited over a glass substrate using the tape casting process under room temperature.

The material collected after the carbothermal reduction synthesis was analyzed by X-ray diffraction (Shimadzu, model XRD 6000) using the Cu Kα radiation in the range of 5 to 75°. The morphology, structure and chemical ratio of the cations in the ITO nanowires were also analyzed by transmission electron microscopy (TEM; Jeol JSM 3010 operated at 300 kV) equipped with energy dispersive X-ray spectroscopy (EDS). The same technique (TEM, using a Philips CM 200 microscope operated at 200 kV) was used to study the percolation of belts inside the polymeric matrix. To carry out electrical measurements of the nanowires, a quartz substrate with gold contacts made by lithograph was used. A micromanipulator (Wentworth, model MP1008) was then used to connect the substrate to a stabilized voltage source (keithley, model 237). For the two and four probe electrical measurements in the composite, a resistivity measurement system (Signatone, model Pro-4) and the Keithley stabilized source were used simultaneously.

RESULTS AND DISCUSSION

After the synthesis of the ITO structures, a wool-like material was collected inside the tube in a region between 450 and 500 °C. Figure 1 presents the XRD pattern of the material collected. All the peaks in the diffractogram can be indexed by the Al, $In_{0,2}Sn_{0,8}$ (JCPDS card #48-1547) and In_2O_3 (JCPDS card #6-416) phases. The all peaks are related to the aluminum sample holder used in the XRD measurements and have no correlation with the ITO sample. Moreover, the relative intensities between the In_2O_3 and the $In_{0,2}Sn_{0,8}$ peaks confirm that the former is present in greater amounts. Relative to the In_2O_3 phase, it is easy to perceive that the peak at $2\theta = 35.5°$ is much higher than it should be for a non-oriented powder, meaning that the sample has a preferential orientation for the (400) planes with an interplanar distance of 2.5 Å.

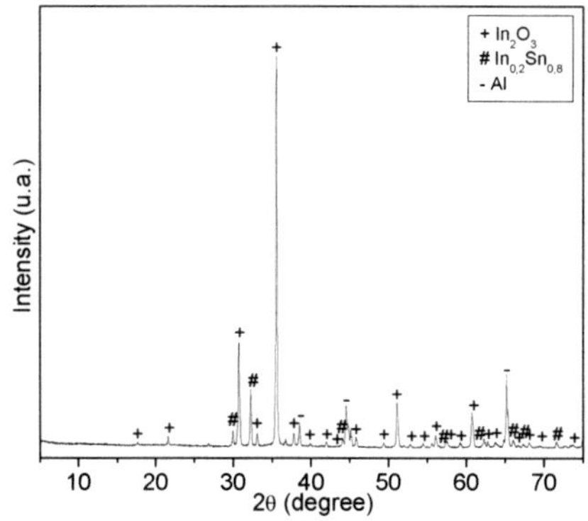

Figure 1: XRD pattern of the collected material after the synthesis.

Figure 2-a illustrate the FEG-SEM characterization of nanostructures. It is possible to observe that the material collected is composed of one-dimensional structures with well defined edges, homogeneous along their lengths and without any macroscopic surface defects, square cross section and have a metallic sphere at one extremity. This result is expected for the In_2O_3 phase, known to be cubic if the growth occurs in (h00) planes together with the fact that the XRD results showed preferential orientation for planes of this family and a metallic sphere at one extremity of the nanowires, thus suggesting that the growth mechanism of wires follows a self-catalytic vapor-liquid-solid (VLS) process.

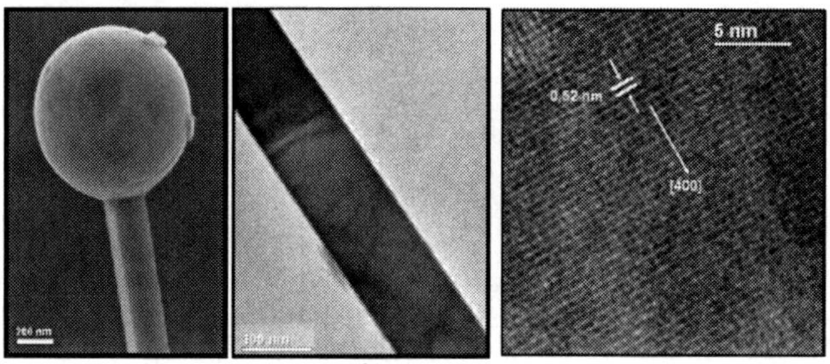

(a) (b) (c)

Figure 2: Images of an ITO nanowire of the: a) FEG-SEM, b) TEM and c) HRTEM.

Figure 2-b presents TEM images of nanowires, the high resolution micrograph in Figure 2-c show that the crystal is single-crystalline and grows in the In_2O_3 structure. The interplanar distance in the HRTEM image and the diffraction indexation both confirm the wires grow in the (200) planes of the indium oxide phase. This result agrees with the XRD analysis which showed preferential orientation for the same family of planes. In order to ascertain that tin-doped In_2O_3 nanowires were synthesized, the energy dispersive X-ray spectroscopy (EDS) technique was used to examine the chemical composition of wires and spheres. The EDS analysis showed that the wires are constituted of indium, tin and oxygen atoms with a In:Sn proportion of 89:11 at%, thus indicating this method was successful in obtaining ITO nanostructures. Meanwhile, the sphere chemical analysis showed the sphere is a Sn rich Sn-In alloy with a In:Sn proportion of 22:78 at%. The EDS results for the metallic spheres support those of the XRD, due to the fact that the proportion between cations obtained by EDS is close to that obtained in the XRD metallic phase.

Figure 3 presents the two probe electrical response for the ITO nanowires. The electrical current is observed to vary linearly with the applied voltage for both positive and negative values of voltage and it is observed that the graph cross the origin. This result indicates that ITO nanowires present an ohmic behavior. The slope of the plot represents the resistance of the wire which in this case is about 10^2 Ω. After carrying out geometric parameters correction, a conductivity of about $1.1x10^6$ S/m was obtained (resistivity of about 90 $\mu\Omega$ cm). This value is found to be far higher than that of pure In_2O_3 nanowires, which is about 10^2 S/m [4], and similar to that obtained in other works [4, 14] for ITO nanowires.

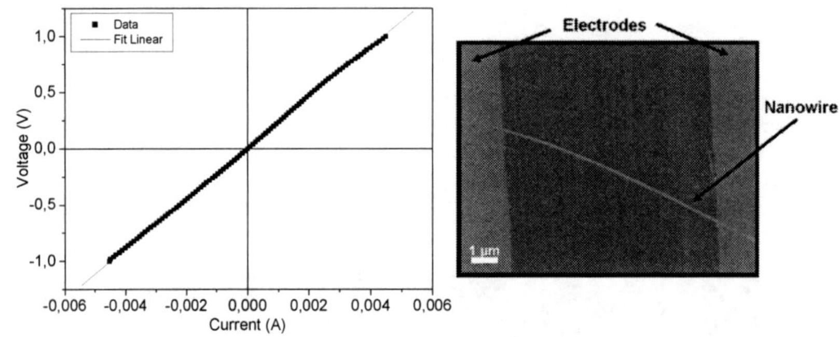

(a) (b)

Figure 3: a) The two probe electrical response for ITO nanowires; b) FEG-SEM image of the approach used to perform the electrical measurements.

Figure 4 shows the electrical resistance of composites as a function of the amount of filler. The electrical resistance decreases abruptly as the amount of filler changes from 2 to 5 wt%, indicating the percolation begins at this range of nanowires insertion. Increasing the amount of filler to 10 wt% enabled an additional reduction in the composite resistance, indicating that a close to total percolation occurred. The electrical resistance of nanocomposites changed by over 7 orders of magnitude by increasing the amount of filler, and using 5 wt% of filler the composite resistance decreased from 10^{10} Ω to about 10^4 Ω. This is an interesting result once for nanocomposites filled with ITO nanoparticles is necessary about 18% in weight to obtain percolation. The addition of filler up to 10 wt% decreased the resistance of the composite to 10^3 Ω, which is a value close to the resistance of wires.

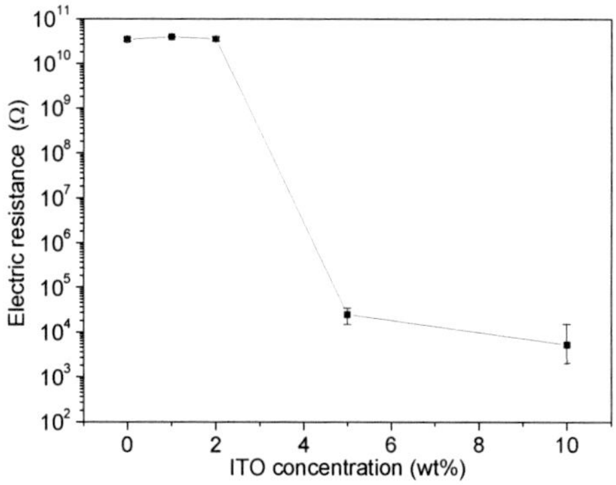

Figure 4: Electrical resistance of composites as function of the amount of nanowires.

Figure 5 presents low magnification TEM images of the composites for different amount of filler. In all images the typical structures of PMMA can be clearly seen [15], while for composites with 1 and 2 wt%, some isolated nanowires can also be observed. This result is in good agreement with the electrical results since isolated wires could not decrease the electrical resistance of the composites.

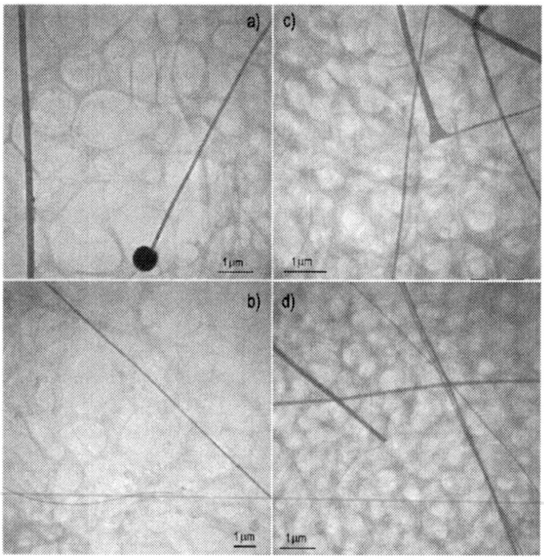

Figure 5: TEM images for the nanocomposite films: a) 1 wt%, b) 2 wt%, c) 5 wt% and d) 10 wt% of ITO nanowires.

Furthermore, it is worth mentioning that although for 1 and 2 wt% filler systems only isolated wires were observed, these wires were more commonly found to occur in 2 wt% filler systems than in 1 wt% filler systems as expected. For the composite with 5wt% of ITO, the TEM image clearly illustrates that percolation was reached in some parts of the composite. This result also supports the electrical results, which indicate the percolation initiated for this amount of filler. For the sample with greater amount of ITO percolation was also observed as shown in Figure 9d. This kind of image is easily obtained during the TEM analysis thus explaining the reason for the smaller resistance obtained for this composite.

CONCLUSION

Based on the exposed analyses and observations, it can be concluded that it is possible to produce conductive composites by inserting conductive ITO nanowires in a polymeric matrix. This result is important because transparent and conductive films have a wide range of potential applications, and practical tests can be performed with this system in order to examine its technological viability.

REFERENCES

1. C. Burda, X. Chen, N. Radha and A. El-Sayed, *Chem. Rev.,* **4**, 1025, (2005).
2. M. C. Johnson, S. Aloni, D. E. Mccready and E. D. Bourret-Courchesne, *Cryst. Growth. Des.,* **6**, 1936 (2006).
3. Y. Yan, L. Zhou, J. Zhang, H. Zeng, Y. Zhang and L. Zhang, J. *Phys. Chem. C Nanomater. Interfaces*, **112**, 10412 (2008).
4. Q. Wan, E. N. Dattoli, W. Y. Fung, W. Guo, Y. Chen, X. Pan and W. Lu, *Nano Lett.*; **6**, 2909 (2006).
5. G. Carotenuto, M. Valente, G. Sciume, T. Valente, G. Pepe, A. Ruotolo and L. Nicolais, *J. Mater. Sci.* **41**, 5587, 2006.
6. K. I. Winey and R. A. Vaia, *MRS Bull.*, **32**, 314 (2007).
7. S. J. Tans, M. H. Devoret, H. Dai, A. Thess, R. E. Smalley, L. J. Geerligs and C.Dekker, *Nature* **386**, 474, 1997.
8. T. Nozaki, T. Karatsu, K. Ohnishi and K. Okazaki, *Carbon N. Y.* **48**, 232, 2010.
9. E. Shanthi, V. Dutta, A. Banerjee and K. L. Chopra, *J. Appl. Phys.* **51**, 6243 (1980).
10. Y. Hoshi, H. Kato and K. Funatsu, *Thin Solid Films*, **445**, 245 (2003).
11. D. S. Ginley and C. Bright, *MRS Bull.* **25,** 15, (2000).
12. M. O. Orlandi, R. Aguiar, A. J. C. Lanfredi, E. Longo, J. A. Varela and E. R. Leite, *Appl. Phys. A*, **8**, 23, 2005.
13. M. O. Orlandi, A. J. C. Lanfredi and E. Longo. In: Méndes-Vilas A, Dias J. Microscopy Book Series - Volume # 4: "Microscopy: Science, Technology, Applications and Education", accepted to publication.
14. S. P. Chiu, H. F. Chung, Y. H. Lin, J. J. Kai, F. R. Chen and J. J. Lin, *Nanotech.*, **20**, 105203, (2009).
15. J. M. Yeh, K. C. Chang, C. W. Peng, M. C. Lai, C. B. Hung, S. C. Hsu, S. S. Hwang and H. R. Lin, *Mater. Chem. Phys*, **115**, 744, (2009).

Mater. Res. Soc. Symp. Proc. Vol. 1312 © 2011 Materials Research Society
DOI: 10.1557/opl.2011.332

Optical Coating by Hybrid Sol

Wei-Hong Wang[1] and Lih-Yue Chen[1]
[1]Chemical Engineering Section, Chemical Systems Research Division,
Chung Shan Institute of Science and Technology, Taiwan 325, R.O.C.

ABSTRACT

The fluorocarbon/SiO_2 hybrid sol (L) for low refractive index coating was made by tetraethoxysilane (TEOS) and polytetrafluoroethylether derivative(PTFEED) with the hydrolysis and condensation in ethanol to get with coating refractive index as (n_L=1.38). It can be coated on glass with single layer by quarter wave, the anti-reflective (AR) coating glass would have less than 3 % reflection. The TiO_2 hybrid sol are prepared by tetraisopropyl orthotitanate(TPOT) with silane coupling agent and acrylic resin. The TiO_2 hybrid sol with the coating refractive index (n_H=1.75) used as high refraction coating sol (H). More acrylic resin added to get less coating refractive index (n_M=1.65) used as middle refractive coating sol (M). Three layers AR coating glass can be made by M/H/L design to get 1.0 % reflection, have 3H hardness, 500g scratch resistance, and 107° water contact angle. These two kinds AR coating with fluorocarbon /SiO_2 hybrid sol (L) coating on surface has anti-stain efficacy, which can be applied on the touch panel screen of flat panel display for mobile phone, PDA and e-book.

INTRODUCTION

In the market, flat panel displays (FPD) with the touching screen for mobile phone, PDA and e-book is basic demand. The FPD with a protect glass for touching screen with optical coating, such as anti-stain[9] and anti-glare[10] with anti-reflection, is indispensable to improve the visibility of FPD. Anti- reflective(AR) coating is the effective to improve the visibility of FPD by optical coating with one-layer, two-layer, three-layer or multilayer. As the anti-stain(AS) or anti-glare(AG) coating on the glass for FPD, many other functions required, such as anti-static [11], conducting[12], heat-insulating[13], UV absorbing[14] properties, which can be made by multiplayer coating. This AR coating is designed by quarter wave with interference phenomenon as the optical thickness, it was coated symmetrically on two-sides of the substrate. In general, one or two layers AR coating can get the V-type or W-type reflective spectrum, three or more layers AR coating with M/H/L design can obtain W-type or U-type reflective spectrum.

Refractive index is the fundamental characteristic of materials for optical coating. By sol-gel technology, It is the advantage to prepare the sol easily with different refractive index coating, such as: low-refractive index coating by SiO_2 sol, high-refractive index coating by TiO_2 sol. and middle refractive index coating sol which can be mixed with SiO_2 and TiO_2 sol by the refractive index to be need with interpolate ratio of them or their precursor for SiO_2-TiO_2 sol. These SiO_2, TiO_2 and SiO_2-TiO_2 sol are coated on glass by dip coating with high temperature sintering for AR coating. Other advantage of sol-gel optical coating, for plastic, those sols can be modified by organic with the functional group for thermal or photo curing [1]. Those organic-inorganic hybrid sols are prepared for AR coating with refractive index as table I.

Table I : Different coating refractive index of those sols

Sol type	SiO_2	SiO_2-TiO_2	TiO_2	SiO_2 hybrid sol	SiO_2-TiO_2 hybrid sol	TiO_2 hybrid sol
Refractive index	1.45	1.45~2.00	2.00	1.38	1.65	1.75

For making the flat panel displays (FPD) with touching screen on mobile phone, PDA and e-book is developed quickly. The FPD with touching screen need not only AR coating, but also AS coating with anti-finger-print efficacy. Now, it was applied on smart phone with PDA, GPS, TV, digital camera, even with computer, all in one pocket secretary. The optical coating with AR and AS coating for touching screen was developed for good visibility and long durability. We are asked to develop this technology, to prepare fluorocarbon/ SiO_2 hybrid sol [2,5-8] as low-refractive coating, and TiO_2 hybrid sol [3,4] as high-refractive coating. To develop the fluorocarbon/SiO_2 hybrid sol, it not only has low refractive index but also anti-stain properties.

EXPERIMENTAL DETAILS

This study provides the preparation of the sol for AR coating to have good abrasion resistance and stain resistance. Three parts of study were included: (1) prepare the fluorocarbon / SiO_2 hybrid sol for low refractive coating, (2) prepare the TiO_2 hybrid sol for high refractive coating, and (3) fabricate the anti-reflective coating glass.

Materials used: Chemicals used in this study including: inorganic precursor as: tetraethylorthosilicate (TEOS, Aldrich) and tetraisopropylorthotitanate (TPOT, Alfa), fluorocarbon as: polytetrafluoroethyl ether derivative (PTFEED, Ausimont), chelating agent as: acetic acid and 2,4-pentanedione, silane coupling agent (SCA) as: methylacryloxypropyl trimethoxyl silane(KBM 503, Shin-Etsu), acrylic monomer as: 2-hydroxylethyl methylacrylate (2-HEMA), tripropylene glycol diacrylate (TPGDA, SR 306) and trimethylolpropyl triacrylate (TMPTA, SR 351), solvent as: ethanol or iso-propyl acohol, catalyst as: hydrochloric acid and nitric acid, photo-initiator and surfactant, etc.

Instrument applied: coater as dip-coating HER-D600, optical spectra measurement as MCPD-2000, hardness measurement as Prema P-247, and Contact angles measurement as APPR#BV-710.

Low refractive coating sol preparation

SiO_2 sol is prepared by the hydrolysis and condensation of TEOS with de-ionized water in anhydrous ethanol, catalyzed by hydrochloric acid. Then the fluorocarbon was added into the SiO_2 sol by weight ratio of PTFEED/TEOS as 1.0, the homogenous and transparent solution of fluorocarbon /SiO_2 hybrid sol (SP-86-1) was obtained for low refractive coating. With the weight ratio of PTFEE/TEOS as 1.5, the fluorocarbon/SiO_2 hybrid sol (SP-87-1) for lower refractive index coating.

High refractive coating sol preparation

TiO_2 sol is prepared by TPOT with acetic acid and de-ionized water in anhydrous ethanol catalyzed by HNO_3 at PH=1.5. After that the TiO_2 sol was mixed with silane coupling agent (SCA), acrylate monomer(AM) and photo initiator (PI). The coating of high refractive index by TiO_2 hybrid sol (H) was prepared with the weight ratio of TPOT: SCA: AM: PI as 100: 25: 25: 5. The coating of middle refractive index by SiO_2-TiO_2 hybrid sol (M) was prepared by same process, mix TPOT with acetic acid, then hydrolyzed, caped with SCA , added AM and PI with the weight ratio about 100: 50: 50: 5.

Anti-reflection coating:

Single layer anti-reflection: Glass substrates were degreased by hydrochloric acid and methanol, washed with water, dried at 100°C, then treated by atmosphere plasma. The single layer AR coating was dip-coated with the fluorocarbon/SiO_2 hybrid sol, dried and baked at 200°C for 3hrs. This coating refractive index (L) is 1.38, with optical thickness of 1/4λ, λ=550 nm. This single layer AR coating glass has good anti-stain efficacy with less than 3.0% reflection.

Three layers anti-reflection: Three layers AR coating glass was fabricated by three different sol with coating refraction as M/H/L in series coating. First, the cleaning glass was coated with middle refraction SiO_2- TiO_2 hybrid sol, then photo cured and baked to get the middle refractive index (M) coating as 1.65 with optical thickness about 1/8λ. After first layer coating, it was coated by TiO_2 hybrid sol as high refractive index (H) coating as n=1.75 with optical thickness about of 3/8λ. After that, it was coated with fluorocarbon/SiO_2 hybrid sol, then baked at 200°C for 1hr. to get the top layer with the low refractive index (L) coating as n=1.38. This three layers AR coating has 1.0% reflection with anti-stain efficacy.

RESULTS AND DISCUSSION

Single layers AR coating

To make the low refractive index coating, fluorocarbon/SiO_2 hybrid sol was selected to be used. The single layer AR coating was dip-coated with the sol, then dried and baked in the oven at 150 °C, 200 °C and 400 °C with optical thickness of 1/4λ, got the reflection spectrum as Figure 1, 2, and 3. By optical thin film, we can estimate the coating thickness and refractive index of SP-86-1 or SP-87-1 coating on glass as table II.

Table II : The coating thickness and refractive index of SP-86-1 or SP-87-1 coating

Baking Temperature	refractive index /coating thickness	
	SP-86-1	SP-87-1
150 °C	1.400 / 95nm	1.390 / 117nm
200 °C	1.395 / 104nm	1.375 / 71nm
400 °C	1.290 / 119nm	1.280 / 98nm

Table III: Contact angle analysis of single and three layer AR coating

Temperature/ Contact angle	SP-86-1	SP-87-1	Three layers AR coating
150 °C	103°	105°	□
200 °C	105°	105°	108°~109°
400 °C	35°	25°	□

From figure 3 showing, the fluorocarbon in the coating is decomposed and volatilized at 400 °C, and the coating refractive index down 1.28 ~ 1.29 (as table II), water contact angle down

25° ~ 35° (as table III) and hardness down <1H (as table IV), so the coating of hybrid sol could not stand at 400 °C high temperature, although fluorocarbon hybrid. For the SP-86-1 or SP-87-1 coating, the SP-87-1 has lower refractive index than SP-86-1 as figure 1, 2, and 3 display. It means to increase the fluorocarbon content could got the lower refractive index.

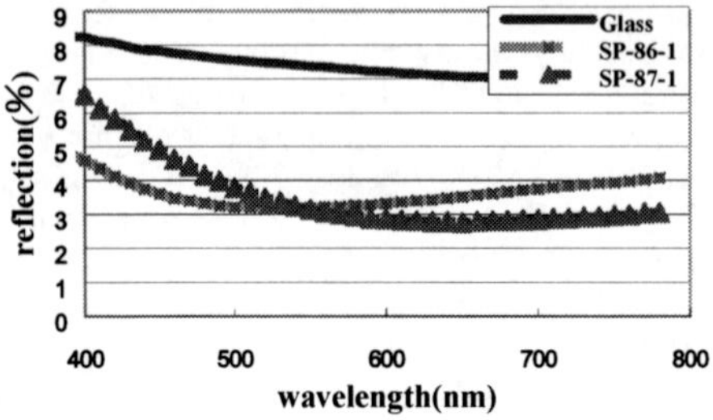

Figure 1: Reflection spectrum of fluorocarbon/SiO_2 hybrid sol coating by 150 °C

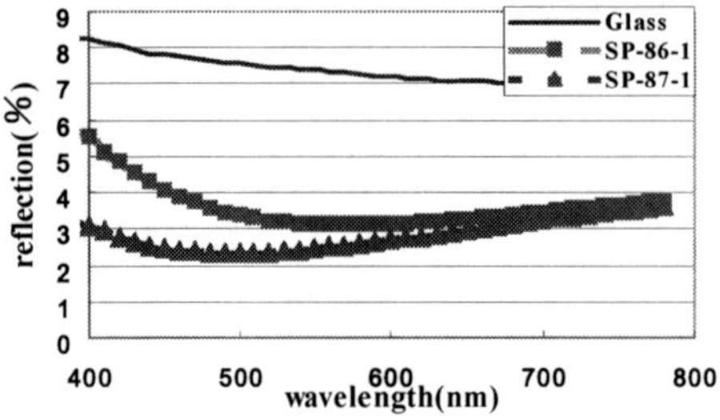

Figure 2: Reflection spectrum of fluorocarbon/SiO_2 hybrid sol coating by 200 °C

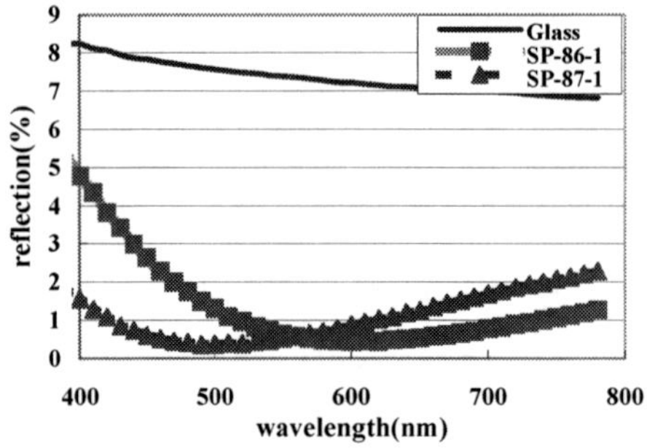

Figure 3: Reflection spectrum of fluorocarbon/SiO_2 hybrid sol coating by 400 °C

Three layers AR coating

Three layers AR coating on glass with 1.0% reflection spectrum, was coated by SiO_2-TiO_2 (ST-54) hybrid sol, which has middle refractive index as n=1.65 with optical thickness about 1/8λ

(λ=550nm) with 15% reflective spectrum. After first layer coating, it was coated with TiO_2 (TS-8) hybrid sol as second layer coating, which had high refractive index as n=1.75 with optical thickness about 3/8λ (λ=550nm) to get about 15~25% reflective spectrum.

The SP-86-1 hybrid sol was chose as the low refractive coating sol for third layer of AR coating. After TiO2 hybrid sol coating, it was coated with fluorocarbon/SiO_2 hybrid sol, then baked in oven at 200 °C for 1hr., the surface layer has the low refractive index as n=1.38 with optical thickness about 1/4λ (λ=550nm) to got 1.0~2.0% reflective spectrum., This three layer AR coating was made on glass as ST-54/TS-8/SP-86-1 sequential coating to get each sequence reflection spectrum showing as figure 4.

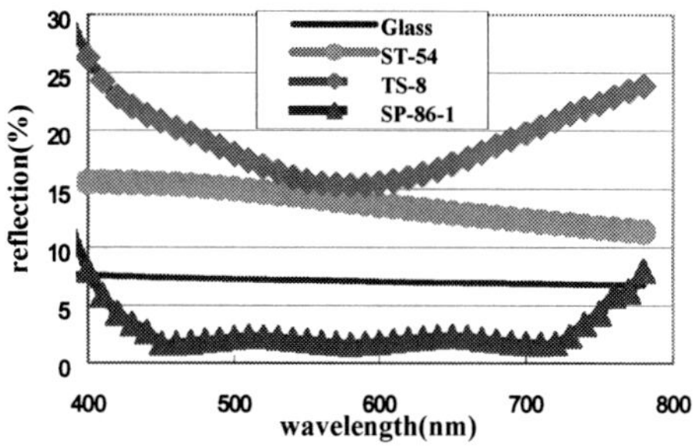

Figure 4: Three-layer anti-reflection coating on a glass surface with different reflection

Contact angle measurement:

The fluorocarbon/SiO_2 hybrid sol was coated on glass with low refractive index as 1.38 with optical thickness of 1/4λ, λ=550 nm to get AR coating. It was baked in 150~200 °C with water contact angle about 103°~105°, examples as table 3, have anti-stain efficacy. But higher temperature baking at 400 °C, fluorocarbon was decomposed and nano-porous presented on surface and change to low water contact angle.

Three layers AR coating glass was fabricated by three different sol with coating refraction as M/H/L. The top coating is low refractive index by the fluorocarbon /SiO_2 hybrid sol. It also has high water contact angle about 108°~109°. This fluorocarbon/SiO_2 hybrid sol coating with low refractive index and high water contact angle, it can be applied on the cover glass of FPD for anti-stain and anti-finger-print.

Hardness measurement:

Table IV: Hardness of single layer and three layers AR coating

Pencil Hardness / Temperature	SP-86-1	SP-87-1	Three layers AR coating
150 °C	≤3H	≤3H	□
200 °C	3H	3H	3H
400 °C	<1H	<1H	□

From table IV indicated, the fluorocarbon in the coating is decomposed and volatilized at 400 °C, although the refractive index of SP-87-1 decrease from 1.375 to 1.28, which caused the porosity of the hybrid coating increase, but this also caused the hardness going down from 3H to less than 1H, the water contact angle decrease from 105° to 25°. So this hybrid sol coating could not bake at temperature higher than 300°C. In single layer or three layer AR coating with 200 °C drying and curing, to get hardness about 3H.

CONCLUSIONS

Use polytetrafluoroethylether derivative (PTFEED) and Tetraethylorthosilicate (TEOS) to prepare the hybrid sol by weight ratio (1:1) for low refractive index coating. Single layer AR coating can be made by the fluorocarbon/SiO_2 hybrid sol with the refractive index of 1.38. This coating will get less about 3.0% reflection, 3H hardness, 105° water contact angle as water repellent as anti-stain coating.

Three layers AR coating on the glass by three kinds sol, such as: middle refraction coating (M) of the SiO_2-TiO_2 hybrid sol (n=1.65), high refraction coating (H) of the TiO_2 hybrid sol (n=1.75), and low refraction coating (L) of the fluorocarbon/SiO_2 hybrid sol (n=1.38). This M/H/L three layer AR coating will have 1.0% reflection, 3H in hardness, 500 g scratch resistance, and 109° in water contact angle for water repellent and anti-stain.

This kinds AR coating has water repellent and anti-stain ability, it could be applied for touching screen on flat panel display, which would applied on smart phone with PDA, GPS, TV, digital camera and computer, as all in one pocket secretary.

REFERENCES

1. W.H. Wang; W. Lo, J.C. Hwang, CSIST-260.3-T401, Tech. Report, 2002.
2. B.T. Liu; W.H. Wang, CSIST-BD97007P, Contract Report, 2008
3. C.C. Chang; W.H. Wang, CSIST-808-V112, Contract Report, 2009
4. S.J. Tong; B.T. Liu, "High refractive index transparent nanocomposites, Master Thesis of National Yunlin University of S&T, 2010.
5. M. Yamaya; K. Sato, US 7,270,887 B2, 2007
6. K. Sato; M. Yamaya; M. Asai and K. Matsumura, US 6,558,804 B2, 2003.
7. T. Nakada; K. Motoyama; R. Gunji; M. Wakabayashi; H. Furusho; H. Fukuro, US 6,472,012 B2, 2002.
8. T. Aoyama; T. Shimizu; N. Otsuki; N. Nagahama, US 6,383,620 B1, 2002
9. X. Talpaert; M. Simonet; M. Azzopardi; A. Durandeau, WO 2001032578, 2001
10. T. Sugou, "Functionalized triacetylcellulose optical film" WO2006101251 A1, 2006
11. W. Kim; T. Kim; C. Park; J. Kim; T. Lee; H. Yoon; K. Suh, e-polymer, 2008.
12. J. Chu; C. Li; C. Shih; S. Chang, US 6441964, 2002.
13. G. Choi, KR 2010046873, 2010
14. A. Schmid; M. Guardalben; T. Chipp, DE 84008381, 1984.

Mater. Res. Soc. Symp. Proc. Vol. 1312 © 2011 Materials Research Society
DOI: 10.1557/opl.2011.110

Characterization on the Electrical Properties of PDMS Nanocomposites by Conducting Polymer Nanowires

Ping Du, Xi Lin and Xin Zhang
Department of Mechanical Engineering, Boston University, Boston, MA 02215, U.S.A.

ABSTRACT

Polydimethylsiloxane (PDMS) is one of the most used materials in bio-applications. However, previous works were mainly focus on the mechanical aspect. In this paper we presented a practical and efficient approach to enhance the electrical properties of PDMS by using conducting polymer nanowires (CPNWs). The nanowires were synthesized using template method and added in PDMS to form nanocomposites. The dielectric constants of the composites were characterized by impedance measurements, and the dielectric relaxation behavior and the volume fraction of CPNWs was investigated. Based on the percolation theory a much lower threshold (5.3 vol%) was achieved.

INTRODUCTION

PDMS plays an essential role in chemical and biological applications due to its optical transparency, mechanical compliance, chemical stability, bio-compatibility, and easy of fabrication [1]. However, being a non-conductive elastomer, PDMS has been limited to a structural material in most applications [2-5]. Hence there is an increasing demand to explore the multifunctionality of PDMS, such as electrical properties, to broaden the range of applications. A common approach is through adding inorganic or organic conductors as fillers to form composites. However, the phase separation of inhomogeneous particles [6] and high percolation threshold for large-sized particles limit the practical applications. Conducting polymer nanowires become a promising filler candidate because of the mechanical flexibility, elongated shape and much lower threshold concentration [7]. In this work, polypyrrole (PPy) nanowires were synthesized and added in PDMS to form the nanocomposites. The enhancement of dielectric constants of the composites was characterized, and the relationship to the volume fraction of CPNWs was investigated based on percolation theory.

EXPERIMENT

In this work the CPNWs were fabricated using porous membrane as the template [8-10]. Anopore porous alumina membranes (Whatman) with a pore diameter of 200 nm were used, and the electrochemical synthesis was carried out at a constant current density of 1 mA/cm^2 (Figure 1(a)). The solution consisted of 0.1 M pyrrole monomer and 0.1 M sodium dodecylbenzene sulfonate (NaDBS, all from Sigma Aldrich). The length of the NWs was controlled by synthesis time. A typical 4 hour synthesis resulted in an average length of 49.12 μm (Figure 1(b)).

After synthesis, the seed layer was etched, followed by etching of the alumina membrane to free the nanowires. Subsequently centrifuge and ultrasonic stirring were applied for several times to replace the etchant solution by ethanol and obtain an evenly distributed CPNW dispersion. This dispersion was then mixed with PDMS and thermal cured to form the composites.

(a)

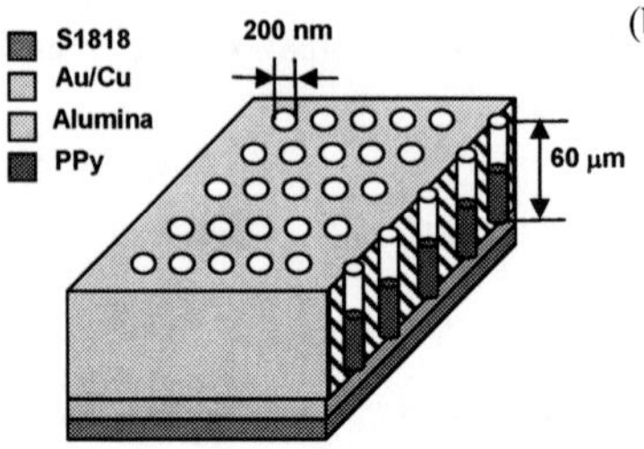

(b)

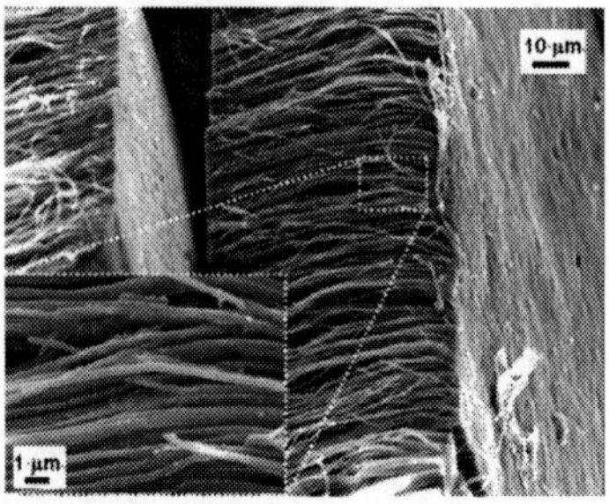

Figure 1. (a) Schematic diagram of CPNW fabrication in the porous alumina membrane. Gold and copper were served as seed layer for electrochemical synthesis. S1818 photoresist was used as the insulating layer. (b) SEM micrograph of PPy nanowires grown for 4 hr. (The alumina membrane was etched away.)

The electrical properties of the composites can be analyzed by impedance technique. The impedance of pristine PDMS and nanocomposites with 1 vol%, 3 vol% and 5 vol% PPy nanowires were measured using a 1260 Impedance analyzer (Solartron). A home-made test fixture was built according to ASTM standard D150 (Figure 2). The parallel plate method was employed and the distance between two electrodes can be precisely controlled by a micrometer with a resolution of 1 µm. Due to the instrument error and stray impedance of cables and test fixture, all measurements are normalized by the corresponding air gap impedance using a matched-capacitance method.

(a)

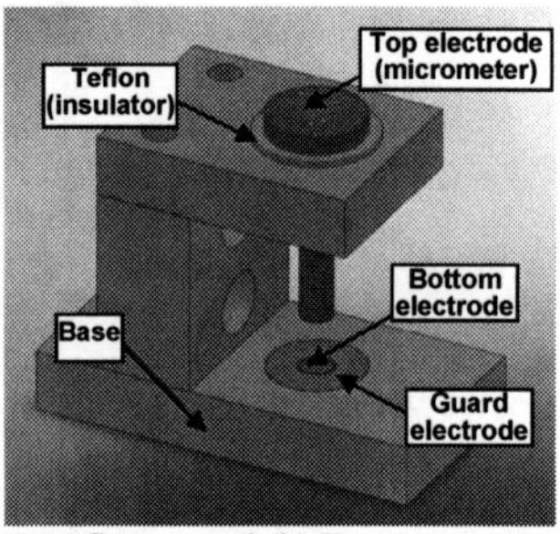

(b)

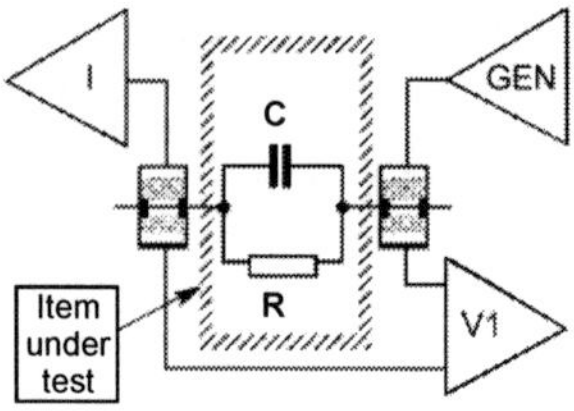

Figure 2. (a) Test fixture and (b) Four-port connection for the impedance measurement.

DISCUSSION

<u>Effect of etching solution pH</u>

In the template synthesis, the alumina membrane needs to be removed to free the NWs for the next mixing step of nanocomposites. Both base [11] and acid [12] can be used to dissolve alumina, but it will have distinct effect on the electroactivity of conducting polymers. It has been reported that deleterious effect will induce by exposure to strong base, while strong acid has only a small effect on the electronic structure of the polymer, causing slightly increase of the conductivity [13]. These are all related to the protonation-deprotonation process on the polymer chain.

To verify this effect, we have conducted cyclic voltammetry on the PPy films of as-deposit state, treated with Potassium hydroxide (KOH) and Phosphoric acid (H_3PO_4). As shown in Figure 3(a), the as-deposit PPy has clear anodic and cathodic peaks. However, exposure to KOH severely decreases the PPy activity, while exposure to H_3PO_4 can maintain or even slightly increase the activity. The good agreement with literatures confirm that acid is better to preserve the conductivity. Therefore alumina etching was carried out only in H_3PO_4 for the PPy nanowires.

After dissolving the alumina, PPy nanowires were ultrasonic stirred and mixed with PDMS to form the nanocomposites. The sonication tends to break the nanowires into shorter wires, which average length was reduced to 16.97 μm (Figure3(b)).

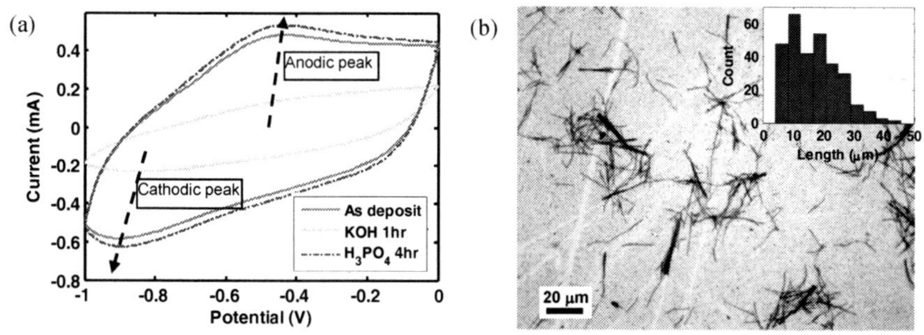

Figure 3. (a) Cyclic voltammetries of PPy thin films at three states: as deposit, after dip in KOH for 1 hr and H_3PO_4 for 4 hr. (Each curve was recorded until stable scan reached.) (b) Optical micrograph of dispersed CPNWs on a silver coated silicon chip. The average length of 300 NWs was 16.97 μm.

Electrical properties measurement

The dielectric constant, dielectric loss and AC conductivity can be readily calculated from the impedance magnitude and phase angle data. For instance, the impedance Z can be written into polar form as $\vec{Z} = |Z|e^{j\theta}$. The Admittance Y is the inverse of Z

$$\vec{Y} = \frac{1}{\vec{Z}} = \frac{1}{|Z|}e^{-j\theta} = Y' + jY'' \quad (1)$$

The relative permittivity ε_r is obtained from Y by

$$\bar{\varepsilon}_r = \frac{\bar{Y}}{j\omega C_0} = \frac{Y' + jY''}{j\omega C_0} = \frac{1}{\omega C_0}(Y'' - jY') = \varepsilon_r' - j\varepsilon_r'' \qquad (2)$$

where $\omega = 2\pi f$ is the angular frequency, and $C_0 = \varepsilon_0 A/t$, with ε_0 the vacuum permittivity, A the cross section area and t the thickness of test sample.

The real part of ε_r is usually referred as the dielectric constant, the dielectric loss D is defined as the ratio of imaginary part to real part of ε_r, and the frequency-dependent AC conductivity σ is related to the imaginary part of ε_r by $\sigma = \omega\varepsilon_0\varepsilon_r''$ [14].

We have done the preliminary impedance measurements of pristine PDMS and nanocomposites with 1 vol%, 3 vol% and 5 vol% PPy nanowires. The real parts of the complex dielectric constants are plotted in Figure 4. Due to the high impedance and low dielectric loss of these samples, the impedance analyzer was working near the limit of the detection range (maximum impedance of 100 MΩ). Therefore the dielectric loss data became much noisier with large scattering, which were not present in this paper.

As shown in Figure 4(a), the dielectric constant of PDMS is in the range of 2.32 to 2.40, which agrees well with existing literature [15]. For the nanocomposites, the dielectric constants show clearly enhancement as the nanowire concentration increases. The high dielectric constants in this conductor-dielectric composite are due to the conductor inclusions. The quasi-1D nanowire shape is one of the structures that enables long-range migration of electronic carriers along the longitudinal axis and gives rise to nanowires with giant polarization.

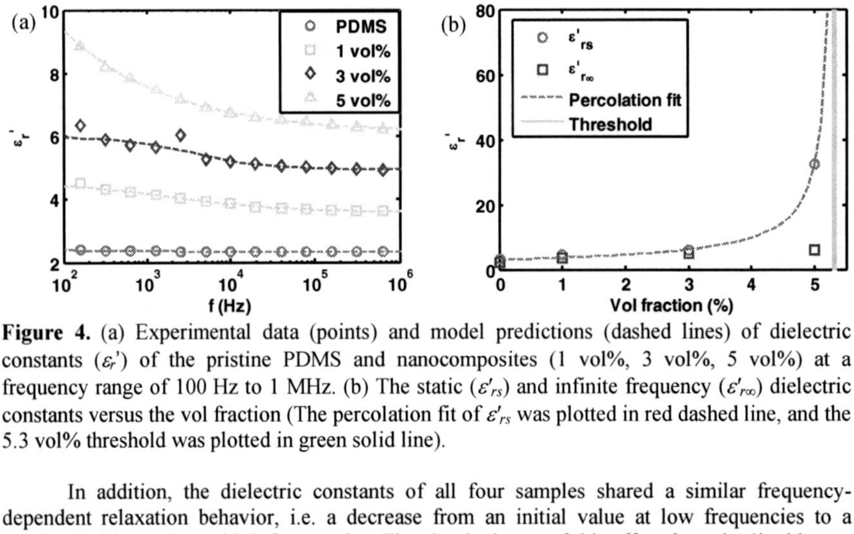

Figure 4. (a) Experimental data (points) and model predictions (dashed lines) of dielectric constants (ε_r') of the pristine PDMS and nanocomposites (1 vol%, 3 vol%, 5 vol%) at a frequency range of 100 Hz to 1 MHz. (b) The static (ε'_{rs}) and infinite frequency ($\varepsilon'_{r\infty}$) dielectric constants versus the vol fraction (The percolation fit of ε'_{rs} was plotted in red dashed line, and the 5.3 vol% threshold was plotted in green solid line).

In addition, the dielectric constants of all four samples shared a similar frequency-dependent relaxation behavior, i.e. a decrease from an initial value at low frequencies to a smaller limiting value at high frequencies. The classic theory of this effect for polar liquids was presented by Debye [16]. However, deviations were found when applying Debye's theory to explain the experimental results for other liquids and solids. We found the similar issue in our system. Therefore a modified equation, i.e. Cole-Cole equation [17] was used in this paper:

$$\varepsilon^* - \varepsilon_\infty = (\varepsilon_s - \varepsilon_\infty)/[(1+(j\omega\tau)^{1-\alpha}] \quad (3)$$

where ε^* is the complex dielectric constant, ε_s and ε_∞ are the "static" and "infinite frequency" dielectric constants, ω is the angular frequency and τ is the relaxation constant. The parameter α, which takes a value between 0 and 1, is an experimentally determined correction factor. When α = 0, the Cole-Cole equation reduces to the Debye equation.

The real part ε' of the complex dielectric constant can be obtained by using the relation $j^{(1-\alpha)}=\exp[j(1-\alpha)\pi/2]$ as

$$\varepsilon' - \varepsilon_\infty = \frac{1}{2}(\varepsilon_s - \varepsilon_\infty)\left[1 - \frac{\sinh(1-\alpha)x}{\cosh(1-\alpha)x + \cos(\frac{\alpha\pi}{2})}\right] \quad (4)$$

where $x = \ln(\omega\tau)$.

The least squares curve fitting method was used to extract all the coefficients by fitting the experimental data with Equation 4. In all calculations, the real part of dielectric constant relative to vacuum dielectric constant was used. The model predictions achieved good agreements with experimental data. Moreover, the static ε'_{rs} and infinite frequency $\varepsilon'_{r\infty}$ dielectric constants show monotonic increases as the vol fraction increases (Figure 4(b)). The static dielectric constant ε'_{rs} shows a dramatic jump when the vol fraction changes from 3 vol% to 5 vol%, indicating that the composite reaches the percolation threshold. Similar phenomenon has also been reported in other literature [18]. A percolation structure can be thought to be made up by randomly packing the two types of granules (not necessarily the same size) together. There is a critical volume fraction below which the electrical properties are dominated by the insulating component and above which the conducting component dominates. Such percolation systems are best analyzed using the two-exponent phenomenological percolation equation (TEPPE) [19]

$$\varepsilon = \varepsilon_1\left(1 - \frac{v}{v_c}\right)^{-q}, \qquad (v < v_c) \quad (5)$$

where ε_1 is the dielectric constant of PDMS matrix, v and v_c are the volume fraction of filler and critical threshold, and q is the critical exponent.

A similar least squares curve fitting was applied to the static dielectric constant using Equation 5. The best-fit shows that v_c equals to 5.3 vol% and q equals to 0.82. The value of q falls in the range of 0.37 to 1.28 [19], indicating the validity of the percolation model. This critical volume fraction is far below the value (14-16 %) predicted for a perfectly random three-dimensional percolation system, which indicates that the scaling effect of nanowire indeed improve the composite properties. However, when the filler are further scaled downward to tens of nanometers in diameter, the interface-to-volume ratios, space charge accumulation and short-range dipole-dipole interactions will become more prominent. Therefore it can be predicted that more improvements in the dielectric constant of PDMS composites could be achieved by using the much smaller scale nanowires as the filler.

CONCLUSIONS

In this paper, we provide a practical and efficient approach to enhance the electrical properties of PDMS by using CPNWs. The synthesis of nanowires and the PDMS nanocomposites were presented with the special consideration of etchant pH value on the electro-activity of nanowire. The enhancement of dielectric constants of the composites was characterized by impedance measurements, and the dielectric relaxation behavior and the volume fraction of CPNWs was investigated. Based on the percolation theory a much lower threshold (5.3 vol%) was achieved. The methodology presented herein can be also applied to other similar material systems. And the multifunctional PDMS nanocomposite will enable new potential application fields, such as energy storage (capacitors) and electromechanical devices (actuators).

ACKNOWLEDGMENTS

This work is supported by National Science Foundation under Grant CMMI-0826191. The authors gratefully acknowledge the assistance from Professor ZhiYong Gu from University of Massachusetts at Lowell for his valuable discussion on the nanowire synthesis process, Jim Mason at Solartron Analytical for the technical advice on the operation of the impedance analyzer, and Tianchen Liu for the synthesis of PPy nanowires.

REFERENCES

1. K. M. Yamada and K. Olden, *Nature* **275**, 179-184 (1978).
2. J. C. McDonald, D. C. Duffy, J. R. Anderson, D. T. Chiu, H. K. Wu, O. J. A. Schueller and G. M. Whitesides, *Electrophoresis* **21**, 27-40 (2000).
3. J. L. Tan, J. Tien, D. M. Pirone, D. S. Gray, K. Bhadriraju and C. S. Chen, *Proc. Natl. Acad. Sci. USA.* **100**, 1484-1489 (2003).
4. Y. Zhao and X. Zhang, *18th IEEE MEMS*, Miami Beach, FL, 398-404 (2005).
5. X. Zheng and X. Zhang, *J. Micromech. Microeng.* **18**, 125006 (2008).
6. J. Hong, J. Lee, C. Hong and S. Shim, *J. Therm. Anal. Calorim.* **101**, 297-302 (2010).
7. H. D. Tran, D. Li and R. B. Kaner, *Adv. Mater.* **21**, 1487-1499 (2009).
8. L. S. Van Dyke and C. R. Martin, *Langmuir* **6**, 1118-1123 (1990).
9. C. R. Martin, *Chem. Mat.* **8**, 1739-1746 (1996).
10. C. R. Martin, *Science* **266**, 1961-1966 (1994).
11. Z. L. Xiao, C. Y. Han, U. Welp, H. H. Wang, W. K. Kwok, G. A. Willing, J. M. Hiller, R. E. Cook, D. J. Miller and G. W. Crabtree, *Nano Lett.* **2**, 1293-1297 (2002).
12. R. Xiao, S. I. Cho, R. Liu and S. B. Lee, *J. Am. Chem. Soc.* **129**, 4483-4489 (2007).
13. Q. Pei and R. Qian, *Synthetic Met.* **45**, 35-48 (1991).
14. J. S. Seybold, *Introduction to RF propagation.* (John Wiley and Sons, 2005).
15. J. E. Mark, *Polymer Data Handbook.* (Oxford University Press, New York, 1999).
16. P. Debye, *Polar Molecules.* (Chemical Catalog Company, New York, 1929).
17. K. S. Cole and R. H. Cole, *J. Chem. Phys.* **9**, 341-351 (1941).
18. C. C. Wang, J. F. Song, H. M. Bao, Q. D. Shen and C. Z. Yang, *Adv. Funct. Mater.* **18**, 1299-1306 (2008).
19. D. S. McLachlan and G. Sauti, *J. Nanomater.* **2007**, 1-9 (2007).

Mater. Res. Soc. Symp. Proc. Vol. 1312 © 2011 Materials Research Society
DOI: 10.1557/opl.2011.480

Nylon 6 reinforced with acrylic polymer nanoparticles. Thermal properties and nano structure

Estefania Huitron-Rattinger[1,2], Bonifacio Alvarado-Tenorio[1,2], Angel Romo-Uribe[2,*]

[1]Departamento de Ingeniería Química Metalúrgica, Facultad de Química, Universidad Nacional Autónoma de México, 04510 México D.F., MEXICO.
[2]Instituto de Ciencias Físicas, Universidad Nacional Autónoma de México, Cuernavaca Mor. 62210, MEXICO.
* To whom all correspondence should be addressed: aromo-uribe@fis.unam.mx

ABSTRACT

The correlation of thermal properties and nanostructure of nylon 6 (denoted PA6) reinforced with polymer nanoparticles (denoted PNP, size~8 nm) has been investigated. PNPs are highly crosslinked acrylic-based polymers synthesized by the Rohm and Haas Co. PNPs and those grafted with maleic anhydride (denoted PNP-g-MA) were each one dispersed into a commercial PA 6 matrix by melt extrusion, at a concentration of 3 wt%. Thermal analysis showed that the PNPs increased the thermal stability of PA6 and reduced the melting and crystallization temperatures as well as the enthalpy. Small-angle light scattering showed that the PNPs crystallize into a spherulitic morphology, typical of PA6. Isothermal crystallization studies showed that the PNPs act as nucleating agents, accelerating the rate of crystallization. Wide-angle X-ray scattering showed that the composites crystallize in the α-form, and the PNPs reduced the degree of crystallinity, in agreement with thermal analysis results. The smaller degree of crystallinity was also reflected in the long range spacing, it was also reduced by the presence of the PNPs as measured directly by small-angle X-ray scattering.

INTRODUCTION

Polymer reinforcement has become a challenging growing field in polymer science. The enhancement of specific properties regarding polymer performance such as toughness, thermal resistance, stiffness, solvent resistance, barrier properties, flame retardation or conductivity can be achieved by carefully choosing the type of filler for the intended reinforcement. The resulting materials are known as polymer composites. However, despite the fact that such changes can be tunable, sometimes unpredictable or undesirable outcomes may arise such as the loss in the elastic modulus. Particularly important are those composites in which at least one component´s characteristic length scale is in the nanometer range, therefore their name polymer nanocomposites. Polymer Nanocomposites typically contain 1-5 vol% of nanoparticles and are formed of a polymeric matrix and a nanoscale reinforcement. Common nanoreinforcements are nanoclays, nanofibers, nanoparticles and polymer nanoparticles [1-4]. The properties displayed by these polymer nanocomposites arise from particle size and shape, as well as particle proximity and distribution throughout the polymer matrix. Polymer nanocomposite structure and properties are affected by the synthetic method, the identity of the polymer matrix, the type of nanoparticle used and any kind of surface treatment performed for compatibilization between the phases. Addition of nanoreinforcements impacts, among other properties, the rheological behavior of the material in the molten state. Even though understanding and optimizing the nanocomposites rheological properties is both challenging and important, the mechanisms of enhancement or attenuation of such properties are not yet completely elucidated. Recently we have reported on the shear and extensional rheological properties of PA6-PNP nanocomposites [5]. The results showed that the PNPs induced melt elasticity in an otherwise viscous melt. This was evidenced by an increase in the shear storage modulus and even the production of strain hardening under extensional deformation [5]. The aim of this work is now to study the impact that the PNPs have on the micro and nanostructure of nylon 6 in the solid state.

EXPERIMENTAL

Nylon 6 (PA6) Capron 8202 was purchased from BASF. This is a low viscosity, general purpose polyamide 6 injection molding homopolymer. The polymer nanoparticles used for this study (denoted hereafter PNP) are highly crosslinked acrylic copolymers with nearly spherical morphology especially synthesized by the Rohm and Haas Co., Research Laboratories (Spring House, Pennsylvania, USA). The synthesis route as well as the exact chemical composition are proprietary information, suffices to say that the PNPs are synthesized from butyl acrylate and methyl methacrylate among other monomers. One of the PNP was additionally copolymerized with maleic anhydride, and it is here denoted as PNP2. The PNPs have average diameter of 8 ± 2 nm, as determined by Rohm and Haas Co. using small-angle X-ray scattering, dynamic light scattering, static light scattering and gel permeation chromatography with subsequent matrix assisted laser desorption time-of-flight mass spectrometry (GPC/MALDI-TOF) techniques. Rohm and Haas Co. also determined the weight average molecular weight of the PNPs, $\overline{M}_w$=2 x 10^4 g/mol. PA6 and PNPs were mixed by melt extrusion and the extrudates were pelletized. The thermal transitions were characterized by differential scanning calorimetry (DSC) using the Pyris1® DSC (Perkin Elmer Inc., CT, USA). Temperature and enthalpy calibration were carried out using analytical grade indium (T_m=156.6 °C) and zinc (T_m=419.5 °C) and corrections were made for the instrument 'baseline'. Samples weighing about 10 mg were loaded into standard aluminum pans. The scans were carried out at 20 °C/min and under nitrogen atmosphere. The decomposition temperatures, T_{dec}, were determined by thermogravimetric analysis (TGA), scanning at 10 °C/min, using the TGA2950 analyzer (manufactured by TA Instruments, New Castle, DE, USA). These experiments were carried out under nitrogen atmosphere.

RESULTS AND DISCUSSION

The decomposition temperatures were determined by thermogravimetric analysis (TGA). Figure 1(a) shows the plot of weight loss and first derivative with respect to temperature as a function of temperature, for the three specimens. The onset for thermal degradation of PA6 (trace i) was found to be 401°C. After the onset the thermal degradation occurred at a rate of 1.97%/°C, and by 475°C 98% of the specimen had been burnt off. On the other hand, the nanocomposite PA6-PNP1 showed essentially the same thermal degradation temperature as the neat PA6. However, there was a significant increase of thermal degradation temperature for the nanocomposite PA6-PNP2. The onset for degradation was increased to 417°C although it showed the same rate of degradation as PA6. Closer inspection (enlargement of weight loss, including first derivative data) showed that the inflection temperatures for (i) PA6, (ii) PA6-PNP1 and (iii) PA6-PNP2 were 441.5, 444.5 and 462.0°C respectively, and the corresponding weight loss values at the given inflection temperatures were 86.5%, 80. 9% and 79.7%. Table 1 summarizes the thermal properties of PA6 and the nanocomposites.

Figure 1 (b) shows DSC scans obtained during first heating. The results show that the melting transition temperature decreased by the presence of nanoparticles. There was also a reduction of the melting enthalpy (see Table 1) and therefore, of the degree of crystallinity of the nanocomposites. The degree of crystallinity, χ, was determined from the experimental heat of fusion values and taking the value of enthalpy at 100% crystallinity for PA6 ΔH_o= 189 J/g, and correcting for the weight percentage of the nanofiller, x, using the expression [6,7]

$$\chi(\%) = \Delta H_m \frac{1-x}{\Delta H_m^o} \times 100 \qquad (1)$$

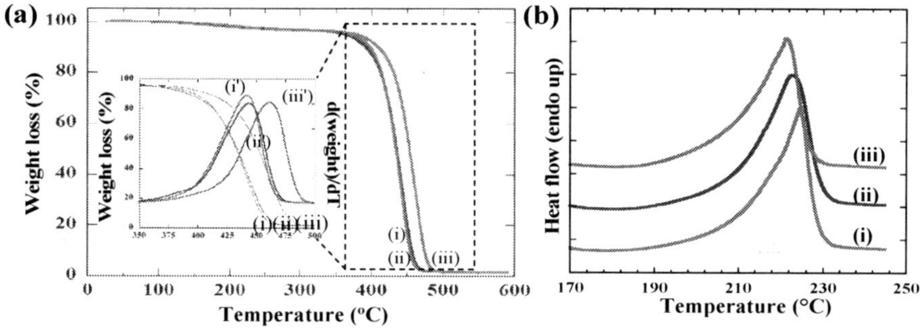

Figure 1. (a) Weight loss as a function of temperature (inset shows enlargement of onset region). (b) DSC first heating scans. i) PA6, ii) PA6-PNP1, iii) PA6-PNP2. Scans carried out at 20°C/min.

Sample	T_{dec} (°C) [a]	T_m (°C) [b]	ΔH_m (J/g) [b]	T_c (°C) [b]	ΔH_c (J/g) [b]	χ (%) [b]	χ (%) [c]	*Long range* (Å) [d]
PA6	401	224.8	53.7	189	-60	27.5	26.1	125
PA6-PNP1	400	222.9	51.5	187	-57	26.5	25.6	102
PA6-PNP2	417	221.4	46.3	187	-58	23.7	6.4	75

Table 1. Thermal and X-ray data for PA6 and its nanocomposites. ***a***. TGA, ***b.*** DSC, ***c.*** data obtained from WAXS, ***d*** data obtained from SAXS.

It is known that PA6 crystallizes with a spherulitic morphology, where the small-angle light scattering pattern gives rise to a 4 lobe pattern [8]. Results from the isothermal crystallization showed that the PA6-PNP nanocomposites also crystallize with the same morphology. Figure 2(a) show a POM micrograph of PA6-PNP2 nanocomposite, and Figure 2(b) show the corresponding SALS H_V pattern. Nylon and its nanocomposites showed variation in spherulite size. The radius of the spherulite, R, was determined from the angular position of the intensity maximum, $2\theta_{max}$, using Stein et al relation [9]

$$\frac{4\pi R}{\lambda}\sin\theta_{max} = 3.9 \tag{2}$$

where λ is the radiation wavelength (He-Ne laser, λ=632.8 nm). A spherulite radius R of 5 μm was determined for PA6, whereas for the nanocomposites PA6-PNP1 and PA6-PNP2 R was determined to be 15 and 8 μm, respectively.

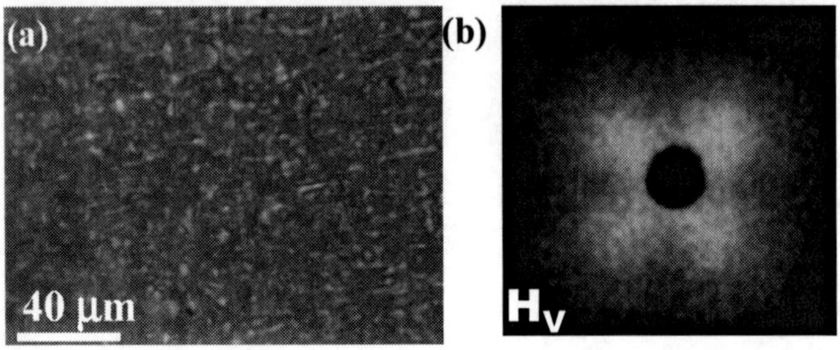

Figure 2. (a) POM micrograph and (b) H_V SALS pattern for PA6-PNP2 nanocomposite film; He-Ne laser.

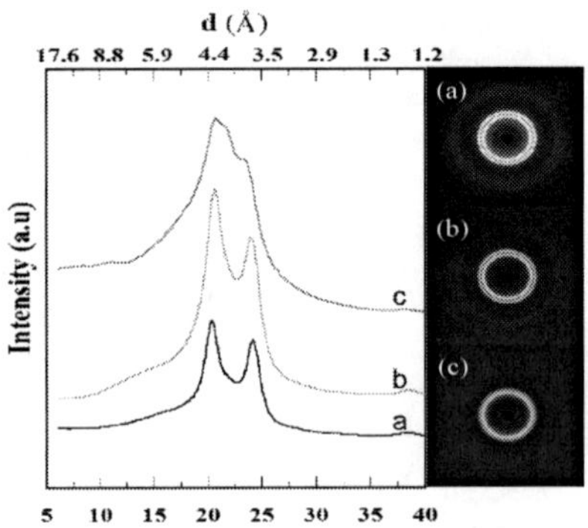

Figure 3. Azimuthally averaged WAXS intensity traces and diffraction patterns of (a) PA6, (b) PA6-PNP1, and (c) PA6-PNP2. Room temperature and CuKα radiation.

The influence of polymer nanoparticles on water absorption of PA6 was also investigated. For this, the samples were dried in a vacuum oven at 90°C overnight, and then placed in Petrie dishes at room temperature and the weight gain monitored using an analytical balance with a resolution of 10^{-2} mg. Weight gain was monitored for a period of 10 hours; the experimental data showed a linear growth. Data showed that the nanocomposites have a rate of mass gain nearly half of PA6. The rate of mass gain was extracted from the straight line slope, this rate was determined to be 9.3×10^{-4} g/hr for neat PA6, and 5.5×10^{-4} g/hr for PA6-PNP1 and PA6-PNP2.

Figure 3 shows the WAXS diffraction patterns of PA6 and the nanocomposites. The samples were hot pressed as thin films, taken into the melt and cooled slowly to allow for recrystalization. The patterns show sharp Debye-Scherrer crystalline rings, corresponding to the *100* and *010* crystalline reflections, indexed according to the α form (triclinic lattice), and some amorphous background. The azimuthally averaged intensity traces are also shown in Figure 3. The patterns show that the PNP2 nanoparticles strongly disrupted the crystallinity in the nanocomposites as the crystalline reflections are very weak, predominating the amorphous background. The degree of crystallinity was determined from the WAXS data using Ruland's method after subtracting the amorphous background [10]. The results, listed in Table 1, are consistent with the reduction in melting enthalpy obtained from DSC analyses.

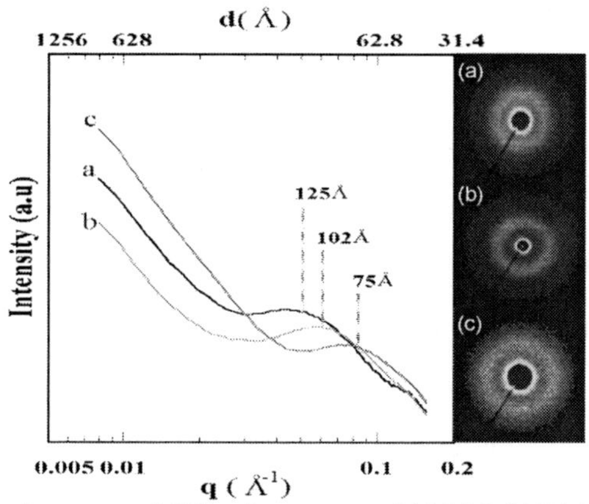

Figure 4. SAXS intensity traces and diffraction patterns of (a) PA6, (b) PA6-PNP1, and (c) PA6-PNP2.

The nanostructure was investigated via SAXS, the results are shown in Figure 4. The SAXS pattern of PA6 shows a well defined amorphous diffraction halo indicative of spatially periodic regions with higher electron density, and it is denoted the long range order. The azimuthally averaged intensity traces for all the samples, as a function of scattering vector q ($=4\pi \text{ Sin } \theta/\lambda$), are also shown in figure 4. The long range spacing for each sample is obtained from the intensity maxima using $D=2\pi/q_{max}$. The long range spacing is the distance between lamellae, including amorphous intra-lamellae regions. D and the spherulite radius R are not related as the former denotes nanoscale regions whereas the latter denotes micron scale regions. It can be seen that the PNPs also have a strong influence on the long range order in the nanocomposites, as the long range spacing has been reduced from 125 Å for PA6 to 102 Å for PA6-PNP1 and 75 Å for

PA6-PNP2 (the intensity maxima shifted to larger q values, meaning smaller distances as scattering and real space are reciprocal [10].

CONCLUSIONS

The results of this investigation have shown that the acrylic PNPs have a significant influence on the thermal properties of PA6, reducing the melting temperature and increasing the thermal stability. Furthermore, the presence of these nanoparticles reduced the rate of moisture absorption, and disrupted the ordering at the molecular and nano-scale, reducing the degree of crystallinity and the long range order. The rheological behavior (shear and extensional rheometry) previously reported had shown that the PNPs induced melt elasticity and strain hardening in an otherwise viscous melt [5]. The rheological results suggest a strong interaction between the PNPs and the PA6 matrix, perhaps molecular jamming, resembling molecular entanglements. The results of this investigation suggest a strong interaction of the PNPs and PA6.

ACKNOWLEDGEMENTS

E Huitron-Rattinger and B Alvarado-Tenorio were supported by graduate scholarships from the Mexican Council for Science and Technology (CONACyT). This research was supported by CONACyT (CIAM2006, grant 58646)

REFERENCES

1. A. Okada, M. Kawasumi, T. Kurauchi and O. Kamigaito, Polym Prep 28, 447 (1987).
2. M. Biswas and S. Sinha, Adv. Polym. Sci. 155, 167 (2001).
3. R.A. Vaia, H. Ishii and E.P. Giannelis, Chem. Mater. 5, 1694 (1993).
4. E.P. Giannelis, Adv Mater 8, 29 (1996).
5. E. Huitron-Rattinger, M.E. Romero-Guzman and A. Romo-Uribe, ACS PMSE 103, 166 (2010).
6. Y. Katoh and M. Okamoto, Polymer 50, 4718 (2009).
7. C. Valerio-Cardenas, A. Romo-Uribe and R. Cruz-Silva, Polym Eng Sci. 51, 254 (2011).
8. Y. Lin, W. Zhong, L. Shen, P. Xu and Q. Du, J. Macromol. Sci., Phys. B44, 161 (2005).
9. R.S. Stein, *Newer Methods of Polymer Characterization*, (Interscience Publishers, New York 1964) p. 155.
10. L. Alexander, *X-ray Diffraction Methods in Polymer Science,* (John Wiley and Sons, London 1969)

Mater. Res. Soc. Symp. Proc. Vol. 1312 © 2011 Materials Research Society
DOI: 10.1557/opl.2011.264

Novel Aster-like ZnO Nanowire Clusters for Nanocomposites

Mikhail Ladanov[1, 2, 3], Manoj Ram[2, 3], Ashok Kumar[2, 3], Garrett Matthews[4]

[1]Department of Electrical Engineering, University of South Florida, Tampa, FL, United States.
[2]Department of Mechanical Engineering, University of South Florida, Tampa, FL, United States.
[3]Nanotechnology Research and Education Center, University of South Florida, Tampa, FL, United States.
[4]Department of Physics, University of South Florida, Tampa, FL, United States.

ABSTRACT

ZnO nanostructures have attracted a great deal of interest because of their biocompatibility and outstanding optical and piezoelectric properties. Their uses are widely varying, including incorporation in sensors, solar cells, and nanogenerators. Biological systems are yet another area of application of ZnO nanowires. Apart from their electrical and optical properties, ZnO nanostructures can be used for the mechanical reinforcement of existing biomimetic scaffolds such as collagen and/or other biodegradable polymers (poly(lactic acid), polyglycolide, poly(alkyene succinate)s or polyhydroxylalkanoates). In this work, we have demonstrated a cheap and comparatively facile hydrothermal growth method for the bulk production of ZnO nanostructures exhibiting an aster-like geometry. The novel nanostructures of ZnO can be used as reinforced material to biopolymers. The aster shape has presented an increased surface area, providing a means for enhancing the stabilization of the gels and\or polymers. With controllable growth of ZnO nanostructures this method allows the geometry which could be tuned for maximal coupling between the two phases of composite and increased mechanical strength.

INTRODUCTION

ZnO is a biocompatible piezoelectric II-VI semiconductor with a wide direct band gap of 3.37 eV and large exiton binding energy of 60 meV. Recently, ZnO nanostructures have attracted great attention as a promising functional material due to its suitability for application in UV lasers[1, 2], light emitting diodes[3], sensors[4-6], nanogenerators[7-9] and solar cells. Due to biocompatibility[10, 11], ZnO can be used as a functional material in implantable devices or even as a reinforcing material for different biopolymers. In this case bulk growth of structures with large surface area to volume ratios while keeping the effective volume large enough for successful incorporation into biopolymers is desired.

Many methods have been used to prepare ZnO nanowires, including using chemical vapor deposition[4, 12], metal–organic chemical deposition[13, 14], pulsed laser deposition[15], physical vapor deposition[13, 16], vapor-liquid-solid methods[9, 17] and hydrothermal methods[1, 18, 19]. The hydrothermal method is one of the most simple and cost effective methods, making the scaling up of production straightforward.

Growth of ZnO nanostructures using the hydrothermal method has been studied extensively as it is a simple way to produce various shapes and geometries of nanostructures. The use of modifying agents, various substrates, wide spectra of seeding techniques and electric field assistance has become common for the modification of resultant properties, such as shape, size and morphology of ZnO nanostructures. However, mostly these techniques are suitable for the growth of ZnO nanostructures on various substrates.

In this paper, we demonstrate the growth of ZnO nanostructures in a bulk solution, where different sizes of ZnO nanoparticles were used as nucleation sites for the hydrothermal growth. During conventional hydrothermal growth ZnO nanowires grow not only on the surface of the substrate, but in the solution as well. This method often leaves debris on the resultant samples, but has the advantage that adding ZnO nanoparticles as growth sites changes the bulk-grown nanostructures greatly. We demonstrate these nanostructures, and discuss possibilities for large scale repeatable growth of such structures.

EXPERIMENTAL

Chemicals and reagents

The zinc nitrate hexahydrate ($Zn(NO_3)_2{\cdot}6H_2O$) and hexamethylenetetramine (HMTA), trisodium citrate ($C_6H_5Na_3O_7$), were all ACS grade and purchased from Sigma-Aldrich (USA). Zinc Oxide Nanoparticles with a nominal size of 10-30 nm were purchased from SkySpring Nanomaterials, Inc. and Zinc Oxide Nanopowder of nominal sizes <50 nm and <100 nm were purchased from Sigma Aldrich. All solvents and materials were employed as purchased without any further purification unless specified.

TEM of the nanoparticles was performed. The method followed was: a small amount of nanoparticles as purchased were mixed with ethanol and actively sonicated to separate bundled particles. A TEM copper mesh was dipped in the solution and dried. TEM was performed on this grid.

ZnO nanostructures growth

ZnO nanostructures were grown using the conventional hydrothermal method. 25 mM L^{-1} of zinc nitrate and hexamine solutions in deionized water (DIW) at 80°C were used as the growth solution. Different amounts of sodium citrate solution in DIW were added into selected samples, as specified.

ZnO nanoparticles in amount of 20-30 mg were vigorously sonicated in 1 mL of DIW to effectively separate and unclog them and were added to the growth solution. They acted as nucleation sites for nanostructure growth.

The solution was kept at 80°C with vigorous stirring for 4 hours and then washed with DIW and ethanol and finally filtered. Samples obtained were studied using SEM technique.

RESULTS AND DISCUSSION

Figure 1a shows a TEM image of as purchased ZnO nanoparticles with nominal size <30 nm. The TEM picture reveals less distribution in size than other samples with nanoparticles size of nearly 200 nm in comparison with the nominal 10-30. Figure 1b shows a TEM image of as purchased ZnO nanoparticles with nominal size <50 nm. ZnO nanoparticles have a large distribution in size and generally are of size much less than 50 nm. Figure 1c shows a TEM image of as purchased ZnO nanoparticles with nominal size <100 nm. ZnO nanoparticles have a large distribution in size and generally are of size ~100 nm. Insets in Figure 1 show atomic planes of the corresponding particles; the scale bar is 2 nm.

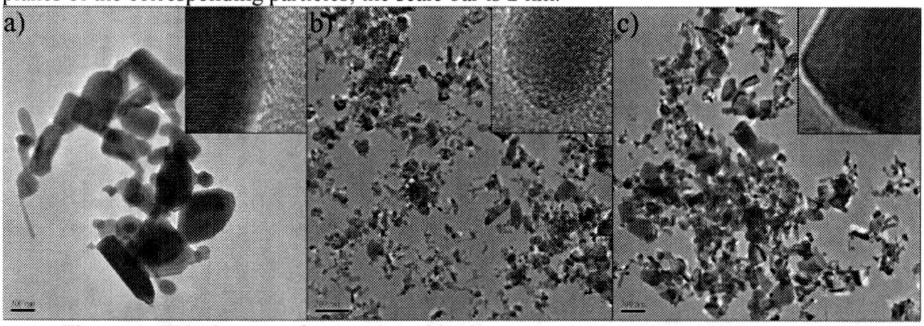

Figure 1. TEM images of as purchased ZnO nanoparticles with nominal sizes of a) 10-30 nm, b) <50 nm, c) <100 nm. Scale bar on the images is 200 nm. Insets: high resolution TEM of atomic planes of the nanoparticles. Scale bar is 2 nm.

Figure 2a shows a TEM image of ZnO nanostructures grown in bulk solution as described before with addition of 2μL of 0.1 mM/L of sodium citrate. Approximately 24 mg of ZnO nanoparticles with nominal size 10-30 nm were added as nucleation seeds. Aster-like shape of resultant ZnO nanostructure has been observed in the image, showing ZnO nanowires growing out of common nucleation site.

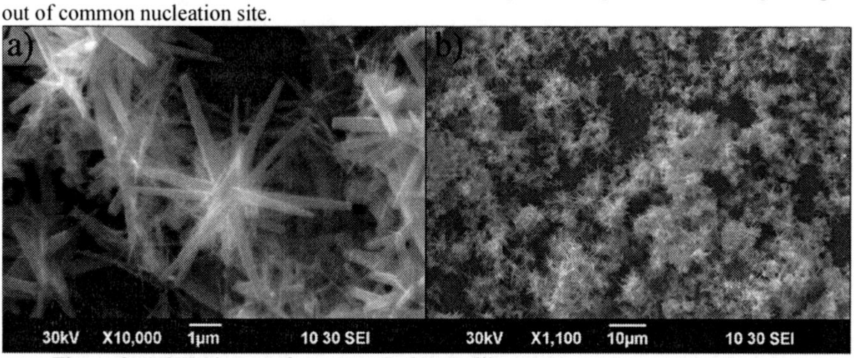

Figure 2. a) SEM image of as grown ZnO aster-like nanostructures with ZnO nanparticles of the nominal size of 10-30 nm acting as nucleation sites, b) Zoomed out SEM image of as grown ZnO aster-like nanostructures with ZnO nanparticles of the nominal size of 10-30 nm acting as nucleation sites. The aster-like structures are the preferable resultant structure with these growth conditions.

A lower magnification image of the same sample is presented in Figure 2b and shows the abundance of aster-like structures in the samples. This indicates that this growth mode is preferable under the present growth conditions.

An SEM image of ZnO aster-like structures grown with ZnO nanoparticles of nominal size of <50 nm are shown in Figure 3a. The nanostructures growth was performed following the general recipe with no sodium citrate added. 25 mg of ZnO nanoparticles of nominal size <50 nm were used as nucleation sites. Aster-like nanostructures are not as pronounced as structures with 30 nm ZnO nanoparticles nucleation, but are in abundance as shown in Figure 3a.

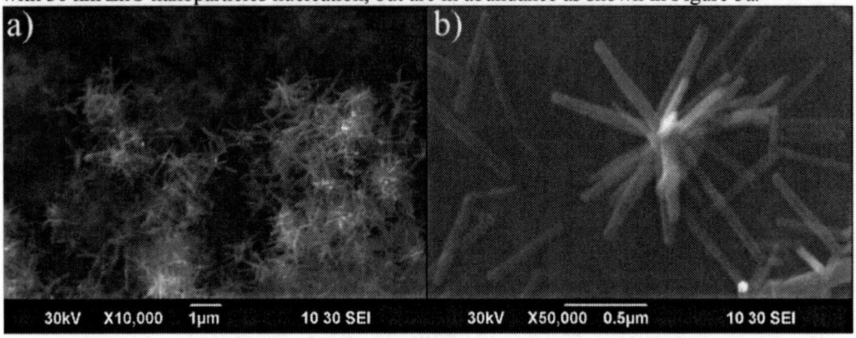

Figure 3. a) SEM image of ZnO aster-like structures grown with ZnO nanparticles of nominal size of <50 nm, b) SEM image of ZnO aster-like structures grown with ZnO nanoparticles of nominal size of <100 nm.

Figure 3b shows ZnO nanostructures grown with the same general recipe, with no sodium citrate added. 27 mg ZnO nanoparticles of <100 nm nominal size were added as nucleation sites in this growth. Although aster-like nanostructures have been observed, the resultant nanostructures are in mass plain ZnO nanowires.

Also performed was a series of experiments in which higher concentrations of sodium citrate, which acts as surfactant agent, were used. Here aster-like structures were not observed in any sample, as can be seen in Figure 4. Instead, rod-like nanowires of increased size (with addition of 20 µL of 10 mM/L sodium citrate) were obtained. The use of even larger concentrations of the surfactant (addition of 2 mL of 10 mM/L of sodium citrate) somewhat cubic-like structures resulted (Figure 4b).

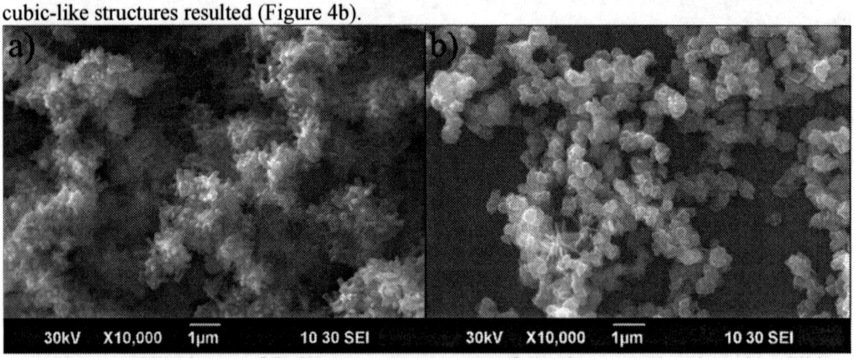

Figure 4. SEM image of ZnO nanostructures grown following general recipe with addition of a) 20 µL of 10 mM/L of sodium citrate per 250 mL of growth solution and b) 2 mL of 10 mM/L of sodium citrate per 250 mL of growth solution.

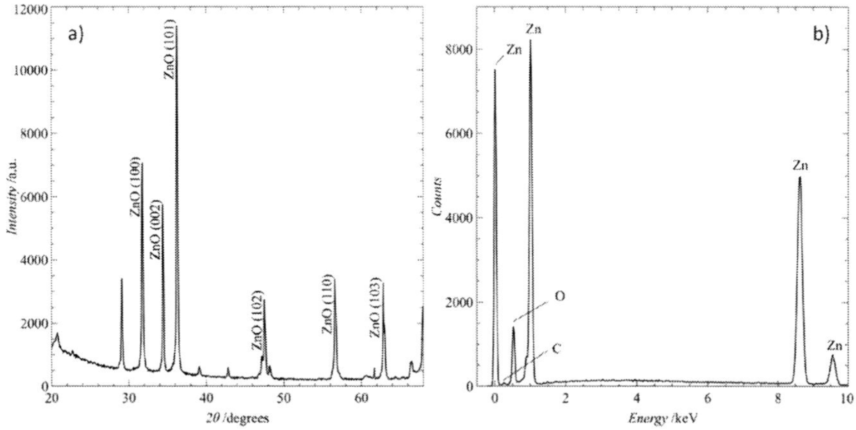

Figure 5. Typical a) XRD spectra and b) EDS spectra of ZnO nanostructures grown following general recipe. Nanostructures were diluted in ethanol, dropcasted on Si substrate and dried for XRD and EDS measurements.

X-ray diffraction (XRD) and Energy-dispersive X-ray spectroscopy (EDS) measurements were also performed to characterize obtained ZnO nanostructures. Figures 5a and 5b show typical XRD and EDS spectra, respectively. In XRD measurements obtained peaks at 31.77°, 34.43°, 36.26°, 47.55°, 56.61° and 62.87° correspond to hexagonal ZnO structure with crystallographic (100), (002), (101), (102), (110) and (103) directions, respectively. EDS measurements reveal that sample generally contains the atomic ratio 35.7% and 53.76% of zinc and oxygen, respectively. Higher atomic percent of oxygen is not uncommon for metal oxides. Meanwhile, EDS measurements support the proper identification of peaks in XRD spectra of ZnO nanostructures.

CONCLUSION

Conventional hydrothermal growth with addition of ZnO nanoparticles as nucleation sites could be used for growth of aster-like nanostructures. Care should be taken regarding the size and quality of nanoparticles. Uniformly sized seeding particles with lateral dimensions of ~200 nm should be used as nucleation sites. In this study the best results were obtained using nanoparticles with nominal size of 10-30 nm.

REFERENCES

1. L. E. Greene, M. Law, J. Goldberger, F. Kim, J. C. Johnson, Y. Zhang, R. J. Saykally and P. Yang, Angew. Chem. Int. Ed. **42** (26), 3031-3034 (2003).
2. Z. K. Tang, G. K. L. Wong, P. Yu, M. Kawasaki, A. Ohtomo, H. Koinuma and Y. Segawa, Appl. Phys. Lett. **72** (25), 3270-3270 (1998).
3. C. H. Liu, J. A. Zapien, Y. Yao, X. M. Meng, C. S. Lee, S. S. Fan, Y. Lifshitz and S. T. Lee, Adv. Mater. (Weinheim, Ger.) **15** (10), 838-841 (2003).

4. Z. Fan, D. Wang, P.-C. Chang, W.-Y. Tseng and J. G. Lu, Appl. Phys. Lett. **85** (24), 5923-5923 (2004).
5. Q. Wan, Q. H. Li, Y. J. Chen, T. H. Wang, X. L. He, J. P. Li and C. L. Lin, Appl. Phys. Lett. **84** (18), 3654-3654 (2004).
6. WangWang, J. Zhou, Song, J. Liu, N. Xu and Z. L. Wang, Nano Lett. **6** (12), 2768-2772 (2006).
7. M.-P. Lu, J. Song, M.-Y. Lu, M.-T. Chen, Y. Gao, L.-J. Chen and Z. L. Wang, Nano Lett. **9** (3), 1223-1227 (2009).
8. X. Wang, J. Song, J. Liu and Z. L. Wang, Science **316** (5821), 102-105 (2007).
9. Z. L. Wang and J. Song, Science **312** (5771), 242-246 (2006).
10. Z. Li, R. Yang, M. Yu, F. Bai, C. Li and Z. L. Wang, The Journal of Physical Chemistry C **112** (51), 20114-20117 (2008).
11. J. Zhou, N. S. Xu and Z. L. Wang, Adv. Mater. (Weinheim, Ger.) **18** (18), 2432-2435 (2006).
12. P.-C. Chang, Z. Fan, D. Wang, W.-Y. Tseng, W.-A. Chiou, J. Hong and J. G. Lu, Chem. Mater. **16** (24), 5133-5137 (2004).
13. W. Lee, M.-C. Jeong and J.-M. Myoung, Acta Mater. **52** (13), 3949-3957 (2004).
14. W. I. Park, D. H. Kim, S. W. Jung and G.-C. Yi, Appl. Phys. Lett. **80** (22), 4232-4232 (2002).
15. Y. Zhang, R. E. Russo and S. S. Mao, Appl. Phys. Lett. **87** (13), 133115-133115 (2005).
16. S. C. Lyu, Y. Zhang, C. J. Lee, H. Ruh and H. J. Lee, Chem. Mater. **15** (17), 3294-3299 (2003).
17. M. H. Huang, Y. Wu, H. Feick, N. Tran, E. Weber and P. Yang, Adv. Mater. (Weinheim, Ger.) **13** (2), 113-116 (2001).
18. L. E. Greene, M. Law, D. H. Tan, M. Montano, J. Goldberger, G. Somorjai and P. Yang, Nano Lett. **5** (7), 1231-1236 (2005).
19. Y. F. Hsu, Y. Y. Xi, K. H. Tam, A. B. Djurišić, J. Luo, C. C. Ling, C. K. Cheung, A. M. C. Ng, W. K. Chan, X. Deng, C. D. Beling, S. Fung, K. W. Cheah, P. W. K. Fong and C. C. Surya, Adv. Funct. Mater. **18** (7), 1020-1030 (2008).

Mater. Res. Soc. Symp. Proc. Vol. 1312 © 2011 Materials Research Society
DOI: 10.1557/opl.2011.169

Thermal and Mechanical Characterization of Jute-Biopol Nanophased Green Composites

Mohammad K Hossain, Mohammad W Dewan, Mahesh Hosur and Shaik Jeelani
Center for Advanced Materials (T-CAM), Tuskegee University
101 Chappie James Center, Tuskegee, AL 36088, U.S.A.

ABSTRACT

Surface modification of jute fibers was accomplished by performing chemical treatments including detergent washing, dewaxing, alkali, and acetic acid treatment. Morphology of modified surfaces examined using scanning electron microscopy (SEM) revealed improved surfaces for better adhesion with matrix. Better thermal performance of treated fibers was found from thermogravimetric analysis (TGA). Enhanced tensile properties of treated fibers were obtained from tensile tests. Using solution intercalation technique and magnetic stirring, 2%, 3%, and 4% by weight Montmorillonite K10 nanoclay were dispersed into a biodegradable polymer, Biopol. Thermal performance of nanoclay infused Biopol characterized using dynamic scanning calorimetry (DSC) showed improved degree of crystallinity by 7%. Jute fiber reinforced Biopol biocomposites with and without nanoclay were manufactured using treated and untreated jute fibers by compression molding process. Thermal and mechanical responses of treated fiber reinforced Biopol composites (TJBC) without nanoclay evaluated using dynamic mechanical analysis (DMA) and flexure tests showed 9% and 12% increase in storage modulus and flexure strength, respectively, compared to untreated jute fiber reinforced composites (UTJBC). The respective values were 100% and 35% for 4% nanoclay infused TJBC compared to UTJBC without nanoclay.

INTRODUCTION

Manufacturing of synthetic fiber composites not only consume huge energy but also their disposal at the end of the life cycle is very difficult since there is virtually no recycling option. Hence, the biocomposite industry is developing at a significant pace to meet growing consumer awareness and follow new environmental regulations [1]. Lignocellulosic bio-fibers derived from leaf, bast, fruit, grass or cane contribute to the strength of bio as well as synthetic polymer composites in various applications [2].

Elements of jute fibers are- cellulose, hemicelluloses, lignin, and pectin. Cellulose is the main element of jute fiber, which is resistant to alkali but hydrolyzed in acid. Hemicellulose is hydrophilic, soluble in alkali, and easily hydrolyzed in acids. Lignins are amorphous and hydrophobic in nature. Pectins are like waxes that provide plant flexibility [3]. Jute fibers consist of –OH group which causes it to be susceptible to moisture and impairs the dimensional stability. These natural fibers do not efficiently adhere to non-polar matrices due to this polar group. To overcome this difficulty, these fibers should be modified chemically or physically [4]. Chemically modified surfaces decrease moisture absorption, and increase tensile strength [5-6] and wettability of fibers by matrix [7]. Various research groups have worked on biodegradable polymeric materials such as Bionolle, Biopol, poly(3-hydroxybutyrate) (PHB) and Polylactic acid (PLA) [8] to use in natural fiber reinforced composite. PHB and Biopol are synthesized by bacteria as macromolecules. The remaining biodegradable polymers are synthetic or semi-synthetic [9]. PHB and its copolymers are highly crystalline and have a melting point, strength, and modulus comparable to those of isotactic polypropylene [10]. Moisture barrier, flammability resistance, thermal, and mechanical properties of polymeric composites can be improved by adding a small amount of nanoclay as filler particles [11]. Nanoclay agglomerates into the composites due to improper dispersion. A stabilization process is used to remove larger clay agglomerates in the nanoclay suspension [12]. To the best of authors' knowledge, no study was reported in the open literatures performed on jute-based nanocomposites that are 100% biodegradable. Hence, the objective of this study is to develop a 100% biodegradable jute-based nanocomposite for thermal and mechanical characterization.

MATERIALS

Poly(3-hydroxybutyrate-co-3-hydroxyvalerate-12%-PHBV) (Biopol), hessian jute fibers, and Montmorillonite K10 nanoclay were used as matrix, reinforcement, and nanofillers in this study. Alcojet detergent, 50% ethanol solution, 50% NaOH solution, and 99% acetic acid solution were used for chemical treatments and 99% chloroform solution was used to dissolve Biopol.

SURFACE MODIFICATION

Detergent washing, dewaxing, alkali treatment, and soaking with acetic acid were performed on hydrophilic jute fibers to improve interfacial bonding with hydrophobic polymer, Biopol. Dirt was removed from fibers by detergent washing keeping fibers into 5% detergent solutions at 30 ^{0}C for 1 hour and washing them subsequently with water. After drying pectin was removed from these detergent washed fibers by keeping them into 5% ethanol solution at 30 ^{0}C for 1 hour followed by washing with water and drying. Dewaxed fibers were kept into 5% NaOH solution at 30^{0}C for 2 hours and washed with distilled water to remove lignins and hemicelluloses. Alkali treatment resulted in major number of –OH group accessible on surfaces of fibers [13], and broke down fiber bundles into single fibers and increased effective surface area available for interacting with matrix. Alkali treated fibers were soaked with distilled water-acetic acid (2%) solution for 1 hour followed by washing with distilled water and drying. Acetic acid neutralized sodium ions which came with fibers during alkali treatment and reacted with -OH group on fiber surfaces to convert hydrophilic surfaces of fibers into hydrophobic for better adhesion with Biopol.

COMPOSITE FABRICATION

Biopol was dissolved into cholorofrom at a ratio of 1:8 at room temperature and stirred by magnetic stirrer for 4 hours to prepare a homogeneous solution. In case of nanophased composites, nanoclay were infused into Biopol using solution intercalation techniques [14]. Measured amount of nanoclay was dissolved into chloroform and solution was then pourd to Biopol chloroform solution and again stirred for 2 hours to prepare a homogeneous mixture. Final solution with/without nanoclay was poured into a mold to prepare a 1.5 mm thick film and dried in a vacuum oven at 60 °C to evaporate all choloroform. Dried thick films were then placed in the hot press and 3 kips force was applied at 166 °C for 10 minutes to prepare 0.50 mm thin films. Jute-Biopol composites were manufactured by stacking prepared films with/without nanoclay and treated/untreated fibers like a sandwich using compression molding process applying 3 kips force at 166 °C for 15 minutes.

EXPERIMENTAL PROCEDURE

DSC scans were performed from 30 to 200 °C using heating and cooling cycles at a rate of 5 °C/min under nitrogen flow. TGA was conducted using nitrogen purge gas in order to acquire information about mass loss and thermal stability of jute fibers, Biopol, and jute-Biopol composites with/without nanoclay. Storage modulus, loss modulus, and tanδ of jute-Biopol samples were obtained from tests under three point bending mode at a heating rate of 5 °C/min from 30 to 120 °C with oscillation frequency of 1 Hz and an amplitude of 15μm according to ASTM D4065-01 standard. Tensile properties of untreated and treated jute fiber bundles were characterized using Zwick Roell testing unit according to ASTM D2256-02 standard under displacement control mode at a strain rate of 0.5/min and on a gauge length of 50 mm. Flexural properties of jute-Biopol composites with/without nanoclay were evaluated according to ASTM D790-02 standard under displacement control mode at a crosshead speed of 2.0 mm/min.

RESULTS AND DISCUSSIONS

Uniaxial tensile properties of treated and untreated jute fiber bindles are shown in figure 1(a). Fibers treated with NaOH for 2 hours exhibited improvements in tensile strength and modulus by 13% and 17%, respectively, compared to untreated fibers. Alkali treatment caused removal of non-cellulosic materials including hemicellulose, lignin, and pectin from interfibriller regions [15] and resulted higher percentage of cellulose which was the main contributor to higher tensile strength and modulus of fibers [16]. Higher concentrations of NaOH or longer exposure to NaOH weaken fibers and make it more brittle. Two hour-exposure times with 5% NaOH showed optimum results.

Surfaces of treated and untreated fibers were investigated using SEM micrographs (Figure 1(b)). These micrographs revealed relatively rougher surfaces in treated fibers compared to untreated fibers. Removal of surface impurities, non-cellulosic substances, inorganic materials, and waxes resulted in cleaner and rougher surfaces in finally treated fibers. Mesh like structure was observed in untreated fibers, which entangled the fibrils. Chemical treatments broke mesh-structure and caused defibrillation of fibers that provided higher strength to fibers. This observation was reported elsewhere [17]. Rougher surface and defibrillation were also attributed to better interaction of fibers with matrix for larger surface area.

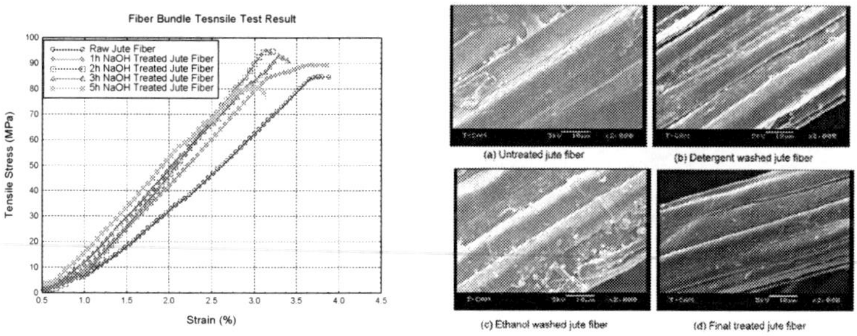

Figure 1. (a) Tensile test curves (b) SEM micrographs of treated and untreated jute fiber

DSC curves are illustrated in figure 2 for Biopol with/without nanoclay. Incorporation of nanoclay decreased the crystallization temperature indicating an enhanced crystallization ability of PHBV. Nanoclay acted as nucleating agent [18] and induced PHBV crystallization at lower temperature. Neat, 2%, 3%, and 4% nanoclay infused Biopol showed 44.36%, 47.46%, 47.47%, and 47.67% degree of crystallinity, respectively. Samples with nanoclay showed improvement by 7% in degree of crystallinity compared to samples without nanoclay. Bimodal melting endothermic peaks were observed during heating cycle due to crystallization phenomena of heterogeneous crystal morphology formed during processing [15]. It is believed that the lower melting temperature (T_{lm}) peaks are attributed to the melting of imperfect crystals formed during the isothermal crystallization process. Some of these crystals further experience melting-recrystallization and melt then at a higher melting temperature (T_{hm}). Most of the samples with nanoclay showed lower peaks at the T_{lm} compared to samples without nanoclay, indicating less crystal perfection in the neat Biopol. On the other hand, T_{hm} was the melting of those recrystallized crystals that was independent of the isothermal crystallization temperature and was almost constant. Thus, nanoclay addition increases not only the crystallization rate but also crystal perfection.

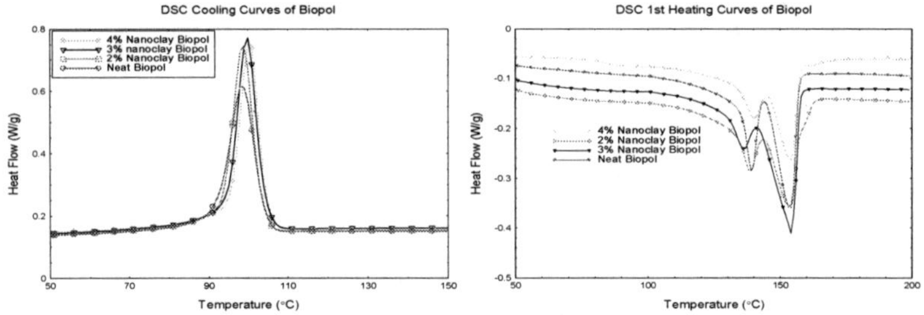

Figure 2. DSC cooling and heating curves for neat and nanoclay infused Biopol

Decomposition temperature of jute fibers and neat Biopol is about 323 °C and 265 °C, respectively. Thermal degradation of biocomposites is a cumulative phenomenon of matrix and fiber alone [19]. Composites started to degrade near to 270 °C. TJBC with nanoclay showed about 5% higher decomposition temperature compared to UTJBC (Figure 3). Untreated fibers may contain some volatile materials, which may lead to decompose at lower temperature. Two distinct decomposition temperature regions were observed in these composites: first region for matrix and second one for fiber.

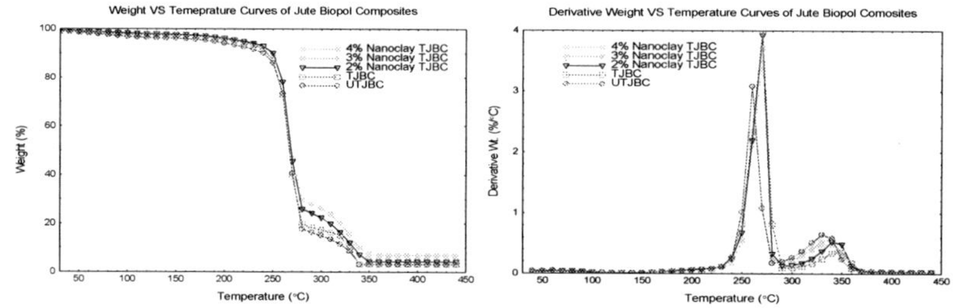

Figure 3. TGA curves of jute Biopol composites

From DMA tests (Figure 4(a)), it was observed that storage modulus was the highest at room temperature and decreased linearly with increasing temperature. Storage modulus (1605 MPa) was higher in TJBC compared to UTJBC (1467 MPa). Jute-Biopol samples with 2%, 3% and 4% nanoclay showed about 2550, 2900 and 3050 MPa storage modulus at 30 °C, respectively. As stiffness increases, tanδ value decreases reflecting reduced energy losses. With increase of temperature, energy losses increase thus the value of tanδ increases (Figure 4(b)). Nanoclay infusion into biocomposites decreased tanδ value indicating higher degree of crystallinity.

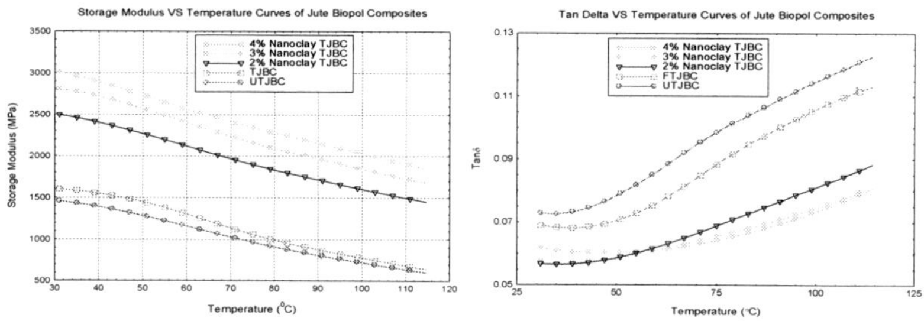

Figure 4. (a) Storage modulus and (b) tanδ VS temperature curves of jute Biopol composite

Flexural test results of TJBC and UTJBC with/without are shown table I. Better flexural properties were found in TJBC compared to UTJBC without nanoclay. Nanophased TJBC had resulted higher strength and modulus. However, addition of nanoclay was attributed to brittleness in composites, which was obvious from reduced percentage of strain at maximum stress.

Table I. Flexure test result of jute Biopol composite

	Strength (MPa	Strength change (%)	Strain (%)	Modulus (GPa)	Modulus change (%)
UTJBC	30.25	0	9.68	1.91	0
TJBC	33.88	12	6.71	2.16	13
2% Nanoclay TJBC	38.17	26	6.56	2.17	13.6
3% Nanoclay TJBC	39.43	30	5.30	3.20	67
4% Nanoclay TJBC	40.87	35	4.85	3.20	67

CONCLUSIONS

The significant conclusions drawn from this investigation are given below:

- Rougher and increased surface areas were observed in treated fibers that are important for proper fiber wetting and better bonding with matrix. Treated fibers also showed better tensile properties for the presence of higher percentage of crystalline celluloses.
- TJBC showed 5% and 9% improvement, respectively, in decomposition temperature and storage modulus compared to UTJBC. These respective values were 5% and 100% for 4% nanoclay infused TJBC compared to UTJBC without nanoclay.
- TJBC resulted in 12% and 13% improvement in flexural strength and modulus, respectively, whereas these respective values were 35% and 67% for TJBC with 4% nanoclay compared to UTJBC without nanoclay.

ACKNOWLEDGMENTS

The authors acknowledge the EPSCoR and NSF-RISE financial support for this research work.

REFERENCES

1. T. Doan, S. Gao, and E. Mader, *Compos. Sci. Technol.* **66**, 952 (2006).
2. E. Zini, M.L. Focarete, I. Noda, and M. Scandola, *Compos. Sci. Technol.* **67**, 2085(2007).
3. M.J. John, and S. Thomas, *Carbohydr. Polym.* **71**, 343 (2008).
4. F. Corrales, F. Vilaseca, M. Llop, J. Girones, J.A. Mendez, and P. Mutje, *J. Hazard. Mater.* **144**, 730 (2007).
5. B.M. Pejic, M.M. Kostic, P.D. Skundric, and J.Z. Praskalo, *Biosour. Technol.* **99**, 2686 (2008).
6. D. Ray, and B.K. Sarkar, *J. Appl. Polym. Sci.* **80**, 1013 (2001).
7. S. Wong, R. Shanks, and A. Hodzic, *Compos. Sci. Technol.* **64**, 1321 (2004).
8. E. Zini, M. Bairado, and M. Scandola, *Macromol. Biosci.* **4**, 286 (2004).
9. K. Sudesh, H. Abe, and Y. Doi, *Progr. Polym. Sci.* **25**, 1503 (2000).
10. N.M. Barkoula, S.K. Garkhail, and T. Peijs, Indus. *Crops. Prod.* **1**, 34 (2010).
11. S. Bruzaud, and A. Bourmaud, *Polym. Test.* **26**, 652 (2007).
12. G. Huang, and A.Netravali, *Compos. Sci. Technol.* **67**, 2005 (2005).
13. F. Vilaseca, J.A. Mendez, A. Pelach, M. Llop, N. Canigueral, J. Girones, X. Turon, and P. Mutje, *Process Biochem.* **42**, 329 (2007).
14. S. Wang, C. Song, G. Chen, G., T. Guo, J. Liu, B. Zhang, and S. Takeuchi, *Polym. Degrad. Stab.* **87**, 69 (2004).
15. S. Singh, A.K. Mohanty, T. Sugie, Y. Takai, and H. Hamada, *Composites Part A.* **39**, 875 (2008).
16. P. Saha, S. Manna, S.R. Chowdhury, R. Sen, D. Roy, and B. Adhikari, *Bioresour. Technol.* **101**, 3182 (2010).
17. M.A. Martins, P.K. Kiyohara, and I. Joekes, *J. Appl. Polym. Sci.* **94**, 2333 (2004).
18. R. Renstad, S. Karlsson, A.C. Albertsson, P.E. Werner, and M. Westdahl, *Polym. Int.* **43**, 201 (1997).
19. M.A. Khan, and J. Ganster, H.P. Fink, *Composites Part A.* **40**, 846 (2009).

Mater. Res. Soc. Symp. Proc. Vol. 1312 © 2011 Materials Research Society
DOI: 10.1557/opl.2011.146

Study on the Mechanical Properties and Creep Behaviour of Carbon Fiber Nano-Composites

Yi-Luen LI[1], Wei-Jen Chen[1], Chin-Lung Chiang[2], Ming-Chuen Yip[1,*]

[1] Department of Power Mechanical Engineering, National Tsing-Hua University, Hsin-Chu 30043, Taiwan, R.O.C.

[2] Department of Safety, Health and Environmental Engineering, Hung-Kuang University, Taichung, 433, Taiwan, R.O.C.

ABSTRACT

The surface modification of carbon nanotubes (CNTs) has been recently observed to influence the distribution of CNTs in epoxy resin and the mechanical properties and electrical conductivities of these CNTs. Accordingly, the treatment of CNTs with organic acids to oxidize them generates functional groups on the surface of CNTs. This investigation studies the consequent enhancement of the mechanical properties and electrical conductivities of CNTs. The influence of adding various proportions of CNTs to the epoxy resin on the mechanical properties and electrical conductivities of the composites thus formed is investigated, and the strength of the material is tested at different temperatures. Moreover, the creep behavior of carbon fiber (CF) /epoxy resin thermosetting composites was tested and analyzed at different temperatures and stresses. The creep exhibits only two stages- primary creep and steady-state creep. The effects of creep stress, creep time, and humidity on the creep of composites that contain various proportion of CNTs were investigated at various temperatures.

Creep strain is believed to increase with applied stress, creep time, humidity, and temperature. It also increases as hardness decrease. The test results also indicate that mechanical strength and electrical conductivity increase with the amount of CNTs added to the composites. Different coefficients of expansion of the matrix, fiber and CNTs, are such that overexpansion of the matrix at high temperature results in cracking in it. An SEM image of the fracture surface reveals debonding and the pulling out of longitudinal fibers because of poor interfacial bonding between fiber and matrix, which reduce overall strength.

INTRODUCTION

The incorporation of nanoparticles or nanotubes into the matrix of fiber-reinforced composites has been extensively investigated in recent years [1]. The creep deformation of engineering materials at elevated temperatures has received significant attention for many years [2-5]. This investigation studies the surface modification of CNTs, and the mechanical strength [6,7] and electrical conductivity of thermosetting composites with various proportions (0.25, 0.5, 0.75 wt%) of CNTs [8,9], The creep behavior of thermosetting composites is tested under various conditions of temperature, load and humidity [10-11]. The results of testing are adopted to investigate the mechanical strength, electrical conductivity and creep behavior.

EXPERIMENTAL

Materials and Experimental Procedure

The MWCNTs that were used for reinforcement in this study were supplied by Desun Mamo Co. Ltd. A MWCNT / epoxy resin prepreg laminate was formed by hot pressing. Epoxy was shaped under a pressure of 1000 psi, at a temperature of 150℃. In the experimental procedure, CNTs were dispersed among isopropyl alcohol, into which titanium (IV) n-butoxide, (3-aminopropyl) triethoxysilane (APTES) and HCl were stirred for 48 hours. The system was then placed in an oven to bake it to modify the CNTs. The mixed solution with dispersed CNTs was poured into a liquid impregnation sink, and then the fixed-size carbon fiber prepreg was placed into the solution. Ultrasonication was performed to distribute uniformly the CNTs among the carbon fiber prepreg. After the impregnation, the carbon fiber prepreg was placed in an oven to remove the residual solution by baking it at 50℃ for six hours to produce the prepreg. Then, the prepregs were piled into a mold to fabricate the prepreg laminates by hot pressing.

Measurement

Tensile strength, flexural strength, impact strength, electrical conductivity and hygroscopicity tests were performed according to the standards ASTM D3039, ASTM D790, ASTM D256, ASTM D257 and ASTM D570-98, respectively. The variation of the flexure strength of the composites with CNT content at the three temperatures 25℃, 85℃ and 175℃ was also examined. Creep strain was tested under various conditions, including of temperature, moisture and stress.

DISCUSSION

Static strength test

Figures 1 (a) and (b) comapre the mechanical strength of unmodified CNT-added composite with that of modified CNT-added composite.

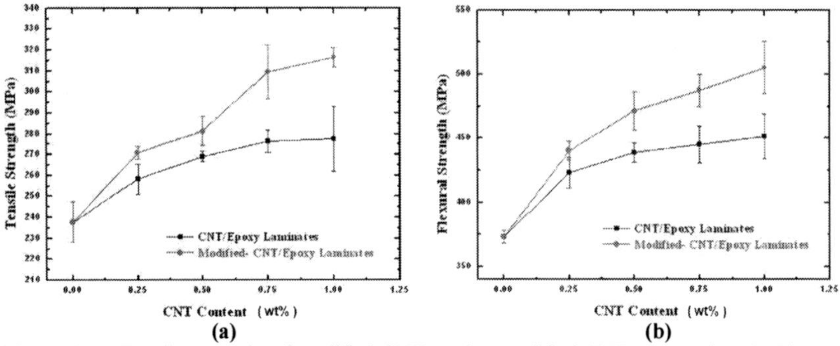

Figure 1. (a) Tensile strengths of modified-CNTs and unmodified-CNTs composites.**(b)** Flexure strengths of modified-CNTs and unmodified-CNTs composites

According to the test results, the mechanical strength of the modified CNT-added composite exceeded that of the unmodified CNT-added composite.

Figures 2 (a) and (b) display SEM images of the distribution of unmodified CNTs and modified CNTs among epoxy resin prepreg laminate, respectively. Figure 2 (a) presents agglomeration and a poorer distribution of CNTs throughout the epoxy resin than that of modified CNTs throughout the epoxy resin in Fig. 2 (b).

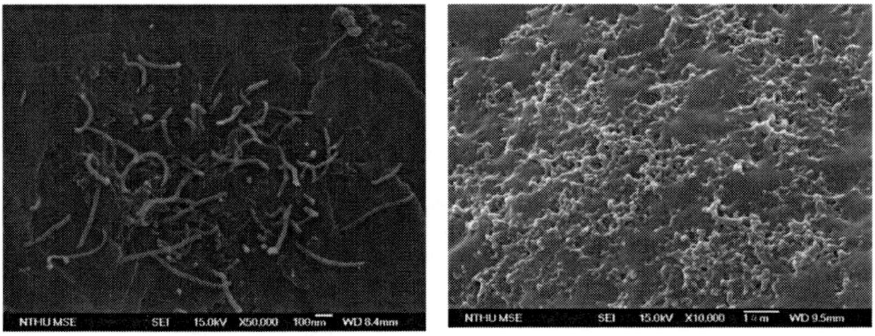

(a) **(b)**

Figure 2. (a). Dispersion of unmodified CNTs among epoxy resin matrix. **(b).** Dispersion of modified CNTs among epoxy resin matrix

Modified CNT-added composite has a very high mechanical strength. The mechanical test results demonstrate that the strength increases with CNT content. This result demonstrates that adding CNTs is essential to the reinforcement of nanocomposite.

Electrical Conductivity

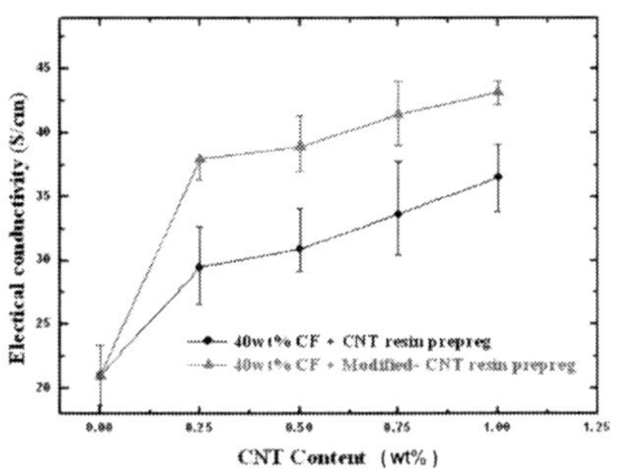

Figure 3. Electrical conductivities of modified-CNTs and unmodified-CNTs/CF/epoxy resin laminates

Figure 3 displays the obtained surface resistivities. The CNTs/epoxy resin prepreg laminates have greater electrical conductivity than the pure epoxy resin prepreg laminates.

Accordingly, the electrical conductivity of CNTs/epoxy resin prepreg laminate increases with CNT content, and the electrical conductivity of modified-CNTs/ epoxy resin prepreg laminate exceeds that of unmodified-CNTs/epoxy resin prepreg laminate.

Creep Behavior

As shown in Figs.4 (a) and (b), the creep behavior of CF/epoxy resin and CNTs/ CF/epoxy resin thermosetting composite were tested at a constant temperature of 55℃, a constant relative humidity (75%RH) and four loadings. The creep strain of the composites increased with stress.

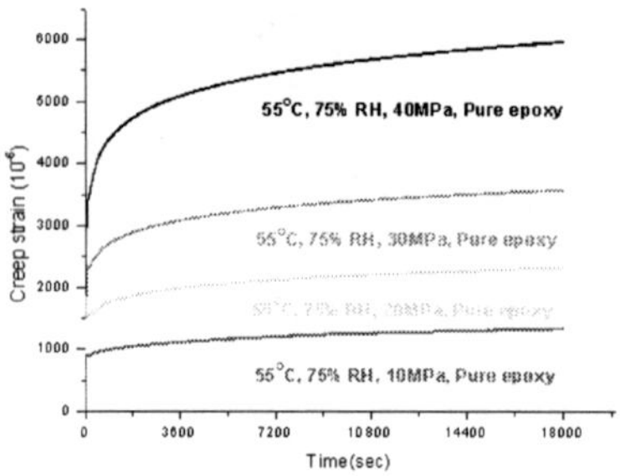

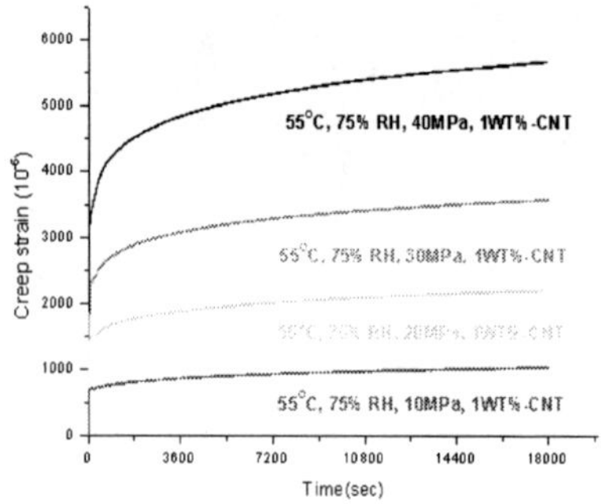

(a) (b)

Figure 4. (a) Creep behavior of CF/epoxy resin laminates tested at constant temperature and relative humidity, under four loadings. **(b)** Creep behavior of CNTs/CF/epoxy resin laminates tested at constant temperature and relative humidity, under four different loadings

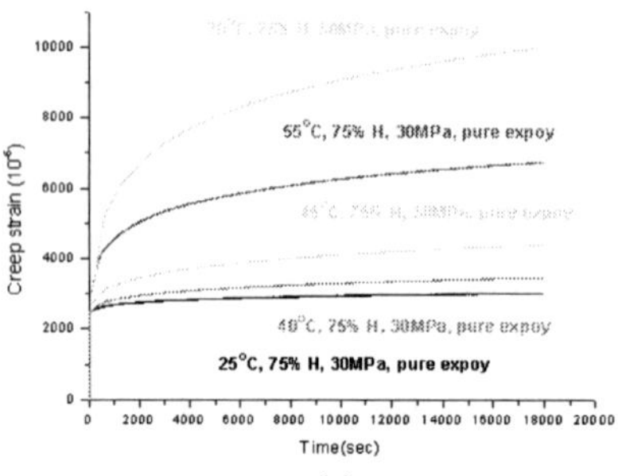

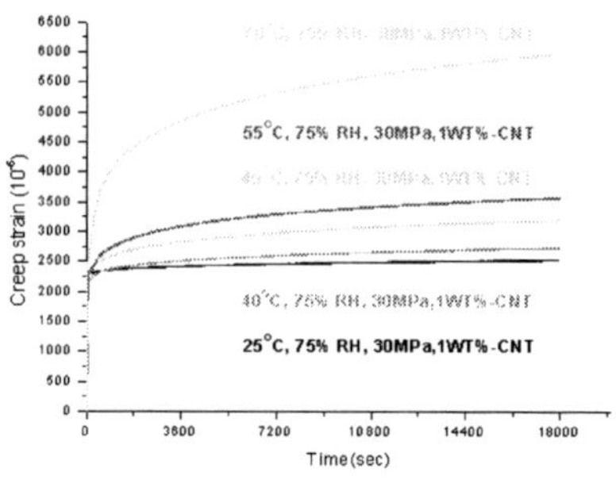

(a) (b)

Figure 5. (a) Creep behavior of CF/epoxy resin laminates tested under constant loading and relative humidity, at five temperatures. **(b)** Creep behavior of CNTs/CF/epoxy resin laminates tested at constant relative humidity and constant loading, at five temperatures.

Figures 5(a) and (b) display the creep behavior of CF/epoxy and CNTs/CF/epoxy resin composite, respectively, at a constant loading of 30MPa, a constant relative humidity of 75%H and five temperatures （25℃, 40℃, 45℃, 55℃, 70℃）.

The creep strain of composites increased with temperature. Creep behavior was especially evident when the specimen was tested at a high temperature of 70℃.

Figure 6 (a) compares the creep behaviors of CNTs/CF/epoxy resin and CF/epoxy resin thermosetting composite at constant temperature, relative humidity and loading. Figure 6 (b) plots the creep behavior of CF/epoxy resin composite at constant loading, constant temperature and two relative humidities.

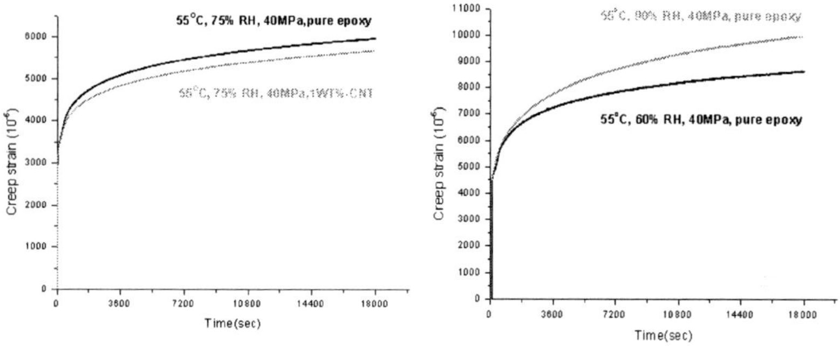

(a) **(b)**

Figure 6. (a) Creep behaviors of CNTs/CF/epoxy resin and CF/epoxy resin thermosetting composite at constant temperature, relative humidity and loading. **(b)** Creep behavior of CF/epoxy resin laminates at a constant loading of 40MPa, a constant temperature at 55℃ and two relative humidities（60%RH, 90%RH）.

CONCLUSIONS

Test results demonstrate that the mechanical strength of modified CNT-added composite exceeds that of CNT-added composite because the distribution of modified CNTs among epoxy resin is better and the interfacial bonding between modified CNTs and epoxy resin is better, without any agglomeration. In fact, modified CNT-added composite has great mechanical strength. Because CNTs have very high electrical conductivity, the electrical conductivity of CNT-added composite exceeds that of the pure resin laminate. Therefore, the electrical conductivity of CNT-added composite increases with CNT content. The test results and comparisons of the creep behaviors of CF/epoxy and CNT/CF/epoxy resin laminates under various conditions of loading, temperature and humidity, demonstrate that these are all factors that critically influence on the creep behavior. According to the above results, CNT can provide great mechanical reinforcement, and adding an appropriate amount of CNTs also can reduce the creep strain rate.

ACKNOWLEDGEMENTS

The authors are grateful for the help of the Center for Nanotechology, Materials Science, and Microsystems at National Tsing-Hua University, Taiwan, R.O.C. Ted Knoy is appreciated for his editorial assistance.

REFERENCES

1. M. Cochet, W. K. Maser, A. M. Benito, M. A. Callejas, M. T. Martinez and J. M. Benoit, "Synthesis of a new polyaniline/nanotube composite: "in-situ" polymerization and charge transfer through siteselective interaction," Chem. Comm. 16 (2001) 1450–1451.
2. Hayhurst DR. J. Mech. Phys. Solids 1972;20:381.
3. Finnie L, Heller WR. Creep of engineering materials. London: McGraw—Hill Book Company, 1959.
4. Bendersky L, Rosen A, Mukherjee AK. Int. Metal. Rev.1985;130:1]15.
5. Mughrabi H. Acta Metall. Mater. 1991;139:3067-3070.
6. S. Kumar, H. Doshi, M. Srinivasarao, J. O. Park and D. A. Schiraldi, "Fibers from polypropylene/nano carbon fiber composites," Polymer 43 (2002) 1701–1703.
7. D. Qian, E. C. Dickey, R. Andrews and T. Rantell, "Load transfer and deformation mechanisms in carbon nanotube-polystyrene composites," Applied Physics Letters 76(20) (2000) 2868–2870.
8. I. Novak., I. Chodak., Investigation of the correlateion between electrical conductiveity and elongateion at break in polyurethane-based adhesives, Synthesis Metal, Vol. 131, (2002), pp.93-98.
9. M. T. Kortschot and R. T. Woodhams, Computer Simulation of the Electrical Conductivity of Polymer composite Containing Matallic Fillers, Polymer Composite, Vol. 9, No.1, February, (1988), pp.60-71.
10. Zhang, S. Y. and Xiang, X. Y., "Creep Characterization of a Fiber Reinforced Plastic Material", J. Reinforced Plastic and Composites , 11,1187-1195,1992.
11. Chen, C. H. and Chen, Y. C., "The Creep Behavior of Solid filled Rubber Composites", J. Polymer Research, 1(1), 75-83, 1994.

Mater. Res. Soc. Symp. Proc. Vol. 1312 © 2011 Materials Research Society
DOI: 10.1557/opl.2011.111

Study of Mechanical Responses and Thermal Expansion of CNF-modified Polyester Nanocomposites Processed by Different Mixing Systems

Muhammad E. Hossain, Mohammad K. Hossain[1], Mahesh Hosur and Shaik Jeelani
Center for Advanced Materials (T-CAM), Tuskegee University
101 Chappie James Center, Tuskegee, AL 36088, U.S.A.

ABSTRACT

In this study, different dispersion techniques such as sonication at high frequency, mechanical mixing, and magnetic stirring methods were employed to infuse 0.1 to 0.4 wt.% carbon nanofiber (CNF) into polyester matrix to study the influence of CNF on mechanical and thermal properties of the polyester nanocomposites. Dispersion of CNF studied using scanning electron microscopy (SEM) micrographs revealed excellent dispersion of CNF using sonication when 0.2 wt.% CNF was mixed in polyester resulting in enhanced mechanical response. On the other hand, agglomerations were observed in samples prepared with other mixing methods. Polyester with 0.2 wt.% CNF samples prepared by sonication resulted in 88% and 16% increase in flexural strength and modulus, respectively, over neat samples. Quasi-static compression tests showed similar increasing trend with addition of 0.2 wt.% CNF. Dynamic mechanical analysis (DMA) showed 35% and 5 °C improvement in the storage modulus and glass transition temperature (T_g), respectively, in the 0.2 wt.% loaded samples. Thermal mechanical analysis (TMA) performed on neat and samples with 0.2 wt.% CNF showed lower coefficient of thermal expansion (CTE) in nanophased sample compared to neat. Fracture morphology evaluated using SEM revealed relatively rougher surface in CNF-loaded polyester compared to neat as a result of better interaction between fiber and matrix due to the presence of CNF.

INTRODUCTION

Polyester resin is widely used for advanced composites where low cost, good strength and stiffness, processability, dimensional stability, and chemical resistance are required. However, mechanical properties of this polymer matrix, such as the strength, modulus, and toughness of polyester resins may not be enough for some end-use applications. In order to utilize polyester resin for such applications, addition of nanosize fillers to the resin which results in enhanced properties is a useful approach [1]. The performance of nanocomposites is highly dependent on uniform distribution of nanoparticles into the host polymer. Maximum performance of nanocomposites can be achieved with homogeneous dispersion of nanoparticles into a polymer matrix while non-homogeneous dispersion causes even degradation of the host polymer's properties. In this rationale, an efficient technique is required to be employed to disperse carbon nanofiber (CNF) into the polyester matrix in order to obtain the enhancement in the properties of polyester composites. Prolongo et al. [2] recommended carrying out several stages of CNF dispersion along chemical modification and functionalization of CNF and combining different dispersing techniques including sonication with high frequency, magnetic stirring, and high shear mixing for homogeneous nanocomposites fabrication. Rana et al. [3] achieved uniform dispersion of CNF into epoxy resin applying a combination of ultrasonication using solvent and surfactants. He et al. [4] concluded that mechanical properties of nanocomposites depended on ultrasonic dispersion time and long term sonication damaged surfaces of both untreated and

[1]Corresponding Author

plasma polymerized CNF. Eskin concluded in his papers [5] that acoustic cavitation is one of the efficient ways to disperse nanoparticles into polymeric materials. Zheng et al. [6] employed high frequency ultrasonic and mechanical mixing methods to disperse spherical SiO_2 in epoxy resin and found that addition of 3 wt.% of SiO_2 in epoxy improved tensile strength, stiffness and impact strength by 114, 13, and 56%, respectively. Xu et al. [7] have confirmed that infusion of vapor grown carbon nanofiber (VGCNF) had little influence on flexural and glass transition temperature but increased storage modulus. Pramoda et al. [8] showed a good dispersion of graphite nanoparticles throughout the PMMA matrix, and the addition of graphite into matrix led to the significant improvement in the thermal stability and storage modulus of the PMMA nanocomposites. CTE of the PMMA/graphite nanocomposites (PMMA/GNC) was decreased with an increase in the graphite content. Tognana et al. [9] evaluated the CTE of the epoxy matrix particulate composite as a function of filler content in both glassy and rubbery state ($293\text{ K} \leq T \leq 493\text{ K}$). In this work, polyester with 0.1-0.4 wt.% CNF loading nanocomposites along with neat were fabricated. Nanocomposites were characterized through flexural and quasi-static compression tests, TMA, and SEM analysis. The experimental results were used to assess the influence of CNF on properties of polyester material system.

SAMPLE FABRICATION

B-440 premium polyester resin, styrene, and heat treated PR-24 CNF were considered as matrix, thinner, and nanoparticle, respectively, in the current study. Polyester resin contains two-part: part-A (polyester resin) and part-B (MEKP- methyl ethyl ketone peroxide) as crosslinking agents. Sonication was performed using a high intensity ultrasonic irradiation (Ti-horn, 20 kHz Sonics Vibra Cell, Sonics Mandmaterials, Inc, USA) for 60, 90, and 120 minutes, respectively, adding 0.1-0.4 wt.% CNF to polyester resin while adding 10 wt.% styrene in a glass beaker [11-12]. The mixing process was carried out in a pulse mode of 30 sec. on and 15 sec. off at an amplitude of 50%. For other types of samples, CNF were mixed using mechanical mixing and magnetic stirring methods. To lessen the void formation, vacuum was applied using Brand Tech Vacuum system for about 90-120 minutes and 0.7 wt.% catalyst was then added to the mixer using a high-speed mechanical stirrer for about 2-3 minutes and vacuum was again applied for about 6-8 minutes to degasify the bubbles produced during the catalyst mixing. The as-prepared resin was poured into the mold and kept at room temperature for 12-15 hours. Controlled polyester samples were fabricated to compare with the nanophased samples. All samples were kept in a mechanical convection oven at 110 °C for 3 hours for post curing.

TEST SETUP

SEM (JEOL JSM 5800) studies were carried out in order to examine the change in microstructure due to CNF infusion. SEM also facilitates to monitor the failure approach at micro level. Flexure tests under three-point bend configuration were performed according to ASTM D790-02 standard in a displacement control mode at a crosshead speed of 2 mm/min. Quasi-static compression tests were carried out according to ASTM D 695-02a standard in a displacement control mode at a crosshead speed of 1.27 mm/min. DMA tests were carried out in the dual cantilever mode at a heating rate of 5 °C/min from 30 °C to 175 °C with an oscillation frequency of 1 Hz and amplitude of 15μm. TMA tests were performed in the expansion mode at a heating rate of 5 °C/min from 30 °C to 175 °C.

RESULTS AND DISCUSSIONS

Micrographic Analysis

To investigate the dispersion properties of CNF in polyester, drops of concentrated HNO_3 acid liquid were added on the cleavage surfaces to partly unveil the CNF formerly covered by the polyester. From the 0.2 wt.% CNF-filled polyester micrographs (Figure 1), excellent dispersion of CNF was found and there was no agglomeration of CNF. In fact, CNF was uniformly embedded in the resin. In 0.2 wt.% CNF-filled matrix, only broken ends of CNF was observed near the surface. It was also revealed that some CNF were broken in a brittle manner and some were pulled out. Etched surfaces revealed that interfacial bonding between the CNF and matrix was intact which would make CNF anchored in the embedding matrix [13]. Addition of CNF enhanced the crosslinking between polymer chains and provided the better chemical and interfacial bonding. Agglomerations in the polyester matrix were observed from the micrographs (Figure 1) of 0.2 wt.% infused polyester samples prepared through mechanical mixing and magnetic stirring methods, respectively. These agglomerates area create stress concentration zones which might act as crack initiator.

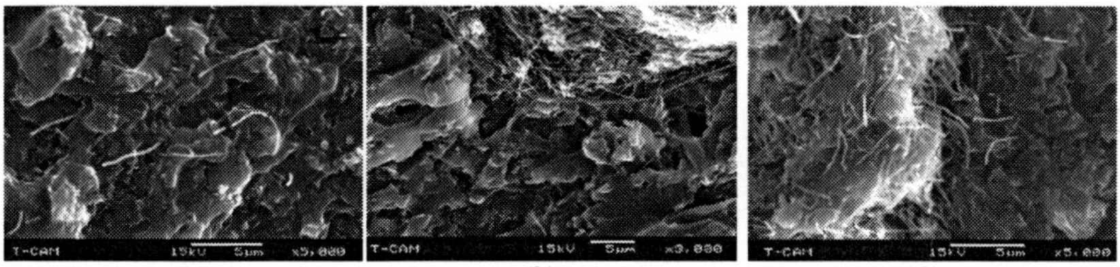

(a) (b) (c)

Figure 1. Micrographs of acid-etched 0.2 wt.% CNF-loaded polyester: (a) sonication, (b) mechanical mixing and (b) magnetic stirring

Flexure Test

Flexural tests performed on neat and nanophased samples showed higher flexural strength, modulus, and strain at maximum strength for all nanophased samples than that of the neat samples (Figure 2) due to the addition of CNF as CNF has high aspect ratio which can prevent crack generation and crack propagation in the matrix. In all cases, the samples failed rapidly after experiencing the maximum load showing brittle nature of failure due to the addition of CNF. From the analysis, it was found that the 0.2 wt.% CNF loading showed the optimal condition resulting in 88% and 16% enhancement in the flexural strength and modulus, respectively, compared to neat. The failure strain also increased significantly with the addition of CNF into the matrix. The flexural properties were slightly decreased at higher CNF content. It might be due to the CNF micro aggregates in various regions of the polymer, which act as areas of weakness. CNF loaded polyester prepared by sonication exhibited better properties than those prepared by other mixing methods like mechanical and magnetic stirring (Table I). Improved properties of nanocomposites are attributed to excellent CNF dispersion by sonication mixing whereas agglomeration of CNF when dispersed by mechanical and magnetic stirring methods resulted in reduced properties.

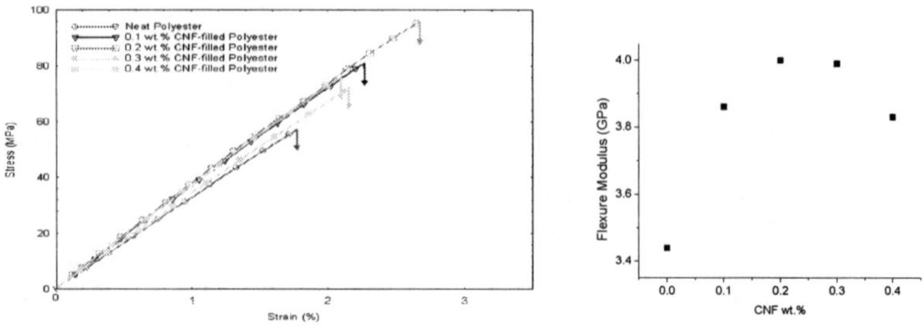

(a) (b)

Figure 2. Flexural (a) stress-strain (b) modulus-CNF wt.% plots of polyester samples with different wt.% of CNF

Table I. Flexural test results of polyester samples with different mixing methods

Type of Mixing	Average Max. Stress (MPa)	Average Flex-Modulus (GPa)
Neat Polyester	50.62 ± 1.85	3.44 ± 0.21
0.2% CNF infused by Magnetic Stirring 5 hrs.	54.78 ± 3.05	3.88 ± 0.09
0.2% CNF infused by Mechanical Mixing 90 m	62.04 ± 6.10	3.53 ± 0.10
0.2% CNF infused by Sonication Mixing 90 m	95.03 ± 4.44	4.00 ± 0.15

Quasi-Static Compression Tests

Nanocomposite with 0.2 wt.% CNF prepared by sonication exhibited better properties than that prepared by mixing methods like mechanical mixing and magnetic stirring (Figure 3).

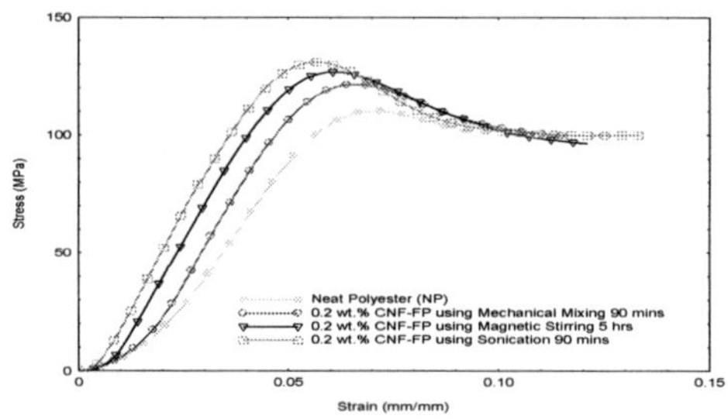

Figure 3. Compressive stress-strain plot of polyester samples with different mixing methods

In all cases, samples were compressed with increasing stress without sudden failure. It was evident that 0.2 wt.% CNF-loaded polyester prepared by sonication showed maximum enhancement in compressive strength and modulus by about 13% and 32%, respectively, compared to neat sample.

Dynamic Mechanical Analysis (DMA)

DMA resulted in improved storage modulus with the addition of CNF content up to 0.2 wt.% and then decreased for 0.3 and 0.4 wt.% (Table II). Modest improvement in T_g was observed due to the addition of CNF.

Table II. DMA results for neat polyester and CNF-loaded polyester samples

Composite Name	Storage Modulus (MPa)	% Change	Glass Transition Temp. (°C)	Change (°C)
Neat polyester	3321 ± 213	------	115.1 ± 0.51	------
0.1%CNF-loaded polyester	3561 ± 128	7.23	120.5 ± 0.68	5.4
0.2%CNF-loaded polyester	4489 ± 124	35.17	120.2 ± 0.46	5.1
0.3%CNF-loaded polyester	3923 ± 226	18.13	120.7 ± 0.54	5.6
0.4%CNF-loaded polyester	3642 ± 201	9.67	119.6 ± 0.37	4.5

Thermo Mechanical Analysis (TMA)

TMA performed on neat and 0.2 wt.% CNF loaded samples exhibited lower CTE than that of neat sample (Table III). The reduction of CTE observed after the addition of CNF was due to its good dispersion in the polyester and the reduced segmental motion of the polyester matrix [8].

Table III. Calculated data of CTE for polyester samples from TMA results

Type	Co-efficient of Thermal Expansion (CTE)	
	Below Glass Transition Temp. (40--80°C)	Above Glass Transition Temp. (120--170°C)
Neat Polyester	106 μm/(m°C)	201 μm/(m°C)
0.2% CNF-Polyester	104 μm/(m°C)	196 μm/(m°C)

Fracture Studies

SEM performed on the fractured samples of flexure tests revealed rough and smooth fracture surfaces in 0.2 CNF-loaded sample and neat sample, respectively (Figure 4).

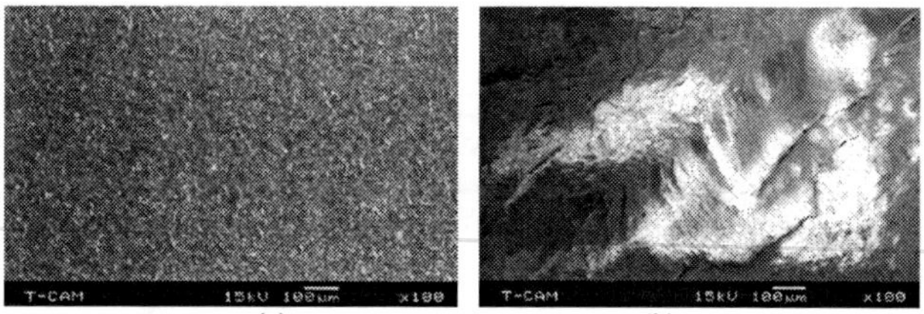

(a) (b)

Figure 4. SEM micrographs of fractured surfaces: (a) neat and (b) 0.2 wt.% loaded samples

CONCLUSIONS

The significant conclusions drawn from this study are given below:

- SEM micrographs revealed better dispersion of CNF in 0.2 wt.% CNF-loaded polyester prepared by sonication mixing and agglomeration in samples prepared by mechanical and magnetic mixing methods.
- Sample with 2 wt.% CNF prepared by sonication showed maximum improvement in flexural strength and modulus by 88% and 16%, respectively, over neat sample.
- Sample with 2 wt.% CNF prepared by sonication showed maximum improvement in compressive strength and modulus by 13% and 32%, respectively, over neat sample.
- The storage modulus was improved by about 35% for 0.2 wt.% CNF-loaded polyester samples.
- There was an increase in the glass transition temperature T_g with addition of CNF.
- Nanophased sample exhibited lower CTE than that of pure matrix which was the indication of the effect of CNF on the thermal expansion of polyester matrix.
- SEM micrographs revealed rougher fracture surface in nanophased sample compared to the neat as a result of better interaction between fiber and matrix due to the presence of CNF.

ACKNOWLEDGEMENTS

The authors greatly acknowledge the funding agency DOE, EPSCoR, NSF-RISE, NSF-CREST.

REFERENCES

1. E.S.A. Rashid, K. Ariffin, C.C. Kooi, and H.M. Akil, *Mater. Des.* **30**, 1(2009).
2. S.G. Prolongo, M. Buron, M.R. Gude, R. Chaos-Moran, M. Campo and A. Urena, *Compos. Sci. Technol.* **68**, 2722(2008).
3. S. Rana, R. Alagirusamy, and M. Joshi, *J. of Appl. Poly. Sci.* **118**, 2276(2010).
4. P. He, Y. Gao, J. Lian, L. Wang, D. Qian, J. Zhao, W. Wang, M. J. Schulz, X. P. Zhou and D. Shi, *Composites Part A.* **37**, 1270 (2006).
5. G.I. Eskin, Ultrason. *Sonochem.* **8**, 319(2001); **2**, 137(1995); **1**, 59(1994).
6. Y. Zheng, and R. Ning, *Mater. Lett.* **57**, 2940(2003).
7. J. Xu, J.P. Donohee, and C.U. Pittman Jr., *Composites Part A.* 35, 693(2004).
8. K.P. Pramoda, N.T.T. Linh, P.S. Tang, W.C. Tjiu, S.H. Goh, and C.B. He, *Comp. Sci. Technol.* **70,** 578(2010).
9. S. Tognana, W. Salgueiro, A. Somoza, J.A. Pomarico and H.F. Ranea-Sandoval, *Mater. Sci. Eng.* **157,** 26(2003).
10. H. Mahfuz, D. Das, F. Powell, and R. Granata, Proceedings of the SAMPE '10 International SAMPE Symposium and Exhibition, Seattle, WA, USA, May 17-20, 2010.
11. www.uscomposites.com/solvents.html (Browse date June 11, 2010).
12. R. Sakin, I. Ay, and R. Yaman, *Mater. Des.* **29**, 212(2008).
13. X.F. Li, K.T. Lau, and Y.S.Yin, *Compos. Sci. Technol.* **68**, 2876(2008).

Mater. Res. Soc. Symp. Proc. Vol. 1312 © 2011 Materials Research Society
DOI: 10.1557/opl.2011.202

Concomitant Channel Cracking and Interfacial Delamination in Polymer/Oxide Nano Hybrid Permeation Barriers in Flexible Electronics

Zheng Jia[1], Matthew B. Tucker[1], Teng Li[1,2,*]
[1] Department of Mechanical Engineering, [2]Maryland NanoCenter, University of Maryland, College Park, MD 20742
[*]Email: LiT@umd.edu

ABSTRACT

Polymer/oxide nano hybrid multilayer permeation barriers are emerging as a promising solution to the stringent barrier requirement of flexible electronics. Yet the mechanical failure of the multilayer permeation barriers could be fatal to their barrier performance. We study two co-evolving failure mechanisms of the multilayer permeation barriers under tension, namely, the cracking of the inorganic oxide layer and the delamination along the oxide-organic interface, using computational modeling. An effective driving force for the oxide layer cracking is determined, which decreases as the oxide-organic interfacial adhesion increases.

INTRODUCTION

The past decade has seen a surge of interests in flexible electronics, an emerging technology with an array of promising applications, such as paper-like displays and organic light emitting diodes (OLEDs) [1-5]. Flexible electronics are attractive for their large deformability, low weight, large surface areas, and low cost. Organic materials (e.g., polymers), used for both active device layers and structural substrates, make these attributes possible. For example, the emissive electroluminescent layer of OLEDs is composed of a thin film of organic compounds; and flexible devices are often fabricated on thin polymer substrates. The functional organic materials used in flexible devices are extremely vulnerable to moisture [6-10], thus result in a rather limited device lifetime — a grand challenge to the future success of flexible electronics technology. A natural choice for long-lasting flexible devices is to protect them with permeation barriers. Organic-inorganic multilayer permeation barriers are emerging as a promising solution to the stringent barrier requirement of flexible electronics [11-13]. The mechanical failure of the multilayer permeation barriers under large deformation, however, could be fatal to the barrier performance [13]. So far, little attention, if any [14, 15], has been given to the failure mechanics of multilayer organic-inorganic permeation barriers. To address this largely unexplored issue, in this paper we study two co-evolving tensile failure mechanisms of the multilayer permeation barriers, namely, channel cracking and interfacial delamination, through systematic computational modeling.

Designing a permeation barrier to allow a reasonable service life of flexible devices is remarkably challenging[13, 16]. For example, to guarantee a service life of about one year for OLEDs, the water vapor permeation rate (WVPR) of the barriers should be no more than 10^{-6} g/m^2/day [17, 18]. In contrast, the WVPR requirements for current electronic devices (e.g., LCD displays) range from 10^{0} g/m^2/day to 10^{-2} g/m^2/day [19], while the best traditional barriers can only achieve a WVPR of 10^{-3} g/m^2/day, still over three orders of magnitude greater than the design criteria for OLEDs. Furthermore, the large and repeated deformation that flexible devices undergo also requires that the permeation barriers be rugged.

Organic materials (e.g., polymers) are often rugged but highly permeable to moisture, while inorganic materials (e.g., ceramics and glass) are almost impermeable to moisture but extremely fragile. Thus, neither by itself is an effective permeation barrier for flexible electronics. Such a conundrum has led to the design of multilayer barrier coatings consisting of alternating thin films of inorganic and organic materials, with the hope to capitalize on both the impermeability of the inorganic layers and the ruggedness of the organic layers [15, 18, 20]. For example, a multilayer barrier made of a stack of alternating layers of polyacrylate and Al_2O_3 has been demonstrated to have a WVPR of about 2×10^{-6} $g/m^2/day$ at ambient conditions (e.g., Fig. 1) [11, 12]. It has been proposed that the organic layers "decouple" the pin-hole defects in neighboring Al_2O_3 layers, resulting in long and meandering moisture diffusion paths, thus significantly impeding the moisture permeation through the barrier [17, 21]. Such a promising barrier performance, however, was achieved in the as-made barriers that are not subject to large mechanical deformation.

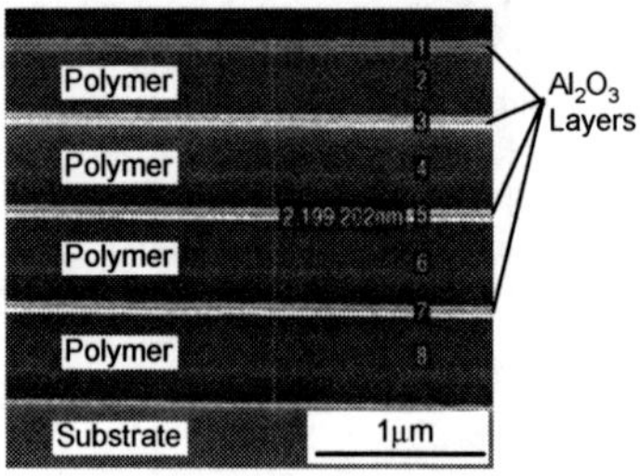

Fig. 1. Scanning electron microscope image of the cross section of an organic-inorganic multilayer permeation barrier (Courtesy of Vitex Sys. Inc.)

Given the stringent barrier requirements for flexible electronics, any mechanical failure in the brittle oxide films of the multilayer barriers would be fatal for the barrier performance. While the organic layers in multilayer barriers are compliant and can recover from large strains, the oxide layers are brittle and fracture at small strains. Typical fracture modes include channel cracks in the oxide layers and delamination along the oxide-organic interfaces. Moisture permeates through an as-made multilayer barrier mainly via the pin-hole defects in the oxide layers [17, 21]. By contrast, a channel crack in the oxide layer can substantially increase moisture permeation through the barrier by opening *a line source of moisture*, rather than the point source from the initial pin-holes. The situation becomes even worse when oxide-organic interfacial delamination accompanies the channel crack propagation, because the moisture wicking along the delaminated interface results in *an area source of moisture* at the delamination site. Therefore, fracture-induced defects (i.e., channel cracks and interfacial delamination) are expected to drastically accelerate moisture permeation through the barrier, and hence are fatal to the flexible devices. As a result, to study the barrier performance under large mechanical deformation, it is imperative to investigate the failure mechanism of the organic-inorganic hybrid layered structures.

Fracture mechanics of thin brittle films on rigid substrates has been a focal topic in the discipline of solid mechanics in the past two decades. Existing studies have largely originated from the context of traditional microelectronics [22-29], thus focused on film-substrate structures with comparable stiffness and subject to small strains (~1%). For flexible devices, the large difference in stiffness of the constituent materials of permeation barriers (e.g., the Young's modulus of Al_2O_3 is more than 100 times higher than that of polyacrylate) and the large mechanical deformation they undergo leads to rich but complicate failure behaviors that have not yet been clearly described in existing studies. On the research front of the fracture of thin brittle films on compliant substrates, Cotterell and Chen [30] investigated the buckling and cracking of thin films on compliant substrate under compression; Ambrico and Begley [31] explored the role of substrate compliance on channel cracking in a brittle film and found that on a compliant substrate the channel crack in the brittle film reaches steady-state propagation at a length much greater than the film thickness. Mei et al. [32] investigated the effect of interfacial delamination on channel cracking of a brittle thin film on an elastic substrate, and showed that interfacial delaminate not only increases the fracture driving force for channel crack growth, but also adds further to the fracture resistance by requiring additional energy for the interfacial fracture. These studies mainly focused on the fracture of film-on-substrate *bi-layer* structures, which shed important lights on but cannot fully capture the fracture characteristics of the *multilayer* organic-inorganic permeation barriers. For example, the channel cracking of an oxide layer in a multilayer permeation barrier can be constrained by the presence of adjacent stiff oxide films. Such a constraint is, in turn, mediated by the thickness of the compliant organic layer in between two oxide films. Cordero et al. [14] studied the critical strains for various configurations of channel cracks in an inorganic-organic-inorganic tri-layer permeation barrier on a polymer substrate and showed that the barrier can sustain the largest strain when the organic layer is of some intermediate thicknesses. In that study, however, the organic-inorganic interfaces were assumed to remain perfectly bonded as channel crack propagates, thus no interfacial delamination was considered.

When a multilayer permeation barrier is under large deformation, it is highly possible that the oxide-organic interfacial delamination and the channel cracking of the oxide layer facilitate each other and thus co-evolve. On one hand, as the delaminated oxide layer becomes freestanding, it induces a larger driving force for channel crack propagation. On the other hand, the channel crack exerts more concentrated traction on the interface at the delamination front, promoting further delamination. The co-evolution of interfacial delamination and necking fracture of a thin Cu film on a compliant polyimide substrate under large tensile deformation has been predicted in computational modeling [33] and further observed in recent experiments [34, 35]. The co-evolution of delamination and buckling of a compressive film on a substrate has also been studied earlier [30].

Aiming to understand the yet unexplored mechanisms that govern the mechanical durability of organic-inorganic multilayer permeation barriers, we study the co-evolving channel cracking and interfacial delamination in such barriers under tension, through computational modeling.

COMPUTATIONAL MODEL

It has been shown that [14], if the organic layers are thicker than the oxide layers in a multilayer permeation barrier, channel cracking in the top oxide layer requires the least driving force when compared with other tensile failure modes (e.g., parallel channel cracking in multiple oxide layers, or channel cracking across the entire thickness of the coating). In the real design of multilayer permeation barriers (e.g., Fig. 1), the organic layers are often much thicker (e.g., 5~7 times) than the oxide layers. Therefore, it is reasonable to assume the dominant tensile failure mode of a real barrier to be the channel cracking in the top oxide layer with possible interfacial delamination originating from the channel root. Since other oxide layers except the top one remains intact in this dominant failure mode, and the oxide layers are more than 100 times stiffer than the organic layers, it is justified to assume that the constraint from other layers of the barrier to the channel cracking and interfacial delamination of the top oxide layer is mainly due to the oxide layer next to the top one, while the contribution from other layers below, be it oxide or organic, is negligible. Above said, we model a tri-layer barrier (two oxide layers sandwiching an organic layer) perfectly bonded to a thick polymer substrate, and analyze the tensile failure of this structure using the finite element code ABAQUS.

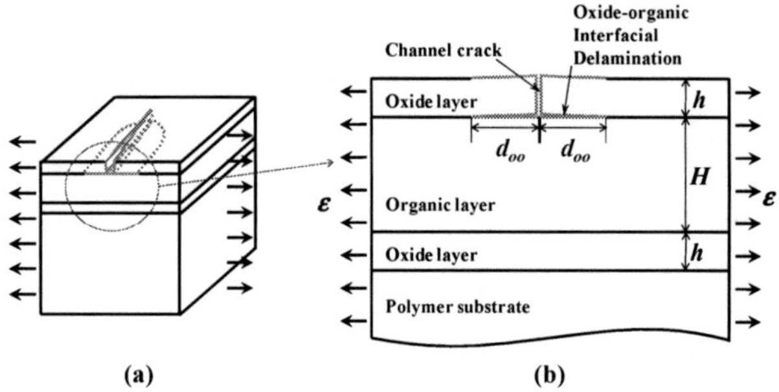

Fig. 2. Schematic of (a) the computational model considering (b) both the channel cracking of the oxide layer and the concomitant oxide-organic interfacial delamination.

Figure 2 illustrates the fracture of a blanket tri-layer barrier on a polymer substrate subject to uniaxial tension. Under sufficiently large tension, cracks initiate from built-in imperfections (e.g., pinholes, edges or corners) in the oxide layer and further propagate to form channel cracks. The stress concentration near the channel root can be severe enough to cause interfacial delamination. As to be shown in next section, under tension, the driving force for interfacial delamination decreases as the delamination advances in the tensile direction. As a result, the interfacial delamination eventually stops advancing in the tensile direction when such a driving force becomes smaller than the oxide-organic interfacial toughness. On the other hand, the channel crack continues propagating in the direction perpendicular to the applied tension and so does the concomitant interfacial delamination near the root of the channel crack front.

The detailed shape and deformation state near the front of the channel crack and the concomitant interfacial delamination can be rather complicated, and thus hard to predict. By contrast, far ahead and far behind the cracking and delamination front, the structure can be taken to deform under the plane strain conditions. To compute the driving force for interfacial delamination in the tensile direction, we simulate a slice of material of unit thickness far behind the cracking and delamination front and calculate the energy release rate at the delaminating front in the tensile direction using contour integral available in ABAQUS. Taking all materials in the structure to be linearly elastic, the driving force for both the channel cracking and the interfacial delamination to propagate in the direction perpendicular to the applied tension can be calculated by the elastic energy stored in a slice of material of unit thickness far ahead of the fracture front minus the elastic energy stored in a slice of material of unit thickness far behind the fracture front.

In the simulation model, the two oxide layers are films of thickness h, the organic layer between two oxide layers is a film of thickness H, and the polymer substrate is a block of thickness $1000h$ and length $1000h$. The horizontal displacement is set to be zero along the centerline of the laminate, and set to be u along both sides of the laminate. The quantity $\varepsilon = u/500h$ will be called the applied strain. Far behind the channel front, the width of the oxide-organic interfacial delamination originated from the channel root is d_{oo} at each side. In simulations, we fix the ratio H/h=5 but vary the ratio d_{oo}/h to study its effect on the driving force for delamination. In the region near the tip of the interfacial delamination, both the oxide film and the organic layer are densely meshed into four-node quadrilateral plane strain concentric-circle elements. The portion outside of this region are meshed with triangle plane strain elements of matching sizes near the regions and coarser elements far away from the regions to reduce the computation time. In the simulations, the Young's modulus and the Poisson's ratio of the oxide layer are $E_{oxide} = 300\,GPa$ and $\nu_{oxide} = 0.3$, respectively, and those of the organic layer are $E_{organic} = 2\,GPa$ and $\nu_{organic} = 0.3$, respectively. The Young's modulus and the Poisson's ratio of the substrate are taken to be the same as that of the organic layer, respectively. These values are representative of an Al_2O_3-polyacrylate multilayer permeation barrier on a polyethylene terephthalate substrate.

RESULTS

The driving force for interfacial delamination *in the tensile direction* far behind the channel front (e.g., inset of Fig.3a) can be represented by the energy release rate at the delamination front. Dimensional considerations lead to this energy release rate taking the form

$$G_d = f\left(\frac{d_{oo}}{h}, \frac{E_{organic}}{E_{oxide}}, \frac{H}{h}\right) E^*_{oxide} \varepsilon^2 h, \tag{1}$$

where $E^*_{oxide} = E_{oxide}/(1-\nu^2_{oxide})$ is the plane strain modulus of the oxide. The dimensionless function f denotes the normalized driving force of interfacial delamination and can be calculated numerically by contour integral using finite element method.

Figure 3a plots normalized driving force of interfacial delamination $G_d / E^*_{oxide}\varepsilon^2 h$ as a function of normalized delamination width d_{oo}/h. When the delamination width is small (e.g., $d_{oo}/h < 1$), the stress concentration near the channel crack root leads to a significantly large driving force of interfacial delamination. As the delamination width increases, the driving force of interfacial delamination drops considerably when $d_{oo}/h < 10$ and then gradually reaches a plateau when $d_{oo}/h > 10$. The initiation and propagation of the interfacial delamination is governed by the competition between driving force G_d and oxide-organic interfacial toughness Γ_{oo}. Figure 3a indicates that, for a modest oxide-organic interfacial toughness, delamination can initiate at the channel crack root. As the delamination advances along the interface in the tensile direction, the driving force decreases. The delamination eventually stops propagating at a width, d^s_{oo}, where the driving force drops to the value of the interfacial toughness, i.e., $G_d = \Gamma_{oo}$.

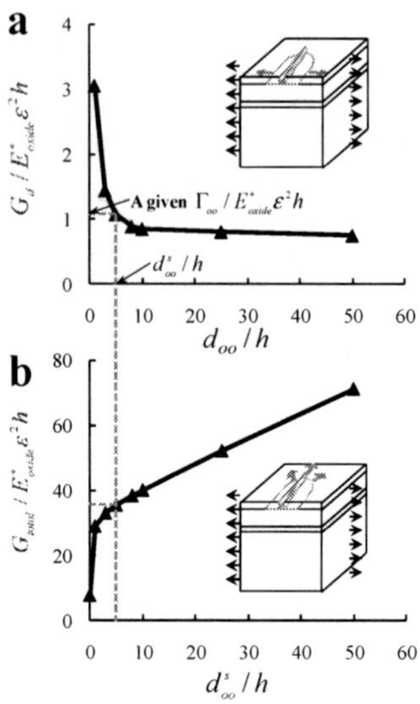

Fig. 3. (a) Normalized driving force of interfacial delamination along the tensile direction $G_d / E^*_{oxide}\varepsilon^2 h$ as a function of normalized delamination width d_{oo} / h; (b) Normalized driving force of oxide cracking and interfacial delamination perpendicular to the tensile direction $G_{total} / E^*_{oxide}\varepsilon^2 h$ as a function of normalized steady-state delamination width d^s_{oo} / h. Dash lines: for a given oxide-organic interfacial toughness Γ_{oo}, the corresponding steady-state delamination width d^s_{oo} can be determined from (a), from which G_{total} can be further determined in (b).

We next compute the total driving force G_{total} for both the channel cracking and the interfacial delamination to propagate *in the direction perpendicular to the applied tension* (e.g., inset of Fig.3b). G_{total} takes a form similar to Eq. (1) but with a different dimensionless function *f*. As discussed above, the stress state and deformation characteristics near the channel cracking front and the concomitant interfacial delamination are rather complicated. Assuming the structure deforms elastically, G_{total} can be calculated by comparing the elastic energy stored in a slice of material of unit thickness far ahead of the fracture front and that stored in a slice of material of unit thickness far behind the fracture front. We consider the steady-state fracture process where far behind the channel cracking front the oxide-organic interfacial delamination reaches a constant width of d^s_{oo}. Figure 3b plots $G_{total} / E^*_{oxide}\varepsilon^2 h$ as a function of the normalized steady-state delamination width d^s_{oo} / h. In the case of perfect bonding between the oxide layer and the organic layer (e.g., no interfacial delamination, $d^s_{oo} / h = 0$), G_{total} corresponds to the driving force for pure channel cracking, which agrees with previous study [14, 32]. Once interfacial delamination occurs and advances, G_{total} increase significantly. This can be explained by the considerable elastic energy released from the delaminated portion of the oxide layer in the wake of the channel cracking due to the loss of substrate constraint. For example, G_{total} triples for $d_{oo}^{s}/h=1$ and increases nearly six-fold for $d_{oo}^{s}/h=25$. By contrast, if there is no interfacial delamination, only a narrow region in the oxide layer along the channel crack surface can be relaxed, leading to a modest elastic energy release. The condition for steady-state channel cracking and interfacial delamination can be described as

$$G_{total} \geq \Gamma_{oxide} + 2\Gamma_{oo} d_{oo}^{s} / h, \tag{2}$$

where Γ_{oxide} is the cohesive fracture toughness of the oxide layer, and the second term on the right denotes the energy required for the oxide-organic interfacial delamination accompanying per unit area advance of the channel crack.

For a given oxide-organic interfacial toughness Γ_{oo}, one can determine the corresponding steady-state delamination width d_{oo}^{s} (e.g., Fig. 3a), from which one can further obtain G_{total} (e.g., Fig. 3b). In this sense, both d_{oo}^{s} and G_{total} are functions of Γ_{oo}. By moving the term of $2\Gamma_{oo}d_{oo}^{s}/h$ in Eq. 2 to the left-hand side, an effective driving force for steady-state channel cracking of the oxide layer can be defined as

$$G_{eff} = G_{total} - 2\Gamma_{oo} d_{oo}^{s} / h, \tag{3}$$

and thus the condition for steady-state channel cracking can be given by

$$G_{eff} \geq \Gamma_{oxide}, \tag{4}$$

in which the right-hand side denotes the resistance that is independent of the interfacial property, and the left-hand side represents the driving force that is a function of the oxide-organic interfacial toughness.

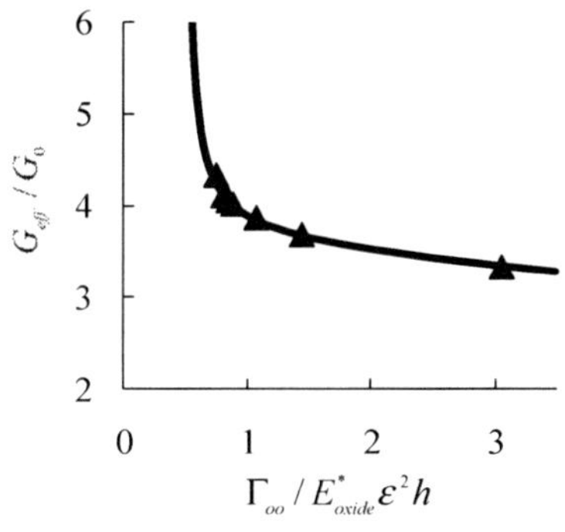

Fig. 4. Normalized effective driving force for channel cracking G_{eff}/G_0 as a function of normalized interfacial toughness $\Gamma_{oo}/E^{*}_{oxide}\varepsilon^2 h$. Here $G_0 = G_{total}(d_{oo}^{s}=0)$.

Figure 4 plots the normalized effective driving force for channel cracking G_{eff}/G_0 as a function of normalized interfacial toughness $\Gamma_{oo}/E^{*}_{oxide}\varepsilon^2 h$. Here G_0 is the driving force for channel cracking without oxide-organic interfacial delamination, i.e., $G_0 = G_{total}(d^s_{\infty} = 0)$ as in Fig. 4b. The smaller the oxide-organic interfacial toughness, the larger the steady-state delamination width, and therefore, the larger the effective driving force for channel cracking of the oxide layer.

CONCLUDING REMARKS

In this paper, we perform comprehensive computational modeling to study the tensile failure of organic-inorganic multilayer permeation barriers in flexible electronics. Major findings of the present study are recapped as follows.

Two failure mechanisms that are fatal for the barrier performance, namely, cracking of the top oxide layer and delamination along the oxide-organic interface, are shown to facilitate each other, and thus to co-evolve. The driving force for the oxide-organic delamination along the tensile direction and that for the oxide layer cracking and the concomitant oxide-organic delamination perpendicular to the tensile direction are quantitatively determined, from which an effective driving force for the oxide layer cracking is defined and shown to decrease as the oxide-organic interfacial adhesion increases. This suggests that a strong interlayer adhesion is crucial for the mechanical durability of the multilayer hybrid permeation barriers. The results from the present study shed light on possible structural design strategy to enhance the mechanical durability of the multilayer permeation barriers. For example, it is expected that a protective coating on the surface of the top oxide layer of a multilayer permeation barrier can potentially offer additional constraint to channel crack opening, leading to a reduction in the driving force for the channel cracking, and therefore, a better mechanical durability of the permeation barriers. More systematic study on this aspect is necessary to quantify the effect of such a protective layer, which will be reported elsewhere in the future.

In summary, the results from the present study are expected to offer mechanistic understanding of the failure mechanics of the organic-inorganic multilayer permeation barriers in flexible electronics. We call for further experimental investigation to validate the research findings in this paper.

ACKNOWLEDGMENTS

This research is supported by National Science Foundation (under grants 0856540, 0928278) and Minta Martin Foundation.

REFERENCES

1. Forrest S (2004) The path to ubiquitous and low-cost organic electronic appliances on plastic. Nature 428: 911-918.
2. Reuss RH, Chalamala BR, Moussessian A, Kane MG, Kumar A, Zhang DC, Rogers JA, Hatalis M, Temple D, Moddel G, Eliasson BJ, Estes MJ, Kunze J, Handy ES, Harmon ES, Salzman DB, Woodall JM, Alam MA, Murthy JY, Jacobsen SC, Olivier M, Markus D, Campbell PM and Snow E (2005) Macroelectronics: perspectives on technology and applications. Proceedings of the IEEE 93: 1239-1256.
3. Nathan A and Chalamala BR (2005) Flexible electronics technology, part II: materials and devices. Proceedings of the IEEE 93: 1391-1393.
4. Crawford GP (2005) Flexible flat panel displays John Wiley & Sons, Chichester ; Hoboken, NJ

5. Rogers JA, Someya T and Huang YG (2010) Materials and Mechanics for Stretchable Electronics. Science 327: 1603-1607.
6. McElvain J, Antoniadis H, Hueschen MR, Miller JN, Roitman DM, Sheats JR and Moon RL (1996) Formation and growth of black spots in organic light-emitting diodes. Journal of Applied Physics 80: 6002-6007.
7. Sutherland DGJ, Carlisle JA, Elliker P, Fox G, Hagler TW, Jimenez I, Lee HW, Pakbaz K, Terminello LJ, Williams SC, Himpsel FJ, Shuh DK, Tong WM, Jia JJ, Callcott TA and Ederer DL (1996) Photo-oxidation of electroluminescent polymers studied by core-level photoabsorption spectroscopy. Applied Physics Letters 68: 2046-2048.
8. Liew YF, Aziz H, Hu NX, Chan HSO, Xu G and Popovic Z (2000) Investigation of the sites of dark spots in organic light-emitting devices. Applied Physics Letters 77: 2650-2652.
9. Papadimitrakopoulos F, Zhang XM and Higginson KA (1998) Chemical and morphological stability of aluminum tris(8-hydroxyquinoline) (Alq(3)): Effects in light-emitting devices. Ieee Journal of Selected Topics in Quantum Electronics 4: 49-57.
10. Schaer M, Nuesch F, Berner D, Leo W and Zuppiroli L (2001) Water vapor and oxygen degradation mechanisms in organic light emitting diodes. Advanced Functional Materials 11: 116-121.
11. Nisato G, Kuilder M, Bouten P, Moro P, Philips LO and Rutherford N (2003) Thin film encapsulation for OLEDs: evaluation of multilayer barriers using the Ca test. SID International Symposium Digest of Technical Papers, pp 550--553.
12. Moro L, Krajewski TA, Rutherford N, Philips O, Visser RJ, Gross M, Bennett WD and Graff G (2004) Integrated encapsulation of bottom and top emission OLED displays. SPIE--The International Society for Optical Engineering, Organic Light-Emitting Materials and Devices VIII.
13. Lewis JS and Weaver MS (2004) Thin-film permeation-barrier technology for flexible organic light-emitting devices. Selected Topics in Quantum Electronics, IEEE Journal of 10: 45-57.
14. Cordero N, Yoon J and Suo ZG (2007) Channel cracks in a hermetic coating consisting of organic and inorganic layers. Applied Physics Letters 90111910.
15. Chwang AB, Rothman MA, Mao SY, Hewitt RH, Weaver MS, Silvernail JA, Rajan K, Hack M, Brown JJ, Chu X, Moro L, Krajewski T and Rutherford N (2003) Thin film encapsulated flexible organic electroluminescent displays. Applied Physics Letters 83: 413-415
16. Erlat AGn, Yan M and R. Duggal A (2009) Substrates and Thin-Film Barrier Technology for Flexible Electronics, pp 413-449.
17. Burrows PE, Graff GL, Gross ME, Martin PM, Hall M, Mast E, Bonham CC, Bennett WD, Michalski LA and Weaver MS (2001) Gas permeation and lifetime tests on polymer-based barrier coatings Proceedings of SPIE, pp 75.
18. Weaver MS, Michalski LA, Rajan K, Rothman MA, Silvernail JA, Brown JJ, Burrows PE, Graff GL, Gross ME, Martin PM, Hall M, Mast E, Bonham C, Bennett W and Zumhoff M (2002) Organic light-emitting devices with extended operating lifetimes on plastic substrates. Applied Physics Letters 81: 2929-2931.
19. Lueder E (1998) Passive and active matrix liquid crystal displays with plastic substrates In: Kuo Yea (ed) Thin-Film Transistor Technologies IV, Electrochem Soc Proc, pp 336-354.
20. Kim TW, Yan M, Erlat G, McConnelee PA, Pellow M, Deluca J, Feist TP, Duggal AR and Schaepkens M (2005) Transparent hybrid inorganic/organic barrier coatings for plastic

organic light-emitting diode substrates. Journal of Vacuum Science & Technology A 23: 971-977.
21. Graff GL, Williford RE and Burrows PE (2004) Mechanisms of vapor permeation through multilayer barrier films: Lag time versus equilibrium permeation. Journal of Applied Physics 96: 1840-1849.
22. Beuth JL (1992) Cracking of thin bonded films in residual tension. International Journal of Solids and Structures 29: 1657-1675.
23. Hutchinson JW and Suo Z (1992) Mixed-mode cracking in layered materials. Advances in Applied Mechanics, 29: 63-191.
24. Huang R, Prevost JH, Huang ZY and Suo Z (2003) Channel-cracking of thin films with the extended finite element method. Engineering Fracture Mechanics 70: 2513-2526.
25. Vlassak JJ (2003) Channel cracking in thin films on substrates of finite thickness. International Journal of Fracture 119: 299-323.
26. Tsui TY, McKerrow AJ and Vlassak JJ (2005) Constraint effects on thin film channel cracking behavior. Journal of Materials Research 20: 2266-2273.
27. Ma Q (1997) A four-point bending technique for studying subcritical crack growth in thin films and at interfaces. Journal of Materials Research 12: 840-845.
28. Ho S and Suo Z (1993) Tunneling cracks in constrained layers. Journal of Applied Mechanics-Transactions of the Asme 60: 890-894.
29. Liu XH, Lane MW, Shaw TM and Simonyi E (2007) Delamination in patterned films. International Journal of Solids and Structures 44: 1706-1718.
30. Cotterell B and Chen Z (2000) Buckling and cracking of thin films on compliant substrates under compression. International Journal of Fracture 104: 169-179.
31. Ambrico JM and Begley MR (2002) The role of initial flaw size, elastic compliance and plasticity in channel cracking of thin films. Thin Solid Films 419: 144-153.
32. Mei HX, Pang YY and Huang R (2007) Influence of interfacial delamination on channel cracking of elastic thin films. International Journal of Fracture 148: 331-342.
33. Li T and Suo Z (2007) Ductility of thin metal films on polymer substrates modulated by interfacial adhesion. International Journal of Solids and Structures 44: 1696-1705.
34. Lu NS, Wang X, Suo ZG and Vlassak J (2007) Metal films on polymer substrates stretched beyond 50%. Applied Physics Letters 91221909.
35. Lu NS, Wang X, Suo Z and Vlassak J (2009) Failure by simultaneous grain growth, strain localization, and interface debonding in metal films on polymer substrates. Journal of Materials Research 24: 379-385.

Mater. Res. Soc. Symp. Proc. Vol. 1312 © 2011 Materials Research Society
DOI: 10.1557/opl.2011.203

Structure and Strength of Silica-PDMS Nanocomposites

Adrian Camenzind[1], Thomas Schweizer[2], Michael Sztucki[3] and Sotiris E. Pratsinis[1]

[1]Particle Technology Laboratory, Institute of Process Engineering, Department of Mechanical and Process Engineering, ETH Zurich, Sonneggstrasse 3, CH-8092 Zurich, Switzerland
[2] Institute of Polymers, Department of Materials, ETH Zurich, Wolfgang-Pauli-Strasse 10, CH-8093 Zurich, Switzerland
[3] European Synchrotron Radiation Facility (ESRF), BP 220, F-38043 Grenoble Cedex, France

ABSTRACT

Commercially available SiO_2 nanoparticles (Aerosil, Degussa) with varying primary particle diameter, specific surface area (SSA), degree of aggregation and structure (fractal dimension) were compounded into PDMS-based nanocomposites. Thin sections of cured nanocomposites were analyzed with TEM and small and ultra-small angle X-ray scattering (U/SAXS) with respect to nanocomposite structure such as: filler primary particle, aggregate (chemically or sinter-bonded particles) and agglomerate (physically-bonded particles). Tensile tests (Young's modulus) were used to determine the nanocomposite strength which increased with increasing filler volume fraction (up to 12 vol%) consistent with "bound rubber" theory.

INTRODUCTION

The processing of flame-made oxide filler nanoparticles (e.g. SiO_2, Al_2O_3 or TiO_2) into polymers was lately summarized [1] focusing on particle surface modification and compounding and the resulting mechanical and optical properties of nanocomposites as a function of particle characteristics. The current understanding of nanoparticle formation in flames and the systematic control of particle size and morphology (aggregate or agglomerate formation) is highlighted in Fig. 1: The evolution of primary particle, d_p (solid line), and collision, d_c (dashed-doted line), diameters in a flame reactor as a function of residence time is shown. The characteristic temperature profile (T) in such reactors can be described with a steep increase (by fuel/precursor combustion) and fast cooling (convection by gas entrainment and radiation). Full coalescence of particles takes place at high flame temperatures. Particle sintering (coalescence) slows down as the flame temperature drops downstream of the combustion zone resulting in aggregated particles (partially coalesced) as indicated by the splitting of particle diameter into d_p and d_c in Fig. 1. Further downstream at even lower temperatures, particle sintering stops and the primary particle size, d_p, reaches an asymptotic value so the aggregate diameter (d_{cH}) no longer increases also. Subsequent coagulation of aggregated particles, however, results in larger agglomerate particles (and d_c) that can be broken (e.g. post-processing [2]) into its building blocks, aggregate and/or primary particles.

Nitrogen adsorption, small angle X-ray scattering (SAXS) or microscopic counting allows to determine the primary particle size, d_p. Experimental determination of aggregate size dimensions, d_{cH}, however, is much harder by microscopic counting [3] and light scattering [4]. Flame-made particles have sizes typically in the order of: d_p = 5-100 nm and d_{cH} = 10-300 nm. On the other hand, the agglomerate size is more challenging to measure (typically by electrical mobility – particle mass analysis [5] or SAXS [6]) and distinguish from aggregate size. The extent of particle aggregation (often described by the ratio of aggregate diameter to primary particle diameter, d_{cH}/d_p) can be related to process parameters. Increasing the flame cooling rate decreases the aggregate collision and primary particle sizes [7] while higher maximum flame temperatures lead to the opposite [7,8].

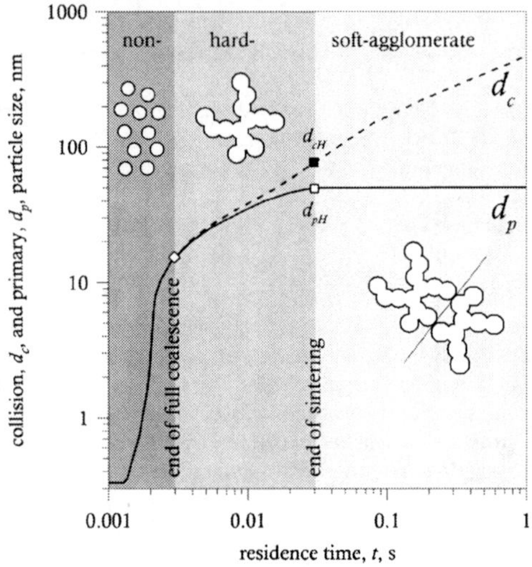

Figure 1 From coalesced to aggregated and agglomerated, flame-made particles: Primary particle size (d_p) and agglomerate collision diameter (d_c). Full coalescence ends when aggregate particles start forming (splitting of solid and dash-doted line). The onset of agglomerate formation occurs when sintering stops and d_p levels off (d_{pH}) while the corresponding d_c at that moment is the largest aggregate diameter (d_{cH}) (adapted from [1,7]).

Here, readily available, flame-made nanoparticles (SiO_2) of different size and morphology were compounded into polydimethyl-siloxane (PDMS) with a dual asymmetric centrifuge (DAC) focusing on the mechanical and structural properties of the resulting nanocomposites as a function of processing conditions.

RESULTS & DISCUSSION

Silica nanoparticles having varying primary particle size (from 7-40 nm) were mixed into a monomer prior to curing with a dual asymmetric centrifuge (DAC) with sufficient small processing volumes of 1-100 ml to afford such systematic investigations. The structural properties of dispersed filler SiO_2 particles in the PDMS composites were investigated with transmission electron microscopy (TEM, Fig. 2) and ultra-small angle X-ray scattering (U/SAXS) with focus on filler primary particle, aggregate (chemically or sinter-bonded particles) and agglomerate (physically-bonded particles) sizes as a function of applied filler concentration and mixing duration. In contrast to the widely applied "scotch" tape method (dry nanoparticles sandwiched between a polymer tape), here, aggregates and agglomerates were distinguishable. Table 1 lists detailed characterization (ϕ_v, $d_{v/s}$, D_f, d_{cH}, n and d_c) for these nanocomposites as well as those filled with 6.2 vol% A150 and A300 and 11.7 vol% OX50 along with d_{BET}.

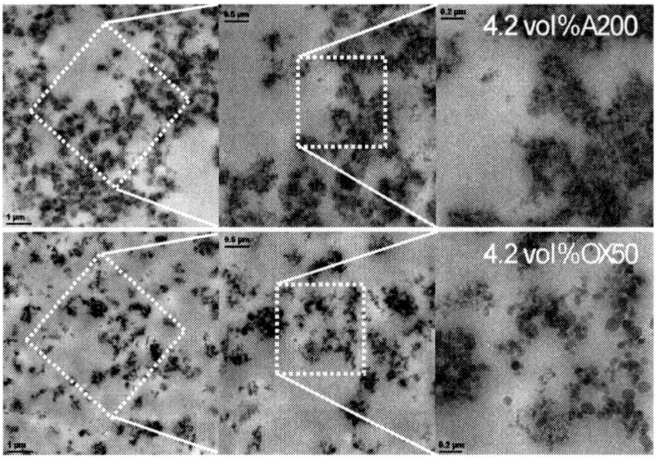

Figure 2 TEM images of SiO_2 (4.2 vol% OX50, A200) filled PDMS nanocomposites mixed for 20 minutes.

Regardless of mixing duration, the primary particle size is 41 nm and close to that of dry powders as expected (Table 1, 11.7 vol% OX50), which is a bit lower than the d_{BET} = 55 nm (as these are two different averages of the distribution [6]). It should be noted that other particle characteristics could not be obtained by SAXS for the dry powders due to fitting limitations for sizes in the order of a few micron although the presence of rather large structures > 1 μm cannot be excluded. For OX50-filled composites, mixing duration hardly affects the filler particle structure in the composite as seen from the small variation of aggregate D_f (1.74-1.75) in agreement to literature D_f = 1.6 ±0.1 (dry OX50 powder [9]) or silica-filled PDMS composites: D_f = 1.8 ±0.2 [10]. The SAXS-measured aggregate size, d_{cH} = 236-281 nm, is in agreement with literature [9,11] and invariant to increasing mixing duration indicating no fragmentation or erosion of aggregate particles with prolonged mixing (Table 1). That is why also the number of primary particles per aggregate, *n*, does not vary and is about 5-6 in all these composites (Table 1). This is not surprising as OX50 silica particles are some of the largest fumed silica particles with the lowest degree of aggregation according to its manufacturer. The agglomerate sizes (d_c) were not affected by mixing duration (t_{mix}), and were larger than 1 μm at the low OX50 concentration (6.2 vol%). However, an increase in OX50 loading (11.7 vol%) resulted in smaller agglomerates (494-746 nm, Table 1) consisting of only a few (2-5) aggregates when comparing d_{cH} to d_c.

In composites containing 6.2 vol% A200 silica particles, mixing duration hardly affects primary particle ($d_{v/s}$ = 13 nm, in close agreement to that measured by nitrogen adsorption, [12]), and aggregate diameters (d_{cH} = 140-146 nm) and n = 16-17 (Table 1). Aggregate diameters obtained by high pressure expansion of A200 aqueous suspensions were in good agreement with SAXS measurements[11]. The more aggregated A200 particle has a narrower primary particle size distribution [13] than OX50 [11] that helps to get a better agreement between $d_{v/s}$ and d_{BET} [6] as sintering narrows the primary particle size distribution in aggregates [13]. The D_f (= 1.78) is close to that measured for A200 powders [6,9] or A200-filled PDMS [10]. While mixing hardly affected primary and aggregate particle size, distinct variations were observed for d_c. Short mixing durations resulted in agglomerate sizes larger than 1 µm, while after 20 minutes of mixing, the d_c was reduced to 325-378 nm (Table 1). This emphasizes that sufficient mixing is needed to establish consistent and uniform agglomerate sizes. In Table 1 composites with 6.2 vol% of A150 and A300 are also listed. Their characteristics as a function of mixing duration were similar to A200 with respect to D_f and their $d_{v/s}$ (as well comparison to d_{BET}). Their n (A150: 27, A300: 82-88), however, were consistently different than that of A200 due to the difference in $d_{v/s}$ and d_{cH} (A150: 178-190 nm, A300: 122-129 nm). These *values* were consistent with high pressure fragmented A150 and A300 suspensions measured by dynamic light scattering (DLS), d_{DLS} = 210 and 180 nm, respectively [14]. Also mixing for more than 10 minutes was needed to break-up agglomerates larger than 1 micron.

Table 1. Particle characteristics by small angle X-ray scattering (SAXS) of nanocomposites made by mixing for 10-40 minutes and curing PDMS with fumed silica of 11.7 vol% or 6.2 vol% OX50 and 6.2 vol% A150 , A200 and A300. BET equivalent primary particle size (d_{BET}) is compared to $d_{v/s}$ obtained from SAXS (adapted from [15]).

Particle	**Mixing duration, min**	**SiO_2 volume fraction,**	**Primary particle size** d_{BET}, nm	$d_{v/s}$, nm		**Mass fractal dimension**	**Aggregate size, nm**	**Number of primary particles**	**Agglomerate size, nm**
	t_{mix}	ϕ_v vol%		dry	composite	D_f	d_{cH}	n	d_c
OX50	10	11.7	55	41.5	42.6-43.3	1.74-1.75	236-251	6	513
	20								647
	30								746
	40								494
	10	6.2			40.5-40.9	1.75	269-281	5	> 1 µm
	20								
	30								
	40								
A150	10	6.2	18	15.4	15.3-15.7	1.82-1.84	178-190	27	507
	20								427
	30								433
	40								491
A200	10	6.2	14	13.4	12.8-13.3	1.78	140-146	17	> 1 µm
	20								325
	30								378
	40								339
A300	10		9	9.2	8.2-8.7	1.81	122-129	82-87	> 1 µm
	20								346
	30								325
	40								335

Finally, besides the structural properties of cured nanocomposites, their mechanical properties were determined through uniaxial tensile measurements. Figure 4 shows the Young's modulus E of OX50 (diamonds), A150 (squares) A200 (triangles) and A300 (circles) filled nanocomposites as a function of particle volume fraction ϕ_v (open symbols). At particle loadings above 2.5 vol% it is seen that A150, A200 and A300 exhibit significantly higher reinforcing ability (larger E) than OX50. However at low particle loading (below 2.5 vol%), the variations between particles are not significant and the evolution of E with decreasing ϕ_v nicely approaches that of unfilled polymers (butterfly), E = 0.8 MPa.

The reinforcing ability of composites filled with aggregated particles having a high *SSA* can not only be explained by the particle volume fraction but also by their large available surface area as it was observed for carbon black - or silica-filled natural rubbers [16]. At the particle surface, a so called "bound rubber" can form, which tends to bind polymer molecules extending the solid volume fraction ϕ_v to an "effective volume fraction" $\phi_{v,eff}$ as seen from Fig. 3.

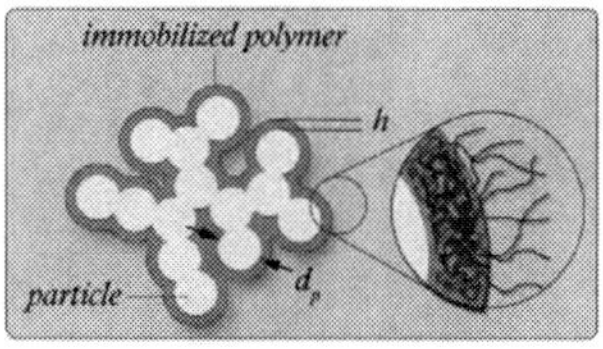

Figure 3 Formation of "bound rubber" layer (immobilized polymer, red) of thickness, h, on top of particles (yellow) of size, d_p.

The effectiveness factor $\beta = \phi_{v,eff} / \phi_v$ describes the ability of a particle to form "bound rubber" at its surface and can be calculated based on particle characteristics as described in more detail elsewhere [15]. The effectiveness factor β equates to 1.28 (OX50), 1.87 (A150), 2.17 (A200) and 2.79 (A300) for the individual particles (see listed values in Fig. 4). Particles consisting of smaller primary particles (like A150, A200 and A300) have a high specific surface area which tends to bind more rubber, resulting in a larger effective volume, hence larger β.

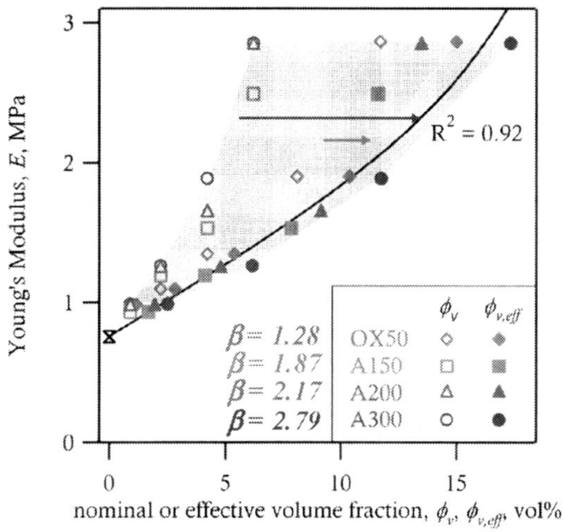

Figure 4 Reinforcing of SiO_2-filled PDMS composites: Young's modulus E as a function of silica volume fraction ϕ_v (open symbols) or effective volume fraction $\phi_{v,eff} = \beta \times \phi_v$ (filled symbols) taking into account the immobilization of polymer on the particle surface (adapted from [15]).

In Figure 4 the Young's modulus E is plotted as a function of the calculated $\phi_{v,eff} = \beta \times \phi_v$ (open symbols). Particles consisting of smaller primary particles (hence higher SSA and higher extent of aggregation, blue) were shifted towards higher "effective" volume fractions as more polymer is trapped on them. For example $\phi_v = 6.2$ vol% becomes $\phi_{v,eff} = 17.3$ vol% for A300 (blue) while a $\phi_v = 11.7$ vol% becomes only $\phi_{v,eff} = 15.0$ vol% for OX50 (green). A better correlation of mechanical strength to particle loading is found using the $\phi_{v,eff}$.

CONCLUSIONS

Agglomerate sizes decrease with increasing mixing duration indicating sufficient mixing for consistent and uniform particle distribution. Aggregated fillers need longer mixing than less aggregated ones. No variation of aggregate sizes and mass fractal dimension of the filler particles in composites with mixing duration was observed by U/SAXS. The less aggregated morphology of OX50 revealed ($n = 5$-6) few primary particles per aggregate in contrast to the more aggregated powders of A150-300 (with $n = 16 - 90$).

The nanocomposite strength was consistent with "bound rubber" theory as the Young's modulus of nanocomposites increased non-linearly with increasing filler volume fraction. The E-modulus is mainly a function of effective filler loading which accounts for the volume uptake by immobilization of polymer at the filler particle surface (Fig. 3 and Fig. 4).

REFERENCES

[1] A. Camenzind, W. R. Caseri and S. E. Pratsinis. "Flame-made nanoparticles for nanocomposites." *Nano Today*, **5**(1), (2010) 48.

[2] A. Teleki, R. Wengeler, L. Wengeler, H. Nirschl and S. E. Pratsinis. "Distinguishing between aggregates and agglomerates of flame-made TiO2 by high-pressure dispersion." *Powder Technol.*, **181**(3), (2008) 292.

[3] R. A. Dobbins and C. M. Megaridis. "Morphology of Flame-Generated Soot as Determined by Thermophoretic Sampling." *Langmuir*, **3**(2), (1987) 254.

[4] D. W. Schaefer and K. D. Keefer, In *Fractal in Physics*, Pietronero, L. and Tosatti, E., Eds. Elsevier Science: Amsterdam, 1986; p 39.

[5] J. H. Scheckman, P. H. McMurry and S. E. Pratsinis. "Rapid Characterization of Agglomerate Aerosols by In Situ Mass-Mobility Measurements." *Langmuir*, **25**(14), (2009) 8248.

[6] H. K. Kammler, G. Beaucage, R. Mueller and S. E. Pratsinis. "Structure of flame-made silica nanoparticles by ultra-small-angle X-ray scattering." *Langmuir*, **20**(5), (2004) 1915.

[7] S. Tsantilis and S. E. Pratsinis. "Soft- and hard-agglomerate aerosols made at high temperatures." *Langmuir*, **20**(14), (2004) 5933.

[8] R. N. Grass, S. Tsantilis and S. E. Pratsinis. "Design of high-temperature, gas-phase synthesis of hard or soft TiO_2 agglomerates." *AIChE J.*, **52**(4), (2006) 1318.

[9] R. Mueller, H. K. Kammler, S. E. Pratsinis, A. Vital, G. Beaucage and P. Burtscher. "Non-agglomerated dry silica nanoparticles." *Powder Technol.*, **140**(1-2), (2004) 40.

[10] J. M. Piau, M. Dorget and J. F. Palierne. "Shear elasticity and yield stress of silica-silicone physical gels: Fractal approach." *J. Rheol.*, **43**(2), (1999) 305.

[11] R. Wengeler, A. Teleki, M. Vetter, S. E. Pratsinis and H. Nirschl. "High-pressure liquid dispersion and fragmentation of flame-made silica agglomerates." *Langmuir*, **22**(11), (2006) 4928.

[12] G. Beaucage, H. K. Kammler and S. E. Pratsinis. "Particle size distributions from small-angle scattering using global scattering functions." *J. Appl. Crystallogr.*, **37**, (2004) 523.

[13] M. C. Heine and S. E. Pratsinis. "Polydispersity of primary particles in agglomerates made by coagulation and sintering." *J. Aerosol Sci.*, **38**(1), (2007) 17.

[14] R. Wengeler and H. Nirschl. "Turbulent hydrodynamic stress induced dispersion and fragmentation of nanoscale agglomerates." *J. Colloid Interf. Sci.*, **306**(2), (2007) 262.

[15] A. Camenzind, T. Schweizer, M. Sztucki and S. E. Pratsinis. "Structure & strength of silica-PDMS nanocomposites." *Polymer*, **51**(8), (2010) 1796.

[16] G. Kraus. "Reinforcement of Elastomers by Carbon-Black." *Rubber Chem. Technol.*, **51**(2), (1978) 297.

Mater. Res. Soc. Symp. Proc. Vol. 1312 © 2011 Materials Research Society
DOI: 10.1557/opl.2011.147

Influence of Preparation Method and Molecular parameters on the Rheology of Model PEO/ Laponite Nanocomposites

Jesmy Jose[1], Abakar Adam Omar[1], Guillaume Brotons[2], Jean-François Tassin[1]

[1]Polymères, Colloïdes, Interfaces, UMR CNRS 6120, Université du Maine
Avenue Olivier Messiaen, 72085 Le Mans Cedex, France

[2] Laboratoire de Physique de l'Etat Condensé, UMR CNRS 6087, Université du Maine
Avenue Olivier Messiaen, 72085 Le Mans Cedex, France

ABSTRACT

Model polymer nanocomposites based on geometrically well defined and protected Laponite particles dispersed in Poly(ethylene oxide) were investigated in order to improve the understanding of the filler dispersion effects on rheology by varying two experimental factors, namely preparation method and PEO matrix molecular weight. Preparation methods are divided into a solution dispersion and a melt dispersion by twin screw extrusion. The linear viscoelastic properties of the samples prepared by solution method revealed an elastic solid like behaviour at Laponite weight fractions as low as 0.1%, dramatically lower than the percolation threshold so far reported for such kind of systems. The sample preparation by melt dispersion, although leading to dispersed particles, does not achieve the same levels of modulus as compared to solution prepared mixtures. We propose a qualitative interpretation of this phenomenon, based on the mixture between a liquid and a dispersed phase of rather solid character. Further experiments using small angle X-ray scattering techniques (SAXS) show that the modulus level is not necessarily related to the height of the correlation peak characteristic of the Laponite stacks. However, for samples prepared with varying PEO matrix molecular weight the fraction of Laponite stacks decreases with increasing PEO molecular weight. The rheology master curve analyses show that confinements of polymer chains arising from high concentrations of particles and high molecular weight matrix chains do not impact the level of the low frequency modulus. However, a slower polymer dynamics, as observed for higher molecular weights, leads to an increase of the modulus at low particle loadings.

INTRODUCTION

Dispersions of nanoparticles, especially anisotropic ones, in polymers offer significant potential for providing a wealth of attractive material properties [1-4]. The interest in these materials stems from the ability to impart huge changes in properties (rheological, mechanical, barrier, thermal etc.) at "fairy dust levels". A prerequisite being the nanofiller dispersion, the dispersal of nanoparticles, which have a very high specific surface, into polymeric matrices remains an outstanding challenge. Dynamic rheology in the linear regime is sensitive to filler dispersion in polymers. At filler concentrations above the percolation value, a sample spanning percolated filler network dominates the viscoelastic behavior in polymer/clay nanocomposites.

We chose to work on a model system consisting of a water soluble polymer, poly (ethylene oxide - PEO), available with different molar masses and particles of synthetic clay, Laponite, which are discs of 25 nm diameter and 1 nm thickness. The purpose of this paper is to investigate the rheological properties of model nanocomposites in the melt state according to the method of preparation and matrix polymer molecular weight.

EXPERIMENT

We used PEO of varying molecular weights (Mw= 10,000, 35,000 and 130,000g/mol) supplied by Sigma Aldrich. The Laponite was obtained from Rockwool (grade RD). Methods of preparation are distinguished between a solution-exfoliation technique [5] and melt intercalation, via a master batch approach. In preparation by solution-exfoliation, the laponite particles are first dispersed in deionized water (milli Q) for 24 hours at a concentration of 2% (w/w), and subsequently subjected to adsorption of free PEO chains by adding an equal volume of a PEO 10 kDa solution. The concentration of PEO is adjusted to obtain the desired Laponite concentration in the mixture. This suspension, in which the particles are still non-aggregated, is lyophilized to preserve as much as possible this organization. The powder is then compressed at 110 °C, to form 1 mm thick discs to be studied in rheometry and SAXS. In melt intercalation, a master batch solution containing 10 % by weight of particles is diluted with pure PEO in a laboratory grade mini extruder, operating at 110 °C, in co-rotation mode, at 50 rpm for 10 min.

The rheological measurements are conducted on molded disks of 25mm diameter and 1mm thickness in a TA ARG2 rheometer by applying very small deformation (pertaining to the linear region). Samples were first equilibrated at 110 °C where the evolution of G 'and G "is followed with time for 6 hours at a frequency of 0.1 rad/s. This is followed by a frequency sweep from 100 to $5x10^{-3}$ rad/s, in the linear domain. X-ray scattering was performed for a wide range of wave vectors ranging from 0.02 Å to 5 $Å^{-1}$ obtained by combining a 2D gas detector and a "flat image plate" for large angles on a RigakuTM copper rotating anode (λ ~ 1.54 Å) with high flux and high resolution (all beam path is under vacuum and the Osmic^{TM} confocal mirrors deliver a cylindrical beam of 2× 10^7 photons/s over 300μm in diameter with a divergence less than 0.5mrad). For reasons of clarity, the intensities are presented up to 1 $Å^{-1}$ (large angle peaks of crystallization of PEO chains are omitted), with absolute scale units (cm^{-1} *i.e.* cm^2 per unit volume of sample).

DISCUSSION

Effect of Preparation Method

Figures 1 a and b compare the frequency dependence of storage modulus for samples obtained by solution (S) and melt (M) preparation techniques. All samples (except M 0.5 %) show a frequency-independent storage modulus indicating a solid-like behavior. At fairly low Laponite loadings, say in M 0.5% and even up to S 0.1 wt % (not shown here), the plateau in the low frequency region reveals a fragile elastic solid with moderate modulus. This reflects a strong connectivity of the structure. Plateau modulus at a given laponite loading is consistently lower for melt intercalated samples. More quantitatively, the behavior of a mixture of 0.5% prepared in solution is very close to a 2% mixture prepared by the melt route, and the modulus G 'of the mixture prepared by solution method at 0.1% requires ten times more Laponite through melt intercalation. The plateau modulus scales with particle concentration with a power law behavior in both cases: $G \approx C^{\alpha}$ with $\alpha = 4.1 \pm 0.3$ for samples of solution type and $\alpha = 2.9 \pm 0.2$ melt intercalated samples as given by figure 1 b. Note that the two lines do not intersect at a concentration of 10 wt%, the original master batch concentration for the dilution in melts. This was confirmed by subjecting a sample prepared by solution to a passage in the extruder where a drop in modulus was found.

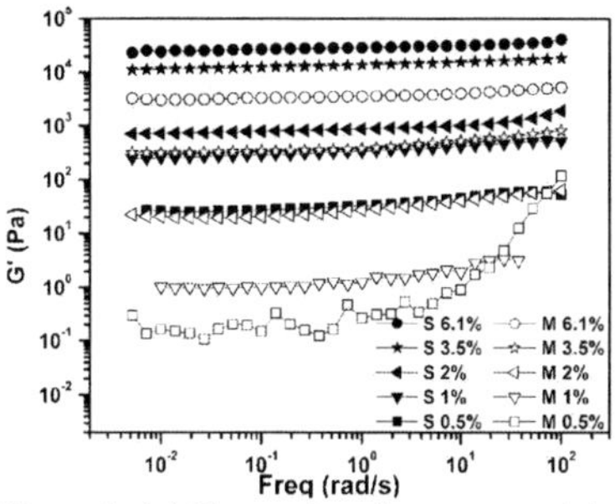

Figure 1: (a) Comparison between modulus of PEO-10k/Laponite samples prepared by Solution(S) and Melt (M).

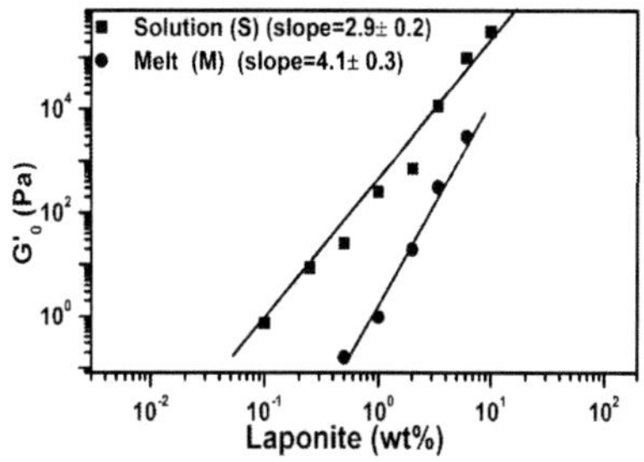

Figure 1 : (b) Concentration dependence of plateau modulus compared for different nanocomposite preparation techniques.

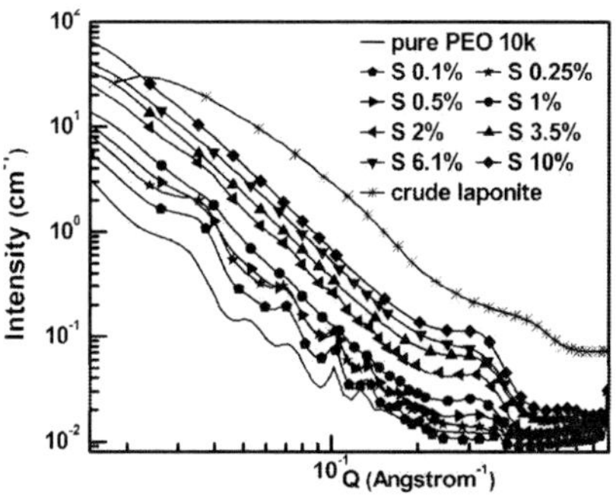

Figure 2: (a) SAXS results for PEO-10k/Laponite samples prepared by solution method (S)[1].

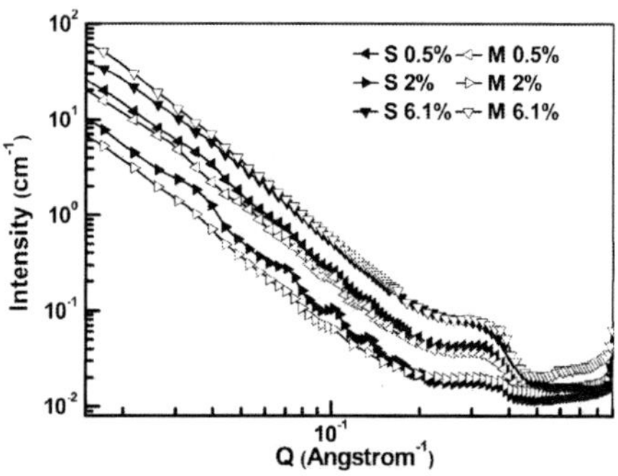

Figure 2: (b) Comparison between solution(S) and melt (M) prepared samples.

To link these observations with structure, scattering measurements were performed on a wide scattering vector window. Unlike rheological measurements carried out in melt, structural characterization was carried out at room temperature, i.e., on crystallized samples. At large angles, the crystallization of PEO is clear (not shown in the figures) and is dependent on the

[1] *The crude laponite curve was obtained from non packed powder hold in a glass capillary and is thus shifted vertically with an arbitrary factor since the sample thickness is not known precisely (in opposition with all other pellet samples).*

concentration of particles in the mixtures studied. Figure 2a shows the I (Q) vs. momentum transfer (Q) curve for the mixtures prepared in solution, and Figure 2-b compares it with those prepared from melt.

The crude laponite powder SAXS curve shows a correlation peak located at 0.5 $Å^{-1}$ that reflects the stacking of laponite particles and indicates a relatively soft smectic order. The highly crystalline pure PEO presents markedly diffuse "wiggles" at small angle, as already reported in the literature [6-8]. These oscillations reflect a degree of periodicity between amorphous regions and crystalline lamellae. Thanks to the relatively low molecular weight of the polymer chains, one PEO chain embedded in a lamellae can form several folds at the surfaces. For instance, the peak at 20.3 nm corresponds to 2 folds. At small angles, the addition of Laponite generates an increase in the SAXS intensity, thereby reflecting an increase in the amount of Laponite-PEO interface per sample unit volume since particles come with a high contrast prior to PEO ($\approx$ 0.4 electrons/ $Å^3$ and $\approx$ 0.8 $Å^3$ for Laponite. The intensity falls along a characteristic slope in log-log scale corresponding to a Q^{-2} law, the same slope obtained by the diffraction of the PEO crystalline smectic sheets. Correspondingly, the oscillations existing in the pure polymer progressively disappear and become virtually invisible above 2% particles, masked by the strong scattering of Laponite discs and the possible increase in PEO disorder. It may be noted that although the system is in the form of a viscoelastic solid, there remains a fraction of Laponite aggregates, giving a correlation peak associated with stacking distances for a few particles of 1.9 nm. This peak is shifted towards lower Q values from that of pure Laponite, showing evidence of intercalation of polymer chains. The intensity of this peak increases with the concentration of Laponite and generally follows the overall increase of the fraction of introduced particles.

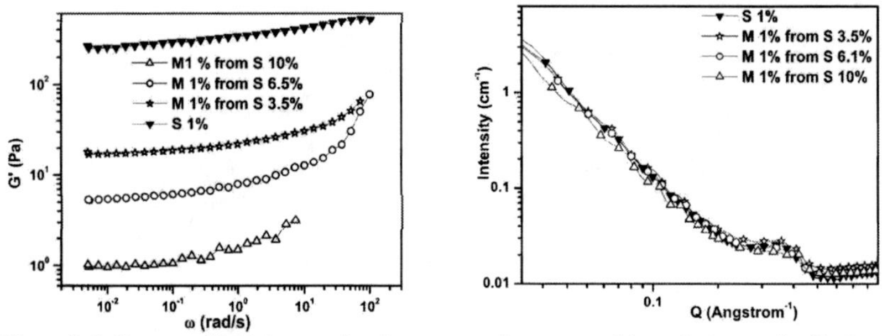

Figure 3: Influence of initial masterbatch concentration prepared by solution on the final melt prepared sample at a 1 % Laponite concentration in PEO 10k matrix: (left) on storage modulus, (right) on SAXS curves.

While rheological measurements show clear differences, SAXS curves, as shown in Figure 2-b are similar. In particular, the peak reflecting the stacking of particles is unaffected. However, there is a noticeable trend toward greater heterogeneity (increased scattering) for the mixtures produced from melts, especially at low Laponite concentrations.

Figure 3a compares the suspensions at 1% particles obtained by dilution in the molten states of solution prepared mixtures ranging from 10 to 3.5% of Laponite. Figure 3b gives the corresponding scattering profiles. It has been noticed that the modulus is greatly reduced when the amount of added pure PEO 10k polymer increases. Quantitatively, the variation of G' is greater than 10-fold when diluted from 10% or 3.5%. As revealed from figure 3b, the structure at large scales probed with X-rays is globally unaffected. The differences seen in X-ray diffraction are subtle, while the mechanical measurements show huge differences. We can conclude that these samples come with almost equal fraction of Laponite dispersed (responsible for the strong scattering that follows a Q^{-2} law as expected for plate like particles observed from 0.02 to 0.2 $Å^{-1}$) and comparable fraction of small Laponite aggregates (peak at Q = 0.35 $Å^{-1}$). Nevertheless, connectivity of solid particles responsible for the behavior at low frequencies is highly dependent on the particle concentration used for preparation of the master batch solution. One can speculate about the existence of assemblies of connected particles, especially since we are dealing with concentrated solutions. From X-ray data we can deduce that the structure that generates such high levels of moduli is produced on a larger scale than probed (greater than 100nm). Note that the subtle PEO connections between separated particles would be non addressable from static scattering techniques on such scales. Dilution by free PEO chains (involving moderate shear) and the influence of annealing at 110 ° C for up to 6-8 hours before the rheological measurement cannot separate domains of connected particles (from solution preparation) on a scale small enough to allow the dispersion of particles and their interactions. This can be explained fairly well by the analogy with the rheology of polymer immiscible blends. The initial nanocomposite with a high particle concentration can be associated with a high modulus phase to be dispersed in a matrix of lower viscosity. Greater the difference in viscosity, lower will be the dispersion, or at least the rupture of the disperse phase. This qualitatively accounts for the observed behavior.

Effect of PEO molecular weight

The solution prepared samples with varying PEO molecular weight exhibited a low frequency plateau behavior with the storage modulus (G') greater than loss modulus (G") even for samples with Laponite concentrations as low as 0.1 wt%. The percolation threshold is thus lower than 0.1 wt%.

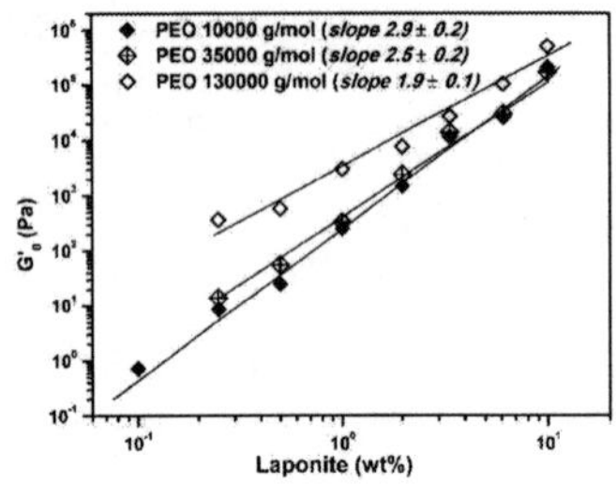

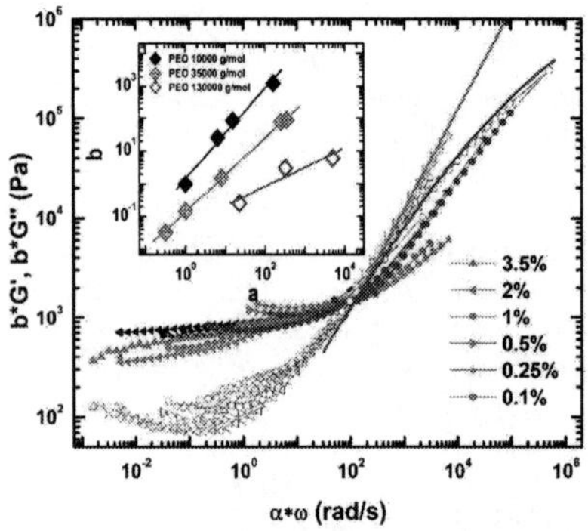

Figure 4: (a) Scaling of plateau modulus with concentration for samples with varying PEO molecular weights.

Figure 4: (b) Master curve showing the scaled moduli for varying particle concentrations and with PEO molecular weights-10,000 (black color), 35,000 (red color) & 130,000 g/mol (blue color) respectively. The inset shows the linear relationship between the scaling factors. The solid lines at high frequency show the loss modulus curves for the pure PEO.

As shown in figure 4a, the scaling exponent of the plateau modulus decreases with the molecular weight of PEO, but at high loadings the low frequency modulus is almost independent on the molecular weight. Remarkably, despite their marked differences, at low Laponite concentrations, all the data for samples showing a crossover between G' and G" could be rescaled to build a single master curve as given in Figure 4b. At high frequencies i.e., at frequencies higher than the G'and G" crossover, the loss modulus curves of the samples varies along the high frequency viscous moduli of the corresponding neat polymers.

At a fixed PEO molecular weight, the horizontal and vertical shift factors depend on each other through a power law (inset of figure 4b). Considering the validated model of a solid filler network being dispersed in a suspending fluid, at particle concentrations above the percolation value, the observed linear relationship between the independent modulus and frequency shift factors implies a shift along the viscosity of the suspending PEO melt [9]. As observed in figure 4b, the separate straight lines, with characteristic slope values determined by the PEO molecular weight would clearly suggest that, at fairly low laponite loadings, (i.e., with a G'and G" cross over region) the plateau modulus of the elastic filler network is influenced by the matrix PEO polymer chain dynamics.

In order to understand the molecular weight effect on the filler network behaviour at higher loadings, a second type of master curve was built by plotting the moduli versus the reduced frequency i.e., the product of the experimental frequency by the terminal relaxation time of the PEO matrix, and is shown in figure 5(a). Interestingly, all data for samples with widely different molecular weights, at high laponite loadings (above 2%), almost superimpose on this very large frequency window (about 9 decades). However, at low loadings, the modulus is affected by the molecular weight of the matrix, so that the superposition of the data obtained with various molecular weights is not observed. These results show that confinements of polymer chains arising from high concentrations of particles and high molecular weight matrix chains do not impact the level of the low frequency modulus. However, a slower polymer dynamics, as observed for higher molecular weights, leads to an increase of the modulus at low particle loadings.

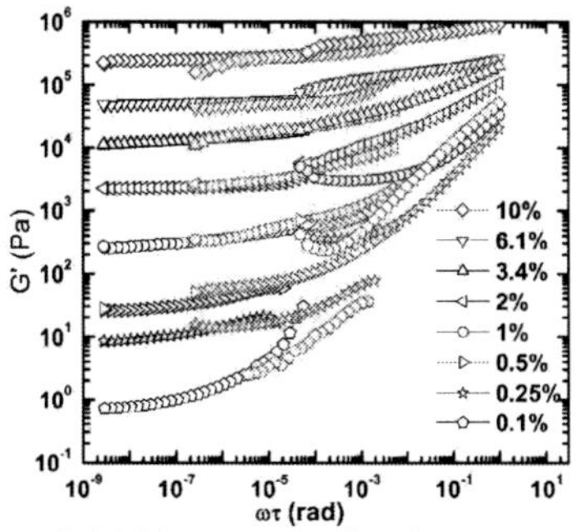

Figure 5: (a) Master curve showing storage modulus vs reduced frequency for solution prepared samples with varying PEO molecular weights-10,000 (black color), 35,000 (red color) & 130,000 g/mol (blue color) respectively.

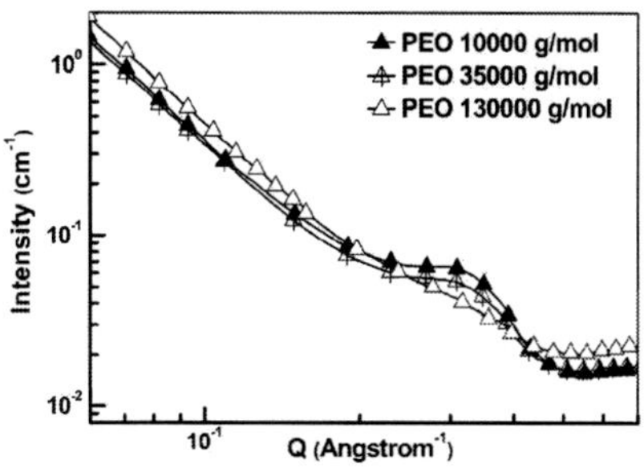

Figure 5: (b) Comparison of the scattering curves for samples at 3.5 % with varying PEO molecular weights.

The scattering analysis of samples with varying PEO molecular weight shows a higher extent of exfoliation at larger PEO molecular weights. For example, figure 4 b compares the laponite X-ray peak intensity for 3.5 % Laponite/PEO samples. The aggregated particles peak intensity decreases with increasing matrix molecular weight. Moreover, an almost complete exfoliation is obtained for PEO at 130,000 g/mol. High molecular weight polymer chains facilitates exfoliation through interparticle bridging or because of the size of the macromolecular coils. Therefore, the fraction of interconnected particle assemblies increases, leading to an enhanced plateau modulus. This further substantiates our previous work that linked the increase of modulus and the reduction of the correlation peak [10].

CONCLUSIONS

The effect of filler dispersion on the linear viscoelastic properties of polymer nanocomposites in the molten state is investigated using a well defined PEO/Laponite model system. The solid like behavior of the filler network is studied using two experimental parameters; preparation route and PEO molecular weight. Small amplitude oscillatory measurements in the linear viscoelastic region revealed a significant disparity in the plateau moduli for the solution and melt prepared samples. However, structural characterization by SAXS showed comparable degree of Laponite stacking and aggregation for both sample types. This behavior was qualitatively explained using the analogy with the dispersion by shearing in polymer blends having wide difference in viscosities. Higher molecular weight PEO facilitates the exfoliation of clay platelets possibly because of chains with larger dimensions or by inter-particle bridging to yield relatively higher plateau modulus values. The effect of molecular parameter is further studied by rheology master curve analyses. The polymer dynamics was found to affect the network behaviour at low Laponite loadings and is independent of matrix molecular weight at higher filler loadings.

ACKNOWLEDGMENTS

The post-doctoral grant for Jesmy Jose from the region Pays de La Loire is gratefully acknowledged.

REFERENCES

[1] H. Kim and C. W. Macosko, "Morphology and properties of polyester/exfoliated graphite nanocomposites," *Macromolecules,* vol. 41, pp. 3317-3327, May 2008.

[2] H. Kim, *et al.*, "Graphene/Polymer Nanocomposites," *Macromolecules,* vol. 43, pp. 6515-6530, Aug 24 2010.

[3] E. P. Giannelis, *et al.*, "Polymer-silicate nanocomposites: Model systems for confined polymers and polymer brushes," *Polymers in Confined Environments,* vol. 138, pp. 107-147, 1999.

[4] M. Moniruzzaman and K. I. Winey, "Polymer nanocomposites containing carbon nanotubes," *Macromolecules,* vol. 39, pp. 5194-5205, Aug 8 2006.

[5] M. Alexandre and P. Dubois, "Polymer-layered silicate nanocomposites: preparation, properties and uses of a new class of materials," *Materials Science & Engineering R-Reports,* vol. 28, pp. 1-63, Jun 15 2000.

[6] S. Z. D. Cheng, *et al.*, "Nonintegral and Integral Folding Crystal-Growth in Low-Molecular Mass Poly(Ethylene Oxide) Fractions .2. End-Group Effect - Alpha,Omega-Methoxy-Poly(Ethylene Oxide)," *Journal of Polymer Science Part B-Polymer Physics,* vol. 29, pp. 299-310, Mar 15 1991.

[7] S. Z. D. Cheng, *et al.*, "Isothermal thickening and thinning processes in low molecular weight poly(ethylene oxide) fractions crystallized from the melt: 2. Crystals involving more than one fold," *Polymer,* vol. 33, pp. 1140-1149, 1992.

[8] S. Z. D. Cheng, *et al.*, "Isothermal Thickening and Thinning Processes in Low-Molecular-Weight Poly(Ethylene Oxide) Fractions .1. From Nonintegral-Folding to Integral-Folding Chain Crystal Transitions," *Macromolecules,* vol. 24, pp. 3937-3944, Jun 24 1991.

[9] V. Trappe and D. A. Weitz, "Scaling of the viscoelasticity of weakly attractive particles," *Physical Review Letters,* vol. 85, pp. 449-452, Jul 2000.

[10] A. Loiseau and J. F. Tassin, "Model nanocomposites based on laponite and poly(ethylene oxide): Preparation and rheology," *Macromolecules,* vol. 39, pp. 9185-9191, Dec 26 2006.

Mater. Res. Soc. Symp. Proc. Vol. 1312 © 2011 Materials Research Society
DOI: 10.1557/opl.2011.160

Rheological Behavior of Nanoclay Containing Nanofluids

Jeffrey C. Munro and YuanQiao Rao
Core R&D Labs, The Dow Chemical Company, 2301 N. Brazosport Blvd., Freeport, TX 77541, U.S.A.

ABSTRACT

A synthetic hectorite nanoclay, Laponite®, with a disc shape (20 – 30 nm in diameter and 1 nm in thickness) was used as a model nanoparticle to prepare dispersions in water and different organic solvents. Although up to 20 wt% of nanoclay can be dispersed in water to form a low-viscosity stable colloidal sol, dispersions in organic solvents behave differently. The effect of the dielectric constant of the medium on the viscosity of the dispersion was studied systematically using two series of water-organic solvent miscible blends. Changes in the rheological behavior of the dispersions suggest that the dielectric constant of the organic solvent is a determining factor in the sol-gel transition of a nanoclay-containing nanofluid.

INTRODUCTION

The incorporation of high aspect ratio nanofillers into organic media is desirable due to the potential for creating high performance materials. One important aspect of these dispersions is the rheological behavior of the nanofluid. In the work described in this paper a synthetic hectorite, Laponite®, with a disc shape (20 – 30 nm in diameter and 1 nm in thickness) was used as a model nanoparticle. Dispersion of this nanofiller in water and other organic solvents were explored.

While the rheology of clay containing nanocomposites has been reported exhaustively in the literature, most of the work has centered on melt rheology and steady high-shear viscosity of nanoclays in polymers.[1,2] Rheological studies and low-shear viscosity of nanoclays in liquid media at room temperature, other than water, are fairly limited.[3,4] In general, rheology has been used, in combination with other techniques such as x-ray scattering, as a means to determine the state of dispersion of clay in a polymer.[5] For clay fillers (aspect ratio ~ 200) in an epoxy monomer, low-viscosity liquids were reported for systems with only 0.7 wt% clay, but a gel state was observed at 5 wt% clay loading.[6] Meanwhile, in rheological studies of dispersions of Laponite RD in polyethylene oxide ($M_w \sim 10{,}000$ g/mol) the critical clay concentration at the liquid-to-gel transition varied from 0.5 to 1 wt% depending on the method used to prepare the dispersion.[5] A transition at lower clay concentration was associated with improved exfoliation of the clay. It is noted that this critical concentration is well below the maximum packing density of clay spheres (where it is assumed each Laponite platelet occupies a spherical volume with a diameter equal to that of the Laponite platelet) of 3.7 wt%.[7] Note that if the maximum packing density is assumed to be the point at which the rotation of the platelets becomes geometrically constrained, as suggested in another study, the weight fraction of Laponite at the concentrated regime threshold is approximately 25 wt%.[4]

EXPERIMENT

Materials

The following materials were used: Laponite S482 was obtained from Southern Clay Products; water at a resistivity of 18.2 MΩ was used; 1,4-butane diol (BDO) and formamide were purchased from Aldrich and used as received. A schematic of the structure of Laponite is shown in **Figure 1**.

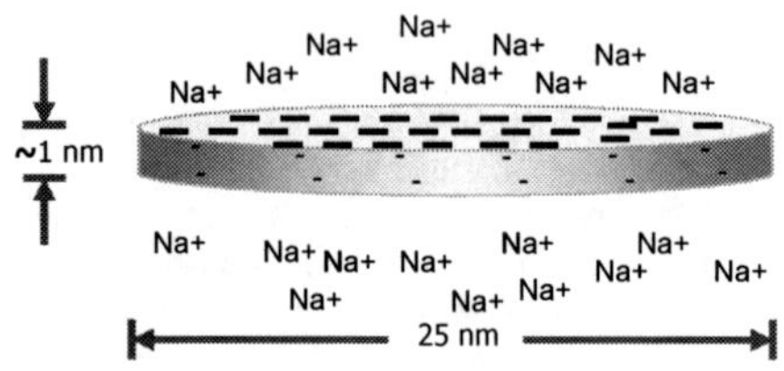

Figure 1: Structure and formula of Laponite.

Dispersion Preparation

Dispersions were prepared using the following methods. For aqueous dispersions, Laponite was added to water while stirring. The dispersions were stirred at room temperature for at least 24 hr, resulting in clear dispersions. For clay dispersions in water-organic mixtures, Laponite was dispersed in water at the desired concentration by adding the clay to water while stirring. Dispersions were stirred at room temperature for at least 24 hr. Then the organic component was added to the clay-water dispersion and stirred at room temperature for 24 hr. The final concentration of clay was kept constant and the water-organic ratio varied. Therefore, the initial concentration of clay in water varied from 4 to 14%. Since the maximum concentration of Laponite in water to get effective exfoliation was 20%, we limited the maximum organic content in the dispersion to about 70% with a desired target of 4 wt% clay in the dispersions.

Rheology Measurements

Rheology data were collected on a stress-controlled AR2000 rheometer (TA Instruments) equipped with a 60 mm stainless steel cone (0.017 rad cone angle) and a temperature-controlled Peltier plate. A stress ramp was performed in which the stress was ramped from 0.1 to 1000 Pa using 10 steps/decade at a temperature of 30 °C. The yield stress was determined by the stress value at which a measurable strain rate (> 0.01 s^{-1}) was first observed. The experiment was stopped if the shear rate exceeded 5000 s^{-1}. Note that the clay dispersions tested were often in the form of structured fluids. Since the rheology of structured fluids is dependent on the shear history of the sample, rheology plots are labeled with apparent viscosity rather than viscosity.

Dynamic Light Scattering (DLS)

DLS data were collected on a Malvern Instruments Zetasizer Nano ZS. Unfiltered clay dispersions (0.1 to 4 wt%) were placed into a disposable cuvette (Malvern Instrument DTS0012). Light scattering data was collected at 25 °C using non-invasive back-scatter (NIBS) mode at 173°. Approximately 4 measurements were collected and averaged for each sample, where each measurement consists of an integrated average of approximately 15 correlation curves (~10 min data collection time).

RESULTS AND DISCUSSION

Laponite dispersions can be considered colloidal systems. There are several concentration, morphology, and rheology regimes in which these dispersions may exist, as illustrated in **Figure 2**. At any concentration, precipitation of aggregated platelets can occur if there are insufficient interactions to maintain colloidal stability. We would like to avoid this region due to lack of dispersion and exfoliation of the platelets in this state. Meanwhile, in the dilute regime the platelets may be present as primary particles or as small aggregates that are small enough to remain suspended. In either of these two cases a stable sol with viscosity close to that of the medium is obtained. In the concentrated region, where the volume fraction of platelets is above the maximum close-packed density, an infinite network of platelets fills the medium resulting in a gel. The intermediate, semi-concentrated regime is of the most interest because it is where the maximum amount of filler that provides a stable, low-viscosity sol can be obtained. In the semi-concentrated regime the volume fraction of filler is high enough that networked platelets could percolate through the medium leading to gel formation, which is not desired. The desired state is for the platelets to behave as non-interacting particles, such that no networking or chaining of particles occurs, resulting in a stable sol with low viscosity. The filler concentrations that define the boundaries between these regimes depend on several factors including the size and shape of the filler, as well as the nature of the inter-particle interactions.

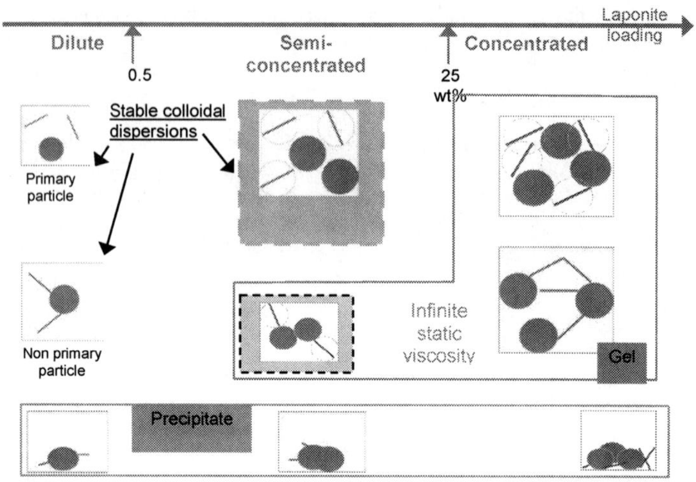

Figure 2. Concentration, morphology, and viscosity regions in Laponite dispersions. Circles and lines represent the faces and edges of Laponite platelets, respectively.

Dispersions of Laponite in Water

The viscosities of aqueous dispersions of Laponite S482 are shown in **Figure 3**. The data plotted correspond to viscosity values at a shear rate of ~ 100 s^{-1}, extracted from a shear stress ramp experiment conducted on a stress-controlled rheometer. Note that the rheological behavior of the aqueous dispersions is Newtonian. Also plotted in the figure is a line representing the Krieger-Dougherty model [8-12]

$$\frac{\eta}{\eta_0}=\left(1-\frac{\phi}{\phi_m}\right)^{-[\eta]\phi_m} \qquad (1)$$

where η is the viscosity of the dispersion, η_0 is the viscosity of the medium (water in this case), ϕ is the volume fraction of filler, ϕ_m is the effective maximum packing volume fraction (percolation threshold) when the free-rotational motion is hindered due to geometric constraints, and $[\eta]$ is the intrinsic viscosity of the platelets, which was estimated based on the Douglas and Garbochi approximation and is related to the aspect ratio of the platelet. The Krieger-Dougherty model is based upon the assumption that the platelets do not interact and that the volume fraction is below the percolation threshold. As can be seen in the plot, the viscosity values of the Laponite S482 dispersions match those predicted by the theory when an aspect ratio of 25 is used. This aspect ratio is consistent with the published size of Laponite (diameter of 25 nm and thickness of 1 nm) as well as the particle size measured by DLS (see **Figure 5** below). The use of existing theory such as the Krieger-Dougherty model to predict the viscosity of a nanofluid containing aspherical nanoparticles when the nanofluid behaves as a stable colloidal sol is validated by this data.

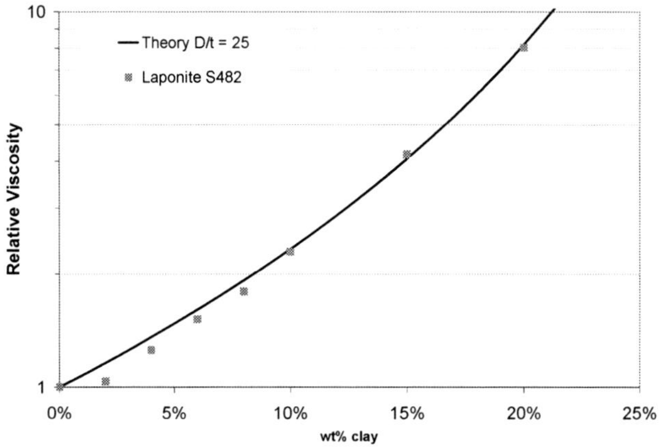

Figure 3. Viscosities of Laponite S482 aqueous sols as a function of clay concentration.

Dispersions of Laponite in Media with Varying Dielectric Constant

When Laponite was dispersed in an organic solvent such as BDO, complex rheological behavior was observed at concentrations significantly lower than in aqueous dispersions. It is speculated that the rheology of such a system is strongly influenced by particle-particle and particle-matrix interactions.

In order to explore the inter-particle interactions in more detail a series of experiments was conducted in which the dielectric constant of the media was changed by adjusting the water/organic content of the dispersant. As shown in **Figure 1**, Laponite is a charged particle and therefore electrostatic interactions are critical to the colloidal stability of Laponite dispersions. Properties such as dielectric constant and ionic strength of the medium, as well as the surface charge of the particle, are important factors in the electrostatic interaction between particles in a medium. Since the charge of Laponite is controlled by the composition of the clay (amount of substitution of Li for Mg in the octahedral layer) and not by the level of protonation of surface groups, we focus on the effects of the medium properties here.

In order to vary the dielectric constant solvent blends of water and water-miscible organics with different dielectric constants were prepared. The dielectric constants for the blends were estimated using Looyenga's formula [14] (see **Figure 4**):

$$\varepsilon_m = \left[\varepsilon_1^{1/3} + V_2\left(\varepsilon_2^{1/3} - \varepsilon_1^{1/3}\right)\right]^3 \tag{2}$$

where ε_m is the dielectric constant of the binary mixture, ε_i is the dielectric constant of pure component i (ε = 78.5 for water, 31.4 for BDO, and 84.0 for formamide), and V_i is the volume fraction of component *i*. For each of the aqueous-organic blends, the liquid-to-gel transition of a 4 wt% dispersion of Laponite S482 in the blends is also shown in **Figure 4.**

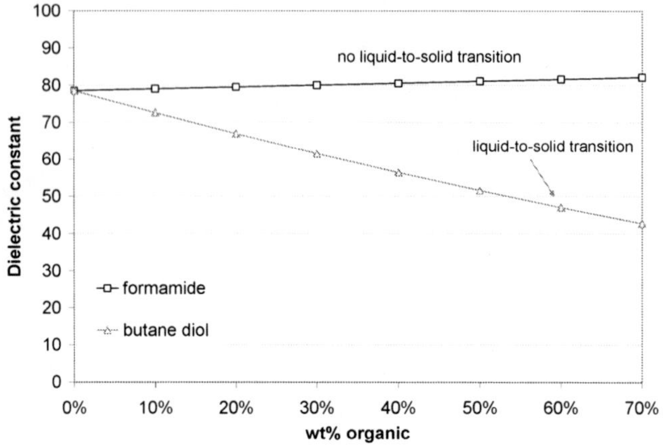

Figure 4: Dielectric constant for aqueous-organic mixtures calculated using Looyenga's formula. Points at which 4 wt% dispersions of Laponite S482 undergo a liquid-to-gel transition in the water-organic pairs are denoted.

The particle sizes of the nanofillers in these dispersions at a concentration below the liquid-to-gel transition were measured by DLS. As shown in **Figure 5,** the measured particle sizes of the clay in the water-formamide blends is consistent with the diameter of single Laponite platelets. In addition, these 4 wt% dispersions of clay in water-formamide blends remain stable, low-viscosity sols. In contrast, the particle sizes of the clay in dilute water-BDO blends increase quickly when the BDO content is 60 wt% or above. This indicates a lose in the stability of the dispersion and growth of clay platelet agglomerates. Note that this increase in particle size coincides with the liquid-to-gel transition observed in semi-concentrated, 4 wt% dispersions in water-BDO indicated in **Figure 4.**

Rheology data for some of the dispersions near the liquid-to-gel transition are shown in **Figure 6**. The viscosity as a function of increasing shear stress is plotted for water-BDO blends. Viscosity values are only recorded once the yield stress of the material is surpassed and there is a measurable strain. For the water-BDO dispersions, the liquid-to-gel transition occurred at 40/60 w/w% water/BDO, where a measurable yield stress was first observed. Viscosity data for the water-formamide dispersions were not collected since no liquid-to-gel transition was observed in this sample series. As depicted in **Figure 4**, as the dielectric constant of the organic decreases, the liquid-to-gel transition shifts to lower organic content in the blend. According to DLVO theory, as the dielectric constant of the medium decreases, the electrostatic repulsive force between like-charged particles decreases. Therefore, the change in the liquid-to-gel transition observed for Laponite dispersions in water-BDO mixtures likely is due in part to changes in the electrostatic interactions of the Laponite platelets in the low-dielectric constant medium.

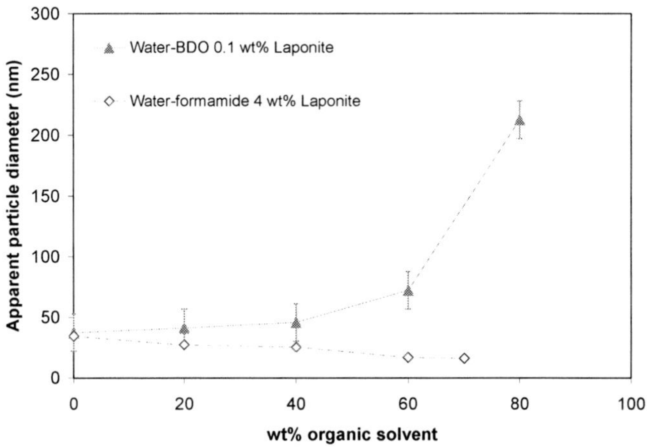

Figure 5. Apparent particle size measured by DLS as a function of the wt% of organic solvent in water-organic solvent blends.

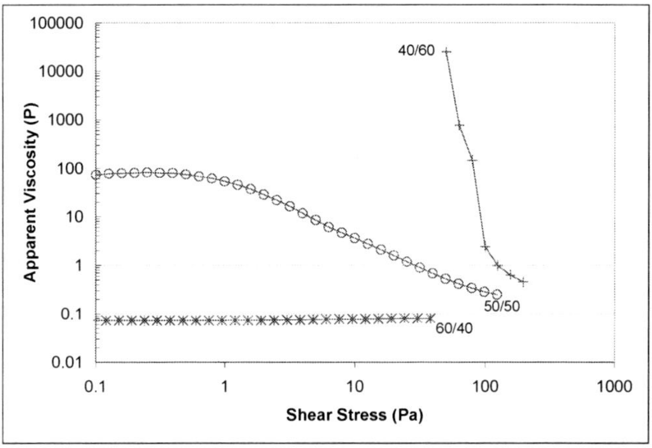

Figure 6: Viscosities of 4 wt% Laponite S482 dispersions in water-BDO blends. Number next to each data series is water/organic content in w/w%.

CONCLUSIONS

The objective of the work presented here was to understand some of the fundamentals of the rheology of nanofluids containing high aspect ratio nanoparticles. Although up to 20 wt% of Laponite nanoclay can be dispersed in water to form a low-viscosity stable colloidal sol, dispersions in organic solvents behave differently. There is a liquid-to-gel transition when the nanofiller concentration is above a critical concentration. This critical concentration is affected significantly by the dielectric constant of the medium. As the dielectric constant of the organic solvent decreases, the critical concentration decreases. When the dispersion is a sol, the viscosity dependence on the nanofiller concentration is well described by the Krieger-Dougherty equation. When the dispersion is a gel, the dispersion behaves as a Bingham fluid, characterized by a yield stress and shear thinning behavior.

ACKNOWLEDGMENTS

The authors would like to acknowledge Mr. Eddie Leyva, Ms. Lisa Allen, and Ms. Abigail Baker for preparing dispersions and performing rheology and DLS measurements. We would also like to thank Drs. Bernie Dems and Valeriy Ginzburg for valuable technical input.

REFERENCES

1. Krishnamoorti, R. and Giannelis, E.P. Macromolecules 1997, 30, 4097 – 4102.
2. Cassagnau, P. Polymer 2008, 49, 2183 – 2196.
3. Plummer, C.J.; Rodlert, M.; Bucaille, J.; Grunbauer, H.J.; and Manson, J.E. Polymer 2005, 46, 6543 – 6553.
4. Sun, L.; Boo, W.J.; Liu, J.; Clearfield, A.; Sue, H.J.; Verghese, N.E. ; Pham, H.Q.; and Bicerano, J. Macromol. Mater. Eng. 2009, 294, 103 – 113.
5. Loiseau, A. and Tassin, J.F. Macromolecules 2006, 39, 9185 – 9191.
6. Wang, H.; Hoa, S.V.; and Wood-Adams, P.M. J. Appl. Polym. Sci. 2006, 100, 4286 – 4296.
7. Estimate is based on the assumptions that each platelet occupies a sphere with diameter equal to the diameter of the platelet, maximum packing density for randomly-packed spheres is approximately 30 vol%, density of Laponite is 2.57 g/cm3, and platelet diameter is 30 nm. Estimate follows analysis found in ref [5].
8. Barnes, H.A. Colloids Surf., A 1994, 91, 89 – 95.
9. Cheng, P.Y. and Schachman, H.K. J. Polym. Sci. 1955, 16, 19 – 30.
10. Haber, S. and Brenner, H. J. Colloid Interface Sci. 1984, 97, 496 – 514.
11. Tadros, T. Adv. Colloid Interface Sci. 2004, 108 – 109, 227 – 258.
12. Thomas, D.G. J. Colloid Sci. 1965, 20, 267 – 277.
13. Wierenga, A.M. and Philipse, A.P. Colloids Surf., A 1998, 137, 355 – 372.
14. Raju, G.G. Dielectrics in Electric Fields. Marcel Dekker, New York, 2003.

Mater. Res. Soc. Symp. Proc. Vol. 1312 © 2011 Materials Research Society
DOI: 10.1557/opl.2011.112

Development of Bacterial Cellulose Nanocomposites

Roberto Benson[1], Hugh M. O'Neill[2], B.R. Evans[2], S. Hutchens[1], C.P. Stephens[3], and R. Hammonds[1]
[1]Materials Science and Engineering, University of Tennessee-Knoxville, Knoxville, TN
[2]Oak Ridge National Laboratories, Oak Ridge, TN
[3]Department of Surgery, Graduate School of Medicine, Knoxville, TN

ABSTRACT
The development of synthetic materials with inherent bone properties would allow the safe restoration of bone function and reduce current risks associated with the use of grafts. This study investigated the development of bacterial cellulose–hydroxyapatite composite (CdHA-BC) as a potential bone substitute material. Composites of bacterial cellulose (BC) and oxidized, degradable, cellulose (OBC) were mineralized by sequential incubation in calcium chloride and aqueous sodium phosphate to form a calcium deficient hydroxyapatite (CdHA). The CdHA produced in BC and OBC is similar in morphology and chemistry to the hydroxyapatite found in natural bone. The formation of CdHA is supported by XRD, and EDS results. The CdHA-BC and CdHA-OBC composites degrade in a simulated aqueous physiological environment.

1. INTRODUCTION

1a. Overall objective

The ideal synthetic material for bone replacement to treat the effects of disease, traumatic injury, and non-unions remains a major concern in biomaterials science and engineering. The drive for using a biomimetic synthetic material over a biologically derived material is that it negates the possibility of disease transmission and adverse immunogenic response. A number of potentially favorable biocompatible materials (ceramics and filled polymers) have been developed and used as bone replacement materials [1-3].

Some clinically used ceramic materials include Bioglass, β- tricalcium phosphate (β-TCP), hydroxyapatite (HA), and biphasic calcium phosphate (BCP, which is a mixture of β-TCP and HA). These ceramics have been shown to support osteoblast adhesion and growth. However, one of the primary drawbacks of utilizing a purely ceramic bone filler/replacement is its inherent brittleness. Creating a polymer-ceramic composite can reduce the brittleness that is characteristic of a purely ceramic scaffold. Thus the composite approach seems appropriate and even biomimetic since natural bone is a composite consisting of 25% water, 15% organic materials, and 60% mineral phase (by weight) [4]. This report will focus on efforts currently underway to develop such a bioactive polymer-ceramic composite.

1b. Development of the polymeric composite matrix

The first component of the composite that must be developed is a nano-fibrous polymeric scaffold that mimics the natural collagen in bone. In vivo, collagen type I makes up 90% of the organic mass in bone [5]. Collagen forms a hierarchical structure starting with the fibril that has a diameter of 1 nm, a fibrillar structure composed of triple helical collagen molecules with a diameter of 100-200 nm and then further organizes into larger lamellar fiber bundles [6, 7]. Our collagen simulant is composed of a similarly hierarchically structured polymer (cellulose) secreted by bacteria, *Gluconacetobacter spp.*

This bacterial species possess the ability to secrete pure cellulose in the form of a nano-fibrous hydrogel [8]. The extruded cellulose subfibrils with a ~1.5 nm width bond together to form ribbons that are typically 3-4 nm thick, 70-80 nm wide, and 1-9 μm long. The ribbons twist and overlap to form the thickness of the pellicle, a gel-like membrane [8]. This polymeric scaffold has found many medical applications in the past including wound dressing [9], periodontal tissue [10], dura mater [11], cartilage [12], and synthetic vascular grafts [13]. In addition to its use as a non-degradable scaffold, it can also be converted into a degradable form of cellulose (OBC) through a periodate ring-opening oxidation. This study will investigate the formation of composites based on both BC and OBC.

1c. Composite formation- Integration of a bioactive ceramic on the polymeric support

These polymeric nano-fiber scaffolds (BC and OBC) lacks bioactivity, therefore a bioactive ceramic must be deposited in the polymeric scaffold to support bone growth. As the ceramic composition increases, the scaffold's mechanical properties will also increase to approach that of native bone. In vivo, the bone modulus increases an order of magnitude when hydroxyapatite (E=100 GPa) is deposited into collagen matrix (E=2-4 GPa) to an average modulus of 20 GPa [5, 7]. In bone, the deposited calcium deficient hydroxyapatite (Cd-HA) takes the form of a nanocrystal that is approximately 5 nm thick, 5-40 nm wide, and 50 to several hundred nanometers long [5]. This nanoceramic and the surrounding organic components of bone can be remodeled by osteocytes. When creating a synthetic bone replacement the commonly used forms of calcium phosphates include β- tricalcium phosphate (β-TCP), hydroxyapatite (HA), and biphasic calcium phosphate (BCP, a mixture of β-TCP and HA).

The main drawback of all of these calcium phosphate forms is their highly synthetic nature, which decreases their bioactivity and bioresorbability. While β-TCP is chemically similar to HA, it is not a natural component of bone mineral [14]. In addition, β-TCP tends to undergo rapid dissolution in the biological environment making it inappropriate for some applications [15]. It is also reported that its resorption rate is variable and dependent on the implant site, the material's structure, porosity, and chemistry. The chemical structure and properties of β-TCP may also be altered after heat sterilization [14]. Synthetic stoichiometric hydroxyapatite (HA), $Ca_{10}(PO_4)_6(OH)_2$, is osteoconductive but it has very low bioresorbability, unlike the physiologic calcium deficient hydroxyapatite (CdHA) [2, 14-17]. Physiological hydroxyapatite can take on the form $Ca_{10-x}(HPO_4)_x(PO_4)_{6-x}(OH)_{2-x}$ with $0< x \leq 1$ [18]. The ceramic that is deposited in this study is the naturally occurring form of calcium phosphate, calcium deficient hydroxyapatite (CdHA), which is both osteoconductive and bioresorbable.

2. EXPERIMENTAL

2.1. Materials preparation

2.1a. Synthesis of bacterial cellulose (BC): The bacterial strain was obtained from the American Type Culture Collection (Manassas, VA, USA) (ATCC 10821). The bacterial cellulose was synthesized using the Gluconacetobacter hansenii bacterial strain, utilizing a modified form of the Schramm and Hestrin method [19]. Discs of cellulose were cultured for 30 days and purified in distilled/deionize water using the method described by Hutchens [20].

2.1b. Bacterial cellulose oxidation (OBC): The oxidation of bacterial cellulose was performed with periodate using a modified method developed by Painter [21]. BC discs were placed in a capped vessel that contained 50 mM $NaIO_4$ in 5% n-propanol. The vessel was covered with aluminum foil and placed on an orbital shaker for 24 h at 23°C. The reaction was terminated by addition of 0.5 mL of glycerol and reduction of temperature by placing the vessel in an ice bath. The oxidized cellulose was purified with several changes of distilled/deionized water.

2.1c. Mineralization of Bacterial Cellulose: Hydroxyapatite crystallites were deposited within the unmodified (CdHA-BC) and oxidized bacterial cellulose (CdHA-OBC) samples by sequentially incubating the samples in aqueous calcium chloride followed by incubation in aqueous sodium phosphate dibasic [22, 23].

2.2. Materials characterization

2.1a. X-Ray Diffraction (XRD): X-ray diffraction was performed on dried polymeric and composite samples mounted on a silicon zero-background holder. Data was obtained on an X'Pert PRO MPD X-ray Diffractometer (PANalytical: Almelo, Netherlands) operated at 45 kV and 40 mA with a CuKα radiation source. Data was collected between 10° and 70° 2θ at a scan rate of 1 degree/min.

2.2b. Fourier Transform Infrared Spectroscopy (FTIR): Dried samples were measured in transmission mode on a BioRad FTS6000 Fourier Transform Infrared Spectrophotometer (Agilent Technology, CA USA). The spectra reported are the result of 256 scans collected with a resolution of 4 cm^{-1}.

2.2c. Scanning Electron Microscopy (SEM): Lyophilized samples were mounted on carbon tape and sputtered with gold on a Spi Module Sputter Coater (Spi Supplies: Westchester, PA, USA) at 20 mA for 10 s. They were then analyzed on a LEO 1525 Scanning Electron Microscope (Zeiss: Oberkochen, Germany). The BC and CdHA-OBC were imaged with a 3 kV accelerating voltage, and the OBC and CdHA-BC were imaged at 5 kV

2.2d. *Mechanical Properties:* Hydrated samples were tensile tested at a strain rate of 12 in/min. These samples had a gage length of 3 in and a width of 1 in.

2.2d. *Degradation Study:* The degradation behavior of the samples was analyzed by immersion in HEPES buffer. Half of the samples from each group were placed in a 37°C Forma Scientific Water Jacketed Incubator (Thermo Electron Corporation: Waltham, MA, U.S.A.) for static incubation. The remaining samples from each group were placed in a 37°C controlled environment incubator shaker (New Brunswick Scientific Co., Inc.: Edison, NJ, U.S.A.) for dynamic incubation. The samples were removed from the HEPES buffer after fourteen days of incubation, rinsed in several changes of distilled/deionized water, dried, and weighed.

3. RESULTS AND DISCUSSION

3.1. Development of the polymeric composite matrix

In development of an optimal polymeric base for a bone tissue engineering scaffold there are a few parameters that become important including porosity, fiber integrity, and degradability. The porosity was visualized using SEM, XRD and tensile testing determined the fiber integrity, and the degradability was investigated by FTIR and hydrolysis studies.

SEM images of the bacterial cellulose before (Figure 1A) and after (Figure 1B) periodate oxidation (Figure 1C) are shown to ensure the morphological integrity of the nanofibers during conversion. XRD was used to study alterations in the crystalline structure of BC due to the periodate oxidation. The XRD pattern of BC (Figure 1D.A) and OBC (Figure 1D.B) showed the typical crystalline structure of unmodified cellulose as defined by defined peaks at 14.7 and 23.1° 2θ indicating that the periodate treatment did not appreciably disrupt the crystalline structure of BC. The sharp peaks are an indication of the high degree of crystallinity of the bacterial cellulose [24].

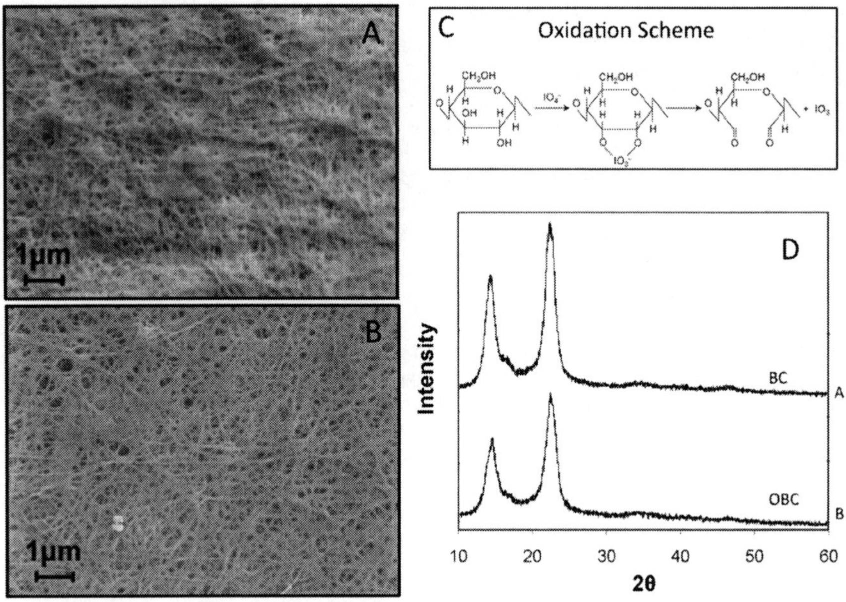

Figure 1: (A) SEM image of BC (B) SEM image of OBC (C) Periodate oxidation scheme (D) XRD pattern for BC and OBC

The mass loss for the 14 day static and dynamic hydrolysis study in both BC and OBC is depicted in Figure 2A. The mass loss for BC was an average of 8% and 20% for BC in static and dynamic incubation, respectively. However, this mass loss was not statistically significant and most likely due to sample variation. The mass loss for OBC was 38% and 36% for BC in static and dynamic incubation, correspondingly. The degradation of OBC is believed to occur by the scheme shown in Figure 2B.

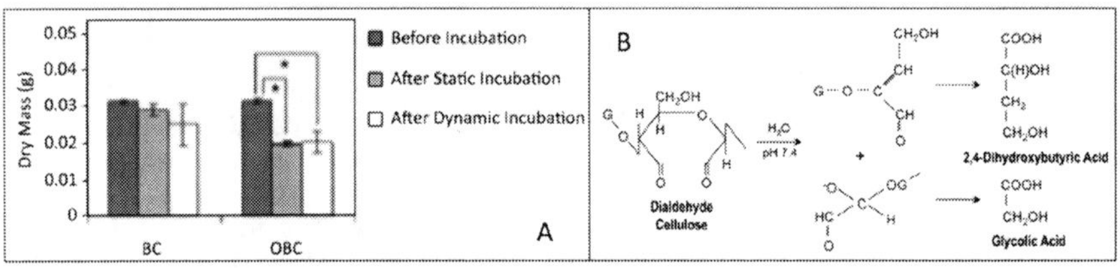

Figure 2: Mass loss due to degradation (A) by the mechanism shown in B

Fourier Transform Infrared Spectroscopy (FTIR) was used to identify structural changes resulting from the periodate oxidation of BC. The spectrum of OBC shown in Figure 3A.A include new absorption bands at 1740, 1652 and 1530 cm^{-1} corresponding to the vibrational stretching mode of aldehyde carbonyl, the hydrogen bonded aldehyde carbonyl, and conjugated aldehydes groups, respectively [25].

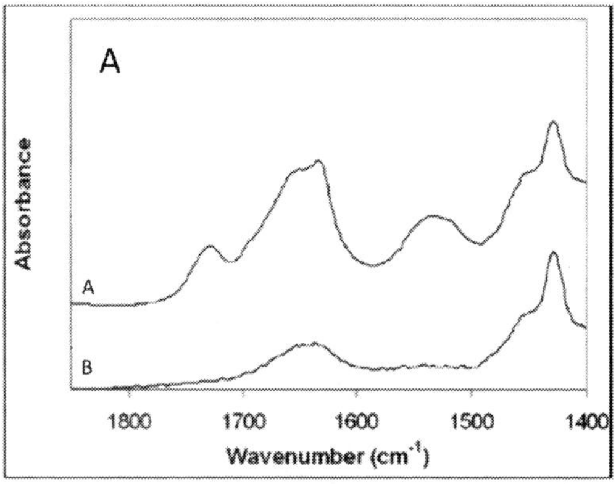

B

Sample	E (Pa)
BC	3910 ± 440
OBC	3900 ± 580

Figure 3: FTIR Spectrum of Bacterial Cellulose, BC, (A.B) and Oxidized Bacterial Cellulose, OBC, (A.A). The Young's modulus is also tabulated in B.

The mechanical properties also show that the fiber integrity remains after periodate oxidation. The results summarized in Figure 3B clearly indicating no differences between the tensile elastic modulus of the unmodified and oxidized bacterial cellulose. The values obtained from the tensile testing support the notion that periodate oxidation did not produce noticeable structural alteration of bacterial cellulose.

3.2. Composite formation- Integration of a bioactive ceramic on the polymeric support

3.2.1. Confirmation of ceramic deposition

The next step in the development of osteoconductive and bioresorbable scaffolding material is the deposition of a biomimetic composite on both BC and OBC. The mass increase with each cycle of calcium chloride and sodium phosphate dibasic is shown in Figure 4. The CdHA-BC increased in mass according to the linear equation 1 (r^2=0.98), where 17 mg is the BC starting mass.

$$Composite\ Mass\,(mg) = 30 * (CycleNumber) + 17 \qquad \textit{(Equation 1)}$$

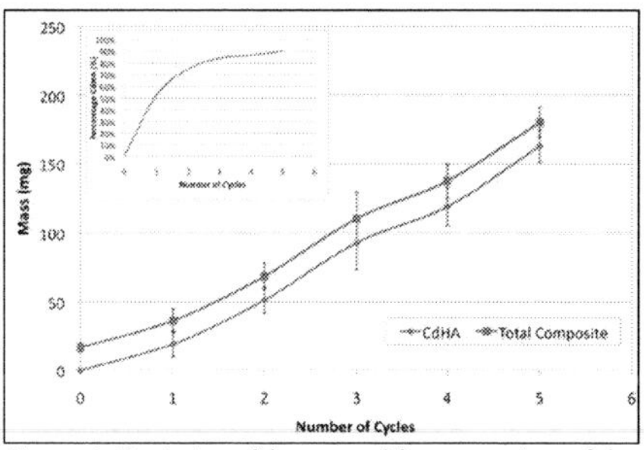

Figure 4: Depiction of the mass of the composite and the mass of the hydroxyapatite at after each cycle of calcium chloride and sodium phosphate dibasic

3.2.2. Determination of ceramic morphology

Once it has been determined that a ceramic has been deposited in the polymeric nanofibrous mesh, the morphology of the ceramic must be investigated; this inquiry was accomplished by SEM and XRD. From SEM it can be seen, Figure 5, that the deposited ceramic forms a calcospherite morphology, which is also seen in vivo in woven bone and the fetal growth plate.

Figure 5: SEM images of bacterial cellulose-hydroxyapatite composite (CdHA(0.53)-BC) at ((a) 1000x, (b) 5000x, (c) 10,000x, (d) 50,000x. A 5 kV accelerating voltage was used for the 1000x-10,000x images while 2 k for the 50,000x image [26].

The CdHA morphology is further investigated using x-ray diffraction. By utilizing the Scherrer equation, the thicknesses of each crystalline diffraction peak can be used to find the crystallite dimensions. The c-dimension was determine from (002) and (004) reflections, and a and b dimensions were obtained from the (100), (200) and (300) reflections, as seen in Figure 6. The crystallite sizes range between 10 and 50 nm. The crystallites appear to be elongated with significantly larger dimension in the c-direction than a or b-directions. These dimensions are comparable in length (20 – 40 nm) to hydroxyapatite crystallites found in bone [27].

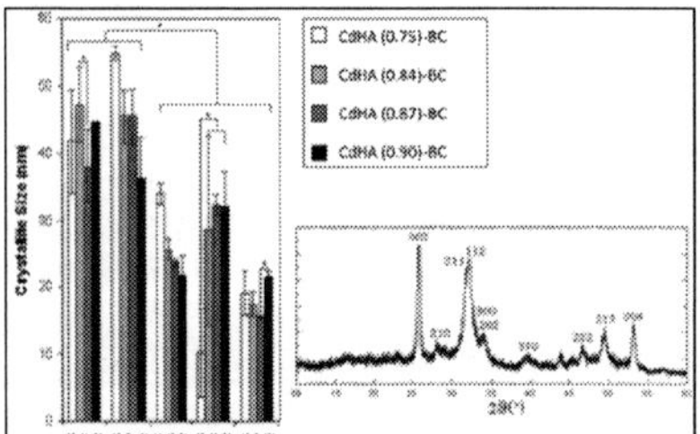

Figure 6: Crystallite thicknesses determined by the Schrerrer equation

3.2.3. Determination of ceramic chemical composition

XRD was also used to ascertain the presence of CdHA in both BC and OBC. XRD patterns of BC and its composites at 75 wt.% CdHA, and 90 wt.% CdHA are shown in Figure 7 where the crystalline diffraction peaks are indexed for hydroxyapatite. These findings support the formation of composites with HA as the dispersed phase.

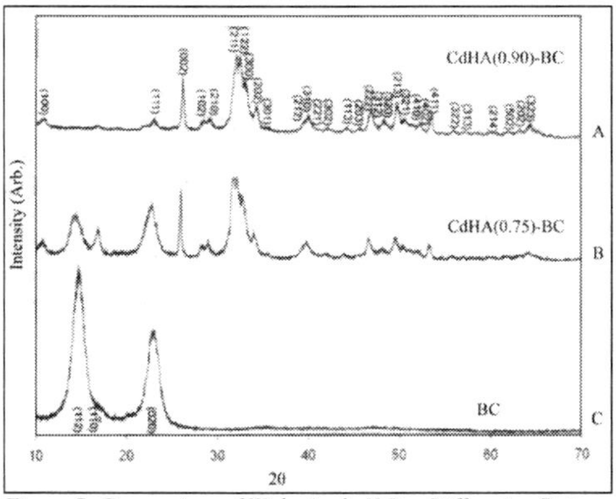

Figure 7: Comparison of Wide-Angle X-Ray Diffraction Pattern with the Addition of CdHA

After XRD demonstrated the deposited ceramic is hydroxyapatite, the next task was to determine its stoichiometry. This was accomplished by elemental analysis using energy dispersive spectroscopy (EDS). A representative energy dispersive electron spectrum of the composites is presented in Figure 8. Analysis of four BC/HA composites revealed calcium to phosphorous ratios (Ca:P) that varied between 1.46 and 1.54, a value less than the 1.67 encountered in stoichiometric HA [27]. These EDS results provide proof that the hydroxyapatite deposited in the BC and OBC is calcium deficient hydroxyapatite (CdHA). It is important to note that Ca:P was independent of the weight percentage of CdHA present in the composite. The carbon and sodium peaks present in the EDS spectra can be attributed to cellulose and sodium to the Na_2HPO_4 used in the mineralization process, respectively.

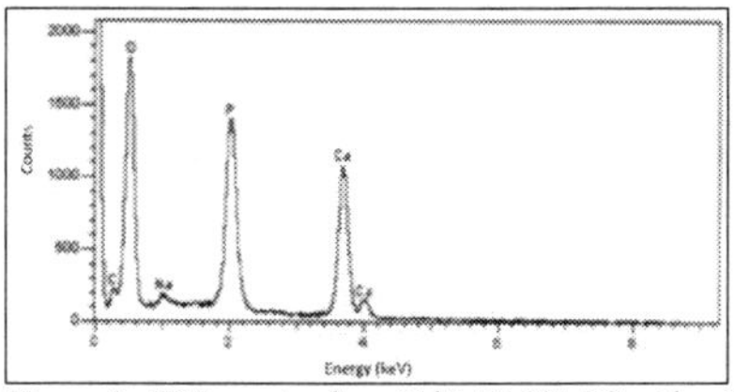

Figure 8: EDS spectra from CdHA(0.90)-BC with a Ca:P of 1.46

3.2.4. Investigation of the polymer-ceramic interface

To obtain bonding information regarding the intra and inter component bonding in CdHA-BC and CdHA-OBC, it was necessary to deconvolute the relevant FTIR absorption bands, i.e. hydroxyl stretching absorption band. The result of the deconvolution of the hydroxyl band for CdHA(0.90)-BC is presented as the insert in Figure 9. The hydroxyl absorption bands at 3350 and 3240 cm^{-1} shifted to 3244 and 3085 cm^{-1}, respectively. These spectral shifts clearly indicate the existence of various degrees of chemical interactions between the BC and CdHA. The formation of chemical bonds between the CdHA and BC serve to stabilizes the composite and maintain the composite mechanical integrity. It is also a necessary material property for bone analogue. Upon implantation in the body, chemical bonding between CdHA particles and BC/OBC would drastically reduce their release to surrounding tissues and initiation of dystrophic calcification [28]. The distribution of interfacially bonded elongated CdHA crystallites throughout the BC matrix would allow effective transfers of stresses to the CdHA crystals [29].

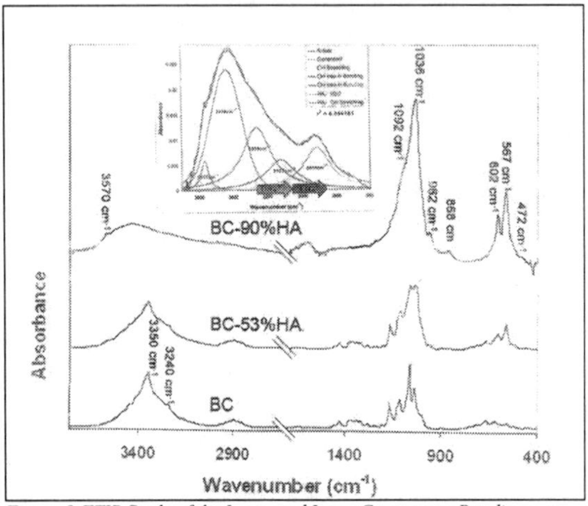

Figure 9:FTIR Study of the Intra- and Inter- Component Bonding

3.2.5. Investigation of the mechanism of CdHA deposition during synthesis

The IR spectra offered information on the formation of CdHA within the BC matrix (Figure 10). CdHA formation from metastable aqueous solutions is usually preceded by a precursor phase, most commonly amorphous calcium phosphate (ACP) ($Ca_2(PO4)_3$) or octacalcium phosphate (OCP) ($Ca_8H_2(PO_4)_6$). The precursor calcium phosphate then hydrolyzes into the more thermodynamically stable hydroxyapatite. The precursor phase depends on the conditions present in the solution during the precipitation reaction [30]. The presence of the v4 PO_4^{3-} doublet at 602 and 567 cm^{-1} in all the composites strongly suggests that the precursor phase of the hydroxyapatite was OCP [30].

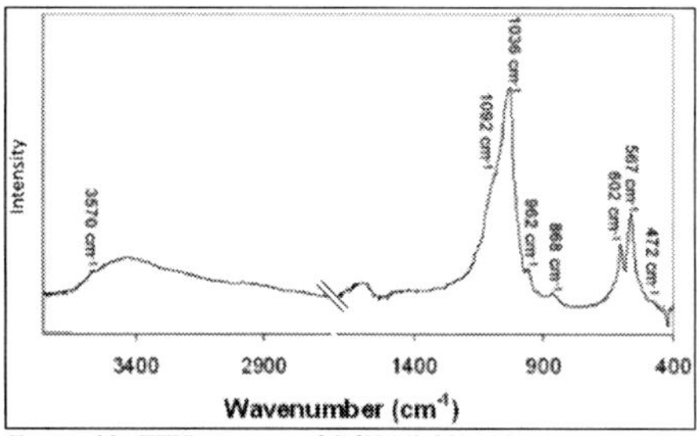

Figure 10: FTIR spectra of CdHA(0.90)-BC

During OCP formation, Ca^{+2} first complexes with another species before combining with phosphate. By organizing the calcium cations onto a matrix prior to mineralization, the OCP precursor ensured a more crystalline hydroxyapatite phase [30]. If the precursor had been ACP, the PO_4^{3-} bands would be a broad singlet instead of a doublet [31]. ACP precipitation usually requires the rapid interaction between $Ca^{2+}$and PO_4^{3-} at high supersaturation instead of precursor complexation with other species. This precludes the spatial ordering of the ions into a crystal lattice resulting in a poorly crystalline (though not amorphous) mineral [31]. The two main theories of bone mineralization are extracellular nucleation of hydroxyapatite crystal onto type I collagen fibers or release of precursor HA crystals formed within matrix vesicles [32]. In both cases, the calcium first associates with a matrix (a functional grin collagen or the matrix vesicle) before forming the calcium phosphate precursor. OCP has been found to be the precursor calcium phosphate phase in an *in-vitro* study of matrix vesicle formation of apatite [30]. In this study, the calcium ions first may first be complexing with the cellulose. This fact is supported by the other IR results that suggested chemical bonding between the apatite and the cellulose hydroxyl groups. The ordered arrangement and negative dipole of the hydroxyl groups in the crystalline BC matrix provides a structured template for the nucleation and formation of the OCP. Thermodynamically unstable OCP then hydrolyzes into crystalline HA which is confirmed by low background and sharp peaks in the XRD pattern. By having OCP precursor, the formation of CdHA in bacterial cellulose mimics the biomineralization of physiological bone.

3.2.6. Determination of mechanical properties for the composite

In addition, the accumulation of the CdHA on BC/OBC produced the expected increase in modulus normally observed for ceramic reinforced composites. Although reported in this document, a decrease was observed in the tensile strength of the CdHA-BC and CdHA-OBC [20].

Table 1: Young's Modulus of BC, OBC, CdHA-BC, and CdHA-OBC

Sample	E (Pa)
BC	3910 ± 440
OBC	3900 ± 580
CdHA-BC	7300 ± 2900
CdHA-OBC	6300 ± 1860

4. CONCLUSIONS

This study has detailed the development of an osteoconductive and bioresorbable polymer/ceramic nanocomposite that could be used for bone tissue engineering, treatment of traumatic injury and remodeling of non-unions. During the formation of the base polymeric nanofibrous mesh, XRD indicated no major structural changes of BC resulting from the periodate treatment that yields oxidized bacterial cellulose. During the mineralization of BC and OBC the weight of CdHA in the composite was linearly controlled by sequential incubations in calcium chloride and aqueous sodium phosphate. The CdHA produced in BC and OBC was determined to be similar in morphology and chemistry to the hydroxyapatite found in natural bone. It was then further determined that the CdHA-BC and CdHA-OBC degrades in a simulated aqueous physiological environment. The summation of these conclusions leads to the development of a composite that contains osteoconductive and bioresorbable CdHA deposited within BC and OBC nanofiber mats.

WORKS CITED

1. Szpalski, M and Gunzburg, R. Orthopedics 202;25 s601 – s609, 2002. **25**(5): 601-s09.
2. Khan, S, Diwan, A, Girardi, F, Cammisa, F, Sandhu, H, and Lane, J. J. Acad. Orthop Surgery, 2005. **13**: 77-86.
3. Betz, R. Orthopedics, 2002. **25**: s561-s70.
4. Hench, L and Best, S, *14 ceramics, glasses, and glass-ceramics*, in *Biomaterials science*, B Ratner, et al, Editors. 2004, Elsevier Academic Press: San Diego, CA. p. 153-70.
5. Currey, J, *Collagen and the mechanical properties of bone and calcified cartilage*, in *Collagen: Structure and mechanics*, P Fratzl, Editor. 2008, Springer. p. 397-420.
6. Wess, T, *Collagen fibrillar structure and hierarchies*, in *Collagen: Structure and mechanics*, P Fratzl, Editor. 2008, Springer. p. 49-80.
7. Gupta, H, *Nanoscale deformation mechanisms in collagen*, in *Collagen: Structure and mechanics*, P Fratzl, Editor. 2008, Springer. p. 155-73.
8. Bielecki, S, Krystoynowicz, E, Turkiewicz, M and Kalinowska, H, *Bacterial cellulose*, in *Biopolymers, volume 5, polysaccharides i; polysaccharides from prokaryotes*, E Vandamme, et al, Editors. 2001, Wiley: Weinham. p. 37-90.
9. Fontana, J, Souza, AD, Fontana, C, Torriani, I, Moresch, J, and Gallotti, B. Appl Biochem Biotechnol, 1990. **24-25**: 253-64.
10. NovaesJr., A, Novaes, A, Grisi, F, Soares, U, and Gabarra, F. Brazilian Dental Journal, 1991. **4**: 65-71.
11. Mello, L, Feltrin, L, Neto, P and Ferraz, F. J Neurosurg, 1997. **86**: 143-50.
12. Azuma, C, Yasuda, K, Tanabe, Y, Taniguro, H, Kanaya, F, Nakayama, A, Chen, Y, Gong, J, and Osada, Y. JBMR, 2007. **81A**: 373-80.
13. Klemm, D, Schumann, D, Udhardt, U and Marsch, S. Prog in Polym Sci., 2001. **26**: 1561-603.
14. Ashman, A and Gross, J, *Synthetic osseous grafting*, in *In biomaterials engineering and devices human applications, vol. 2: Orthopedic, dental and bone graft applications*, D Wise, et al, Editors. 2000, Humana Press: Totowa, NJ. p. 133-54.
15. Parikh, S. J Postgrad Med, 2002. **48**: 142-48.

16. Ley, J, Cranin, A and Katzap, M, *Biomaterials used in implant dentistry*, in *In biomaterials engineering and devices human applications, vol. 2: Orthopedic, dental and bone graft applications.* , D Wise, et al, Editors. 2000, Humana Press: Ttowa, NJ. p. 3-24.

17. Kamitakahara, M, Ohtsuki, C and Miyazaki, T. Journal of Biomaterials Applications, 2008. **23**(3): 197-212.

18. Vallet-Regi, M, Rodriguez-Lorenzo, L and Salinas, A. Solid State Ionics, 1997. **101-103**: 1279-85.

19. Hutchens, S, Leon, R, O'Neill, H and Evans, B. Letters in applied microbiology, 2007. **44**(2): 175-80.

20. Hutchens, S, *Characterization of a calcium-deficient hydroxyapatite-bacterial cellulose composite* in *Biomedical Engineering.* 2007, The University of Tennessee: Knoxville, TN.

21. Painter, T. Carbohydr. Res, 1988. **179**: 259-68.

22. Hutchens, S, *Synthesis and initial characterization of a calcium-deficient hydroxyapatite-bacterial cellulose*, in *Biomedical Engineering.* 2004, The University of Tennessee: Knoxville, TN.

23. Hutchens, S, Woodward, J, Evans, B and O'Neill, H, *Composite material* in *USPTO.*

24. White, D and BrownJr, R, *Prospects for the commercialization of the biosynthesis of microbial cellulose*, in *Cellulose and wood – chemistry and technology.* , C Schuerch, Editor. 1989, Wiley: New York. p. 573-90.

25. Bellamy, L, *The infra-red spectra of complex molecules.* 1966, New York: John Wiley and Sons.

26. Hutchens, S, Benson, R, Evans, B, O'Neill, H, and Rawn, C. Biomaterials, 2006. **27**(26): 4661-70.

27. Park, J, *Biomaterials science and engineering.* 1984, New York: Plenum Publishing.

28. Murugan, R and Ramakrishna, S. Biomaterials, 2004. **25**: 3829-35.

29. Alexander, H, *Composites*, in *Biomaterials science.* , B Ratner, et al, Editors. 1996, Academic Press: San Diego: . p. 94-105.

30. Sauer, G and Wuthier, R. J. Biol Chem, 1988. **27**: 13718-24.

31. Granja, P and Barbosa, M. J Mat. Sci., 2001. **36**: 2163-72.

32. Anderson, H and Morris, D, *Mineralization*, in *Physiology andpharmacology of bone*, G Mundy, et al, Editors. 1993, Springer-Verlag: Berlin Heidelberg.

Mater. Res. Soc. Symp. Proc. Vol. 1312 © 2011 Materials Research Society
DOI: 10.1557/opl.2011.113

Self-Assembled Nano-Needles of Polyaniline, Efficient Structures in Controlling Electrical Conductivity

Michael I. Ibrahim[1,2], Maria J. Bassil[1], Umit B. Demirci[2], Georges El Haj Moussa[1], Mario R. El Tahchi[1] and Philippe Miele[2]

[1] LPA-GBMI, Department of Physics, Lebanese University - Faculty of Sciences II, PO Box 90656 Jdeidet, Lebanon, email: gbmi@ul.edu.lb, Tel: +961 3 209688, Fax: +961 1 681553.
[2] Université Lyon 1, CNRS, UMR 5615, Laboratoire des Multimatériaux et Interfaces, 43 boulevard du 11 Novembre 1918, F-69622 Villeurbanne, France, email: umit.demirci@univ-lyon1.fr.

ABSTRACT

Polyaniline (PANI) is one of the most interesting conducting polymers with a wide and controllable conductivity range, synthesized easily via chemical or electrical route, stable chemically and environmentally, having high absorption in the visible range and high mobility of charge carriers. Under different conditions, PANI morphology can be controlled yielding to the creation of nano-tubes, belts, rods, fibers and particles.

In this study, the chemical oxidative polymerization which consists of mixing aniline hydrochloride (A-HCl) with ammonium peroxydisulfate (APS) was used to synthesize HCl doped PANI. Fixing the weight ratio A-HCl/APS defined by the IUPAC while varying their quantities leads to the formation of PANI nanoparticles with variable diameters. In addition, PANI nano-needles of 60 nm average diameter at the center are also obtained. Different methods are used to investigate of 1-D morphologies. The electrical conductivity of bulk PANI pellets is measured using the four-point probe technique. The absorption in the visible range of PANI particles and nano-needles is determined by UV-Vis spectroscopy. XRD analysis was performed to study the effect of PANI particle size and morphology on the crystallinity of the powder. Such structures could be used in hybrid solar cells for higher conversion efficiencies.

INTRODUCTION

Since the discovery of conductive polymers, new applications such as plastic flexible circuits, hybrid photovoltaics and many others were achieved. Polyaniline (PANI) is one of the most interesting conducting polymers being widely used for its easiness of synthesis, chemical and environmental stability, high absorption in the visible range and high mobility of charge carriers [1]. For these reasons PANI has entered the wide door for many applications from which we cite organic electrodes for supercapacitors [2], gas sensors [3], hybrid solar cells [4] and corrosion protection [5]. Highly crystalline PANI films have proved their suitability in humidity sensing [6].

Varying the polymerization process, PANI morphology could be controlled resulting in the formation of nanofibers, nanoparticles, nanowires, needle-like shapes, and so on. Wang et al have proved that the rapid mixing polymerization of low aniline and APS concentrations leads to the formation of PANI nanofibers with diameter of 30-50 nm [7]. Due to PANI nanofibers high surface area, their response to gas vapors is faster and more intense than that of bulk PANI [8]. The morphology of PANI can be varied from spherical particles to micro and nanofibers using electrochemical preparation route [9, 10]. Carrying out the chemical oxidation polymerization of aniline at 0°C, PANI nanowires acting as a photoswitch under illumination were obtained [11]. Needle-like PANI was synthesized using the emulsion polymerization where APS was dropwise added thus increasing the degree of polymerization and crystallinity [12, 13]. In this study, we investigate the effect of aniline hydrochloride (A-HCl) and APS concentrations, polymerization temperature and stirring speed of the mixture on the morphology and crystallinity of the formed PANI powder. The optical and morphological properties of the PANI powder are investigated using several techniques such as FTIR, UV-Vis, SEM and XRD. Finally the electrical conductivity of PANI pellets is measured using the four-point probe technique.

EXPERIMENT

The standard preparation of PANI defined by the IUPAC [15] was used to prepare PANI. In a typical procedure, 1:1 mole ratio of aniline hydrochloride (A-HCl) to ammonium peroxydisulfate (APS) is used, thus 2.59 g of A-HCl (Sigma-Aldrich purum ≥99%) are dissolved in distilled water to 50 ml of solution. Simultaneously, 5.71 g of APS (Sigma-Aldrich ACS reagent ≥98%) are also dissolved in distilled water to 50 ml of solution. The two aqueous solutions are then mixed briefly together and left to polymerize. After ten minutes, green HCl doped emeraldine salt PANI is formed. Afterwards PANI solution is dried at room temperature for 48 hours, then the powder is collected and ground. Residual monomers, low molecular weight polymers, oligomers and carcinogenic byproducts were removed by washing the collected powder several times with 0.2 M HCl and acetone.

Several procedures were followed in order to explain the formation of PANI needles and other morphologies. Therefore different parameters were varied such as the quantities of the monomer and oxidant, the addition time of the oxidant and the polymerization temperature.

Three methods were mainly used for the synthesis of PANI 1-D morphologies (nano-needles and rods) besides nano-particles:

I. The A-HCl to APS ratio was fixed at 1:1 and the quantities of the monomer and the oxidant were varied in the range of 5 to 50 wt% of the standard quantities defined by the IUPAC. APS aqueous solution was poured over the A-HCl solution instantly then the mixture was briefly stirred for 10 s.
II. The second method consists on fixing the quantities of the monomer and the oxidant to 10% of the quantities defined by the IUPAC and the APS solution is added dropwise during a time period between 1 and 1.5 hours (average flow rate of 0.7 ml/min).
III. Polymerization conducted at -10°C, 0°C and room temperature. APS solution is directly poured over pre-cooled A-HCl solution and the mixture was left for 24 hours to complete the polymerization reaction without any stirring.

The as-formed powder is then morphologically, structurally and optically compared to standard PANI. Using a Scanning Electron Microscope (Hitachi S-800), the particle size of different PANI samples is examined. XRD analysis was performed within the range $2\theta = 5–90°$ for PANI powder on a PANalytical X'Pert Pro, using Cu K_α=1.54184 Å as an X-ray source and operating at 45 kV and 30 mA. UV-Vis spectroscopy was performed using Safas-Monaco spectrometer in the range 185–1050 nm.

The synthesized powders pressed at 7 Kpsi of pressure for 2 min formed the pellets for the surface electrical conductivity study using the four-point probe technique. The probe is a 2205C type from A. & M. Fell ltd Company connected to a Keithley 487/Voltage source and to a Keithley 2000 digital multimeter.

DISCUSSION

SEM imaging

Under SEM, it is noticed that the particles of neat PANI powder (average particle size of 400 nm) are aggregated to form bigger particles (figure 1 d). Using method I for PANI formation, it can be observed that as the A-HCl and APS concentrations decreases PANI particle size decreases reaching an average diameter of 120 nm. Besides the decreasing PANI particles size, nanoneedles are obtained at A-HCl and APS concentrations between 1.2 mM and 4 mM (6 and 20% of IUPAC standard quantities) (figure 1 b & c). At lower concentrations (figure 1 a), lower number of needles is obtained due to the decreasing number of A-HCl monomers making it difficult for unidirectional polymerization.

(a) (b) (c) (d)

Figure 1: SEM images of PANI samples at different concentrations P5 (a), P10 (b), P20 (c) and P100 (d) following method I.

The addition rate of APS strongly affects the formation of PANI particles and needle-like shape. As it can be observed from the SEM images below (figure 2), when APS is added instantly long PANI rods besides particles and short rods interconnected together are formed. However when APS solution is added dropwise with or without stirring, particles are formed easier than other morphologies. This is due to the low quantity of APS which is not enough for a unidirectional polymerization. Thus as monomers start a 1-D polymerization with low APS quantity, the formation of 1-D morphologies is stopped to continue with a 0-D ones.

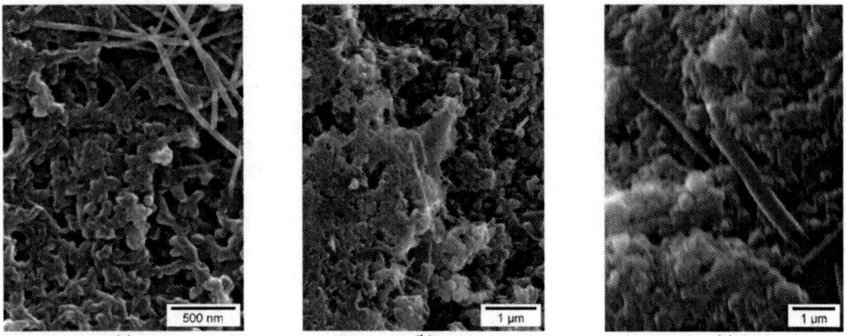

(a) (b) (c)

Figure 2: SEM images of 10% PANI where APS was poured instantly (a), dropwise without stirring (b) and dropwise with brief stirring (c).

When A-HCl is polymerized at temperatures below 0°C, big tubes are obtained with diameters between 1 and 2 μm (figure 3 a & b). In our opinion these tubes are longer than the ones obtained at room temperature having that at low temperatures monomers can undergo easily a 1-D polymerization.

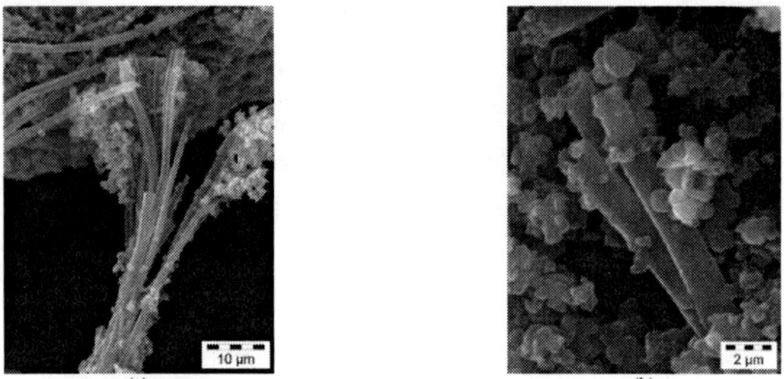

(a) (b)

Figure 3: SEM images of 10% PANI polymerized at 0°C (a) and -10°C (b).

XRD

The XRD pattern of HCl doped PANI show four peaks at 2θ = 8.5°, 15.4°, 20.4° and 25.1° for all PANI samples except for P10, showing its crystalline nature with variable d spacing and Miller indices (h k l) [16]. As it is shown in figure 4, the peaks of P20-50 are stronger than P100 due to the increasing crystallinity of PANI. For P10 and with such low monomer and oxidant concentration, disordered morphologies of nano-needles and particles are formed resulting in the perturbation of the crystalline structure thus amorphous polymer is obtained.

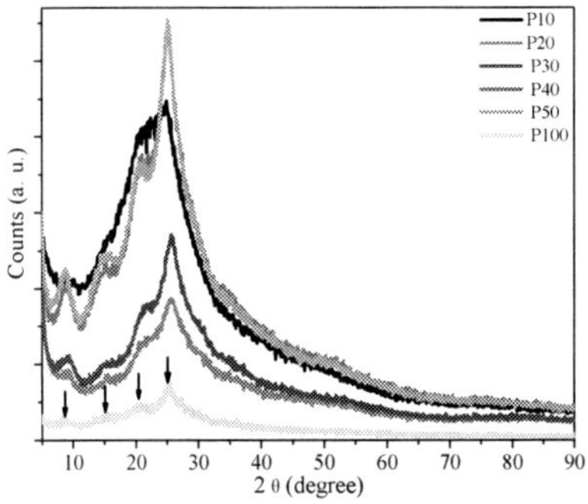

Figure 4: XRD patterns of different PANI concentrations synthesized using method I.

UV-Vis spectroscopy

For HCl doped PANI, three peaks appeared at 349, 440 and 711 nm which are assigned respectively to π-π^* transition of benzenoid and quinoid rings and the polaron band transition [17] (figure 5). It is noticed that the two absorption peaks at 349 and 440 nm do not shift proving that transitions of both benzenoid and quinoid rings occur in PANI particles and needles and that all PANI samples have the same structure. However the polaron band transition peak has blue shifted, which could be referred to the decreasing PANI particle size and to the decreasing number of H^+ and Cl^- ions originating from the dopant.

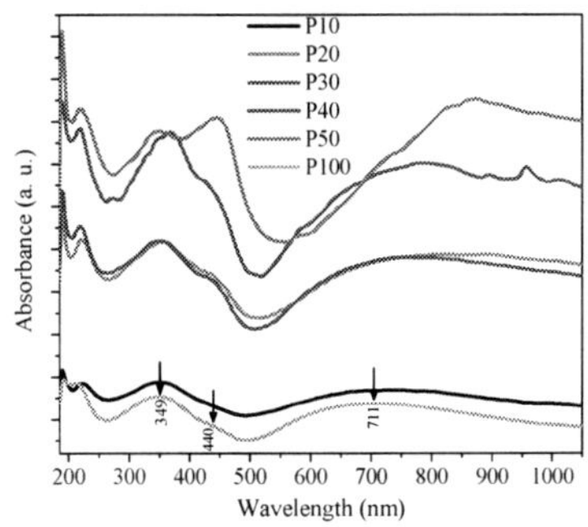

Figure 5: UV-Vis spectra of different PANI concentrations synthesized using method I.

Electrical Conductivity

When shifting the preparation method from the standard one and as the PANI content increases, the electrical conductivity of the PANI samples increases to a maximum value of about 11 S/cm (figure 6). For PANI content percentage between 10 and 50% a fast increase of approximately 400 times in the conductivity is observed then a slower increase (two times) for higher concentrations. Even though the same monomer to oxidant ratio is used to synthesize the conducting emeraldine salt, the low quantity of dopant present inside the monomer leads to a decrease in the conductivity of PANI by a factor of 800.

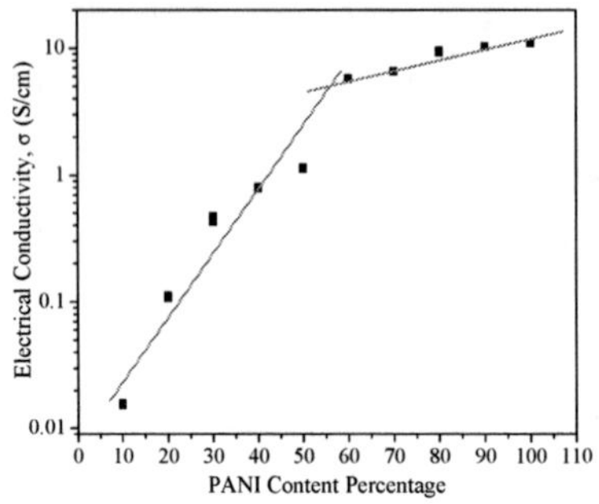

Figure 6: Electrical conductivity of PANI pellets in semi-logarithmic scale of different PANI content using the four-point probe technique.

CONCLUSION

In this study, it was shown that by varying the monomer and oxidant quantities, the addition time of the oxidant and the polymerization temperature a mixture of 1-D and 0-D PANI morphologies were obtained. The characterization techniques showed the effect of PANI structure and dopant quantity on the crystallinity, absorption in the visible range and the electrical conductivity of PANI samples.

ACKNOWLEDGMENTS

This work was financially supported by the Lebanese National Council for Scientific Research (LNCSR) and the Lebanese University. M. Ibrahim thanks the Agence Universitaire de la Francophonie (AUF) and the LNCSR for supporting his PhD studies. Authors acknowledge CTμ (Centre technologique des Microstructures) of University of Lyon 1 for SEM observations.

REFERENCES

[1] M. Roe, J. Ginder, P. Wigen, A. Epstein, M. Angelopoulos and A. MacDiarmid, *Phys. Rev. Lett.* **60,** 2789 (1988).
[2] K. Ryu, K. Kim, N. Park, Y. Park and S. Chang, *J Power Sources* **103**, 305 (2002).
[3] A. Kukla, Y. Shirshov and S. Piletsky, *Sens Actuators B Chem* **37**, 135 (1996).
[4] Z. Liu, J. Zhou, H. Xue, L. Shen , H. Zang and W. Chen, *Synth Met* **156**, 721 (2006).
[5] A. Mirmohseni and A. Oladegaragoze, *Synth Met* **114**, 105 (2000).
[6] S. Shukla, A. Bharadvaja, A. Tiwari, S. Pilla, G. Parashar and G. Dubey, *Adv Mat Lett* **1**, 129 (2010).
[7] Y. Wang and X. Jing, *J Phys Chem B* **112**, 1157 (2008).
[8] Z. Li, F. Blum, M. Bertino, C. Kim and S. Pillalamarri, *Sens Actuators B Chem* **134**, 31 (2008).
[9] G. Rimbu, I. Stamatin, C. Jackson and K. Scott, *J Optoelectronics Advanced Materials* **8**, 670 (2006).
[10] S. Roy, K. Kargupta, S. Chakraborty and S. Ganguly, *Mater Lett* **61**, 2535 (2008).
[11] X. Wang, M. Shao, G. Shao, Z. Wu and S. Wang, *J Colloid Interface Sci* **332**, 74 (2009).
[12] W. Jung, D. Kim, Y. Lee and S. McCarthy, *Mater Res Soc Symp Proc* **949** C07-05, 2007.
[13] W. Jung, Y. Lee and S. McCarthy, *J Vinyl Addit Technol* **13**, 76 (2007).
[15] J. Stejskal and R. G. Gilbert, *Pure Appl Chem* **74**, 857 (2002).
[16] J. Pouget, M. Jozefowicz, A. Epstein, X. Tang, and A. MacDiarmid, *Macromolecules* **24,** 779 (1991).
[17] J. Li, X. Tang, H. Li, Y. Yan and Q. Zhang, *Synth Met* **160**, 1153 (2010).

Mater. Res. Soc. Symp. Proc. Vol. 1312 © 2011 Materials Research Society
DOI: 10.1557/opl.2011.114

The experimental determination of the onset of electrical and thermal conductivity percolation thresholds in carbon nanotube-polymer composites

Byung-wook Kim[1], Steven Pfeifer[2], Sung-Hoon Park[2], and Prabhakar R. Bandaru[2]

[1]Department of Electrical and Computer Engineering, University of California, San Diego, USA

[2]Materials Science Program, Department of Mechanical Engineering, University of California, San Diego, USA

ABSTRACT

We show evidence of electrical and thermal conductivity percolation in polymer based carbon nanotube (CNT) composites, which follow power law variations with respect to the CNT concentrations in the matrix. The experimentally obtained percolation thresholds, *i.e.*, ~ 0.074 vol % for single walled CNTs and ~ 2.0 vol % for multi-walled CNTs, were found to be aspect ratio dependent and in accordance with those determined theoretically from excluded volume percolation theory. A much greater enhancement, over 10 orders of magnitude, was obtained in the electrical conductivity at the percolation threshold, while a smaller increase of ~ 100 % was obtained in the thermal conductivity values. Such a difference is qualitatively explained on the basis of the respective conductivity contrast between the CNT filler and the polymer matrix.

INTRODUCTION

Polymer composites containing conducting fillers [1] such as carbon black [2], carbon fiber, and metal fibers have been extensively investigated for various applications such as electromagnetic interference (EMI) shielding [3], electronic packaging [4], radar absorption [5], *etc.*, However, presently used composites require high filler to polymer loading ratios which deteriorates the overall mechanical properties, through a compromise of intrinsic matrix morphology [6]. A possible way to ameliorate the above problems, through using *low* filler volume fractions, incorporates carbon nanotubes (CNTs) in composites [7-10]. A concomitant large aspect ratio and tunable electrical conductivity would enable electrical percolation to be achieved at very small CNT volume fractions [9]. The use of small CNT volume fractions would also reduce the economic costs and clustering of the CNTs within the polymer matrix, important for large scale application. In this paper, we discuss the issue of the minimum volume fraction necessary, *i.e.*, the percolation threshold (ϕ_c) for the CNT-polymer composites to serve as electrically or thermally conductive materials through percolation theory [11]. Such a study could then lead to an understanding of the viability of such composites for a large number of applications.

EXPERIMENTAL DETAILS

We have sought to understand issues associated with the uniform dispersion of CNTs into polymer matrices through optimized synthesis/processing, involving enhanced nanotube-polymer interactions through chemical functionalization schemes. We chose for our initial experiments, a composite of CNTs and a Reactive Ethylene Terpolymer (RET: Elvaloy 4170) – Figure 1.

Both pristine and carboxyl functionalized SWNTs (average diameter 1-2 nm, and length 5-20 μm) and MWNTs (average diameter 40 nm, length 5-9 μm) were used and dispersed in toluene with sonication. The exact location of the functional groups would depend on the defect density on the nanotubes which can be manipulated [12]. If the defects are randomly dispersed, isotropic bonding of the nanotubes with the polymer matrix is implied and would yield uniform dispersion and mixing. It was typically seen that sonication reduced the average length of the SWNTs to ~ 4.3 μm, with a bundle diameter ~ 4.8 nm, resulting in an aspect ratio of ~ 880. On the other hand, the MWNTs have an average length of 1.4 μm, with a bundle diameter ~ 40 nm, yielding an aspect ratio of ~ 33.

The nanotube dispersion was added to the RET solution (in toluene), and then the mixture was ultrasonicated for 5 hours. To remove excess solvent, the mixture was stirred and evacuated in vacuum (10^{-3} Torr) for 12 hours. Subsequently, a hot press was used to press the composites into the desired thickness, of ~ 2 mm, reported here.

The four-wire resistance method was used to measure the resistance (R) for the nanocomposites with R < 1GΩ, while for higher resistance (> 1 GΩ) composites, two point measurements using the Agilent B1500A semiconductor device analyzer, with triaxial probes were used. The composites were treated with oxygen plasma and subsequently, 5-10 nm of Ti, along with ~ 100 nm of Au were sputtered for electrical contacts. The thermal conductivity (κ) was measured through the 3ω principle [13], using a Wheatstone bridge based methodology as shown in Figure 2.

The principle involves the periodic heating of metal/contact lines (10 mm long, 90 μm wide) which serve as both the heater and thermometer, deposited on the composites through alternating current modulated at a frequency ω which induces a change in the resistance with a 2ω dependence. The κ can then be interpreted through measurement of the third harmonic component of the voltage, $V(3\omega)$.

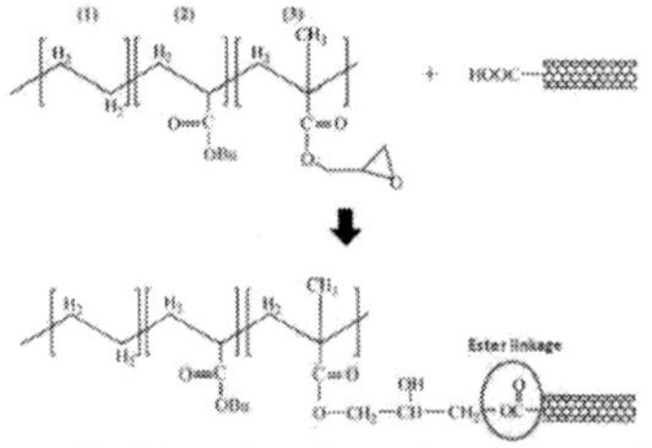

Figure 1. Schematic of reaction between CNT functional groups with epoxy groups on the RET constituted from (1) polyethylene, (2) methyl-methacrylate, and (3) epoxide groups. While (1) and (2) contribute to mechanical robustness, (3) is used for forming ester linkages to –COOH groups on the CNTs for enhanced bonding and dispersion.

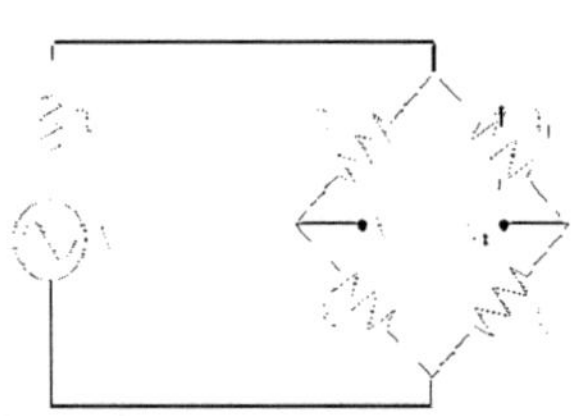

Figure 2. Schematic of the Wheatstone bridge based apparatus for the measurement of the thermal conductivity (κ) of the polymer composites. The 3ω component of across the sample (R_s) was obtained from measuring the voltage difference (V_A-V_B) using a Lock-in amplifier, SR830. The bridge serves to nullify the larger first harmonic component, $V(\omega)$.

RESULTS AND DISCUSSION

We indeed observed a *more uniform dispersion* for both *low* - Figure 3 (a) and *high*- Figure 3 (b) COOH-CNT filling fractions, as seen in the fracture surface images of the composites, in comparison to the clumped fillers observed for pristine CNT based composites in Figure 3 (c).

The DC conductivity (σ_{DC}) of the polymer composites as a function of the CNT (SWNT and MWNT) filler concentration (ϕ) is shown in Figure 4. It is immediately apparent that there is a much greater enhancement in the electrical conductivity (over ten orders of magnitude) compared to the approximately two-fold increase in the thermal conductivity values. The σ_{DC} were fit to the relations of the following form, where ϕ_c is the threshold volume fraction for electrical percolation, while s and t are critical exponents at the subcritical and overcritical region, respectively.

$$\sigma_{DC} \sim \begin{cases} (\phi_c - \phi)^{-s}, \phi_c < \phi \\ (\phi - \phi_c)^{t}, \ \phi > \phi_c \end{cases} \quad (1)$$

Given that the aspect ratio (r) was ~ 880 for the SWNTs and ~ 33 for the MWNTs, the theoretical ϕ_c was calculated from excluded volume percolation theory [14], to be ~ 0.08 and ~ 1.9 vol %, respectively, using

$$\phi_c(r) = \frac{1.4}{\frac{4\pi}{3}r^3 + 2\pi r^2 + \frac{\pi}{2}r}\left(\frac{\pi}{6}r^3 + \frac{\pi}{4}r^2\right). \quad (2)$$

We have also shown how the variability in CNT lengths on the aspect ratio and the electrical percolation threshold could be dealt with through using statistical analyses [14]. Such modeling

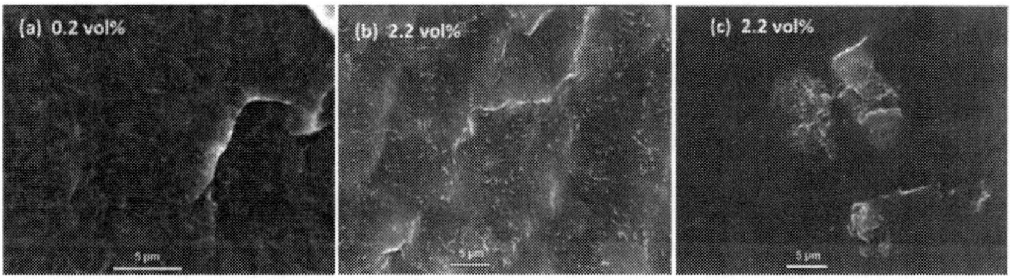

Figure 3. Scanning Electron Microscope (SEM) micrographs illustrating uniform dispersion of functionalized SWNTs in the RET polymer at (a) 0.2 and (b) 2.2 volume % CNT filling fractions while (c) non-functionalized SWNTs exhibit clumping.

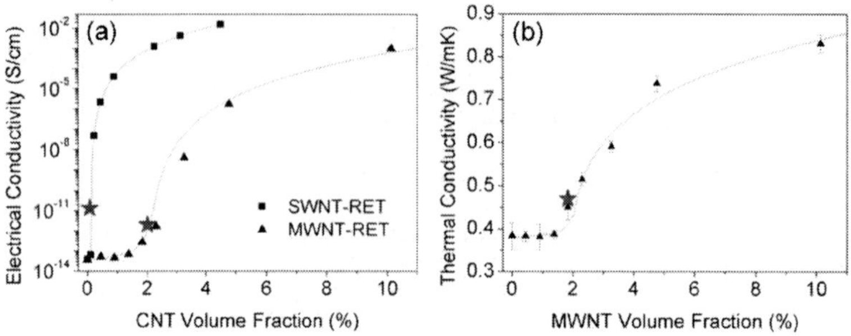

Figure 4. The enhancement of the (a) electrical conductivity of SWNT and MWNT based composites, and (b) the thermal conductivity of MWNT based composites, with increasing CNT volume fractions could be fit (solid lines) to expressions based on excluded volume percolation theory. The effective electrical and thermal conductivities (represented by *) at the threshold were estimated based on a percolation theory with the CNTs and the polymer being modeled as resistive elements [15].

could be used to *a priori* determine the threshold and considers realistic process variability which could then be extended to other mutable CNT characteristics such as diameter, agglomeration, curvature, *etc*. Consequently, for the carbon nanotube-polymer composites, both $\phi_c < \phi$ and $\phi > \phi_c$ regimes were accessible and fit, resulting in a $\phi_c \sim$ 0.074 and 2.0 vol % of SWNT and MWNT filling fractions of the percolation threshold, respectively. For $\phi_c < \phi$, the s was ~ 0.7 for the both SWNT and MWNT based composites, while for $\phi > \phi_c$ t was determined to be ~ 3.5 and ~ 5.2 for the SWNT and MWNT based composites, respectively. We see a remarkable correspondence of the experimentally observed ϕ_c with the theoretically computed values from Figure 4. We then attempted to fit the thermal conductivity (κ) variation in the case of MWNT-RET composites, using expressions similar in form to (1),

$$\kappa \sim \begin{cases} (\phi_c - \phi)^{-p}, \phi_c < \phi \\ (\phi - \phi_c)^{q}, \ \phi > \phi_c \end{cases} \tag{3}$$

where ϕ_c is now the threshold volume fraction for the onset of thermal percolation, and p and q are critical exponents in the respective regimes. We obtained, experimentally, ϕ_c ~1.83 vol %, which is quite close to the theoretically expected value ~1.9 vol % using equation (2), where p ~ 0.02 and q ~ 0.17.

The much larger variation in the increase of the electrical conductivity of the CNT-polymer composites compared to the thermal conductivity may be tentatively attributed to the greater contrast in the conductivity values of the matrix and the filler in both cases. In the former case, one has to consider the much lower electrical resistance of the CNTs compared to the polymers, while there is a much smaller variation in the κ between disparate materials. Invoking a model to determine the critical exponents [15], where the filler (R_f) and matrix (R_m) are represented as individual resistors, it has been determined that the effective conductivity (σ/κ) at the percolation threshold fraction, could be related to $R_f^u R_m^{1-u}$, where $u = t/_{t+s}$ or $u = q/_{p+q}$ for σ/κ, respectively, and the computed value have been indicated (through * in Figure 4).

CONCLUSIONS

We have shown that the onset of electrical and thermal conductivity in nanotube-polymer composites, as determined through the percolation thresholds, can be fitted with power law equations. The close correspondence of the experimental values to those determined from theory is quite remarkable and suggests that similar processes are operative in two widely differing phenomena. Further research should probe the relationships between the exponents, the influence of the respective conductivity contrast ratio of the fillers (nanotubes) and the matrix (polymer), *etc.* Our investigations could yield better insight into percolation phenomena, of wide importance in the theory of networks, and the optimal conditions for nanotube connectivity and dispersion for wide scale application.

ACKNOWLEDGMENTS

We gratefully acknowledge partial support from the National Science Foundation (Grants ECS-05-08514). Discussions with Prof. V. Karbhari and Dr. C. Love are appreciated.

REFERENCES

1. D. D. L. Chung, *Carbon* **39,** 279-285 (2001).
2. M. Al-Saleh and U. Sundararaj, *Macromolecular Materials and Engineering* **293,** 621-630 (2008).
3. S.-H. Park, P. T. Thielemann, P. M. Asbeck, and P. R. Bandaru, *IEEE Transactions on Nanotechnology* **9** 464 (2010).
4. E. S. A. Rashid, K. Ariffin, H. M. Akil, and C. C. Kooi, *Journal of Reinforced Plastics and Composites* **27,** 1573-1584 (2008).
5. Z. Liu, G. Bai, Y. Huang, F. Li, Y. Ma, T. Guo, X. He, X. Lin, H. Gao, and Y. Chen, *Journal of Physical Chemistry C* **111,** 13696-13700 (2007).
6. Y. Lin, M. J. Meziani, and Y.-P. Sun, *Journal of Materials Chemistry* **17,** 1143-1148 (2007).
7. M. Moniruzzaman and K. I. Winey, *Macromolecules* **39,** 5194-5205 (2006).
8. N. Li, Y. Huang, F. Du, X. He, X. Lin, H. Gao, Y. Ma, F. Li, Y. Chen, and P. C. Eklund, *Nanoletters* **6,** 1141-1145 (2006).
9. A. Saib, L. Bednarz, R. Daussin, C. Bailly, X. Lou, J. M. Thomassin, C. Pagnoulle, C. Detrembleur, R. Jérôme, and I. Huynen, *IEEE Transactions on Microwave Theory and Techniques* **54,** 2745-2754 (2006).
10. P. Ajayan, L. S. Schadler, C. Giannaris, and A. Rubio, *Advanced Materials* **12,** 750-753 (2000).
11. S. Kirkpatrick, *Reviews of Modern Physics* **45,** 574-588 (1973).

12. J. A. Nichols, H. Saito, C. Deck, and P. R. Bandaru, *Journal of Appled Physics* **102,** 064306 (2007).
13. T. Borca-Tasciuc, A. R. Kumar, and G. Chen, *Review of Scientific Instruments* **72,** 2139-2147 (2001).
14. S. Pfeifer, S. H. Park, and P. R. Bandaru, *Journal of Applied Physics* **108,** 024305 (2010).
15. J. P. Straley, *Physical Review B* **15,** 5733-5737 (1977).

Mater. Res. Soc. Symp. Proc. Vol. 1312 © 2011 Materials Research Society
DOI: 10.1557/opl.2011.115

In Situ Synthesis and Integration of Polymer Electrolytes in Nanostructured Electrodes for Photovoltaic Applications

Siamak Nejati and Kenneth K. S. Lau
Department of Chemical and Biological Engineering, Drexel University
Philadelphia, PA 19104, U.S.A.

ABSTRACT

The conventional dye sensitized solar cell (DSSC) is limited by the use of a liquid electrolyte that is prone to leakage and evaporation. Efforts to replace the liquid with a solid equivalent have been met with difficulties in penetrating the mesoporous TiO_2 nanostructured photoanode with liquid processing, particularly for photoanode layer thickness greater than 2 μm. Here, initiated chemical vapor deposition (iCVD) is successfully applied to directly synthesize and fill the pores of the mesoporous TiO_2 network of up to 12 μm thickness with poly(2-hydroxyethyl methacrylate) (PHEMA) polymer electrolyte. Comparing with equivalent liquid electrolyte cells, DSSCs integrated with PHEMA polymer electrolyte showed consistently higher open circuit voltage, which is attributed to a decrease in electron recombination with the redox couple at the electrode-electrolyte interface.

INTRODUCTION

The dye sensitized solar cell (DSSC) first proposed by O'Regan and Gratzel makes use of a photosensitizer dye for absorbing sunlight and injecting electrons into a titanium dioxide (TiO_2) anode consisting of a mesoporous network of TiO_2 nanoparticles [1]. The photo-oxidized dye is regenerated by the presence of an iodide/triiodide redox couple in a liquid electrolyte that shuttles between the photoanode and the platinized counter electrode. However, the use of a liquid electrolyte makes the cell prone to leakage and evaporation, reducing its operability and durability [2-4]. Thus, there is significant effort in trying to replace the liquid electrolyte with solid materials, including hole transport materials [5-7], conjugated polymers [8,9], and solid and gel polymer electrolytes [10-13]. To utilize a solid material however requires suitable processing that can provide effective infiltration into the mesoscopic pore network of the TiO_2 nanostructure. Methods such as drop casting, spin coating, and vacuum infiltration from solutions of these materials are ineffective in penetrating the tiny pores, especially with mesoporous layers thicker than 2 μm, due to viscous and steric effects [14-16]. In addition, wet processing typically leaves residual solvent which are not entirely removed and will most likely reduce cell performance. We therefore present a liquid-free initiated chemical vapor deposition (iCVD) process for enabling a one-step synthesis and deposition of polymer electrolyte within mesoporous TiO_2 that is effective in achieving pore filling up to 12 μm in layer thickness. By enabling tight integration of the polymer electrolyte, in this case poly(2-hydroxyethyl methacrylate) (PHEMA), with the nanostructured TiO_2 photoanode, we were able to obtain higher open circuit voltages in fabricated DSSCs that yielded higher power conversion efficiencies compared to equivalent liquid electrolyte cells. Based on impedance spectroscopy, we attribute the increase in open circuit voltage to a lower charge recombination rate with the redox couple in the polymer electrolyte.

EXPERIMENT

iCVD synthesis. As reported in detail elsewhere [17,18], iCVD was utilized to directly form poly(2-hydroxyethyl methacrylate) (PHEMA) inside the mesoporous TiO_2 network of the DSSC photoanode. The monomer, 2-hydroxyethyl methacrylate (HEMA, Aldrich), and a free radical initiator, *di-tert*-amyl peroxide (TAPO, Aldrich), were metered into an iCVD reactor as vapors using precision needle valves (0.2-0.8 sccm). The initiator selectively decomposes to form active radicals in the gas phase as a result of contact with a thermal zone heated by a series of filament wires (350 °C). These radicals together with the monomer vapor penetrate the pore network under a low pressure environment (60-200 mtorr) and surface polymerization is initiated at the surfaces of the pores of the TiO_2 maintained at near room temperature (27 °C). Pore filling was evaluated by looking at SEM cross-sections.

DSSC assembly. The TiO_2 photoanode was fabricated according to established procedures [19]. A TiO_2 slurry was first prepared from TiO_2 nanoparticles (P25, Evonik), and then spin coated onto fluorine doped tin oxide (FTO) conductive glass (Hartford Glass) to obtain layer thicknesses between 2 and 12 μm. The mesoporous layer was then annealed at 500 °C, and a ruthenium dye (N3, Solaronix) was subsequently adsorbed by immersing in an ethanolic solution after cooling to 100 °C. The substrate was thoroughly rinsed and dried for pore filling by iCVD. After iCVD pore filling with PHEMA, a redox couple was then incorporated by soaking in a redox solution of lithium iodide and iodine in either 50:50 or 0:100 vol% γ-butyrolactone (GBL, Aldrich) and propylene carbonate (PC, Aldrich). A platinized FTO glass counter electrode was then place on top to complete the cell.

DSSC characterization. Current-voltage behavior was traced by using a Gamry Reference 600. Samples were illuminated using a custom solar simulator under 100 mW/cm^2 irradiance at AM 1.5 spectral conditions over an area of 0.26 cm^2 with external potential bias. In addition, electrochemical impedance spectroscopy (EIS) was performed under 1 sun conditions over a frequency range of 800 kHz to 20 mHz, and an AC amplitude of 10 mV at the open circuit voltage. The platinized FTO was used as both the counter and reference electrodes. Cells were compared with liquid cells containing the corresponding redox solvent of 50:50 or 0:100 vol% GBL and PC.

DISCUSSION

iCVD was successfully applied to grow PHEMA via a radically initiated addition polymerization of HEMA monomer. Previous FTIR and NMR studies have shown PHEMA made by iCVD to be spectroscopically identical to bulk PHEMA synthesized in the liquid phase [17,18]. In our work here, FTIR shows that iCVD allows thermal decomposition of the radical initiator without harming the monomer and as a result iCVD leads to clean polymer formation that retains the chemistry of the starting monomer, including the ester and hydroxyl groups, see Figure 1. In addition, by GPC, the molecular weight of PHEMA was measured to be around 2300 g/mol, and, by DSC, PHEMA was found to be a glassy polymer with a glass transition temperature between 70-90 C, depending on molecular weight [17]. PHEMA was chosen as a potential polymer electrolyte to replace the liquid electrolyte in DSSCs as its chemical functionality is expected to facilitate ion transport of the redox couple and methacrylate polymers are known to form stable gel systems with carbonate dielectrics.

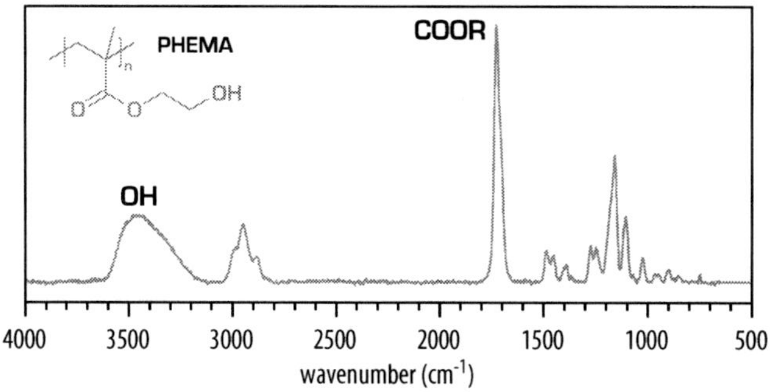

Figure 1. FTIR spectrum of poly(2-hydroxyethyl methacrylate) (PHEMA) showing the retention of ester and hydroxyl groups of the starting monomer.

By controlling iCVD parameters, we were able to achieve complete pore filling of PHEMA within mesoporous TiO_2 layers up to 12 μm thick, as shown in Figure 2. By running iCVD under monomer-limited conditions, we found that the critical process parameter which controls the quality of fill is the relative pressure of the monomer, defined as the ratio of its partial pressure in the vapor phase to its vapor pressure at the temperature of the substrate surface ($z = P_M/P_{M,sat}$). The z parameter affects both mass transport and reaction kinetics. For mass transport, gas phase diffusion inside mesopores is dominated by Knudsen diffusion (Kn>10,000). However, mass transport in mesoporous materials also has a significant surface diffusion component, which is dependent on monomer adsorption behavior that is a function of z [20,21]. Experimentally, it has been observed that diffusivity shows a maximum at some intermediate z condition [22]. For reaction kinetics, the polymerization rate is dependent on monomer

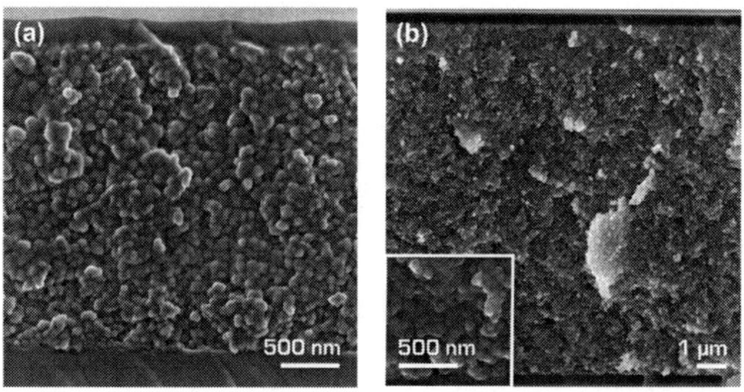

Figure 2. Cross-sectional SEM images of (a) 4 μm, and (b) 12 μm thick mesoporous TiO_2 nanostructured DSSC electrodes completely pore filled with poly(2-hydroxyethyl methacrylate) (PHEMA) polymer electrolyte using initiated chemical vapor deposition (iCVD).

concentration at the surface, which again is a function of monomer adsorption behavior or z [23]. For normal reaction kinetics, which has been found for iCVD free radical polymerization [24,25], reaction rate typically increases with increasing z or greater surface availability of the monomer. Thus, we should expect at an intermediate z region in which reaction is limiting i.e., mass transport resistance is negligible, and that should allow efficient reactant delivery inside the mesopores and enable effective pore filling to occur, see Figure 2. Indeed, by running at an intermediate z between 0.03-0.09, we obtained complete pore filling of PHEMA within the mesoporous TiO_2 nanostructures.

Comparing DSSCs that integrated PHEMA polymer electrolyte with liquid electrolyte cells that used the same redox couple of lithium iodide and iodine in either 50:50 or 0:100 vol% GBL and PC, the polymer cells show consistently higher open circuit voltage while the short current density did not change significantly, see Figure 3. This leads to higher power conversion efficiencies of polymer cells that reach about 2.3% compared with liquid cells with efficiencies of around 1.2%. These efficiencies have been achieved without optimizing the entire cell, and since the only difference between the polymer and liquid cells is the presence of the PHEMA polymer electrolyte, the comparison serves to highlight the effect of the polymer. To elucidate the effect of the polymer, EIS measurements were made. From the Bode plots presented in Figure 3 comparing DSSCs containing the polymer and the corresponding liquid electrolyte, we observe a shift in the mid-frequency peak to lower frequency with the polymer cell. Since the peak is directly related to the recombination time constant of the electrons with the redox couple (triiodide) [26], this shift corresponds to an increase in electron lifetime from 12 ms in the liquid cell to 50 ms in the polymer cell. It is reasonable to believe that the tight integration of the PHEMA polymer electrolyte at the TiO_2 electrode interface reduces surface trap states [27] or limit the amount of lithium adsorption on the TiO_2 surface [28], factors that would lead to an increase in the observed open circuit voltage.

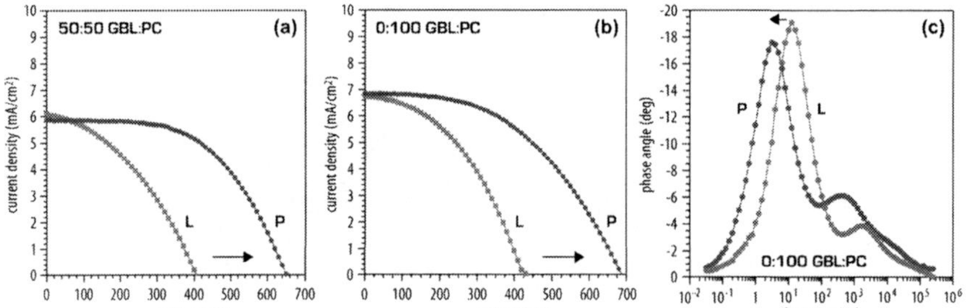

Figure 3. Current-voltage behavior of dye sensitized solar cells integrated with PHEMA polymer electrolyte (P) and the equivalent liquid electrolyte (L) containing (a) 50:50, and (b) 0:100 vol% γ-butyrolactone and propylene carbonate, showing an increase in open circuit voltage when the liquid is replaced with the polymer. For pure propylene carbonate, (c) Bode plots from electrochemical impedance spectroscopy measurements of the polymer and liquid cells show a decrease in mid-frequency peak as a result of a decrease in electron recombination with the redox couple at the electrolyte-electrode interface.

CONCLUSIONS

Initiated chemical vapor deposition (iCVD) has been shown as an effective processing method for directly growing polymer electrolytes within mesoporous TiO_2 nanostructured electrodes of DSSCs of up to 12 μm in thickness. By carefully tuning iCVD processing parameters, effective pore filling of poly(2-hydroxyethyl methacrylate) (PHEMA) polymer electrolyte has been achieved in a regime where the surface polymerization reaction was limiting compared with gas and surface diffusion. DSSCs that effectively incorporated the PHEMA polymer electrolyte showed higher open circuit voltage compared with equivalent liquid electrolyte cells, which is attributed to a decrease in the rate of electron recombination with the contacting redox couple at the electrolyte-electrode interface.

ACKNOWLEDGMENTS

The authors would like to thank the National Science Foundation for support through the NSF CAREER Award (CBET-0846245) and NSF SGER grant (CBET-0820608) as well as the American Chemical Society through the ACS PRF grant (48160-G5). We also acknowledge the use of Drexel University's Centralized Research Facilities.

REFERENCES

1. B. O'Regan, and M. Grätzel, *Nature* **353**, 737 (1991).
2. M. Grätzel, *J. Photochem. Photobio. A* **164**, 3 (2004).
3. P. Wang, S. M. Zakeeruddin, J. E. Moser, M. K. Nazeeruddin, T. Sekiguchi, and M. Grätzel, *Nat. Mater.* **2**, 402 (2003).
4. M. Matsumoto, H. Miyazaki, K. Matsuhiro, Y. Kumashiro, and Y. Takaoka, *Solid State Ionics* **89**, 263 (1996).
5. U. Bach, D. Lupo, P. Comte, J. E. Moser, F. Weissortel, J. Salbeck, H. Spreitzer, and M. Grätzel, *Nature* **395**, 583 (1998).
6. Q. B. Meng, K. Takahashi, X. T. Zhang, I. Sutanto, T. N. Rao, O. Sato, A. Fujishima, H. Watanabe, T. Nakamori, and M. Uragami, *Langmuir* **19**, 3572 (2003).
7. J. Kruger, R. Plass, M. Grätzel, and H. J. Matthieu, *Appl. Phys. Lett.* **81**, 367 (2002).
8. D. Gebeyehu, C. J. Brabec, and N. S. Sariciftci, *Thin Solid Films* **403**, 271 (2002).
9. Y. Saito, N. Fukuri, R. Senadeera, T. Kitamura, Y. Wada, and S. Yanagida, *Electrochem. Commun.* **6**, 71 (2004).
10. F. Cao, G. Oskam, and P. C. Searson, *J. Phys. Chem.* **99**, 17071 (1995).
11. A. F. Nogueira, J. R. Durrant, and M. A. De Paoli *Adv. Mater.* **13**, 826 (2001).
12. J. H. Wu, Z. Lan, J. M. Lin, M. L. Huang, S. C. Hao, T. Sato, and S. Yin, *Adv. Mater.* **19**, 4006 (2007).
13. R. Komiya, L. Y. Han, R. Yamanaka, A. Islam, and T. Mitate, *J. Photochem. Photobio. A* **164**, 123 (2004).
14. L. Schmidt-Mende, and M. Grätzel, *Thin Solid Films* **500**, 296 (2006).
15. B. O'Regan, F. Lenzmann, R. Muis, and J. Wienke, *Chem. Mater.* **14**, 5023 (2002).
16. J. H. Yum, P. Chen, M. Grätzel, and M. K. Nazeeruddin, *ChemSusChem* **1**, 699 (2008).

17. R. K. Bose, and K. K. S. Lau, *Chem. Vap. Deposition* **15**, 150 (2009).
18. R. K. Bose, and K. K. S. Lau, *Biomacromolecules* **11**, 2116 (2010).
19. M. K. Nazeeruddin, A. Kay, I. Rodicio, R. Humphrybaker, E. Muller, P. Liska, N. Vlachopoulos, and M. Grätzel, *J. Am. Chem. Soc.* **115**, 6382 (1993).
20. R. Valiullin, P. Kortunov, J. Karger, and V. Timoshenko, *J. Chem. Phys.* **120**, 11804 (2004).
21. R. Valiullin, J. Karger, and R. Glaser, *Phys. Chem. Chem. Phys.* **11**, 2833 (2009).
22. J. G. Choi, D. D. Do, and H. D. Do, *Ind. Eng. Chem. Res.* **40**, 4005 (2001).
23. P. J. Flory, *Principles of Polymerization* (Cornell University Press, Ithaca, NY, 1953).
24. K. K. S. Lau, and K. K. Gleason, *Macromolecules* **39**, 3688 (2006).
25. K. K. S. Lau, and K. K. Gleason, *Macromolecules* **39**, 3695 (2006).
26. R. Kern, R. Sastrawan, J. Ferber, R. Stangl, and J. Luther, *Electrochim. Acta* **47**, 4213 (2002).
27. J. Moser, S. Punchihewa, P. P. Infelta, and M. Grätzel, *Langmuir* **7**, 3012 (1991).
28. G. Redmond, and D. Fitzmaurice, *J. Phys. Chem.* **97**, 1426 (1993).

Mater. Res. Soc. Symp. Proc. Vol. 1312 © 2011 Materials Research Society
DOI: 10.1557/opl.2011.481

Magnetic properties of acrylic UV-cured films containing magnetite nanoparticles

Alessandro Chiolerio[1], Paolo Allia[2], Paola Tiberto[3], Lorenza Suber[4], Giada Marchegiani[2,4] and Marco Sangermano[2]
[1]Physics Department, Politecnico di Torino, Corso Duca degli Abruzzi 24, IT-10129, Torino, Italy
[2]Materials Science and Chemical Engineering Department, Politecnico di Torino, Corso Duca degli Abruzzi 24, IT-10129, Torino, Italy
[3]Electromagnetism Division, INRiM, Strada delle Cacce 91 IT-10135, Torino, Italy
[4]Istituto di Struttura della Materia, CNR-Area della ricerca di Roma 1, Via Salaria km 29.500 IT-00015, Monterotondo Stazione, Italy

ABSTRACT

Acrylic based films containing thermo-chemically synthesized magnetite nanoparticles (NPs) were prepared by UV-curing. A stable dispersion of Fe_3O_4 NPs in n-hexane was added to polyethylene glycol diacrylate (PEGDA) oligomer or to hexanediol diacrylate (HDDA) oligomer, producing a blend whose viscosity matches the processing requirements for inkjet printing technology. Morphologic characterization is provided by means of Field Effect SEM on a representative nanocomposite section.
By real-time FT-IR analysis it was shown that Fe_3O_4 NPs are able to initiate radical chain-grown polymerization under UV light, for what concerns the HDDA matrix. Tight cross-linked transparent polymeric films were obtained after 1 minute of UV irradiation.
The magnetic properties of the produced films were studied by means of an Alternating-Gradient Force Magnetometer (AGFM) in the temperature range 10 – 300 K and up to 18 kOe. The isothermal magnetization curves of both HDDA and PEGDA -based nanocomposites showed that these hybrid systems must be described as interacting superparamagnets (ISP) characterized by inter-particle magnetic interactions dominating over intra-particle effects.

INTRODUCTION

Inorganic nanoparticles (NPs) dispersed within polymer matrices have been potentially exploited for a number of applications, thanks to the new properties originating from the combination of inorganic components and polymers. Polymeric Matrix Nanocomposite materials (PMNCs) containing either metallic or insulating NPs have been prepared to obtain interesting physical (optical [1], conductive [2-4], and magnetic [5-8] properties for use in micro and nanoelectronics. In addition, important contributions to fundamental studies in nanoscale physics have been brought by studying those materials, such as quantum tunneling of magnetization, spin reversal mechanism in single-domain particles and quantum size effects[9–11], while we foresee advanced use of these materials for their magnetoresistance properties.

In this case, a liquid ferrofluid containing an oligomer resin and a dispersion of magnetic NPs is made to polymerize by UV curing to form a solid film. Magnetite is a cheap material with a lower degree of toxicity than metallic ferromagnets and can be easily synthesized through a variety of low-cost techniques, such as the thermo-chemical route [12].

DISCUSSION

Preparation of the Magnetite NPs

Fe_3O_4 NPs were prepared according to the method proposed by Sun et al. [12]. Fe(III) acetylacetonate,1,2-hexadecanediol, oleic acid, oleylamine, benzyl ether and n-hexane were purchased from Aldrich and used as received. In a typical procedure, Fe(acac)3 (2 mmol), 1,2-hexadecanediol (10 mmol), oleic acid (6 mmol), oleylamine (6mmol), and benzyl ether (20 mL) were mixed and mechanically stirred under a flow of nitrogen. The mixture was heated under a nitrogen blanket to 200 °C for 2 h and then heated up to reflux (300 °C) for 1 h. The black-colored mixture was cooled to room temperature by removing the heat source and treated with ethanol under air. A dark-brown material precipitated from the solution. The product was dissolved in n-hexane in the presence of oleic acid and oleylamine and re-precipitated with ethanol. Finally the product was dispersed in n-hexane.

Preparation of the PMNCs

The magnetite NPs dispersion in n-hexane was added to the PEGDA acrylic resin in two amounts, i.e., 5 and 10 phr and to the HDDA acrylic resin in four amounts, i.e. 5, 10, 20 and 40 phr. The mixtures were stirred reaching a uniform dispersion and the radical photoinitiator (Irgacure 2959 Ciba) was added at a concentration of 2 wt% only to the PEGDA samples, while those based on HDDA were cured without any photoinitiator. The formulations were coated on glass substrates and the curing reaction was performed by irradiation with UV lamp (Hg medium pressure lamp, Italquartz, Milano, Italy) with a light intensity of about 30 mW/cm^2 (measured with EIT radiometer) for one minute under nitrogen. UV cured transparent films about 100 micrometers thick were obtained and exploited in subsequent measurements. Samples with different magnetite content will be referred to as PEGDA *n* and HDDA *m*, where *n* and *m* are the phr content of ferrofluid. In Figure 1 (right) we report the Field Effect Secondary Electron Microscope image of a section of PMNC. On the left the equivalent diameter histogram, extracted with the Matlab ® script as described in [2]. Notice that the cross-correlation between the magnetic NPs and the electron beam may result in features having a higher diameter than the effective NP diameter.

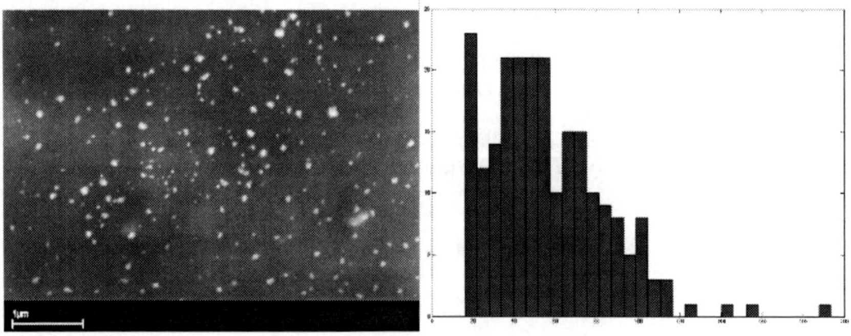

Figure 1. Left: Field Effect SEM image of a section of PEGDA10 film, showing the homogeneous distribution of ferrofluid NPs. Right: histogram representing the equivalent diameter distribution (X axis: diameter in nm, Y axis: counts).

Measurements

Acrylic double bond conversion was evaluated as a function of irradiation time by means of real-time FT-IR employing a Thermo-Nicolet 5700. The formulations were coated onto a silicon wafer. The sample was exposed simultaneously to the UV beam, which induces the polymerization, and to the IR beam for in-situ analysis of the extent of the reaction. Acrylic double bond conversion was followed by monitoring the decrease in the absorbance, due to C=C double bonds, of the peak centered at around 1450 cm^{-1} (the acrylic double bond conversion was normalized with the carbonyl peak centered at 1700 cm^{-1}). A medium pressure mercury lamp (Hamamatsu) equipped with an optical guide was used to induce the photopolymerization (light intensity on the surface of the sample of about 30 mW/cm^2).

Magnetization curves were measured at fixed temperatures from 10 to 300 K on the as-received NP dried powder and PMNCs samples by using an alternating-gradient field magnetometer (AGFM) equipped with a liquid He continuous-flux cryostat and operating in the magnetic field range -10 kOe $< H < +10$ kOe. Field-cooled (FC) and zero-field-cooled (ZFC) curves were obtained by recording sample magnetization under a small constant field ($H = 200$ Oe) while continuously varying the measurement temperature from 10 to 250 K.

RESULTS

In Figure 2 we report the conversion versus magnetite NPs content for the different PMNCs based on HDDA. For what concerns PEGDA the conversion curves as functions of irradiation time (not shown) were practically unmodified both for the pristine PEGDA resin and for the UV-curable formulations containing an increasing amount of magnetite NPs. It can be safely assumed that the magnetic NPs do not significantly influence the rate of conversion of PEGDA; the acrylic double bond formation was almost complete after one minute of irradiation time. On the contrary, for what concerns HDDA it is quite evident an increase of both rate of photopolymerization and acrylic double bond conversion by increasing the concentration of the magnetite dispersion. This is a clear indication of the photoinitiator activity of the magnetite NPs. A slight acrylic double bond conversion is achieved even for the pristine HDDA resin without any photoinitiator, which shows that a certain amount of radicals are self-generated from the hydrogen abstraction of the monomer during UV irradiation.

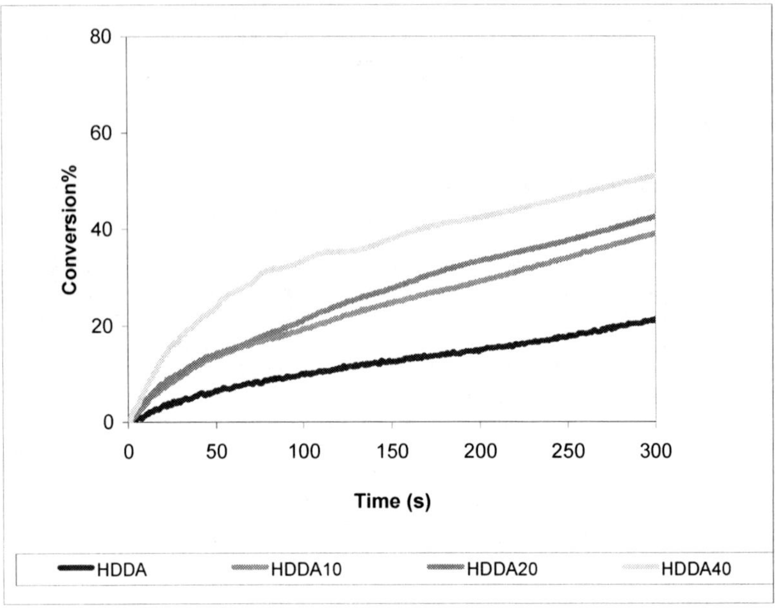

Figure 2. Conversion curves relative to the HDDA-based PMNCs with different dispersoid content.

Isothermal magnetization curves of cured-PMNCs are reported in Figure 3, for two representative samples produced with HDDA and PEGDA resins and for both 300 and 10 K. The magnetization of as-received NPs was calculated to be about 68 emu/g or 375 emu/cm^3 at room temperature, to be compared with 92 emu/g for bulk magnetite [13]. For what concerns PEGDA-based PMNCs the curves taken at room temperature exhibit small hysteresis and a typical unsaturating behaviour. A magnetic hysteresis is observed at T = 10 K. The behavior is slightly different for HDDA-based PMNCs, since in spite of the different nominal concentration of NPs in the precursor dispersion, magnetization curves taken at the same temperature have identical shape within the experimental errors. This applies to all measurement temperatures in the range 10–300K. At room temperature, magnetization curves of HDDA-PMNCs exhibit no hysteretic features and are fitted by a single Langevin function, indicating a superparamagnetic (SPM)-like behavior. Instead, PEGDA-PMNCs magnetization curves reveal that the materials are not in the standard SPM regime. As known [14] this property indicates the gradual emergence of the interacting superparamagnet (ISP) regime of a NP system, where long-range interactions among isolated magnetic moments are no longer completely negligible and bring about a collective magnetic behavior. Figure 4 shows FC and ZFC curves taken on PEGDA PMNC samples with a static field of 50 Oe. From those curves we can estimate the blocking temperatures for the composites: 11.5 and 7 K, for PEGDA5 and PEGDA10 samples, respectively. For HDDA composites, a blocking temperature of 15 K was estimated [8]. These values suggest a strong interaction between polymeric matrix and dispersoids.

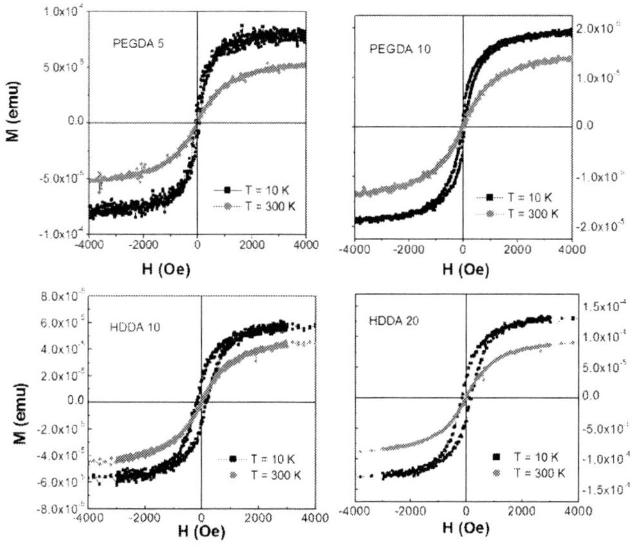

Figure 3. Isothermal magnetization curves of different PMNCs at 10 and 300 K; from top left in clockwise direction: PEGDA5, PEGDA10, HDDA20 and HDDA10.

In conclusion, both single-particle effects (responsible for the measured hysteretic properties) and collective effects (the ISP regime) have been observed in the examined nanocomposite systems, with a strong influence of the polymeric matrix host, indicating that its influence on the dispersoid is fundamental.

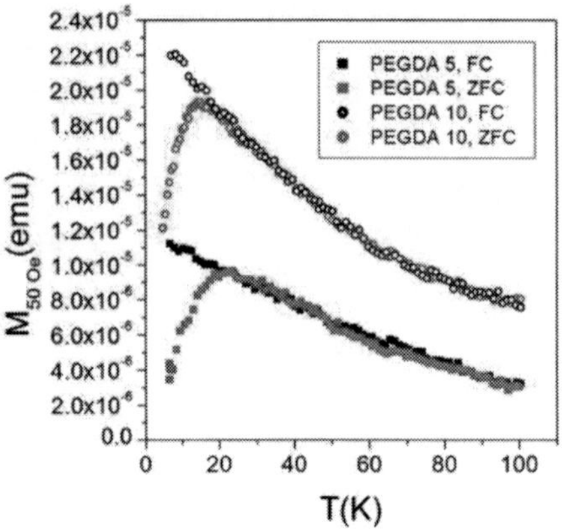

Figure 4. FC and ZFC curves for samples PEGDA5 and PEGDA10.

ACKNOWLEDGMENTS

The Authors would like to thank Regione Piemonte for funding this research within the printTAG project (''Sistemi Avanzati di Produzione'' 2008).

REFERENCES

1. Li S., Lin M. M., Toprak M. S., Kim D. K., Muhammed M. (2010) Nano Rev. 1: 5214-1-5214-19. doi:10.3402/nano.v1i0.5214.
2. Chiolerio A., Vescovo L., Sangermano M. (2010) Macromol. Chem. Phys. 211: 2008-2016.
3. Chiolerio A., Sangermano M. (2011) submitted to European Physical Journal E.
4. Gangopadhyay R., De A. (2000) Chem. Mater. 12: 608–622. doi: 10.1021/cm990537f.
5. Shull R. D., Kerch H. M., Ritter J. J. (1994) J. Appl. Phys. 75: 6840-6842. doi:10.1063/1.356802.
6. Sohn B. H., Cohen R. E., Papaefthymiou G. C. (1998) J. Magn. Magn. Mater. 182: 216-224.
7. Allia P., Tiberto P., Coisson M., Chiolerio A., Celegato F., Vinai F., Sangermano M., Suber L., Marchegiani G. (2010) J. Nanopart. Res. accepted for publication.
8. Sangermano M., Vescovo L., Pepino N., Chiolerio A., Allia P., Tiberto P., Coisson M., Suber L., Marchegiani G. (2010) Macromol. Chem. Phys. 211 2530-2535.
9. Zhang X. X., Hernandez J. M., Tejeda J., Phys. Rev. 1996, B54, 1.
10. Awschalom D. D., Di Vincenzo D. P., Smith J. F., Science 1992, 258, 414.
11. Papaefthymiou G. C., Phys. Rev. 1992, B46, 10366.
12. Sun S., Zeng H., Robinson D. B., Raoux S., Rice P. M., Wang S. X., Li G., J. Am. Chem. Soc. 2002, 124, 8204.
13. Cullity B. D. (1972) "Introduction to Magnetic Materials", Addison-Wesley, Reading.

14. Allia P., Coisson M., Tiberto P., Vinai F., Knobel M., Novak M. A., Nunes W. C. (2001) Phys. Rev. B 64: 144420-1 – 144420-12. doi: 10.1103/PhysRevB.64.144420.

Mater. Res. Soc. Symp. Proc. Vol. 1312 © 2011 Materials Research Society
DOI: 10.1557/opl.2011.116

Magnetic Polypropylene Nanocomposites Reinforced with In-situ Fabricated Iron Oxide Nanoparticles

Jiahua Zhu[1], Suying Wei[2] and Zhanhu Guo[1*]

[1]Integrated Composites Laboratory (ICL), Dan F. Smith Department of Chemical Engineering, Lamar University, Beaumont, TX, 77710 USA zhanhu.guo@lamar.edu

[2]Department of Chemistry and Biochemistry, Lamar University, Beaumont, TX 77710, USA

ABSTRACT

Polypropylene(PP)/Fe_2O_3 nanocomposites are fabricated using an in-situ method to uniformly disperse the magnetic nanoparticles (NPs) in polymer matrix. Maleic anhydride functionalized PP (f-PP) with different molecular weight is used as surfactant to stabilize the in-situ produced nanoparticles. The thermal behavior of PP and its nanocomposites with the incorporation of small amount f-PP is studied with thermal gravimetric analysis (TGA). The results show that the onset degradation temperature is increased by ~117 °C with the addition of NPs. Both melt rheology and transmission electron microscopy are used to investigate the NPs dispersion. Strong saturated magnetization (M_s) is observed after introducing f-PP to the nanocomposites through protecting the as-formed NPs from oxidation.

INTRODUCTION

Polymer nanocomposites (PNCs) are now being extensively studied because of their cost effective processability, lightweight and tunable physical properties, such as mechanical, magnetic, optical and electronic properties.[1-4] With all these unparallel advantages, PNCs have found wide potential applications in solar cells,[5] microwave absorption layers,[6,7] supercapacitors[8] and electrochromic devices.[9] Magnetic PNCs are now being extensively studied for their potential applications in high-capacity magnetic storage media and elements of nanoscale integral circuits. The major challenges for obtaining high performance magnetic units arise from the magnetically induced particle agglomeration. Techniques to stabilize and disperse magnetic NPs by encapsulating the magnetic core within polymers,[10] silica[11] and carbon[12] have been reported. Mono-dispersed magnetic NPs are able to be produced via different appraoches.[13,14] However, it is still a challenge to reach a high quality dispersion of these mono-dispersed NPs in polymer matrix due to the limited surface compatibility between the two phases. Right now, most of the current research work on fabricating PNCs starts from the as prepared NPs and polymers (or monomer) with a direct blending[15] or surface initiated polymerization method.[16] In this work, the PNCs are produced with an in-situ method. Maleic anhydride functionalized polypropylene (f-PP) is introduced as a surfactant to stabilized and disperse the in-situ formed NPs with the advantage of covalent bonding between carboxylic group and NPs. Meanwhile, excellent interfacial compatibility between f-PP and PP are expected owing to the similar backbone structure.

EXPERIMENT

The isotactic Polypropelene (PP) used in this study is supplied by Total Petrochemicals USA, Inc (M_n≈40500, M_w≈155000, ρ=0.9 g/cm^3). The maleic anhydride functionalized PP with different molecular weight (X-10127 (lf-PP), M_w≈800, ρ=0.85 g/cm^3; X-10128 (hf-PP), M_w≈8000, ρ=0.90 g/cm^3) is supplied by Baker Hughes Company. Iron(0) pentacarbonyl ($Fe(CO)_5$) is commercially obtained from Sigma Aldrich. The xylene (laboratory grade) is purchased from Fisher Scientific.

20 g PP and 0.5 g lf-PP (hf-PP) is dissolved in 207 ml xylene and refluxed at ~140 °C for 2 hours. Then 17.48 g $Fe(CO)_5$ is injected into the mixture solution and further refluxed for 3 hours. After cooling down, pour the solution to a large glass plate to allow the solvent evaporation overnight. Meanwhile, the as formed Fe NPs are slowly oxidized to Fe_2O_3. Pure PP powders are also prepared following the above procedures without adding $Fe(CO)_5$.

The desired samples are prepared using hot press (Carver 3853-0, USA). The rheological behavior of the composites is studied using TA Instruments AR 2000ex Rheometer. An ETC steel parallel-plate geometry (25 mm in diameter) is used to perform the measurement at 200 °C, with dynamic oscillation frequency sweeping from 100 Hz to 0.1 Hz in the linear viscoelastic range (strain 1%) under a nitrogen atmosphere to prevent the oxidation of PP. The thermal stability is studied with a thermo-gravimetric analysis (TGA, TA instruments TGA Q-500) from 25 to 600 °C with an air flow rate of 60 ml/min and a heating rate of 10 °C/min.

The particle distribution is examined by a transmission electron microscope (TEM). The samples are microtomed into thin sections with a thickness of less than 100 nm and then observed in a FEI Tecnai G2 F20 with a field emission gun at a working voltage of 200 kV.

The magnetic property measurements are carried out in a 9 T physical properties measurement system (PPMS) by Quantum Design at room temperature.

DISCUSSION

Melt Rheology

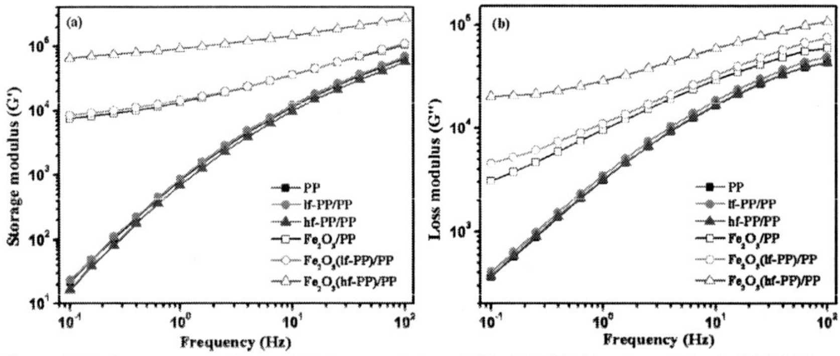

Figure 1 (a) Storage modulus and (b) loss modulus of PP, f-PP/PP blends and Fe_2O_3/PP PNCs.

Figure 1 shows the storage modulus (G') and loss modulus (G") as a function of frequency. Both G' and G" increases with the increase of oscillation frequency. It is clearly observed that the G' (Figure 1a) and G" (Figure 1b) curves of pure PP and f-PP/PP blends are quite similar to each other, while the PNCs show significantly higher G' and G" within the whole frequency range, especially at low frequencies. The G' versus frequency curve for the PNCs approaches a plateau at low frequencies. This "plateau" suggests an interconnected structure of nanofillers and strong interaction between the filler and polymer, which derived from the apparent yield stress in either G' or G" versus frequency.[17] It is worth noting that the Fe_2O_3(hf-PP)/PP PNCs shows the G' value an order of magnitude higher than Fe_2O_3(lf-PP)/PP PNCs at the frequency of 0.1 Hz. With the maleic functional groups reacted and attached to the particle surface, both lf-PP and hf-PP are able to protect NPs from agglomeration. Meanwhile, the excellent compatibility PP and f-PP helps to disperse the NPs more efficiently. The hf-PP with long polymer chains provides the advantage of chain-chain entanglement between themselves and the PP chains. Thus, strong interactions of the two phase materials could be obtained as evidenced by the high storage modulus. The Fe_2O_3(lf-PP)/PP shows slightly enhanced G' and G" values as compared to the Fe_2O_3/PP, which is arising from the weak interfacial interactions.

TGA Analysis

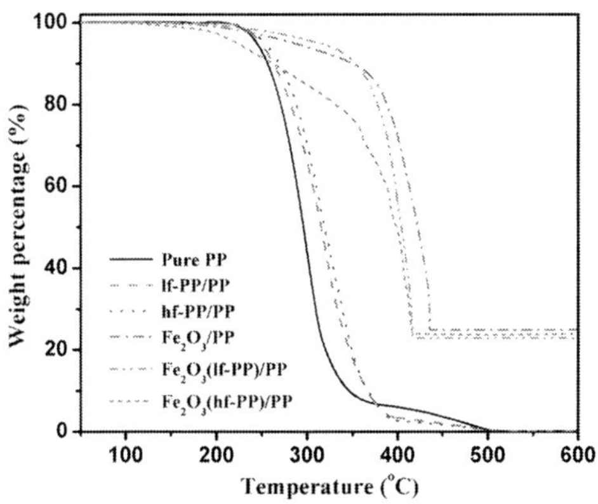

Figure 2 TGA curves of pure PP, f-PP/PP blends and Fe_2O_3/PP PNCs.

Figure 2 shows the thermal decomposition curves of the pure PP, f-PP/PP blends and Fe_2O_3/PP PNCs from room temperature up to 600 °C. The pure PP shows the thermal decomposition temperature (T_d, record on 20% weight loss) at 270.6 °C, and the PP blends with either lf-PP or hf-PP simultaneously contribute to the higher thermal stability, which improves the T_d by 11.7 °C and 18.3 °C, respectively. After introducing 20 wt% NPs with an in-situ thermal decomposition method, the PNCs exhibit a significantly enhanced thermal stability. The Fe_2O_3/PP obtains the highest T_d of 388.2 °C, which is 117.6 °C larger than that of pure PP. Followed by the Fe_2O_3(lf-PP)/PP and Fe_2O_3(hf-PP)/PP PNCs show their Td at 378.6 and 332.6 °C, respectively. The uniformly dispersed NPs formed by this in-situ method show a strong interaction with the organic polymer phase and exhibit great restriction on the polymer chains. Therefore, a tremendous improvement in the thermal stability is observed in the PNCs. Meanwhile, we could not exclude the possibility that the enhanced thermal conductivity after introducing metal oxide NPs may facilitate the heat dissipation within the composites, which is observed in polypropylene/carbon nanotube(CNT)[18] and epoxy/Single-wall carbon nanotubes PNCs.[19] Though a lot of nanomaterials such as carbon nanotubes[20,21] and nanoclays[22,23] have been used to improve the thermal stability of polymers, still only limited enhancement (less than 50 °C) was observed in most of the reported observations. The unexpected high thermal stability of the PNCs synthesized from this in-situ method may provide some guidance to design high performance fire retardant materials.

Magnetic Property

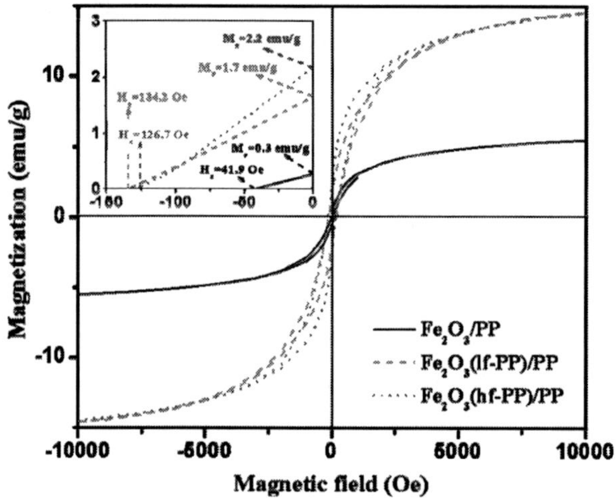

Figure 3 Hysteresis loops of the Fe_2O_3/PP PNCs with/without surfactant at room temperature.

Figure 3 shows the magnetic hysteresis loops of Fe_2O_3/PP PNCs with and without f-PP. The saturation magnetization (M_s) of the PNCs without f-PP is 5.5 emu/g at the magnetic field of 10000 Oe with NPs loading of 20 wt%. Composite at the same particle loading and containing surfactants (lf-PP or hf-PP) is observed to have larger M_s of 14.5 emu/g, which is primarily due to less oxidation of NPs with a protective f-PP layer covalently bonded to the NPs surface. The coercive force (coercivity, H_c; 41.9 Oe) indicates the soft magnetic behavior of the Fe_2O_3/PP PNCs. The H_c is 134.2 and 126.7 Oe for the PNCs with lf-PP and hf-PP, respectively, which is much larger than that of Fe_2O_3/PP PNCs. This is due to the enlarged interparticle distance in the PNCs arising from the good dispersion of single-domain NPs with the assistance of f-PP, which is consistent with particle-loading-dependent coercivity in NPs assembly.[24] The f-PP serves as an effective surfactant to protect the in-situ produced NPs from oxidation and separate the NPs from agglomeration, together with the superior compatibility with PP matrix, these materials are of great potential to the various applications such as electronics, magneto fluids and etc.

The NPs dispersion and identification is characterized by TEM and Mössbauer spectra, respectively, Figure 4. The NPs tend to agglomerate and interconnect to form a network structure without f-PP, Figure 4a. After introducing lf-PP and hf-PP, slight agglomeration is still observed while the crystalline structure of NPs is improved as evidenced by the clear electron diffraction rings, Figure 4d~f. Figure 4g shows the fringe spacing is about 2.6 Å and 2.2 Å, corresponding to the (104) and (113) crystal planes of α-Fe_2O_3. The paramagnetic doublet observed in the center of the Mössbauer spectra (Figure 4h) confirms the existence of Fe^{3+}, which corresponds to the superparamagnetic behavior of Fe_2O_3 at this temperature.[25] The fitting results show a main component at isomer shift (IS)=0.35 mm/s and quadrupole splitting (QS)=0.80 mm/s, which is Fe^{3+} in a paramagnetic state in the distorted oxygen octahedral site. Figure 4 (h) depicts the secondary component with IS=0 and correspondent H=335 kOe, which represents a spectral contribution of 5% α-Fe metallic magnetically ordered state. This is reasonable since Fe NPs were formed by thermal decomposition of $Fe(CO)_5$ and then partially oxidized.

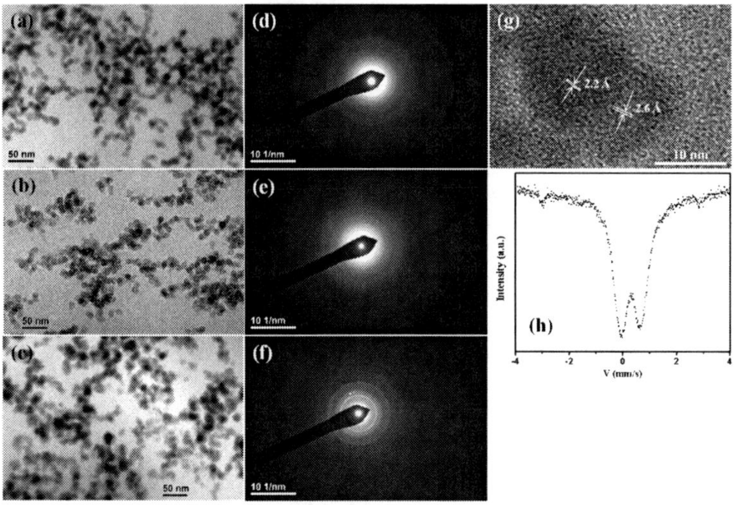

Figure 4 TEM images of (a) Fe_2O_3/PP, (b) Fe_2O_3(lf-PP)/PP and (c) Fe_2O_3(hf-PP)/PP. (d~f) corresponds to the electron diffraction pattern of (a~c), respectively. (g) HRTEM and (h) Mössbauer spectra of Fe_2O_3/PP PNCs.

CONCLUSIONS

Fe_2O_3/PP PNCs have been prepared with an in-situ method. The thermal degradation temperature has been improved by ~120 °C after incorporating NPs. The NPs in the PNCs containing f-PP become magnetically harder with a much larger coercivity than that of the PNCs without f-PP. HRTEM results confirms the monodispersed particles with fairly uniform particle distribution. The Mössbauer spectra prove the formation of Fe_2O_3 from this method.

REFERENCES

1. Z. Guo, L. Henry, V. Palshin and E. J. Podlaha, J. Mater. Chem. 16, 1772-1777 (2006).
2. J. J. Mack, L. M. Viculis, A. Ali, R. Luoh, G. Yang, H. T. Hahn, F. K. Ko and R. B. Kaner, Adv. Mater. 17, 77-80 (2005).
3. J. Zhu, S. Wei, X. Chen, A. B. Karki, D. Rutman, D. P. Young and Z. Guo, J. Phys. Chem. C 114, 8844-8850 (2010).
4. J. Zhu, S. Wei, J. Ryu, M. Budhathoki, G. Liang and Z. Guo, J. Mater. Chem. 20, 4937-4948(2010).
5. W. Huynh, J. Dittmer, W. Libby, G. Whiting and A. Alivisatos, Adv. Func. Mater. 13, 73-79 (2003).
6. Z. Guo, S. E. Lee, H. Kim, S. Park, H. T. Hahn, A. B. Karki and D. P. Young, Acta Mater. 57, 267-277 (2009).
7. J. Gass, P. Poddar, J. Almand, S. Srinath and H. Srikanth, Adv. Func. Mater. 16, 71-75 (2006).
8. J. Li, J. Claude, L. E. Norena-Franco, S. I. Seok and Q. Wang, Chem. Mater. 20, 6304-6306 (2008).
9. J. Zhu, S. Wei, M. J. Alexander, T. D. Dang, T. C. Ho and Z. Guo, Adv. Func. Mater. 18, 3076-3084 (2010).
10. C. Boyer, V. Bulmus, P. Priyanto, W. Y. Teoh, R. Amal and T. P. Davis, J. Mater. Chem. 19, 111-123 (2009).
11. Y. Lu, Y. Yin, B. T. Mayers and Y. Xia, Nano Lett. 2, 183-186 (2002).
12. D. Zhang, S. Wei, C. Kaila, X. Su, J. Wu, A. B. Karki, D. P. Young and Z. Guo, Nanoscale 2, 917-919 (2010).
13. Z. Guo, C. Kumar, L. L. Henry, E. E. Doomes, J. Hormes, and E. J. Podlaha, J. Electrochem. Soc., **152,** D1-D5 (2005).
14. Z. Guo, M. Moldovan, D. P. Young, L. L. Henry, and E. J. Podlaha, Electrochem. Solid State Lett., **10,** E31-E35 (2007).
15. J. Zhu, S. Wei, J. Ryu, L. Sun, Z. Luo and Z. Guo, ACS Appl. Mater. Interfaces 2, 2100-2107 (2010).
16. Z. Guo, S. Park, S. Wei, T. Pereira, M. Moldovan, A. B. Karki, D. P. Young and H. T. Hahn, Nanotechnology, 18, 335704 (2007).
17. P. Pöschke, M. Abdel, I. Alig, S. Dudkin and D. Lellinger, Polymer, 45, 8863 (2004).
18. T. Kashiwagi, E. Grulke, J. Hilding, K. Groth, R. Harris, K. Butler, J. Shields, S. Kharchenko and J. Douglas, Polymer 45, 4227-4239 (2004).
19. M. J. Biercuk, M. C. Llaguno, M. Radosavljevic, J. K. Hyun, A. T. Johnson and J. E. Fischer, Appl. Phys. Lett. 80, 2767-2769 (2002).
20. J. J. Ge, H. Hou, Q. Li, M. J. Graham, A. Greiner, D. H. Reneker, F. W. Harris and S. Z. D. Cheng, J. Am. Chem. Soc. 126, 15754-15761 (2004).
21. T. Kashiwagi, F. Du, J. F. Douglas, K. I. Winey, R. H. Harris and J. R. Shields, Nat. Mater. 4, 928-933 (2005).
22. T. Agag, T. Koga and T. Takeichi, Polymer 42, 3399-3408 (2001).
23. J. Zhu, A. Morgan, F. J. Lamelas and C. A. Wilkie, Chem. Mater. 13, 3774-3780 (2001).
24. D. Kechrakos and K. N. Trohidou, Phys. Rev. B 58, 12169(1998).
25. C. Mikutta, R. Mikutta, S. Bonneville, F. Wagner, A. Voegelin, I. Christl and R. Kretzschmar, Geochim. Cosmochim. Acta 72, 1111-1127 (2008).

Mater. Res. Soc. Symp. Proc. Vol. 1312 © 2011 Materials Research Society
DOI: 10.1557/opl.2011.117

Gelation, Electrical Conductivity and Elasticity of PAM- MWNT

Gulsen A. Evingur[1] and Önder Pekcan[2]

[1]İstanbul Technical University, 34469, Maslak, İstanbul, TURKEY
[2]Kadir Has University, 34320, Cibali, İstanbul, TURKEY

ABSTRACT

Polyacrylamide- Multiwalled carbonnanotube (PAM- MWNT) composites were prepared via free radical crosslinking copolymerization with different amounts of MWNT varying in the range between 0.1 and 15 wt. %. PAM- MWNT composite gels were characterized by fluorescence, dielectric spectroscopy and the tensile testing technique. A small content of doped nanotubes dramatically changed gelation time, conductivity and young modulus, respectively. The gel fraction exponent, β of PAM- MWNT composite gels were measured for various monomer and MWNT concentrations and observed that the gel fraction exponent β agrees best with the percolation theory for various amounts of PAM- MWNT. These polymer systems which are initially of an isolator character are doped with carbon nanotubes of nano dimensions and when the amount of this addition exceeds a critical value (0.3 wt. % MWNT) known as the percolation threshold, then composite gel systems with carbon nanotubes become electrically conducting structures with a critical exponent around r=2 which is close to the theoretical prediction of this value in 3D percolated system as known random resistor network. The observed elasticities are decreased above 3 wt. %MWNT with critical exponent around y=0.72 which is indicative of a transition from liquid-like to solid-like viscoelastic behavior.

INTRODUCTION

Polymer composites with carbon nanotube additions are one of the research subjects which have attracted a lot of attention in recent years. The first polymer nanocomposites using carbon nanotubes as filler were reported in 1994 by Ajayan et al.[1].

Hydrogel- carbon nanotubes (CNT) nanocomposites were demonstrated for potential applications such as sensors and actuators with DC electric field as the stimulus. They are being pursued in sensor applications for the detection of a variety of biomolecules including ethanol, glutamate, and glucose[2].

In this work, the universalities on gelation, electrical conductivity and elasticity of PAM-MWNT composites were presented.

THEORETICAL

Gelation

Percolation offers a particularly simple and yet detailed picture through which one may seek to understand gelation [3-5]. In the language of percolation, one may think of monomers as occupying the vertices of a periodic lattice, and the chemical bonds as corresponding to the edges joining these vertices, at any given moment, with some probability p. Then, the gel point can be identified with the percolation threshold p_c, where, in the thermodynamic limit, the incipient infinite cluster starts to form. Identifying the weight average degree of polymerization DP_W with the average cluster size S_{av} and the gel fraction G with the probability P_∞ of an occupied site belonging to the incipient infinite cluster, one can predict the scaling behavior of these and related quantities near the gel point, as a function of $|p - p_c|$,

$$DP_w \propto (p_c - p)^{-\gamma} \qquad p \to p_c^- \tag{1}$$

$$G \propto (p - p_c)^{\beta} \qquad p \to p_c^+ \tag{2}$$

If p approaches p_c from below it is denoted as $p \to p_c^-$; in contrast, $p \to p_c^+$ denotes when p approaches p_c from above. The critical exponents, β and γ in percolation theory, β=0.41 and γ=1.80, differ from those found in Flory- Stockmayer, β=1 and γ=1.

Electrical conductivity

Two classes of models have been developed for the description of the frequency dependence of the complex conductivity

$$\sigma^*(w) = \sigma'(w) - i\sigma''(w) \tag{3}$$

where σ' *(w)* forms the real part, and σ'' *(w)* the imaginary part of the complex conductivity, i the imaginary unit, and w, the angular frequency in the percolating systems.

(i) The equivalent circuit model which treats a percolation system as a random mixture of resistors and capacitors[4, 6], and *(ii)* the model based on charge carrier diffusion on percolating clusters [4, 7]. For the frequency dependence of AC conductivity at the percolation threshold, Bergman and Imry[6] have derived the following law:

$$\sigma'(w) \propto w^{\frac{r}{r+s}} \tag{4}$$

where σ' is the real part of equation 3; w, the angular frequency; and r and s critical exponents for conductivity.

Elasticity

The mechanical properties of hydrogels with nano materials are best understood using the theories of rubber elasticity and viscoelasticity. Stress- strain analysis of the energetic and entropic contributions to the refractive force indicates that entropy accounts for more than 90% of the stress. After some statistical analysis, shear modulus can be defined by

$$G = \frac{\tau}{\lambda} \tag{5}$$

where τ, the force per unit area and λ, is the extension ratio. While the thermodynamic and statistical thermodynamic approaches describe observed rubber-elastic behavior at low extensions quite well, the equation is invalid higher elongations (λ>3)[8]. Near the percolation threshold G obeys the following scaling law[7],

$$G_s(p) \approx (p_c - p)^{-y} \qquad p \to p_c^{-} \tag{6}$$

$$G_e(p) \approx (p - p_c)^{x} \qquad p \to p_c^{+} \tag{7}$$

where G_s is effective shear modulus of a superelastic percolation network, G_e, is effective shear modulus of the network, p_c is critical concentration, x and y is critical exponent for elasticity.

EXPERIMENTALS DETAILS

Composite gels were prepared via free radical crosslinking copolymerization with different amounts of MWNT varying in the range between 0.1 and 15 wt%. The fluorescence intensity measurements were carried out using a Model LS-50 spectrometer from Perkin-Elmer, equipped with a temperature controller. AC dielectric spectroscopy was performed with a dielectric spectrometer HP- 4192 A impedance analyzer controlling a computer. The mechanical properties of PAM-MWNT composite gels were characterized by the tensile testing machine.

RESULTS AND DISCUSSION

The typical fluorescence spectra of pyranine at two different stages of gelation for PAM-MWNT composite are shown in figure 1. At the beginning of the reaction, only the 512 nm peak exists, and then the intensity of the new (short-wavelength) peak around 380 nm started to increase as the intensity of the 512 nm peak (long-wavelength) peak decreased during the course of gelation of PAM-MWNT composite gels.

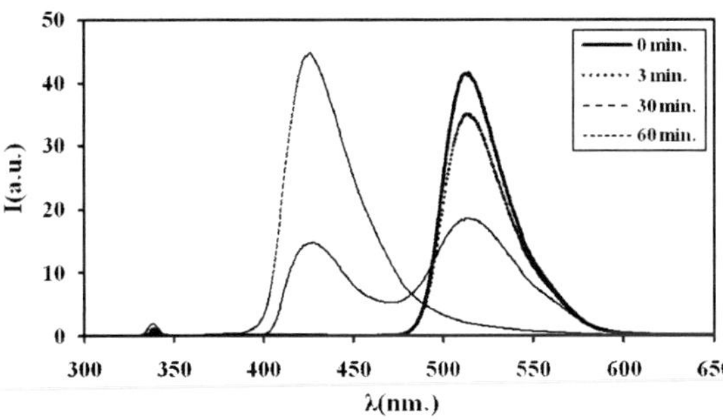

Figure 1. Typical fluorescence spectra of pyranine at different stages of gelation for PAM-MWNT composite during the whole course of gelation presenting the binding of pyranine.

Since the maxima of the spectra shifts from 380 nm to 427 nm as the gelation progresses, we monitored the fluorescence spectra in relatively large periods of time and plotted the intensity I_{em} corresponding to the maxima of the spectra as a function of time. These data were used to evaluate the critical behavior of the sol- gel phase transition. Above t_c, if the intensity from finite clusters distributed through the infinite network I_{ct} is subtracted from the maximum fluorescence intensity, then, the corrected intensity $I - I_{ct}$ measures solely the gel fraction G, the fraction of the monomers that belong to the macroscopic network.

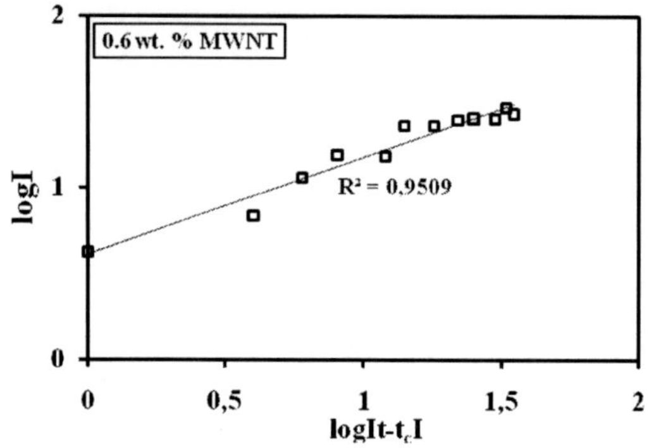

Figure 2. Double logarithmic plot of the intensity I versus time curves above t_c (gelation critical time) for 0.6 wt. % MWNT. The β exponents are determined from the slope of the straight lines.

Figure 2 represents the log-log plots of the typical intensity-time data above the gel point, for 0.6 wt. % MWNT concentration where the slope produced β values which are listed in Table I for various PAM-MWNT.

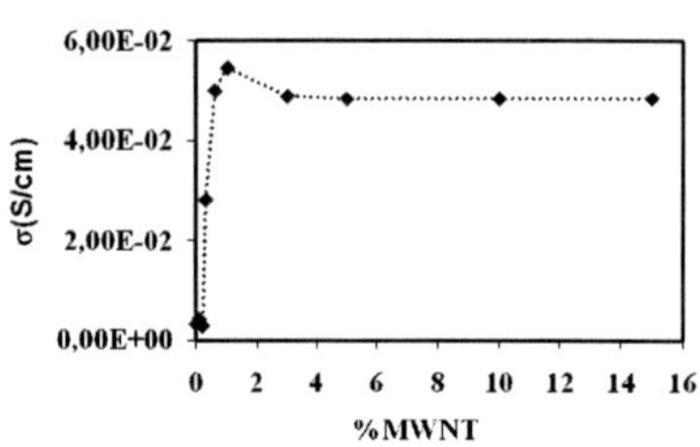

Figure 3. Conductivity, σ as a function MWNTs content at 5 kHz. frequency.

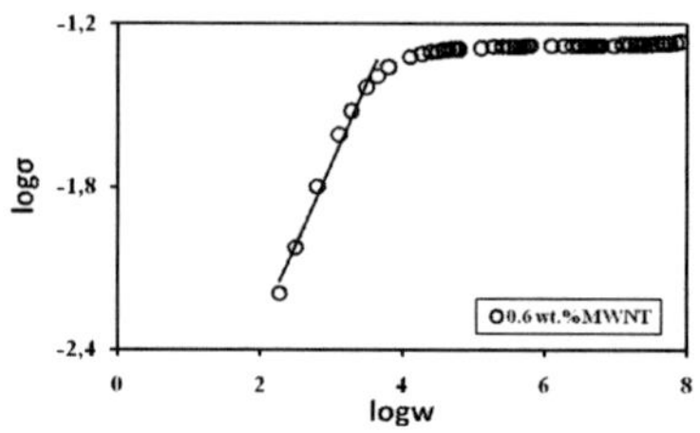

Figure 4. Logarithmic plot of the conductivity versus frequency curves of 0.6 wt. %MWNT.

Figure 3 shows the conductivity as a function of alternative current frequency for composite gels with different content of PAM and/or MWNT. It is observed that the conductivity behavior of the composite gels increases exponentially by increasing the AC frequency. PAM doped by lower MWNT composite system converted into a conductive character at a higher frequency than the frequency value needed for PAM doped by higher MWNT composite system. From Fig. 3, one can seen that the conductivity, (σ) of 0.3 % MWNT gel was stable up to 4 kHz then suddenly increases to some point, and becomes stable. Such a phenomenon confirms the insulator to conductor transition at the critical concentration of conductive filler and/or above the percolation threshold the electrical properties of composites are dominated by the percolating paths of the conductive MWNT.

The value of the fitting exponent r in equation 4 was estimated from the slope of the linear relation between log σ and log w as shown in figure 3. s is the critical exponent which taken from literature by $s \approx 0.7 \pm 0.05$. The electrical percolation occurs above 0.3 wt. % MWNT with a critical exponent around r=2 which is close to the theoretical prediction of this value in 3D percolated system as known random resistor network.

Table I. Experimentally measured parameters for gelation of PAM-MWNT composite [3].

PAM (%)	MWNT (%)	t_c (min)	β	y
100	0	5	0.52	0.72
99	1	7	0.45	
97	3	7	0.47	
95	5	8	0.60	
90	10	9	0.60	
85	15	11	0.41	

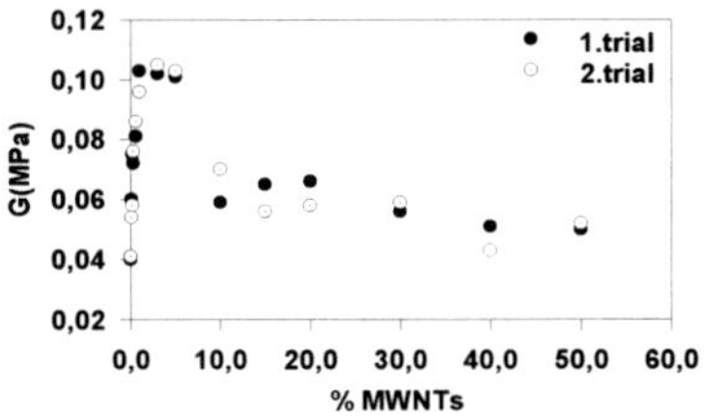

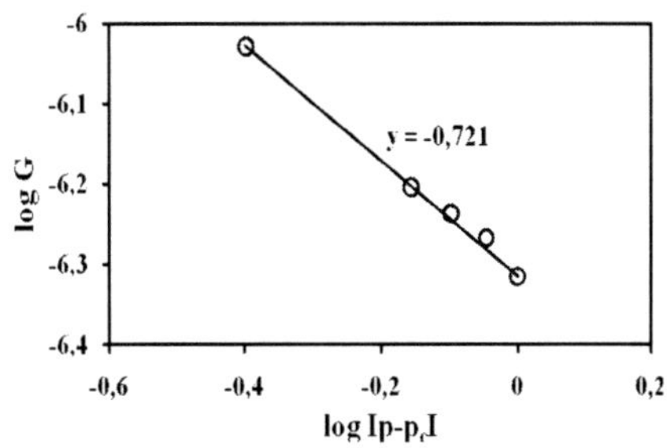

Figure 5. Shear modulus, G dependence of on the MWNT content in composite gels.

Figure 6. Logarithmic plot of the shear modulus versus MWNT concentrations below p_c.

Figure 5 shows that young modulus depends on the content of MWNT in PAM, respectively. Young modulus increases progressively until 3 wt. % MWNT with increasing nanotube content. At contents above 3wt. %MWNT, the young modulus is decreased marginally with increasing MWNTs content in Figure 5. Mechanical measurements show that a decrease in length brings about an increase in entropy because of changes in the end to end distances of the network chains of PAM-MWNT composite gels, thermodynamically. Thus, the entropic model for composite elasticity is a reasonable approximation. Figure 6 shows that double logarithmic plot of the shear modulus, G versus MWNT concentrations curves below p_c. The y exponent is determined from the slope of the straight lines. The critical exponent of elasticity, y obtained from fitting the composition dependence of the shear modulus below the percolation threshold is consistent with the suggestions of a transition from liquid-like to solid-like viscoelastic behavior.

CONCLUSIONS

In this work, we have been able to test the critical phenomena of the gelation, conductivity and elasticity as a function of MWNT concentrations. It was found that the gel fraction exponent β obeyed the percolation theory for PAM-MWNT composite gels. It was also shown that if polymer systems which are initially of an isolator character are doped with carbon nanotubes of nano dimensions and when the amount of this addition exceeds a critical value known as the percolation threshold, then composite gel systems with carbon nanotubes added become electrically conducting structures.

The electrical percolation threshold occurs above 0.3 wt. % MWNT with the critical exponent of r=1.8 - 2.3 which is close to the theoretical prediction of this value in a 3D percolated system (r=2). The percolation critical exponent, r obtained from fitting the composition dependence of the AC conductivity above the percolation threshold is consistent with the suggestions of percolation for a random resistor network.

The shear modulus increase after 3 wt. % MWNT loading indicating that there is a sudden change in the PAM- MWNT composite gels. The behavior is defined by a transition from liquid-like to solid-like viscoelastic behavior. The critical exponent of elasticity, y below critical MWNT concentration is found by 0.72.

ACKNOWLEDGMENTS

We thank Prof. Ferid Salehli for dielectric measurement and Dr. Argun Talat Gökçeören for tensile test measurements.

REFERENCES

1. P. M. Ajayan, O. Stephan, C. Colliex, D. Trauth, *Science*, **265**, 1212, (1994).
2. N. S. Satarkar, D. Johnson, B. Marrs, R. Andrews, C. Poh, B. Gharaibeh, K. Saito, K. W. Anderson, J. Z. Hilt, *J. App. Poly. Sci.*, **117**, 1813, (2010).
3. D. Stauffer, A.Coniglio, M. Adam, *Adv. Poly. Sci.*,. **44**, 103, (1982).
4. D. Stauffer, A. Aharony, *Introduction to Percolation Theory*. (Taylor and Francis, 1994).
5. P. G.de Gennes, *Scaling Concepts in Polymer Physics*.(Cornell University Press., 1979).
6. D. J. Bergman, Y.Imry, *Phys. Rev. Lett.*, **39**, 1222, (1977).
7. M. Sahimi, *Application of Percolation Theory*. (Taylor and Francis.1994).
8. K. S. Anseth, C. N. Bowman, L. B. Peppas, Biomaterials, **17**, 1647, (1996).

Mater. Res. Soc. Symp. Proc. Vol. 1312 © 2011 Materials Research Society
DOI: 10.1557/opl.2011.118

Microwave Metamaterials Containing Magnetically Soft Microwires

L. V. Panina[1,2], M. Ipatov[2], V. Zhukova[2], J. Estevez[2] and A. Zhukov[2]
[1]School of Comp., Comm. and Electr., Univ. of Plymouth, Drake Circus, PL4 AA, Plymouth, UK.
[2] Dpto. de Física de Materiales, Fac. Químicas, UPV/EHU San Sebastián 20009, Spain.

ABSTRACT

This paper discusses a new type of wire media based on amorphous ferromagnetic microwires. The combination of two effects, namely, a strong dispersion of the effective permittivity in metallic wire composites (resonance or plasmonic type) and giant magnetoimpedance effect in wires will result in unusual property that an effective dielectric response may strongly depend on the wire magnetization which can be changed with different external stimuli. We have demonstrated the effect of the external magnetic field on microwave response from composites containing CoFeSiBCr amorphous wires in free space at microwave frequency.

INTRODUCTION

Composites with embedded metallic wires may demonstrate a strong dispersion of the effective permittivity ε_{ef} in the microwave range [1-4]. This dispersion depends on the impedance of the wires that gives the possibility to sensitively tune the dispersion by changing the impedance of the wires. The possibility to control or monitor electromagnetic permittivity (and therefore scattering and absorption) is important for many applications such as remote non-destructive testing, structural health monitoring, tuneable coatings and absorbers.

In our approach we use the ferromagnetic wires with high magnetoimpedance (MI) effect to control the electromagnetic permittivity (and therefore scattering and absorption) of composite metamaterials by changing the magnetic properties of embedded microwires with external stimuli (magnetic field, current, stress, temperature). The high frequency impedance of a soft magnetic conductor may experience enormous changes when its static magnetic structure undergoes transformation due to application of a magnetic field, stress or temperature. This effect is intrinsically related to magnetic softness and requires low magnetic anisotropy constant and high circumferential magnetic permeability. In particular, these conditions are fulfilled in magnetically soft amorphous microwires with low and negative magnetostriction constant ($\lambda_s \approx -10^{-7}$) where GMI effect reaching several hundred percents is often observed. The characteristic magnetic fields required to cause this impedance change could be quite small, in the range of few Oe.

In this paper we have discussed the development of metamaterials with arrays of continuous wires and of short-cut ferromagnetic microwires, exploiting unique magnetic properties of wires to tune the effective electromagnetic parameters in the microwave frequency band. We made the measurements of the impedance in a single wire as a function of the external dc field at microwave frequencies, and free space measurements of reflection/transmission spectra with extracting the effective parameters. We have demonstrated that the application of a magnetic field to the whole composite strongly increases the dielectric losses, which affects the reflection/transmission spectra.

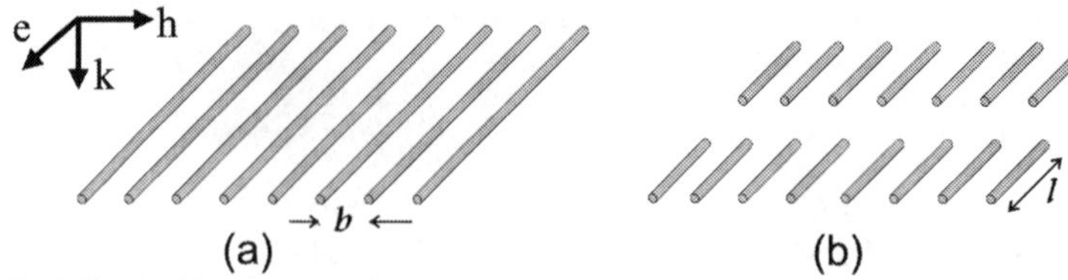

Fig. 1. Sketch of the wire composites.

EFFECTIVE PERMITTIVITY OF WIRE MEDIA

We consider two types of wire composites as shown in Figure 1. Composites containing long parallel wires as shown in figure 1(a), the dispersion of the effective permittivity ε_{ef} corresponds to that for a diluted plasma [1-2] with a negative value of the real part of the permittivity below the characteristic plasma frequency, f_p:

$$\varepsilon_{ef}^2 = 1 - \frac{\omega_p^2}{\omega^2(1+i\gamma)}, \quad \omega_p = 2\pi f_p, \quad f_P^2 = \frac{c^2}{2\pi b^2 \ln(b/a)} \tag{1}$$

Here, γ is the relaxation parameter and c is the velocity of light. For wire radius a in the micron range and spacing b between them of about 1 cm, the characteristic plasma frequency f_p is about 4 GHz. The relaxation parameter γ is defined by the wire surface impedance ς_{zz}:

$$\gamma = \frac{c\varsigma_{zz}}{\omega a \ln(b/a)} \tag{2}$$

The composites containing short pieces of conducting wires are characterized by a resonance type of the effective permittivity as the wires behave as dipole antennas with the resonance at half wave length condition: $f_R = c/2l\sqrt{\varepsilon_d}$, where ε_d is the permittivity of the supporting matrix. If the interaction between the wires is neglected, the effective permittivity is composed of the averaged dipole polarization χ and may be expressed analytically for an important case of not very strong skin effect [7]:

$$\varepsilon_{eff} = \varepsilon + 4\pi p\chi, \quad \chi = \frac{1}{2\pi \ln(l/a)(\tilde{k}a)^2}\left(\frac{2}{\tilde{k}l}\tan(\tilde{k}l/2) - 1\right), \tag{3}$$

$$\tilde{k} = k\left(1 + \frac{ic\varsigma_{zz}}{\omega a \ln(l/a)}\right)^{1/2}, \quad k = \omega\sqrt{\varepsilon}/c \tag{4}$$

Here p is the wire volume concentration and, $\tilde{k}$ is the renormalized wave number. Comparing equations (2) and (4) it is seen, that in both cases the dependence on the wire surface impedance occurs in a similar way, controlling the dielectric losses in the case of a moderate skin effect.

MI IN AMORPHOUS WIRES

As it was demonstrated above, the MI effect can be applied for achieving the tuneability of composite. Amorphous ferromagnetic microwires with negative magnetostriction having a circumferential anisotropy are characterized by large change in impedance when subjected to an axial magnetic field which rotates the magnetisation away from the circular direction. This effect remains essential even at GHz frequencies, that allows to apply them in dielectrics with tuneable microwave properties. As the wire impedance depends on the dc magnetisation (via the ac permeability) it is expected that any physical effect (magnetic field, mechanical stress,temperature) that results in change in the magnetic structure (whilst the ac permeability remains high) will affect the dispersion of the effective permittivity.

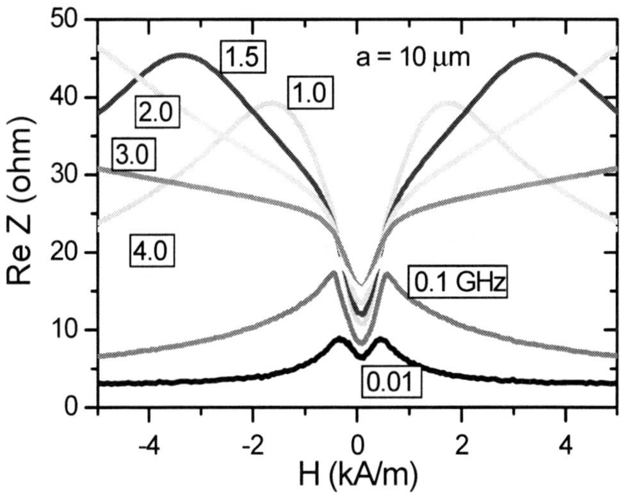

Fig. 2. Real part of the impedance of $Co_{66}Fe_{3.5}B_{16}Si_{11}Cr_{3.5}$ wire as a function of the field for different frequencies. The wire parameters: radius a =10 μm, glass thickness t=2.4 μm , wire length l = 3 mm.

A careful selection of the alloy composition and/or annealing conditions has been performed to developed microwires with a high MI effect at GHz frequencies. Figure 3 shows a dependence of wire impedance on axial external magnetic field. Here we can see a rather high change of the sample impedance with magnetic field at GHz frequency. Besides the magnetic field, the permeability spectra can be sensitive to any other physical stimuli that considerably affect the impedance of the embedded wires. Stress-impedance effect [6] is characterized by the change of GMI spectra with application of external stress. This effect can be used for development of stress-sensitive materials. Temperature-sensitive composites can be constructed on wires with compositions Co-Fe-Ni-Si-B and Co-Fe-Cr- Si-B that are demonstrated to have a low Curie temperature (below 100oC) and pronounced temperature dependent magnetic properties and GMI effect [7]. Further, we will see how the transmission/reflection spectra of the composite containing array of these microwires are affected by magnetic field.

TUNABLE PERMITTIVITY SPECTRA

Combining the dispersion properties of wire media and MI effect it is possible to actively tune the permittivity spectra of arrays of microwires by application of a small magnetic field or a stress. The modeling results with the external field as a parameter are shown in Figure 3. For continuous wire composite, as it is seen that in Figure 3(a) the application of magnetic field, as a result of impedance increasing, the imaginary part of permittivity, that define losses, increases, while the real part decreases its high negative value. For the cut-wire composite (Figure 3(b)), the dispersion region broadens.

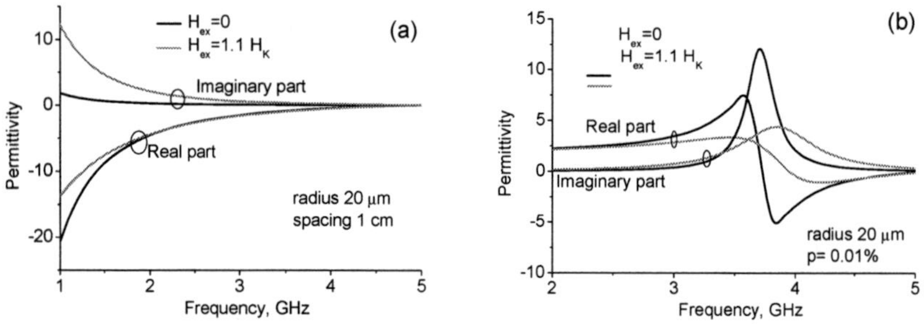

Fig. 3. Effective permittivity spectra in composites depicted in Figs.1a,b, respectively, with the external field as a parameter. Modelling is performed for wires with a circumferential anisotropy (anisotropy field H_k=500A/m). The other parameters are: resistivity 130μΩcm, magnetisation 0.05T, wire radius 20 μm. For (a), *b*=1cm. For (b), *l*=4cm, *p*=0.01%.

The transmission/reflection spectra of composite by free space methods. $Co_{66}Fe_{3.5}B_{16}Si_{11}Cr_{3.5}$ glass coated amorphous wires with radius of 20 μm were glued in paper to form wire-lattices of 50x50 cm^2 with separation *b* of 10 mm, as shown in Figure 1a. After measurements, such wire-lattice was sequentially cut in stripes of 40, 20 and 10 mm wide to form the composites with short dipole wires as depicted in Figure1b. The S-parameters were measured in the frequency range of 0.9-17 GHz in the presence of external field ranging up to 3 kA/m applied through a plane coil with turns perpendicular to the electrical field in the incident wave.

Figure 4 shows the spectra of the reflection *R*, transmission *T*, and the effective permittivity ε_{ef} for long-wire composites with the field H_{ex} as a parameter. The relative change in *R* and *T* is about 10% at lower frequencies while the phase of transmission shifts about 40 degrees at 1 GHz with the change of the field. The permittivity spectra deduced from *R* and *T* plots are consistent with the theoretical plots seen in Fig. 2a. The effective thickness was taken equal to the lattice period of 1 cm. The imaginary part of the permittivity increases with the field due to the increase in the wire impedance resulting in decrease in the transmission amplitude (although the reflection amplitude also decreases).

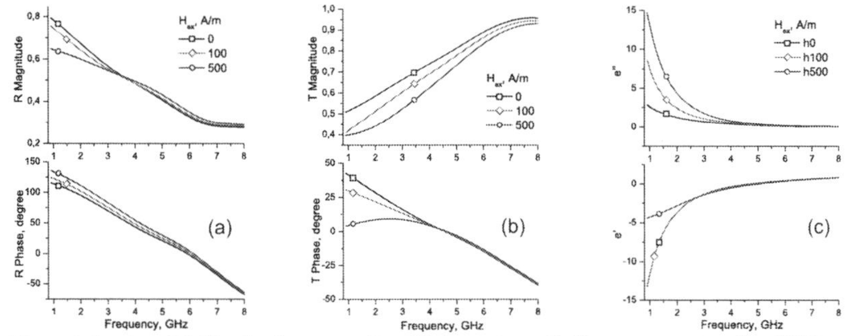

Figure 4. Spectra of *R*, *T* and ε_{ef} for composites with long wires with H_{ex} as a parameter (H_{ex}=0,100, 500A/m).

Figure 5 shows the spectra for cut-wire composites with different wire length of 40, 20 and 10 mm and with the field as a parameter. The transmission spectra have a deep minimum near a resonance demonstrating a stop filter behaviour. The magnitude of this minimum depends strongly on the field for longer wires with lower resonance frequency. For shorter wires with the dispersion at a higher frequency band the field dependence is not noticeable since the wire ac permeability is nearly unity and the impedance becomes insensitive to the magnetic properties. The resonance frequency corresponds to that for wires in the medium with the permittivity of unity.

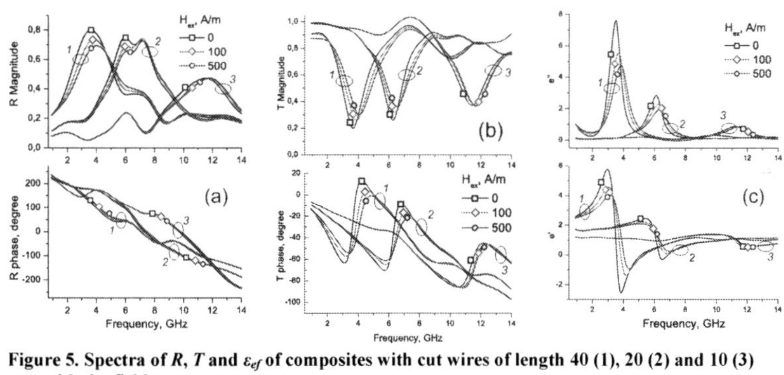

Figure 5. Spectra of *R*, *T* and ε_{ef} of composites with cut wires of length 40 (1), 20 (2) and 10 (3) mm with the field as a parameter.

CONCLUSIONS

We report novel results on the magnetic field effect on the dielectric response in composites with arrays of parallel magnetic wires, continuous and short-cut, in the frequency region of 0.9-17 GHz utilising free-space measurement method and analysis in terms of the effective permittivity depending on the wire surface impedance. Both the real and imaginary parts of ε_{ef} show strong variations with increasing the field owing to the field dependence of the wire impedance which controls the losses in the dielectric response. Long-wire composite has a plasmonic type dispersion of ε_{ef} with negative values of its real part below the plasma frequency which is in the GHz range for wire spacing of about 1 cm and wire diameter of few microns. The presence of the external magnetic field suppresses low-frequency plasmons increasing the value of the real part of the permittivity. In the case of cut-wire composites, it is confirmed that their effective permittivity has a resonance type dispersion due to the dipole resonance in wires at half wavelength condition. The application of the field broadens the resonance and shifts it towards the higher frequencies. Therefore, both types of wire composites possess a strong dependence of the effective permittivity on the external magnetic field and are suitable for large scale applications as tuneable microwave materials.

ACKNOWLEDGMENTS

We acknowledge support under projects MAT2007-66798-CO3-01, MAT2010-18914 (MICINN), by the EU ERA-NET programme under project DEVMAGMIWIRTEC (MANUNET-2007-Basque-3) and by the Basque Government under Saiotek 08 METAMAT (SPRI). Some of the authors, A.Zh and V.Zh wish to acknowledge support of the Basque Government under Program of Mobility of the Investigating Personnel of the Department of Education, Universities and Investigation for stay in Moscow (grants MV-2009-2-21 and MV-2009-2-24).

REFERENCES

1. Rotman W., “Plasma simulation by artificial dielectrics and parallel plate media,” *IRE Trans. Antennas Propagat.*, vol. 10, 82–95, 1962.
2. Pendry, J. B., A. J. Holden, W. J. Stewart, and I. Youngs, “Extremely Low Frequency Plasmons in Metallic Mesostructures,” *Phys. Rev. Lett.*, Vol. 76, No. 25, 4773-4776, 1996.
3. Lagarkov A. N. and A. K. Sarychev, “Electromagnetic properties of composites containing elongated conducting inclusions,” *Phys. Rev. B* 53, 6318, 1996.
4. Liu L., S. M. Matitsine, Y. B. Gan, and K. N. Rozanov, “Effective permittivity of planar composites with randomly or periodically distributed conducting fibers,” *J.Appl.Phys.* 98, 63512, 2005.
5. Makhnovskiy D.P., L. V. Panina, C. Garcia, A. P. Zhukov, and J. Gonzalez “Experimental demonstration of tunable scattering spectra at microwave frequencies in composite media containing CoFeCrSiB glass-coated amorphous ferromagnetic wires and comparison with theory,” *Phys. Rev. B*, Vol. 74, 064205-1–064205-11, 2006.
6. C. García, A. Zhukov, V. Zhukova, M. Ipatov, J. M. Blanco, and J. Gonzalez. Effect of tensile stresses on GMI of Co-rich amorphous microwires. *IEEE Trans. Magn.*, 41:3688–3690, 2005.
7. V. Zhukova, M. Ipatov, A. Zhukov, R. Varga, A. Torcunov, J. Gonzalez, and J. M. Blanco. Studies of magnetic properties of thin microwires with low Curie temperature. *J. Magn. Magn. Mat.*, 300:16–23, 2006

Mater. Res. Soc. Symp. Proc. Vol. 1312 © 2011 Materials Research Society
DOI: 10.1557/opl.2011.119

Photo-polymerizable gold nanorods / methyl methacrylate composite for plasmonic optical application

Kyoko Masui[1,2], Satoru Shoji[1], Xuan-Ming Duan[2], and Satoshi Kawata[1]

[1]Department of Applied Physics, Osaka University, Osaka 565-0871, Japan,
[2]Technical Institute of Physics and Chemistry, Chinese Academy of Sciences, China
masui@ap.eng.osaka-u.ac.jp / http://lasie.ap.eng.osaka-u.ac.jp/

ABSTRACT

We prepared a photo-polymerizable compound of Au nanorods and methyl methacrylate (MMA). An irradiation of ultraviolet (UV) light onto the compound induces photo-polymerization of MMA, resulting in a formation of an Au nanorods/PMMA composite. Au nanorods were synthesized by seed-mediated growth method in aqueous solution. Hydrophobically functionalized Au nanorods were dispersed into MMA with a small amount of photosensitizer and initiator. We investigated the stability and the uniformity of the dispersion of Au nanorods during the photo-polymerization process through VIS-NIR absorption spectroscopy. We also demonstrated the use of the compound for two-photon micro/nanolithography.

INTRODUCTION

Recently noble metal nanoparticles and their arrangement are recognized as fundamental platforms of novel optical/photonic technologies. Varieties of applications, such as optical sensors, super-resolution imaging, metamaterials, optical nano-waveguides, etc. are proposed based on the excitation, transformation, and detection of local surface plasmon polaritons (LSPPs) through an interaction of photons and metal nanoparticles. Currently, for the achievement of such plasmonic technologies, one of the urgent demands is to develop novel micro/nano-fabrication methods to provide three-dimensional (3D) complex microstructures in which metal nanoparticles are embedded in desired orientation and arrangement. Aiming this purpose, we developed a gold nanorods/methyl methacrylate (MMA) compound, which makes it possible to produce arbitrary 3D micro/nano-structures of Au nanorods/PMMA composite. Our produced compound shows good stability of the dispersion of Au nanorods in MMA before / after photo-polymerization. Fabrication of Au nanorod/PMMA composite microstructures is demonstrated by means of two-photon lithography. Our approach could be a variable method for the development of novel plasmonic devices and metamaterials.

PREPARATION OF GOLD NANORODS / METHYL METHACRYLATE COMPOUND

The procedure of the preparation of Au nanorods/MMA compound is shown in Figure 1. Au nanorods were prepared by seed-mediated method [1,2]. The synthesis started from small gold seeds, and by the anisotropic protection of the surface of the gold particles by the surfactant hexadecyltrimethylammonium bromide (CTAB), the individual gold particle grew into a rod shape. Figure 2 shows scanning electron microscope (SEM) images of the Au nanorods dried on a glass substrate. Nanorods with a quite uniform geometrical shape and size (a thickness of 15 nm and a length of 60 nm) were obtained. Because of their shape uniformity, aggregations of self-aligned nanorods were often seen in SEM observation. In order to disperse the Au nanorods into MMA, we performed hydrophobation of the CTAB-wrapped gold nanorods by using mercaptopropyltrimethoxysilane (MPS) and octadecyltrimethoxysilane (ODS) [3]. First, 0.01 M MPS ethanol solution was added to Au nanorods dispersion. With being stirred vigorously for 30 min, the mixture were further mixed with ODS chloroform solution and NaOH aqueous solution. During vigorous stirring for 1 hour, the pink color of the aqueous phase was transferred to the chloroform phase. The photos in Figure 1 (d) show the bottle before / after the transformation, indicating successful hydrophobation of the Au nanorods. The organic phase was collected by a pipette and dried to extract the hydrophobically functionalized nanorods.

We prepared a MMA(methyl-methacrylate)-based photopolymerizable resin, consisting of MMA monomer, crosslinker, initiator, and photosensitizer. The recipe of the resin was described in the references [4-6]. We have been used this recipe for two-photon micro/nanolithography. The absorption spectrum of the photopolymerizable resin is shown in Figure 3 (a). We added hydrophobically functionalized Au nanorods into the photopolymerizable resin. A mild sonication of the mixture by an ultrasonication bath for 10 minutes provides stable monodispersion of Au nanorods in the resin without significant aggregation.

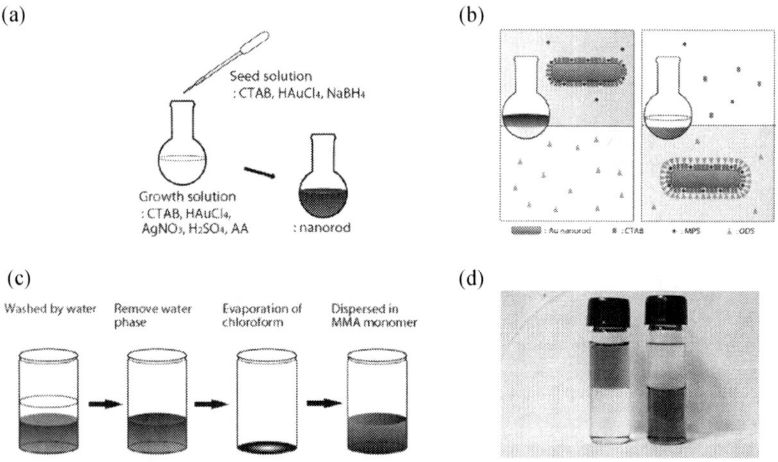

Figure 1 (a) Synthesis of Au nanorods. (b) Hydrophobic modification of Au nanorods. (c) Procedure of Solvent change. (d) The photo of the bottles containing two solution phases (upper layer: aqueous, bottom layer: chloroform) before (left) / after (right) hydrophobation of Au nanorods.

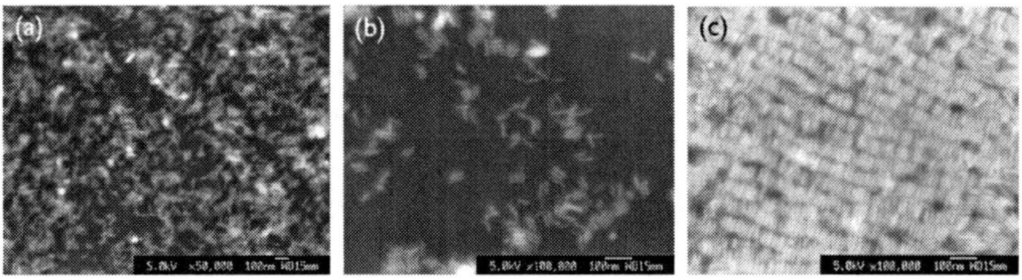

Figure 2 Scanning electron microscope images of Au nanorods. (a)Randomly dispersed Au nanorods, (b) with higher magnification, (c) self-aligned Au nanorods.

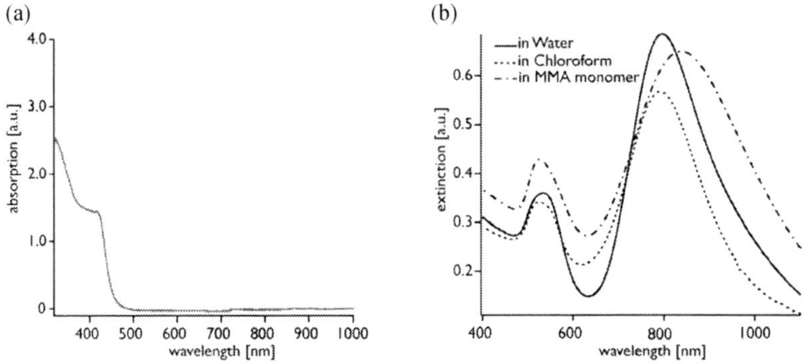

Figure 3 (a) Absorption spectrum of the photopolymerizable resin. (b) Absorption spectra of Au nanorods dispersed in different solvent.

Figure 3 (b) shows typical VIS-NIR absorption spectra of Au nanorods in different solvent. As seen in the spectra, there are two noticeable peaks at around 550 nm and 800 nm, those are representing transverse and longitudinal modes of local surface plasmon polaritons (LSPP), respectively. It is known that the peak at 800 nm, corresponding to the longitudinal LSPP mode of isolated individual gold nanorods, sensitively reflects the shape, surrounding dielectric constant, as well as the aggregation of Au nanorods in the wavelength and the line width of the peak [7]. In Figure 3 (b), the longitudinal mode shows a slight shift in the peak wavelength according to the difference of the dielectric constants of different solvents. However, the peak does not show significant broadening and large shift, meaning the dispersion of Au nanorods was stably maintained during hydrophobation process.

PHOTOPOLYMERIZATION OF GOLD NANORODS/MMA COMPOUND

In order to check the capability of photo-polymerization, we performed UV light irradiation onto the gold nanorods/MMA compound. We prepared a thin layer of the compound between two coverslips, and exposed it to UV light of a mercury lamp. Figure 4 (a) shows the absorption spectra of the compound before and after light irradiation. Although broadening of the longitudinal LSPP peak was seen, the peak wavelength did not show any shift. Figure 4 (b) shows the photo of the polymerized compound as well as the normal photopolymerizable resin without Au nanorods. The polymerized film was colored in pink mixed with yellow of photosensitizer, and optically transparent. These results indicate that the polymerization of MMA does not cause significant destabilization of the dispersion of Au nanorods, and the nanorods were immobilized in PMMA matrix with monodispersed condition.

(a)

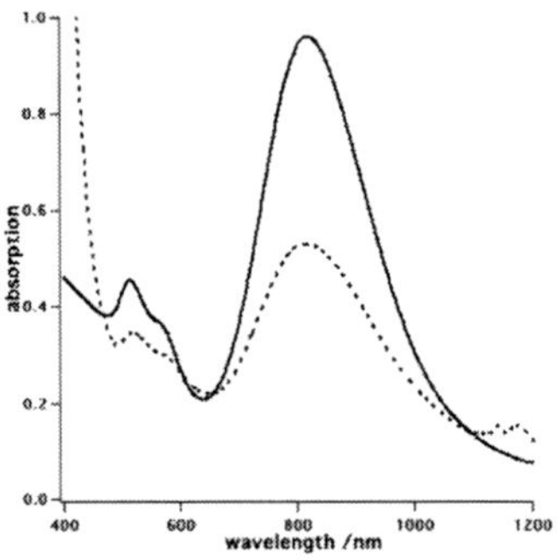

(b)

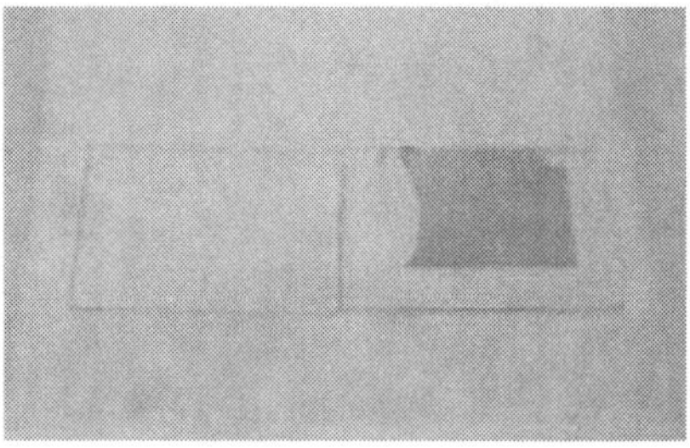

Figure 4 (a) VIS-NIR absorption spectrum of the Au nanorods/MMA compound before/after UV light irradiation. Solid line: before irradiation, dotted line: after irradiation. (b) The photo of (left) the photopolymerizable resin and (right) Au nanorods/MMA compound after UV irradiation.

TWO-PHOTON LITHOGRAPHY WITH GOLD NANORODS/MMA COMPOUND

Laser lithography based on two-photon absorption induced polymerization, called two-photon lithography, possesses intrinsic 3D spatial resolution and super-resolution capabilities beyond the optical diffraction limit [4-6]. Because of these features, two-photon lithography has been applied for producing many kinds of 3D micro/nano devices such as micro-fluidics, photonic crystals, micro-optics, MEMS, biological scaffolds, and so on. Near infrared femtosecond laser light being tightly focused into UV sensitive photopolymerizable resins by a microscope objective lens excites two-photon absorption, initiating polymerization reaction at the center of the focus spot. Since the efficiency of two-photon absorption intrinsically exhibits nonlinear manner to the intensity of light, the photo-polymerized volume, defining the spatial resolution, can be much smaller than the volume of the focus spot of the laser light. By this means we achieved super-resolution 3D laser-lithography with a spatial resolution of less than 100 nm. By scanning the focus spot in resins, we demonstrated a variety of 3D polymer microstructures including animal statues, elastic springs, photonic crystals, cell culturing scaffolds [4-6]. The idea here is to use the Au nanorod/MMA compound for the two-photon lithography to demonstrate the fabrication of Au nanorods/PMMA composite micro/nanostructures.

In the experiment, a femtosecond Ti:Sapphire laser (Tsunami, Spectraphysics Corp.) at 780 nm was used for the excitation of two-photon absorption. The laser light was focused by an oil immersion objective lens with a NA of 1.4 into the Au nanorods/MMA compound on a coverslip. In order to check the effect of the existence of Au nanorods on the property of two-photon polymerization, we fabricated a couple of test patterns consisting of arrays of single dots or single stroke lines on glass substrates. Figure 5 show SEM images of the fabricated test patterns made of Au nanorods/PMMA composite. Similar to the case of an ordinary photopolymerizable resin, polymerized structures were formed. A high magnification image in Figure 5 (c) shows that Au nanorods were embedded in the PMMA matrix of the structures. This result indicates that both of the laser irradiation and the polymerization reaction of MMA do not cause any phase separation or aggregation of Au nanorods in the compound.

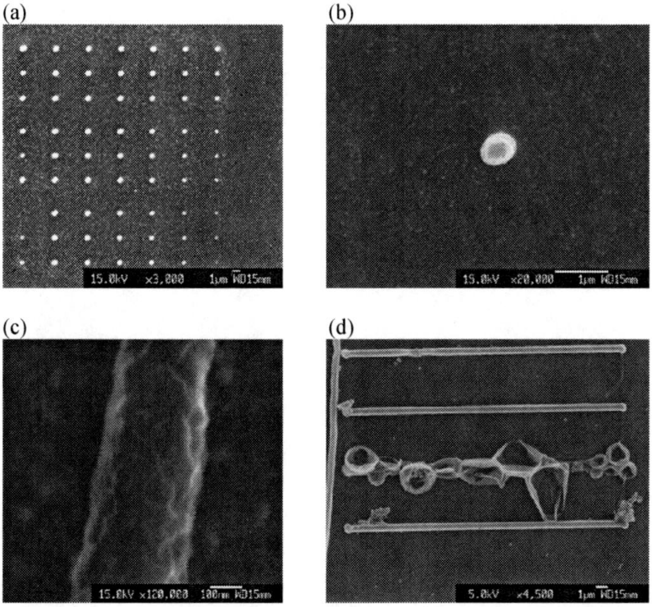

Figure 5 SEM images of test patterns fabricated by two-photon lithography. (a)dot array, (b)single dot with a diameter of 200 nm, (c) magnified images of the surface of a single stroke line, and (d) line array. The result of a laser induced bubbling of the compound is also shown in the third line from the top in (d).

The wavelength of the femtosecond laser is not only resonant to two-photon absorption of the photosensitizer, but also considerably near to the resonance of the longitudinal LSPP of the nanorods. This fact naturally suggests the possibility that the Au nanorods also absorb the laser light quite efficiently, resulting in the disturbance of efficient two-photon polymerization. Moreover, the resonant absorption could induce a local heating surrounding the nanorods.

Indeed, we sometimes observed a transient bubbling of the composite during the lithography process. Figure 5 (d) shows the result of the bubbling. The laser induced bubbling occurred only from the focus spot and caused balloon-like hollow structures. The occurrence of the bubbling was quite random and transient. However, the bubbling effect can be preventable effectively by choosing moderate laser intensity. Especially, with the condition for the finest spatial resolution (i.e. the laser intensity slightly higher than the threshold of two photon absorption), the bubbling effect was hardly occurred. Since the LSPP exhibits sharp resonant condition, detuning of the LSPP mode by controlling the growth of the length of nanorods is also effective method to prevent the composite from the bubbling.

CONCLUSIONS

As shown above, we developed a photo-polymerizable compound of Au nanorods and MMA-based photopolymerizable resin. Our produced compound shows good stability of the dispersion of Au nanorods in the photopolymerizable resin before / after photo-polymerization, and does not show any phase separation or aggregation of Au nanorods during the laser irradiation. In order to apply this material for plasmonic/photonic application, there are several tasks to be solved. Especially, the control of the distance and the arrangement of Au nanorods during photo-polymerization is the most important issue in order to apply this approach for the development of super-resolution imaging, plasmonic waveguides, or metamaterials [8-11]. At this moment, Au nanorods are randomly oriented in the matrix of PMMA. The cooperation of optical gradient force, i.e. optical tweezers, would be one of the possible solutions for this task.

ACKNOWLEDGMENTS

This research is supported by Grant-in-Aid for Young Scientists (no.21686010, and no.19810012), MEXT, Japan.

REFERENCES

1. N. R. Jana, L. Gearheart, and C. J. Murphy, Adv. Mater. **13,** 1389 (2001).
2. B. Nikoobakht and M. A. El-Sayed, Chem. Mater. **15,** 1957 (2003).
3. K. Mitamura, T. Imae, N. Saito, and O. Takai, J. Phys. Chem. B **111,** 8891 (2007).
4. S. Kawata, H.-B. Sun, T. Tanaka, and K. Takada, Nature, **412,** 697 (2001).
5. S. Nakanishi, S. Shoji, H. Yoshikawa, Z. Sekkat, and S. Kawata, J. Phys. Chem. B **112,** 3586-3589 (2008).
6. S. Nakanishi, S. Shoji, S. Kawata, and H.-B. Sun, Appl. Phys. Lett. **91,** 063112 (2007).
7. J. P´erez-Juste, I. Pastoriza-Santos, L. M. Liz-Marz´an, P. Mulvaney, Coord. Chem. Rev. 249 1870 (2005)
8. V. Podolskiy, A. K. Sarychev, V. M. Shalaev, Opt. Express **11,** 735 (2003).
9. A. V. Kobashin, P. Evans, S. Pastkovsky, W. Hendren, G. A. Wurtz, R. Atkinson, R. Pollard, V. A. Podolskiy, and A. V. Zayats, Nature Mater. **8,** 867 (2009).
10. S. Kawata, A. Ono, and P. Verma, Nature Photonics, **2,** 438 (2008).
11. X. Huang, S. Neretina, and M. A. El-Sayed, Adv. Mater. **21,** 4880 (2009).

Mater. Res. Soc. Symp. Proc. Vol. 1312 © 2011 Materials Research Society
DOI: 10.1557/opl.2011.265

Conjugated Polymer:TiO_2 Nanocomposite Solar Cells Based on P3HT Nanoparticles

B. Harihara Venkatraman[1], Akshay Kokil[2], Soumitra Satapathi[2], Jayant Kumar[2], and Dhandapani Venkataraman[1*].
[1]Department of Chemistry, University of Massachusetts Amherst, Amherst, MA 01003, U.S.A.
[2]Center for Advanced Materials and Department of Physics & Applied Physics, University of Massachusetts Lowell, Lowell, MA 01854, U.S.A.

ABSTRACT

There is widespread interest in developing efficient solar cells derived from conjugated polymers and TiO_2. The conjugated polymer can act as a light harvesting dye as well as a hole transport material, and can potentially replace both the ruthenium dye and the I_3^-/I^- couple in the DSSCs. Herein, we report a novel and facile approach of using conjugated polymer nanoparticles to make conjugated polymer:TiO_2 nanocomposite based solar cell. Nanoparticles from poly(3-hexylthiophene) (P3HT) were made using mini-emulsion technique. In this work we report on incorporation of these P3HT nanoparticles into nanoporous titania. Device characteristics made using P3HT nanoparticle sensitized solar cells were measured. These devices showed a short-circuit current density (J_{sc}) of 0.207 mA/cm^2, open-circuit voltage (V_{oc}) of 0.62 V and 0.07% (η) efficiency.

INTRODUCTION

Dye-sensitized solar cells (DSSC) with liquid electrolytes have been considered as potential alternative for expensive silicon based solar cells [1]. Power conversion efficiencies of about 11% have been reported in DSSCs [2]. Recently, a lot of research has been focused on improving the performance of DSSC using dyes derived from organic molecules and exploiting conjugated polymer as a hole transport material, instead of the corrosive liquid electrolyte [3]. It has also been recognized that complete filling of the pores in TiO_2 is important for higher efficiency[4, 5]. Typically, organometallic ruthenium complexes have been utilized as sensitizing dyes. Conjugated polymers have also been utilized as sensitizers for DSSCs and encouraging results have been reported [6-11]. However, since the diffusion of the conjugated polymers in the nanoporous electrode and their adsorption on the nanocrystalline TiO_2 is not adequate, the efficiencies obtained using the conjugated polymer sensitizer have been limited. Therefore, it is necessary to devise a general strategy for filling the pores in titania that can be applied to any conjugated polymers or dyes.

Herein, we discuss a new strategy of using dye nanoparticles (NP), instead of just dye molecules for infiltration (Figure 1). As a proof of concept, we show here the performance of solar cells sensitized with conjugated polymer nanoparticles since solar cells sensitized with conjugated polymer have been widely reported [6-11]. For the initial experiments, we utilized poly(3-hexylthiophene) (P3HT) due to its oxidative stability as well as its large exciton migration length (~ 10 nm).

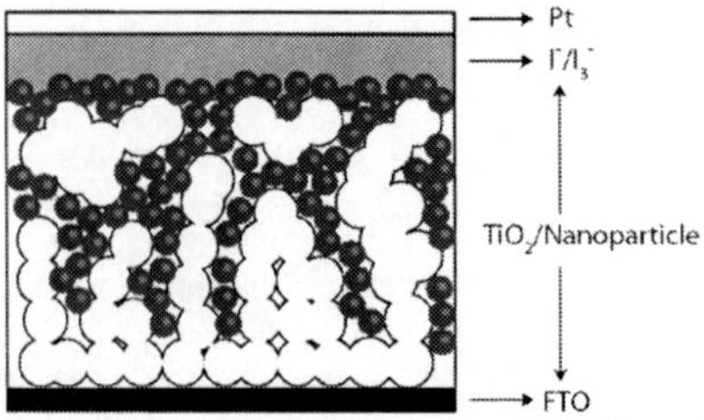

Figure 1: Representation of sensitized solar cell with infiltrated dye nanoparticle

Nanocrystalline TiO_2 with different particle sizes (25 nm and 100 nm) was utilized to obtain nanoporous electrodes with varying pore size. The electrodes were sensitized using an aqueous P3HT NP dispersion followed by fabrication of the DSSC using reported protocols. The testing of the fabricated solar cells was performed under standard conditions and the results of our initial experiments are reported here.

EXPERIMENT

Materials and Reagents Used

Electronic grade poly(3-hexylthiophene) with regioregularity between 90% and 93% was obtained from Rieke Metals, Inc. Surfactant (SDS-sodium dodecyl sulfate) was purchased from Scientific Products. All solvents were purchased from EMD (through VWR). Synthesis of P3HT nanoparticles were carried out using distilled water. The sonication was performed using Aquasonic, VWR Scientific Products (Model 50T). Dynamic light scattering (DLS) was performed using 50 mW He-Ne Laser, an available photodiode detector (BI-APD), a digital time correlator (BI-9000) and software from Brookhaven Instruments Corporation. UV-visible spectroscopy of the nanoparticles was performed using an Agilent 8453 photodiode array UV-Vis spectrometer. TiO_2 nanoparticles used for this study were acquired from Degussa (P25) and Aldrich (~100 nm, 634662). Fluorine doped tin oxide (FTO) coated glass slides were obtained from Hartford Glass Company, Hartford City, Indiana. I^-/I_3^- based redox liquid electrolyte was prepared by dissolving 0.0254 g of I_2 and 0.1358 g of LiI in 2 mL acetonitrile and 0.3 mL 4-tert-butylpyridine.

DISCUSSION

Synthesis of P3HT nanoparticles

Nanoparticles of either low molecular weight dyes or conjugated polymers, with size range of 30 nm to 500 nm, can be made using a variety of techniques [12-16]. In addition, the wetting properties of nanoparticles can be tuned by using sulfate, carboxylate or phosphate surfactant, during nanoparticle synthesis. The diffusion parameter of the nanoparticle can be adjusted by controlling the size of the nanoparticle. We chose to synthesize nanoparticles from poly(3-hexylthiophene) using mini-emulsion technique since the solar cells sensitized with P3HT shows efficiency of around 0.18-0.42 [6]. The detailed synthesis of P3HT nanoparticles will be published elsewhere. Briefly, the polymer dissolved in chloroform was suspended in an aqueous surfactant solution. The mixture was sonicated and the chloroform was evaporated yielding nanoparticles. The size of the nanoparticles was determined using dynamic light scattering (DLS).

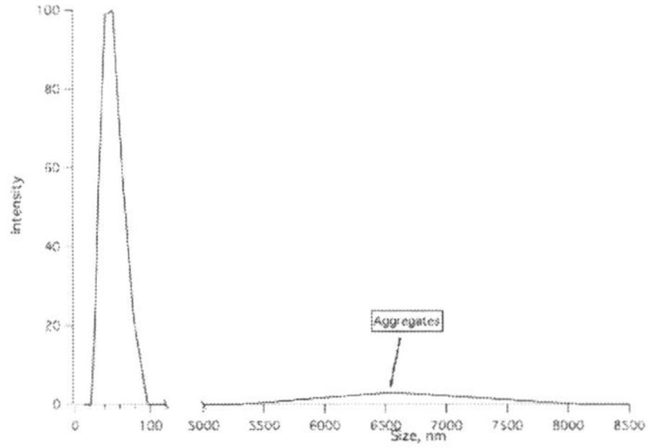

Figure 2: Particle size distribution determined using DLS for P3HT nanoparticles

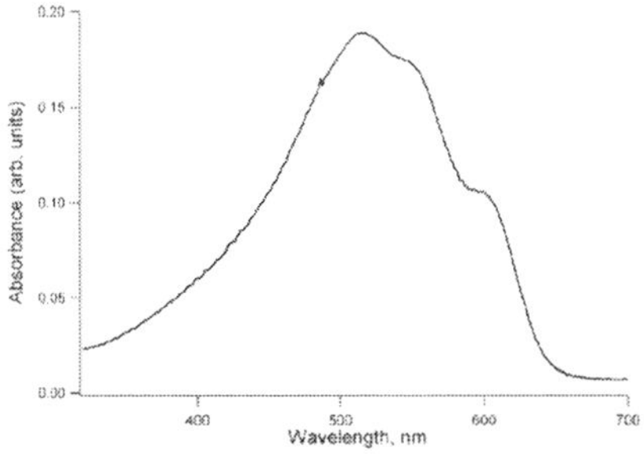

Figure 3: UV-visible absorption spectrum of P3HT NP

Before performing DLS, the aqueous NP dispersions were filtered multiple times through 0.2 μm PTFE filters. The observed major particle size was approximately 50 nm (see Figure 2). The DLS spectrum also showed a small peak around 6 microns, which we attribute to the nanoparticle aggregates. Figure 3 shows the absorption spectrum of the P3HT nanoparticles. The λ_{max} of the nanoparticle is around 510 nm and also shoulder peak around 600 nm corresponding to the π-π interaction between polymer chains. This shows the packing of polymer chains within the nanoparticle.

P3HT sensitized solar cell

FTO coated glass slides were cleaned through successive sonication for 15 min. in deionized water, ethanol and acetone. The slides were then dried under a stream of dry N_2. The TiO_2 nanoparticle slurry was obtained by grinding the dry TiO_2 powder in a pH 3 aqueous acetic acid solution. TiO_2 films were then prepared by spin-coating the TiO_2 nanoparticle slurry onto the cleaned FTO glass substrates. TiO_2 films were sintered at 400 °C for 3 h. For sensitizing the nanoporous electrode, the P3HT nanoparticle dispersion was then added drop-wise on top of the sintered TiO_2 layer. Thick platinum (100 nm) coated FTO glass was used as the counter electrodes. The DSSC was assembled by clamping the counter electrode on top of the sensitized TiO_2 layer and the I^-/I_3^- electrolyte was then infiltrated in the cells. The photovoltaic performance of the solar cells was evaluated under simulated AM1.5 G illumination (100 mW/cm^2) in conjunction with a Keithley SMU2400 source meter. The active area of the DSSCs was controlled to be 0.25 cm^2 with a mask. The current-voltage (I-V) curve of the P3HT NP sensitized DSSC is shown in Figure 4. The solar cell yielded a short-circuit current density (J_{SC}) of 0.175 mA/cm^2, open-circuit voltage (V_{OC}) of 0.35 V and 0.03% (η) efficiency.

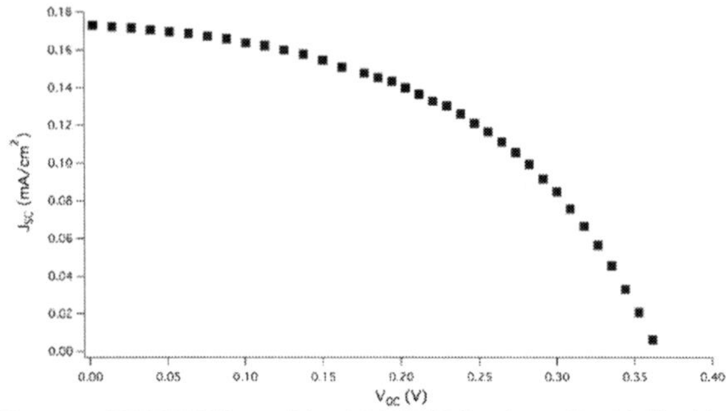

Figure 4: I-V curve of P3HT NP sensitized P25 TiO2 solar cell with liquid electrolyte

We believed that the low efficiency of P3HT nanoparticle sensitized solar cell is due to poor infiltration of P3HT nanoparticle within the titania pores. The size of the pores yielded by sintering P25 titania nanoparticles will be smaller than 20 nm. However, the size of the P3HT nanoparticles used for infiltration was 50 nm. We believe that the large P3HT nanoparticles cannot infiltrate in to the smaller TiO_2 nanopores

Figure 5: Scanning electron micrographs of P3HT NP sensitized P25/100 nm TiO_2 film.

In order to improve the infiltration of P3HT NP within the titania pores, we spin-coated another layer of 100 nm TiO_2 nanoparticle on the sintered P25 titania layer and sintered at 480 °C for 8 h. This process resulted in larger pore size TiO_2 for the P3HT nanoparticle to infiltrate (see Figure 5). Indeed, P3HT nanoparticle sensitized P25/100 nm TiO_2 solar cells showed better performance compared to P25 TiO_2 solar cells. The power (I-V) curve of the P3HT NP sensitized P25/100 nm TiO_2 solar cell is shown in Figure 6. Devices made from this architecture showed an improved performance with a short-circuit current density (J_{sc}) of 0.207 mA/cm^2, open-circuit voltage (V_{oc}) of 0.62 V and 0.07% (η) efficiency. As compared to P25 TiO_2 solar cell this device structure resulted in more than a 100 percent increase in the power conversion efficiency. The optimization of the DSSCs is currently underway and the results will be reported elsewhere.

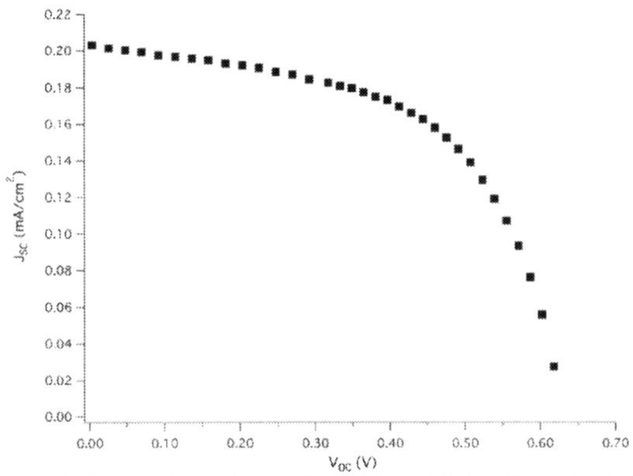

Figure 6: I-V curve of P3HT NP sensitized P25/100 nm TiO_2 solar cell with liquid electrolyte

CONCLUSIONS

P3HT nanoparticles of 50 nm size were synthesized using mini-emulsion technique. The strategy of using nanoparticles to infiltrate the titania pores was explored. Our initial results show that the P3HT nanoparticle sensitized solar cells show a short-circuit current density (J_{sc}) of 0.207 mA/cm^2, open-circuit voltage (V_{oc}) of 0.62 V and 0.07% (η) efficiency. The technique for complete dye infiltration the polymer nanoparticle to improve the power conversion efficiencies is still being optimized. In addition, tuning the size of the P3HT nanoparticle can increase the efficiency of infiltration. Overall, the strategy of using nanoparticles for infiltration seems promising.

ACKNOWLEDGMENTS

This work was supported by the US Department of Energy, Office of Basic Energy Sciences as part of the PHaSE Energy Frontier Research Center under Award Number DE-SC0001087.

REFERENCES

1. Schmidt-Mende, L. and Grätzel, M., *TiO2 pore-filling and its effect on the efficiency of solid-state dye-sensitized solar cells.* Thin Solid Films, 2006. **500**(1-2): p. 296-301.
2. Grätzel, M., *Conversion of sunlight to electric power by nanocrystalline dye-sensitized solar cells.* Journal of Photochemistry and Photobiology a-Chemistry, 2004. **164**(1-3): p. 3-14.
3. Hagfeldt, A., et al., *Dye-Sensitized Solar Cells.* Chemical Reviews. **110**(11): p. 6595-6663.
4. Bhongale, C.J. and Thelakkat, M., *Efficient hybrid polymer/titania solar cells sensitized with carboxylated polymer dye.* Solar Energy Materials and Solar Cells. **94**(5): p. 817-822.
5. Coakley, K.M. and McGehee, M. D., *Photovoltaic cells made from conjugated polymers infiltrated into mesoporous titania.* Applied Physics Letters, 2003. **83**(16): p. 3380-3382.
6. Kim, Y.G., et al., *Efficient light harvesting polymers for nanocrystalline TiO2 photovoltaic cells.* Nano Letters, 2003. **3**(4): p. 523-525.
7. Liu, J.S., et al., *Polythiophene containing thermally removable solubilizing groups enhances the interface and the performance of polymer-titania hybrid solar cells.* Journal of the American Chemical Society, 2004. **126**(31): p. 9486-9487.
8. Sergawie, A., Yohannes, T., and Solomon, T., *A comparative study on liquid-state photoelectrochemical cells based on poly(3-hexylthiophene) and a composite film of poly(3-hexylthiophene) and nanocrystalline titanium dioxide.* Synthetic Metals, 2007. **157**(2-3): p. 75-79.
9. van Hal, P.A., et al., *TiO2 sensitized with an oligo(p-phenylenevinylene) carboxylic acid: a new model compound for a hybrid solar cell.* Journal of Materials Chemistry, 2003. **13**(5): p. 1054-1057.
10. Fatuch, J.C., et al., *Synthesis and characterization of aniline copolymers containing carboxylic groups and their application as sensitizer and hole conductor in solar cells.* Synthetic Metals, 2009. **159**(21-22): p. 2348-2354.
11. Haeldermans, I., et al., *Water based preparation method for 'green' solid-state polythiophene solar cells.* Thin Solid Films, 2008. **516**(20): p. 7245-7250.
12. Piok, T., et al., *Organic light-emitting devices fabricated from semiconducting nanospheres.* Advanced Materials, 2003. **15**(10): p. 800-804.
13. Marie, E., et al., *Synthesis of Polyaniline Particles via Inverse and Direct Miniemulsion.* Macromolecules, 2003. **36**(11): p. 3967-3973.
14. Landfester, K., et al., *Semiconducting polymer nanospheres in aqueous dispersion prepared by a miniemulsion process.* Advanced Materials, 2002. **14**(9): p. 651-655.
15. Pecher, J. and Mecking, S., *Nanoparticles of Conjugated Polymers.* Chemical Reviews, 110(10): p. 6260-6279.
16. Hittinger, E., Kokil, A., and Weder, C., Synthesis and characterization of cross-linked conjugated polymer milli-, micro-, and nanoparticles. Angewandte Chemie-International Edition, 2004. 43(14): p. 1808-1811.

Mater. Res. Soc. Symp. Proc. Vol. 1312 © 2011 Materials Research Society
DOI: 10.1557/opl.2011.120

Melt-Quench Formed Smectic Phase in iPP/CNT Nanocomposites and its Re-Crystallization

Georgi Georgiev[1,2], Yaniel Cabrera[2], Lauren Wielgus[2], Scott Schoen[2], Devin Ivy[2] and Peggy Cebe[2]

[1]Department of Natural Sciences, Assumption College, Worcester, MA 01609, U.S.A.
[2]Department of Physics and Astronomy, Tufts University, Medford, MA 02155, U.S.A.

ABSTRACT

The largest commercial application of carbon nanotubes (CNT) are their polymer nanocomposites (PNCs). This motivates detailed studies of the interactions between CNTs and polymers and the ways the CNTs influence the crystallization behavior and structure of PNCs. We have chosen isotactic polypropylene (iPP) as one of the best model systems because of its ability to form a smectic liquid crystal phase. We studied iPP/CNT nanocomposites with CNT concentrations ranging from 0.01 to 5% per weight. We prepared films by compression molding and quenching in a mixture of isopropyl alcohol and dry ice. Some of the formed smectic liquid crystalline (LC) phase of iPP/CNT PNCs persisted to temperatures higher than the last melting temperature for iPP crystals. By means of differential scanning calorimetry we corroborated the existence of the LC phase in the nanocomposites and studied the impact of multi-walled carbon nanotubes on its melting and re-crystallization. We used a polarizing optical microscopy to reconfirm our DSC data.

INTRODUCTION

Isotactic polypropylene (iPP) is one of the best model systems to study in the field of polymer nanocomposites because iPP/CNT PNCs can form a liquid crystalline smectic phase and alpha, beta, and gamma crystallographic phases under a variety of crystallization conditions [1]. We prepared iPP/CNT nanocomposites from solution with CNT concentrations ranging 0.01 - 5% per weight by co-precipitation from a non-solvent. Films were made by compression molding and quenched in a mixture of isopropyl alcohol and dry ice at -70°C. The quenching of the samples induces a smectic liquid crystalline (LC) phase [1]. One of the questions addressed in this study are: How the CNTs affect the smectic phase of iPP and subsequent re-crystallization? We found that the perfection of the smectic phase is slightly hindered by measuring a decrease of the melting temperature of the melt-quench formed smectic phase. On the other hand we found that the re-crystallization is improved by measuring a downshift of the exotherm corresponding to this transition. The significance of this downshift is that the CNTs surfaces serve as nucleation centers and induce crystallization at lower temperatures. Similar result on increasing crystallization rates we have observed before in iPP/CNTs PNCs in the absence of smectic phase [3]. The melting temperature of thus formed re-crystallized iPP samples was not affected significantly by the presence of CNTs. CNTs form liquid crystal phase in their lyotropic solutions [4] and interact strongly with liquid crystal molecules [5,6,7]. In our research we are exploring whether the CNTs interact in a similar way with polymer liquid crystalline structures as with small molecule liquid crystals.

MATERIALS AND METHODS

iPP and CNTs

Isotactic polypropylene was obtained from Scientific Polymer Products, Inc. in powder form as Catalog #130. Mutli-walled carbon nanotubes came from MER Corporation. MWCNT's were produced by chemical vapor deposition (CVD). Catalytic CVD MWNTs had 140 ± 30 nm diameters and 7 ± 2 micron length, with purity of >90% (Catalog # MRCSD).

Purifying MWCNTs by Oxidation

One gram of the MWNT's was suspended in a mixture of concentrated sulfuric acid and nitric acid (3:1 vol. ratio). This solution was sonicated in a Misonix water bath sonicator for 24h at 50°C. The resultant suspension was diluted with 400mL of deionized water and filtered through a 400 nm pore membrane (PTFE) until the water passing through the filter had a pH between 6 and 7. The dispersions were subsequently filtered or diluted to the desired concentration. The resulting MWNTs had a pH of between 3 and 3.5 and were stable. Scanning Electron Microscopy (SEM) analysis showed that the CNTs were shortened between 30% and 40% of their original length, as expected.

Preparation of iPP/CNT Nanocomposites

Nanocomposites were prepared by sonicating MWCNTs in Xylene (75ml) at 50°C in a flask for 30min. IPP was added to the Xylene containing the MWCNTs to form the different percent concentrations of the nanocomposites for the study. The solution was heated and stirred by placing the flask in an oil bath at 130°C on a hot plate. Once iPP was completely dissolved, the solution was mixed slowly into methanol (a polar compound, non-solvent) with about a 5x volume dilution. The resulting powder precipitate was dried and then pressed into a film using a compression molding hot press at temperature of 200°C and pressure of 3-6000 lbs/sq in.

Thermal treatment

The molten films at 200°C were transferred from the hot press within less than 1sec. into a mixture of isopropyl alcohol and dry ice at -70°C.

Differential scanning calorimetry

DSC analysis was performed on the TA Instruments calorimeter, Model # Q100. We heated the quenched samples from room temperature to 200°C at 10°C/min.

RESULTS AND DISCUSSION

In this section we present the polarized optical microscopy and DSC results for the melt-quench formed smectic phase in iPP/CNTs PNCs. On figure 1 we show the result for melt-quench formed smectic phase, which upon subsequent heating persists above the melting temperature of iPP crystals of 160ºC to 180ºC. The larger portion of the melt-quench formed smectic phase melts around 50°C and then re-crystallizes at around 100°C , according to our DSC measurements (figure 2). The melt-quench formed smectic phase and nanotubes can further be oriented using zone drawing.

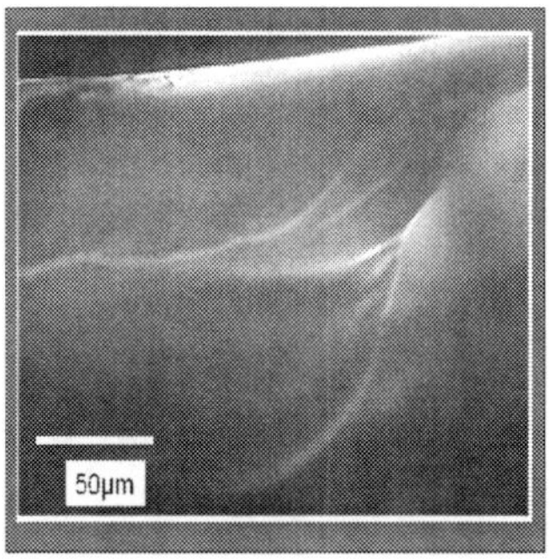

Figure 1: Smectic phase of iPP produced under the conditions in this study. The image represents birefringence in neat iPP, melted in Instec hot stage and quenched from 200°C to -70°C in mixture of solid CO_2 and methanol (using an established procedure [8]). The image is taken upon subsequent heating to 180°C on a glass substrate under crossed linear polarizers. No crystals remain above the crystal melting temperature of 170°C, which indicates that the birefringence is produced solely by the smectic phase. The lines are created by the flow of the sample upon melting and indicate the flow direction.

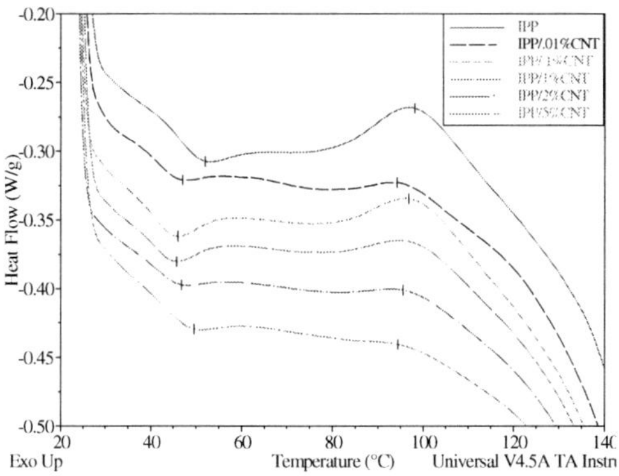

Figure 2. Heat flow vs. temperature DSC heating traces for different CNT concentrations in iPP. The heating rate was 10°C/min. The increasing percent of CNTs in the iPP/CNT nanocomposites is in order from top to bottom, where the topmost curve is the one with 0% CNTs and the lowest curve is the one with 5% CNTs. The order for the rest of the curves is as shown in the legend.

Figure 2 shows the DSC heat traces for the different nanocomposites as the temperature was increased at 10°C/min. All curves show two peaks before the main iPP crystal melting endotherm at ~160°C. The first peak (endotherm 1) is due to melting of the LC smectic phase. This is explained with the fact that liquid crystals have lower degree of order than the fully crystallized polymer, therefore melting occurs earlier [1]. The second peak (exotherm 1) is due to re-crystallizing of the polymer chains freed by the smectic phase melting [1]. The position of the first endotherm shifts down in temperature as the CNT concentration increases from 0% to 1%, and then starts shifting to higher temperatures as the concentration increases further from 1% to 5%. The position of the exotherm does not show a clear trend, but on average slightly decreases. It is possible that during quenching stresses are frozen in the films which when released can have a slight effect on lowering the temperature of the glass transition. This will be a subject for further investigation.

Table I. Peak positions of the peaks from the DSC signal for iPP/CNT nanocomposites for different concentrations of CNTs. T_m is the maximum of the melting endotherm upon continuing heating of the samples, which is almost constant.

CNTs %	Positions of peaks (°C)		
	Endothermic peak	Exothermic peak	T_m
0	51.77	103.21	161.11
0.01	46.05	95.33	159.48
0.1	45.15	98.54	159.8
1	45.12	96.66	159.33
2	46.65	94.29	159.96
5	48.41	92.84	160.39

We answered positively to the question, whether smectic phase can form in iPP when CNTs are introduced. In our DSC experiment the position of the smectic melting endotherm at around 50°C shifts down in temperature by a few degrees as the CNT concentration increases from 0% to 1%, and then starts shifting to higher temperatures as the concentration increases further from 1% to 5%. This shows that in a melt-quench formed unoriented smectic phase, the CNTs, which may span several smectic domains, have weak disordering effect, as expected. The position of the re-crystallization exotherm on average slightly decreases.

CONCLUSIONS

The significance of this study is that the result that smectic phase is shown to form in nanocomposites of iPP with addition of CNTs. Therefore we conclude that the nanotubes do not prevent the formation of smectic phase. Another result is that the presence of CNTs aids the re-crystallization of iPP by shifting the peak temperature of the re-crystallization exotherm to lower temperatures. This increase of the rate of iPP crystallization we have observed in a previous study which measured crystallization in the absence of smectic phase [3]. The melting of the smectic phase is less influenced by the CNTs but we observe a general trend of decreased peak temperature for the melting endotherm at around 50°C in the presence of CNTs. This decrease of the melting temperature can be attributed to reduced order parameter of the smectic phase in the presence of CNTs. After 1% CNTs this trend of peak temperature decrease does not continue and even reverses slightly. The crystal melting temperature of iPP is not significantly affected by the presence of CNTs in those treatments. We conclude that the CNTs are a nucleation agent for the re-crystallized iPP which was released by melting of its smectic phase. Further studies will include exploring more concentrations to test the reversal trend in the shift of the smectic endotherm at around 50°C. We will complement the work with small and wide angle X-ray scattering (SAXS and WAXS), and two dimensional microscopic transmission ellipsometry (2D-MTE).

ACKNOWLEDGMENTS

For support of this research, the authors thank: Assumption College for a Faculty Development Grant; The Natural Science Department at Assumption College for the summer student stipends; The Petroleum Research Fund, 44149-AC7; The National Science Foundation, Polymers Program of the Division of Materials Research, DMR-0602473, and MRI Program under DMR-0520655 for thermal analysis instrumentation.

REFERENCES

1. J. Varga, *Journal of Materials Science* **27,** 2557-2579 (1992).
2. I. Dierking, G. Scalia and P. Morales, *J. Appl. Phys.* **92**, 8 (2004).
3. G. Georgiev, Y. Cabrera, L. Wielgus, Z. Iftikhar, M. Mattera, P. Gati, A. Potter and P. Cebe, in *Artificially Induced Grain Alignment in Thin Films*, edited by V. Matias, R. Hammond, S.-H. Moon, R. Hühne (Mater. Res. Soc. Symp. Proc. **1150**, Warrendale, PA, 2009), RR04-16, p. 185-190.
4. W. Song, I.A. Kinloch, A.H. Windle, Science **302**, 1363 (2003).
5. A. Dawid, W. Gwizdała, *Journal of Non-Crystalline Solids* **355,** 1302–1306 (2009).
6. W. Gwizdała, K. Górny, Z. Gburski, *Journal of Molecular Structure* **887,** 148–151 (2008).
7. G. Georgiev, E. A. Gombos, M. McIntyre, M. Mattera, P. Gati, Y. Cabrera and P. Cebe, in *Nanoscale Pattern Formation*, edited by E. Chason, R. Cuerno, J. Gray, K.-H. Heinig, (Mater. Res. Soc. Symp. Proc. **1228E**, Warrendale, PA, 2010), KK11-81.
8. F. de Candia, P. Iannelli, G. Staulo, V. Vittoria, *Colloid and Polymer Science* **266,** 608-613 (1988).

Mater. Res. Soc. Symp. Proc. Vol. 1312 © 2011 Materials Research Society
DOI: 10.1557/opl.2011.121

Nucleation of the electroactive phase of poly(vinylidene fluoride) by ferrite nanoparticles: surface versus size effects

P. Martins, C. M. Costa, M. Benelmekki, S. Lanceros-Mendez

Centro/Departamento de Física, Universidade do Minho, 4710-057 Braga, Portugal.

ABSTRACT

Multiferroics and magnetoelectric materials show interesting scientific challenges and technnologial applications in sensors, acuators and data storage. In view of the fact that only a small number of materials show this kind of properties, exhaustive research activity is being pursued towards the development of new composite materials. Multiferroic nanocomposites films composed of piezoelectric poly(vinylidene fluoride) (PVDF) and magnetostrictive nano-size $CoFe_2O_4$, $NiFe_2O_4$ or $NiZnFe_2O_4$ ferrites were prepared by a solution method. Those ferrite nanoparticles have the ability to nucleate the electroactive β-phase of the polymer, providing in this way an easy route for the preparation of magnetoelectric particulate composites. The fact that the different nanoparticles promotes different amount of β-phase nucleation for different concentrations of nanoparticles indicates that filler size is not the most important parameter determining phase nucleation but the filler-matrix surface interaction. Further, when the polymer-ferrite surface interaction is modified through surfactation, the electroactive phase is not nucleated.

INTRODUCTION

The Magnetoelectric (ME) effect (induction of polarization by an applied magnetic field or induction of magnetization by an applied electric field) was first observed in singe phase materials (e.g. Cr_2O_3) [1], [2], [3]. Since the ME response of single phase materials is too weak for device applications, intensive studies were focused on the development of multiferroic composites combining piezoelectric and magnetostrictive materials [2], [4]. Multilayer and particulate piezoelectric/magnetostrictive films have been reported over the past few years [4], [5]. Laminated composites nevertheless are difficult to produce in large scale [5]. On the other hand, ceramic based composites are limited by deleterious reactions at the interface regions leading to low resistivities and high dielectric losses making such composites not suitable for device applications [6], [7].

One possible and efficient way to simplify the preparation process as well to avoid the problems of laminated and ceramic composites is to use polymer based composites with magnetostrictive particles. There composites are also promising candidates for outstanding magnetoelectric performance [8], [9].

PVDF is one of the most used piezoelectric polymer materials [10]. It shows five different crystalline phases known has α, β, γ, δ and ε that can be obtained depending on the processing conditions. The α and β phases are the most important and also the most studied. The α-phase is non-polar and it is obtained directly by cooling from the melt. The β-phase is the polymer phase with the highest piezoelectric, pyroelectric, and ferroelectric properties [10]. The polar β-phase is usually obtained by stretching α-phase films at temperatures between 70°C and 100°C and stretch rations larger than four. Un-oriented β-phase films can be also obtained by crystallization from dimethyl formamide (DMF) or dimethyl acetamide (DMAc) solutions at temperatures below 70°C, higher temperatures will usually result in a mixture of the α and β-phase, with the α-phase fraction increasing with increasing temperatures [11], [12].

The preparation of the β-phase by stretchin from the α-phase is not suitable for magnetoelectric composites once it will reduce the coupling between the polymer and the magnetostrictive phase.

Nano-size ferrites such as $CoFe_2O_4$, $NiFe_2O_4$ or $NiZnFe_2O_4$ can be incorporated into a poly(vinilidene fluoride), PVDF, matrix and have the ability to nucleate the electroactive beta phase of the polymer, providing in this way an easy route for the preparation of magnetoelectric particulate composites.

EXPERIMENTAL

$CoFe_2O_4$, $NiFe_2O_4$ and $NiZnFe_2O_4$ ferrite nanoparticles with dimensions between 35-55, 20-30 and 10-30 nm respectively were purchased from *NanoA*mor: Nanopowders & Dispersions for Nanotechnology, Houston, USA. *N,N*-dimethylformamide (DMF, pure grade) and citric acid were supplied by Sigma-Aldrich, St. Louis, USA and PVDF was supplied by Solvay Solexis, Bollate, Italy. All the chemicals and nanoparticles were used as received from the suppliers.

Towards the preparation of the nanocomposites, two methods were applied to ensure a good dispersion of ferrite nanoparticles and to avoid loose aggregates:

a) the desired amount of nanoparticles were added to DMF and then placed in an ultrasound bath during 6 hours;

b) nanoparticles were surfactated with citric acid and later added to DMF [13].

Then, PVDF powder was added to the mixture of DMF and nanoparticles. The obtained blend was mixed in a Teflon mechanical stirrer for complete dissolution of the polymer powder during 1h. Samples with thickness of ±50 μm were obtained by spreading the solution on a very well clean glass substrate at room temperature. The thickness of the films was controlled by a spreader. Solvent evaporation was obtained inside an oven at a controlled temperature of 210ºC. After this step, samples were cooled down to room temperature. The weight percentage of ferrite nanoparticles varied from 0.001 to 50. Fourier transform infrared spectroscopy (FTIR) has been proved to be suitable to identify and quantify phase content in PVDF [14]. Fourier transformed infrared spectroscopy (FTIR) spectra of the films were recorded on a Perkin-Elmer Spectrum 100 in ATR mode over a range of 650-1150cm^{-1} with a resolution of 4 cm^{-1}. 32 scans were performed to each sample.

Scanning electron microscopy (SEM) measurements were performed in a *FEI Nova NanoSEM 200* with an accelerating voltage of 20 kV, in previously gold coated samples (*Polaron, model SC502 sputter coater*). The samples analysed in cross-section were previously immersed in liquid nitrogen for 180 s in order to ensure that the sample temperature decreases below glass transition (*Tg*), and then were mechanically broken.

TGA was carried out to using Pyris 1 TGA – Perkin-Elmer under nitrogen atmosphere supplied at a constant 50 mL min^{-1} flow rate. The sample holders used were crucibles of ceramic with capacity of 60 μL. The samples were subjected to same heating rate of 10 ± 0.2°C.min^{-1}, between 50 and 850 °C, in order.

DISCUSSION

Figure 1 shows a representative SEM image of the samples with 20% of NiZn ferrite content prepared by ultrasound dispersion (a) and surfactated with citric acid (b).

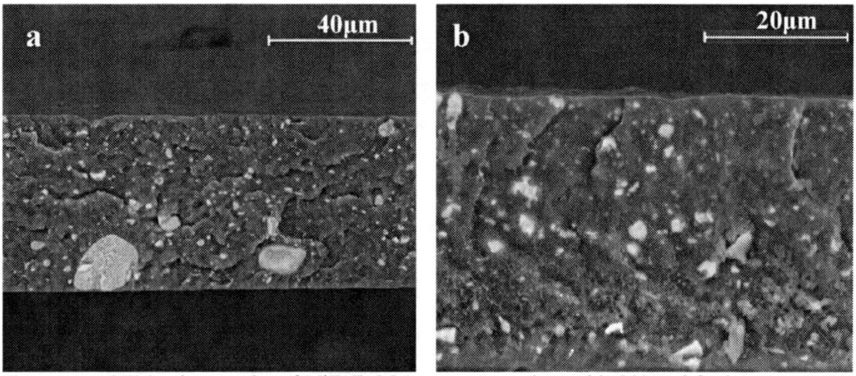

Figure 1 – SEM micrographs of NiZnFe2O4 nanocomposites with 20% of ferrite content: a) dispersed by ultrasound; b) using citric acid as surfactant

Agglomerates are observed in figure 1a. Due to the higher density of the ferrite agglomerates, they appear at the bottom of the samples, leading to a bad distribution of the ferrite particles along the composites. Some of the agglomerates have sizes of 30 μm each, which difficult or even makes impossible the electric poling of the composites. On the other hand, citric acid surfactation of the samples lead to a homogeneous distribution of the ferrite nanoparticles by avoiding large agglomerates, 1b.

For the quantification of the polymer phase, specific bands such as 766 and 840 cm^{-1} have been identified to correspond to the α and β-phase respectively. The relative amount *β* phase (*F(β)*) present in the samples was calculated applying equation 1 [15]:

$$F(\beta) = \frac{X_\beta}{X_\alpha + X_\beta} = \frac{A_\beta}{(K_\beta / K_\alpha)A_\alpha + A_\beta} \quad \text{Eq. 1}$$

Here, *F(β)* represents the *β*-phase content; A_α and A_β the absorbencies at 766 and 840 cm^{-1}, corresponding to the *α* and *β* phase material; K_α and K_β are the absorption coefficients at the respective wave number and X_α and X_β the degree of crystallinity of each phase. The value of

K_α is 6.1×10^4 and K_β is 7.7×10^4 cm^2/mol. A similar procedure was used for the calculation of the β-phase content.

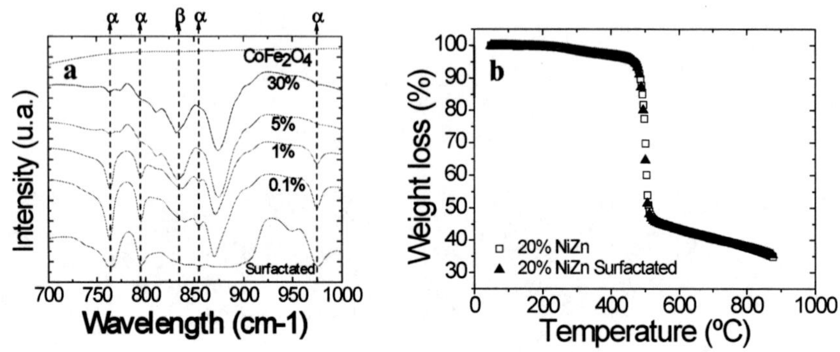

Figure 2 – *a)* Infrared measurements for the PVDF/CoFe2O4 composites samples *b)* Thermal degradation of PVDF/$NiZnFe_2O_4$ (80/20 wt) composites obtained at 10°C.min^{-1} without and with surfactated nanoparticles.

The evolution of the α- and β- phases for composites with increasing ferrite concentration is presented in Figure 2a and in Table1. The β-phase increases and correspondingly the α-phase content decreases with increasing nanoparticle concentration.

Table 1. Phase content of all nanocomposites.

Ferrite	**$CoFe_2O_4$**		**$NiFe_2O_4$**		**$NiZnFe_2O_4$**	
Ferrite Content (%)	**α phase content (%)**	**β phase content (%)**	**α phase content (%)**	**β phase content (%)**	**α phase content (%)**	**β phase content (%)**
0.01	86.5	3,5	96.4	3,6	96	4
0.1	83	7	96.8	3,2	95	5
0.5	81	19	96.4	3.6	80	20
1	67	23	96.6	3,4	70	30
5	12.5	87,5	96.7	3,3	12.5	87.5
10	13	87	93.3	6,7	13	87
20	13.5	86,5	70	30	11.5	88.5
30	13	87	44	56	12.3	87.7
40	12.5	87,5	29	71	13	87
50	12	88	16	84	21.1	87.9

The fact that the β-phase of PVDF forms preferentially in this PVDF/ferrite nanocomposites indicates that the ferrite nanoparticles are nucleating β-phase epitaxially on their surfaces, this is supported by the FTIR spectra of the composites prepared from surfactated particles: modifying the surface properties of the particles improves dispersion but also hinders phase transformation resulting just in non-electroactive α-phase composites (Figure 2a). Further there seems to be no particular relationship between the size of the particles and the concentration rate at which the β-phase is nucleated: the larger (Co) and smallest (NiZn Fe_2O_4) nucleate similar amounts of β-phase for similar concentrations.

Since the surface interactions between ferrite nanopaticles and the PVDF polymer induces the nucleation process of the β-phase, TGA was used to verify if the thermal stability and degradation of the composite is affected by the surfactation process. The TGA curves of the composites with and without citric acid surfactation are shown in Figure 2 b. It is observed that two degradation mechanism occur, one around 250ºC due to the presence of the nanoparticles (it does not exists for pure PVDF) [16] and the main degradation of PVDF at higher temperatures, that they are not affected by the surfactation process.

CONCLUSIONS

The nonfunctionalized ferrite nanoparticles used in this work nucleate the electroactive phase of PVDF, but at a different rate: whereas beta phase contents larger than 80% are obtained for 5%wt of $CoFe_2O_4$ and $NiZnFe_2O_4$, 50%wt of $NiFe_2O_4$ is needed for obtaining those values of β-PVDF. This fact indicates that filler size is not the most important parameter determining phase nucleation but the filler-matrix surface interaction. These results do not support the theory of nanoparticles promoting phase nucleation when their radius is less than the radius of gyration, R_g of the polymer [17]. The R_g value for PVDF is 27,5 nm, and the average radius of nanoparticles is 45 nm for $CoFe_2O_4$, 25nm for $NiFe_2O_4$ and 20 nm for. $NiZnFe_2O_4$. Nanoparticle funtionalization, i.e. modification of the polymer-ferrite surface interaction, on the other hand, promotes dispersion but prevents β-phase nucleation.

ACKNOWLEDGMENTS

The authors thank the Foundation for Science and Technology (FCT) through the 3º Quadro Comunitario de Apoio, the POCTI and FEDER programs and the PTDC/CTM/69316/2006 and NANO/NMed-SD/0156/2007 projects. P. Martins thanks the support of the FCT (grant SFRH/BD/45265/2008)

REFERENCES

1. Mathur, N. D.et al, *Multiferroic and magnetoelectric materials*. Nature, 2006. **442** (7104): p. 759-765.
2. Spaldin, N. et al, T*he renaissance of magnetoelectric multiferroics*. Science. 2005. **309** (573): p. 391-392.
3. Rado, G. et al, *Observation of the Magnetically Induced Magnetoelectric Effect and*

Evidence for Antiferromagnetic Domains. Physical Review Letters,1961. **7**(8): p. 310-311.

4. Nan, C. et al, *Multiferroic magnetoelectric composites: Historical perspective, status, and future directions*. Journal of Applied Physics, 2008. **103**(3): p. 1-35.
5. Guo, Y. et al, *Giant Magnetodielectric Effect in 0-3 Ni0.5Zn0.5Fe2O4-Poly(vinylidene-fluoride) Nanocomposite Films*. Journal of Physical Chemistry C, 2010. **114** (32): p. 13861-13866 .
6. Li, Y et al, *Magnetoelectric effect of Ni0.8Zn0.2Fe2O4/Sr0.5Ba0.5Nb2O6 composites*. Journal of the European Ceramic Society, 2006. **26**(13): p. 2839-2844.
7. Mitoseriu, L. et al, *BaTiO3-(Ni0.5Zn0.5)Fe2O4 ceramic composites with ferroelectric and magnetic properties*. Journal of the European Ceramic Society, 2007. **27**(13-15): p. 4379-4382.
8. Martins, P et al, *Nucleation of electroactive β-phase poly(vinilidene fluoride) with CoFe2O4 and NiFe2O4 nanofillers: a new method for the preparation of multiferroic nanocomposites*. Applied Physics A, Materials Science & Processing, 2010. DOI: 10.1007/s00339-010-6003-7.
9. Zhang, Z. et al, *The effect of magnetic nanoparticles on the morphology, ferroelectric, and magnetoelectric behaviors of CFO/P(VDF-TrFE) 0-3 nanocomposites*. Journal of Applied Physics, 2009. **105**(5): p. 1-6.
10. Fukada, E. et al, *History and recent progress in piezoelectric polymers*. IEEE Transactions on ultrasonics ferroelectrics and frequency control, 2000. **47**(6) p. 1277-1290.
11. Branciforti, M. et al, *New Technique of Processing Highly Oriented Poly(vinylidene fluoride) Films Exclusively in the β Phase*. Journal of polymer science Part B-Polymer Physics, 2007. **45**(19): p. 2793-2801.
12. Martins, P. et al, *Local variation of the dielectric properties of poly(vinylidene fluoride) during the alpha- to beta-phase transformation*. Physics Letters A, 2009. **373**(2): p. 177-180.
13. Campelj, S. et al, P*reparation and properties of water-based magnetic fluids*. Journal of Physics: Condensed Matter, 2008. **20**(20): p. 1-5.
14. Kobayashi, M et al, *Molecular Vibrations of Three Crystal Forms of Poly(vinylidene fluoride)*. Macromolecules, 1975. **8**(2): p. 158–171.
15. Miranda, D. et al, *Influence of Silver Nanoparticles Concentration on the alpha- to beta-Phase Transformation and the Physical Properties of Silver Nanoparticles Doped Poly(vinylidene fluoride) Nanocomposites*. Journal of nanoscience and nanotechnology, 2009. **9**(5): p. 2910-2916.
16. Botelho, G et al, *Relationship between processing conditions, defects and thermal degradation of poly(vinylidene fluoride) in the β-phase*. Journal of non-crystalline solids, 2008, **354**(1): p. 72-78.
17. Andrew, J. et al, *Enhanced Ferroelectric Phase Content of Polyvinylidene Difluoride Fibers with the Addition of Magnetic Nanoparticles*. Langmuir, 2008. **24**(16): 8435-8438.

Mater. Res. Soc. Symp. Proc. Vol. 1312 © 2011 Materials Research Society
DOI: 10.1557/opl.2011.122

Laser assisted fabrication of porous polymer MEMS with nano structured additives

Igor V. Shishkovsky[1] and Yuri G. Morozov[2]
[1]P.N. Lebedev Physics Institute of Russian Academy of Sciences, Samara branch,
Novo-Sadovaja st. 221, Samara 443011, Russia. shiv@fian.smr.ru
[2]Institute of Structural Macrokinetics and Materials Science (ISMMS), RAS,
Chernogolovka, Moscow, 142432 Russia.

ABSTRACT

Selective laser sintering (SLS) process was used to obtain nanostructured porous 3D objects consisting of Ni or/and Cu nanoclusters enveloped with a polycarbonate (PC) matrix for M/NEMS applications. The liquid phase SLS process was performed in air or Ar. The powder mixtures of Ni + PC in proportions 1:1, 1:2 (Ni particle size ranged from 27 to 184 nm) and Cu + PC = 1:9, 1:4, 3:7 (Cu particle size ranged from 76 to 90 nm) and a cw YAG:Nd^{+3} laser (P = 4–10 W) with defocused laser beam were used. Optimal regimes of laser treatment as monolayer as layerwise 3D part manufacturing in above mention nano compositions were determined. The scanning electronic microscopy (SEM) equipped EDX and qualitative XRD analysis showed that content and size of nanoparticles are remained unaffected after the liquid-phase sintering. It was shown a principle possibility of functional graded 3D parts fabrication via the interleaving of the metal - polymer powdered compositions with Ni and/or Cu additives, which ensures the nano particle sizes conservation. Proposed approach is opened a new way by deliberated M/NEMS device synthesis. Alternation of ferromagnetic and non-magnetic layers with Ni and/or Cu core – polymer shell structures was provided the interesting electro-physical properties of such devices.

INTRODUCTION

Nanoparticles are promising candidate as the solid lubricant, in conducting and magnetic materials, and in pressure cells and hydrogen storage devices also. Multilayers with giant magnetoresistance, which could be realized in the layers interleaving ferromagnetic and non-magnetic metals, have attractive applications [1-3]. With the decrease of particle sizes from the micro- to the nano- level, the specific surface area to volume ratios grows and, consequently, the chemical activity of nanoparticles follows to this behavior. The direct multilayer fabrication of the nanoparticles is a difficult task. It is known, that the sintering is the thermally activated process, which compulsorily must be accompanied by the nanoparticle coagulation in the micrometer conglomerations. To stabilize the nanoparticles in the polymeric matrix is attractive [4], because of this makes possible the organization of beforehand determined distribution of nanoparticles by porous polymer structure. It is allow to protect them from the undesirable oxidation and even to attempt the functional graded structures design on this way.

Therefore, the polymer mixture sintering is represented as more optimum due to the temperature for the polymer sintering is deliberately lower than temperatures under which the nanoparticles will begin actively to consolidate. Nano size core/polymer shell structure can change hydrophobic properties of the filter porous structure, ensure its colloidal stability also.

Selective Laser Sintering (SLS) is appropriate technique for functional-graded structure fabrication with nano impurities. Actually, it is possible to realize the "bottom-up" paradigm of nanotechnologies by the layerwise synthesis of the smart micro and nano- objects (M/NEMS devices). Earlier [5], we showed the possibility of the laser sintering of the functional- graded filters from the powdered Metal (nickel or brass) plus Polymer (polycarbonate /PC/ or polyamide /PA/) Compositions (MPC).

At present report, the possibilities of a layer-by-layer laser sintering of the porous samples from polycarbonate powder with an interleaving of the nano - nickel and copper additives and laser parameter optimization are discussed. Phase-structural and electro-physical properties of Cu + PC – Ni + PC layers were investigated.

EXPERIMENT

The nickel and copper nanoparticles were prepared at the ISMMS of RAS by means of the levitation jet technique analogous to that described in [6]. The mean size values of thus prepared Ni particles were - 27.8, 32.2, 119, and 184 nm. The Ni percentage and specific surface area of the powders were 86.6/ 25.1; 11.2/ 26.9; 94.2/ 5.73; and 98.3 at. %/ 3.68 m^2/g, respectively (the rest of the mass was NiO). The mean size values of Cu particles were (1.- 76-100; 2.- 90-120) nm. The Cu percentage and specific surface area of the powders were measured (1. – 96.8/3.4; 2. – 98.2/2.7) w.% Cu (bal., CuO) / m^2/g, respectively.

The thermostable PC powder was used as a binder. A commercially available PC powder (LET-7 grade, Russia) had the particle size ~ 20-40 μm. The starting Ni + PC blend powders were prepared in the weight ratios 1:1 or 1:2 and Cu + PC = 1:9, 1:4, 3:7.

The cw YAG:Nd^{+3} laser was used to carry out the SLS process. The beam power P ranged from 4 to 10 W. The laser beam was scanning the powder surface at the velocity v = 8.7–34.8 cm/s, the optic facilities was used with the beam focal length of 147 mm and shift of 25 mm from the bottleneck. Layer-by-layer laser sintering was performed in a specially designed camera in Ar or in the air. The three-dimensional model is built via CAD file.

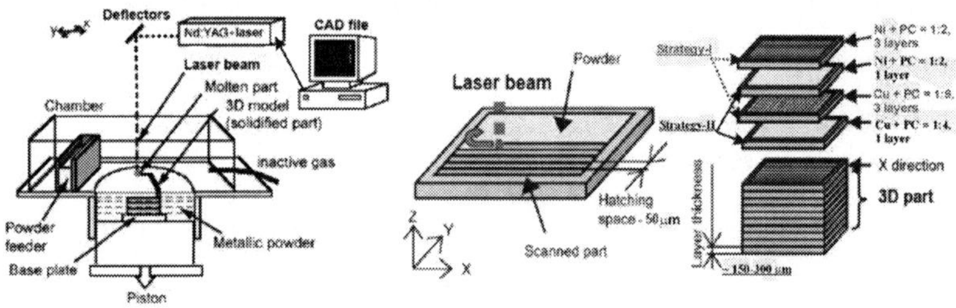

Figure 1. SLS setup (left) and strategy (right) of 3D synthesis.

The delivery of the metal powders was conducted using a double powder feeder system in chamber. The first hopper contained nano Ni based MPC, while the second hopper contained nano Cu based MPC. The laser beam is carried through the deflector system (Fig. 1a), and it is focused on the powder bed. During the SLS process, a powder layer is deposited onto the base plate attached to the piston. The laser beam scans by the powder bed according to the slice data of the model, and the first solid layer was made on the base plate. Then, the piston is lowered, and the next powder layer is deposited. Successive scanning and lowering of the powder bed are carried out until the model is completed.

Fig. 1b shows the two main synthesis strategies of the laser fabrication of 3D models carried out at present study. Each cube sample was created layerwise and had the 10x10 mm square. By the first strategy the FG was synthesized in the nano Ni + nano Cu based MPC, where a Cu based MPC was delivered as the first layer, a Ni based MPC was second and the mixture interchange was accomplished in the each layer (case on the fig. 3c). By the second strategy the possibility of the FG -structures was fabricated in the nano Ni + nano Cu based MPC with interleaving through 3 layers (case on the fig. 3b).

The structure of the sintered porous PC matrices with incorporated nanoparticles was studied by SEM/EDX (LEO 1450 microscope, Carl Zeiss), equipped with an EDX accessory (INCA ENERGY 300, Oxford Instruments) and XRD (DRON-3 diffractometer, Cu-K_α radiation). The electro physical properties of sintered samples were evaluated on laboratory digital equipment (magnetometer and LRC E7-12). Magnetic measurements were carried out by M4500 EG&G Princeton Applied Research vibrating sample magnetometer in magnetic field strengths up to 10 kOe at the room temperature.

RESULTS AND DISCUSSION

In the laser-assisted M/NEMS fabrication, the polymer matrix is normally expected to provide the contact between the surface Ni and/or Cu particles and the flow of a reactive or purified medium. The major aim of our experiments was the laser beam parameters optimization in order to achieve the maximum strength of the porous structure and minimum nanoparticle recrystallization or coagulation during the sintering procedure. It is believed [4] that the hydrophobic polycarbonate ties itself with the nanoparticles surface and can also interact with the reactive media through specific hydrophobic interactions.

Figure 2 SEM image of the sintered mixtures: a) Ni+PC (1 : 1), $P = 6$ W, $v = 17.4$ cm/s; b) insert of S1 area under high magnification; c) Cu+PC (1 : 9), P = 6.2 W, $v = 40$ cm/s.

Fig. 2 shows the inner structure of material obtained by the liquid-phase laser sintering of PC–Ni and PC-Cu mixtures. More dark areas on the fig. 2(a-c) are corresponded to the pure PC while bright areas are presented as the Ni and Cu nanoparticles with the polymer coating. Sub nano structure of nickel (see an insert - fig. 2b) is well-marked present on the common black background of polymer. The estimations gave for the small grain size ~ 43–48 nm (fig. 2b) that is slightly higher as compared to the initial value (27–32 nm). However, there are exist conglomerates with bigger then initial size on the order, for account of thermal coagulation.

The SEM results suggest that the growth mechanism for the metal core–polymer shell structures is also applicable in our case. The formation of the PC shell structure is clearly seen in Fig. 2a,c. It seems reasonable to assume that the polymer macro molecules dissolve in the Ni or/and Cu particles and, and if once the particle becomes supersaturated with carbon, then the nano size particles starts to precipitate on the metal surface thus forming graphitic layers, that normally occurs at lower temperatures.

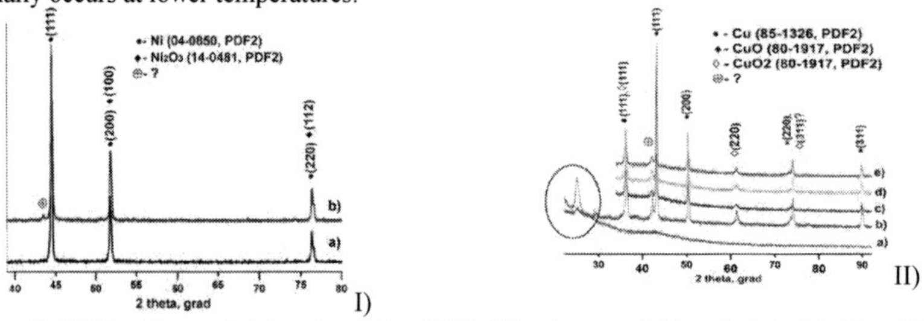

Figure 3. XRD patterns of sintered samples: I) Ni–PC mixtures: 1 (a) and 1 : 2 (b); P = 6 W, v = 17.4 cm/s. II) PC + Cu mixtures: a) pure PC before of the LI; b) PC + nano Cu = 7:3 without LI; c) P = 11 W, v =20 cm/s, PC + Cu = 1:4; d) 8.7/160/1:4; e) 6.2/80/3:7, respectively.

On the fig. 3 (I), (curves a, b) the XRD patterns after PC + Ni sintering allowed to detect diffraction peaks (111, 200, 220) typical of the Ni crystal with a face-centered *Fm*3*m* structure. Some peaks (100, 112) can be attributed to metastable Ni_2O_3 oxide. There were no detectable impurities arising from another nickel oxide (NiO), nickel carbide or nickel nitride. Variations in the laser parameters and powder composition had no influence on the XRD patterns. The polymer-encapsulated Ni nanoparticles were invented as the way to prevent them from the undesirable oxidation, carburization or nitridation during the sintering process.

Fig. 3(II) (a-e) shows the X-ray diffraction patterns obtained for PC + Cu compositions. The two curves (a) and b)) on the fig. 3(II) are shown the powder content before a laser treatment. Pure PC and initial PC + nano Cu mixture had two clear peaks on 2Θ~ 25 and 29 degrees, which we connect with polymer crystallinity (remarked on fig.3(II)a as a circle). After SLS those peaks disappear. For nanosized copper + PC (fig. 3(II) c-e), the patterns were included the standard Cu and CuO cubic and monoclinic phases, respectively. Also the XRD patterns for PC + nano Cu reported about CuO2 oxide also (diffraction peaks (220, 311?)). Actually, near the main Cu peak (111), it is clear determined an unresolved peak, which is keeping on diffractograms after laser treatment (marked as "⊕"). By our opinion, it is may be amorphous part of nano copper powder.

Fig. 4 present typical appearance of functional graded 3D porous samples, which were manufactured by SLS process. Powder specifications, alternation methods and LI regimes are shown at the figure caption. The liquid-phase laser sintering with interlining of PC–Ni and/or PC-Cu layer was successfully realized. We had similar results by SLS polyamide with nano additives. It is open the great possibilities by management of future electro physical properties in M/NEMS devices. Turn-on of external magnet field during SLS process was used also for the structure ordering of the ferromagnetic component in polymer with nano particles compositions. This additional degree of freedom by nanophase substructure design as magnetoresistance measurements demands future examinations.

a) b) c)

Figure 4 Typical appearance of the 3D functional graded sintered samples (surface square – 10x10 mm, height ~ 4 mm): a) Ni + PC = 1:2 – 5 layers on the top, and bottom Cu + PC = 1:4 - 5 layers; b) Ni + PC = 1:2 – by 3 layers, then Cu + PC = 1:9 – by 3 layers and again; c) Cu + PC = 1:4, then Ni + PC = 1:2 – and alternation per layer.

Temperature dependence of electrophysical properties in laser synthesized 3D samples was studied on example of 3D samples from PC + Cu powdered MPC. By fig. 5a it is seen, the temperature dependencies of a real part of a dielectric permeability (as a no presented here dielectric loss tangent) has hysteresis behaviour. On the stage of heating a dielectric permeability is growth (fig. 5a), and on the cooling stage this value is saving for same time but then was decreased until initial level. In our measurements the heating values were below appropriated cooling values. In the paper [7] similar behaviour of copper - polymer nanocomposite film was explained by the thermo stimulated charge redistribution on the copper-polymer boundary. Charges are accumulated on long-living traps or in the metastable phase structures. Those reasons could be take place in our synthesized samples.

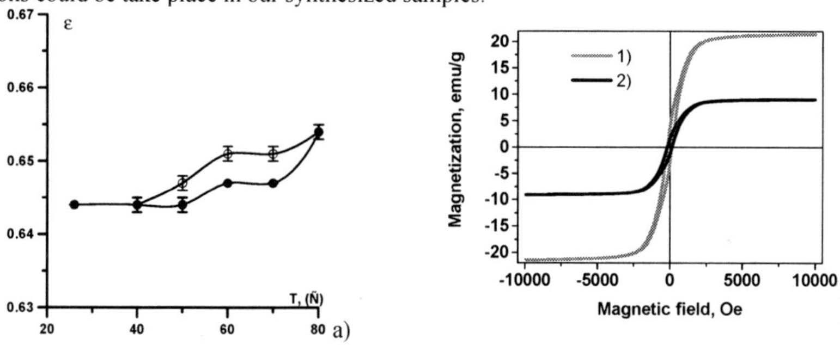

a) b)

Figure 5 Electro physical properties of laser sintered 3D samples: a) temperature effect of real part of dielectric permeability in laser synthesized nano composition base Cu + PC = 1:9: 1) black circles are heating stage; 2) white circles are cooling. LI regime – P = 6.2 W, V = 40 mm/s; b) hysteresis loop for 3D SLS samples from Ni + PC = 1:2 mixture: 1) laser sintering in external magnet field; 2) SLS without external magnet field. LI regime – P = 6 W, V = 36.7 cm/s.

Fig. 5b is presented magnet hysteresis loop which was evaluated in the nano Ni + PC = 1:2 mixture after SLS. It is need to remark a significant growth of magnetization (σ_s - 21.44 contrast 9.08 emu/g and σ_r - 4.14 vs. 1.75 emu/g) in the sintered sample with external magnet field. Another word, SLS process allows create 3D samples with control magnetization and ratio of Ni/NiO phases, which will determined catalytically properties such smart objects.

Earlier [8] it has been found that giant magnetoresistance of the multilayers is strongly dependent on the relative amount of ferromagnetic and paramagnetic phases. Magnetoresistance is strongly depend on the thickness of both ferromagnetic and non-magnetic layers and increase linearly with increasing number of interfaces also. SLS process allows to fabricate such interleaving ferromagnetic (Ni, Fe) and non-magnetic (Cu, PC) layers via natural course (see fig. 4). On the stage of computer modeling of future 3D part the magnetoresistance properties could numerical calculated and optimized. Then laser assisted nanofabrication process control can additional manage layerwise synthesis smart M/NEMS devices.

CONCLUSIONS

The laser-assisted technique of the 3D synthesis was used to prepare porous polycarbonate structures containing encapsulated nickel and / or copper nanoparticles, chemical active and uniformly distributed over the sintered polymer. The obtained SEM and XRD data confirmed the formation of the Ni /Cu core –PC shell structures around the nanoparticles. The incorporation of Ni nanoparticles into the polymer matrix and the high speed heating–cooling cycles were chosen as the ways to decelerate the polymer degradation during the SLS process.

It was shown a principle possibility of functional graded three - dimensional parts fabrication via the interleaving of the metal - polymer powdered compositions with Ni and/or Cu additives, which ensures nano particle sizes conservation. The optimal regimes of LI were determined as for single monolayers as layer - by - layer SLS process.

Electrophysical properties of synthesized 3D samples had hysteresis phenomena that can be useful for MEMS-NEMS applications: sensors, magnetic recordings, filters with catalyst additives, tissue-engineering scaffolds, and microfluidic chips. In addition, our technique can be extended for the encapsulation in the polymer matrix of iron, aluminium, titanium, tin, zirconium or cobalt nanoparticles and those oxides.

ACKNOWLEDGMENTS

This work was supported by the Russian Foundation for Basic Research (Project no. 10-08-00208-a).

REFERENCES

1. V. Sunny, D.S. Kumar, Y. Yoshida, M. Makarewicz, W. Tabis, M.R. Anantharaman, *Carbon,* **48,** 1643 (2010).
2. L. Joly-Pottuz, B. Vacher, et al, *Tribol. Lett.,* **29,** 213, (2008).
3. Y.X. Yang, R.K. Singh, P.A. Webley, *Adsorption,* **14** 265, (2008).
4. C. Boyer, M. R. Whittaker, V. Bulmus, J. Liu, T. P. Davis, *NPG Asia Mater.,* **2**(1) 23, (2010).
5. I. Shishkovsky, V. Sherbakov, A. Pitrov, *Proc. SPIE,* **6732,** 43, (2007).
6. A.P. Krasnov, Yu.G. Morozov, E.A. Chernov, *Powder Technology,* **81,** 93, (1994).
7. N.M. Ushakov, A.N. Ulzuev, I.D. Kosobudski, *Russian Journal of Technical Physics,* **78** (12), 65, (2008).
8. S.M. Dulal, E.A. Charles, *Transactions of the Institute of Metal Finishing,* **86** (5), 260, (2008).

Mater. Res. Soc. Symp. Proc. Vol. 1312 © 2011 Materials Research Society
DOI: 10.1557/opl.2011.425

Microwave Synthesis of Metallic Nanoparticles Supported on Porous Coordination Polymers: A Bi-functional Catalyst Design for CO_2 Activation and Conversion

Victor Abdelsayed[1,2], Yueying Fan[1,2] and Todd Gardner[1*]

[1]National Energy Technology Laboratory, 3610 Collins Ferry Rd., Morgantown, WV 26507, USA

[2]URS Corporation, 3610 Collins Ferry Rd., Morgantown, WV 26507, USA

ABSTRACT

The efficient separation and conversion of CO_2 from power plant flue gases remains a significant and far reaching global goal. Partially, because of the high energy cost associated with CO_2 separation which makes technologies, such as amine scrubbing and conventional sorbent materials, less attractive. In this work we have used a microwave-assisted method to prepare Mg-based porous coordination polymers (MOF-74) with open metal sites. The material shows a high CO_2 uptake at room temperature mainly due to the high binding energy between the open Mg site (in the coordination polymer) and the oxygen atom of the adsorbed CO_2 molecule. Our results show that the microwave-based Mg-MOF-74 shows superior properties compared to thermally prepared samples such as higher surface area, crystallinity and CO_2 uptake. Pt nanoparticles (<5-10 nm) were deposited on the Mg-MOF-74 as manifested from the TEM results. The Pt nanoparticles were uniformly dispersed across the sample with more aggregations at higher Pt loadings. Different techniques were used to characterize the materials such as (BET surface area, XRD, SEM, EDS, TEM and TGA). These materials function as a bi-functional catalyst and sorbent material with the Mg-MOF-74 material responsible for CO_2 capture and supported Pt nanoparticles responsible for its catalytic activity.

INTRODUCTION

Recently, metal organic frameworks (MOFs) have attracted enormous attention due to its various applications in gas storage, separation, catalysis and sensors. MOFs have ordered microporous structures, large surface areas and pore volumes. MOFs are hybrid materials constructed from metal ions or clusters covalently bonded to organic molecules called linkers. The design of this porous material is theoretically limitless and hence can be engineered to address specific industrial problems.

Unlike the classical MOF materials with their metal ions coordinatively saturated with ligand which adds a screening effect for gas adsorption properties, MOF-74 is an open-metal-site coordination polymer with exceptional CO_2 adsorption properties. Low density Mg-based MOF with an isostructure similar to Zn-based MOF-74 were prepared previously using solvothermal and room temperature synthesis approaches which require long preparation time and energy [1,2]. To our knowledge, Mg-based MOF-74 was never prepared by microwave assisted solvothermal technique. In this paper we have investigated this method as a rapid and energy efficient route that can be scale-up economically. We have also employed the microwave technique to incorporate high-quality; nearly monodisperse nanoscale Pt particles without using any reducing agents with total metal loading from 0.5 to 10 wt%.

* Corresponding author.

EXPERIMENT

Materials: Chloroplatinic acid (H_2PtCl_6, 8wt.% solution in water) and 2,5-dihydroxyterephthalic acid (DHTP, 98%), were purchased from Sigma Aldrich. Propylene oxide (PO, 99+%), magnesium nitrate hexahydrate ($Mg(NO^3)^2.6H^2O$, 98-102%), and anhydrous ethanol (94-96%), were purchased from Alfa Aesar. N,N-dimethylformamide (DMF, 99.8% extra dry) and dichloromethane (DCM, 99.5%) were purchased from Acros Organics. Methanol (99.9%) was purchased from Fisher Chemical. All reagents were used without further purification. Ultra pure carbon dioxide (CO^2, 99.999%) gas was purchased from Matheson Tri Gas. Nanopure water (~18MΩ) was used for all experiments.

Microwave Synthesis of Activated Mg-MOF-74: Mg-based MOF-74 crystals were prepared by microwave-assisted solvothermal rapid technique where 1.4g of magnesium nitrate hexahydrate and 0.337g of 2,5-dihydroxyterephthalic acid were dissolved in a mixture of 135 ml DMF, 9ml ethanol and 9ml water. The reaction mixture was decanted into eight 100ml-closed high pressure Teflon vessels then sealed and placed on a turntable of a microwave reaction system operating at 2.45 GHz (Synthos 3000 from Anton Paar) equipped with pressure and temperature sensors. The microwave system was operated under the power mode at 500 watt for 30 min with the maximum allowed reaction temperature of 183°C and corresponding pressure of 11.2 bar. After the reaction temperature cooled down, the yellow Mg-MOF-74 crystals were collected, combined and washed with methanol. The sample was exchanged into methanol daily for 4 days to remove the guest DMF. Finally, the methanol was decanted and the sample was dried under vacuum (10-3 torr) for another 2 days.

Solvothermal Synthesis of Activated Mg-MOF-74: Mg-MOF samples were prepared according to a modified synthesis procedure reported previously [1].

Synthesis of Pt Catalysts Supported on Mg-MOF-74: 1 gram of Mg-MOF-74 was dispersed in a 100ml methanol-chloroplatinic acid solution. The solution was prepared by adding the calculated volume of 8wt.% chloroplatinic acid solution to methanol equivalent to 0.5, 2.0, 5.0, and 10.0 wt% of Pt metal loading on Mg-MOF-74. The dispersion was placed inside the microwave, stirred for 10 min then the temperature was ramped to 130oC in 1 min and kept at this temperature for 3 min while under stirring. The Pt/Mg-MOF-74 catalysts were then filtered, washed with methanol then dried under vacuum for 2 days. The yellow color of Mg-MOF-74 gets darker as the Pt loading increases, indicating the deposition of Pt nanoparticles.

RESULTS AND DISCUSSION

The morphology of Mg-based MOF74 samples prepared by microwave-assisted solvothermal method is displayed in Figure 1(a-b). The SEM image (Figure 1a) shows uniform hexagonal crystals with rod-like shapes that are obtained after only 30 min of irradiation time. The crystals are about ~5μm long by ~1μm width as also observed from the TEM image (Figure 1b). For comparison, the Mg-MOF-74 sample was prepared according to the literature [1] by a thermal conventional method in a Teflon lined vessel for 20 hour at 125 °C. It is clear that no definable crystalline structure can be observed as shown from the SEM image (Figure 1c) with the corresponding EDX spectrum (Figure 1d) confirming the presence of C, O, and Mg in the crystals. The crystals are highly aggregated with a high level of crystal defects and facets. The homogeneous dielectric heating employed by microwave irradiation ensures fast nucleation and growth rates. In contrast, thermal methods are always entailing thermal gradient effects that could produce a wide size-distribution of crystals especially if applied for long reaction time that could amplify that effect as observed in thermal Mg-MOF-74 samples.

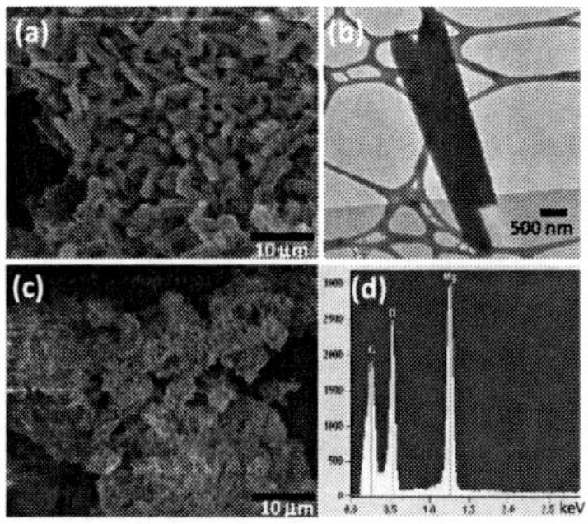

Figure 1. (a) SEM and (b) TEM images for Mg-MOF-74 prepared by microwave-assisted method. (c) SEM and (b) EDX spectrum for thermally-prepared Mg-MOF-74 crystals.

The BET surface area analysis by nitrogen adsorption-desorption isotherms was performed at 77 K on Mg-MOF-74 samples to give 1096 m^2/g and 1053 m^2/g for microwave and thermal Mg-MOF-74 samples using the BET model, respectively. It is obvious that the microwave sample is slightly higher than the thermal sample. The CO_2 adsorption-desorption isotherms were measured at 273 K as shown in Figure 2. The microwave sample shows a higher CO_2 uptake of 35 wt% compared to 31 wt% for the solvothermal sample at 1.0 atm.

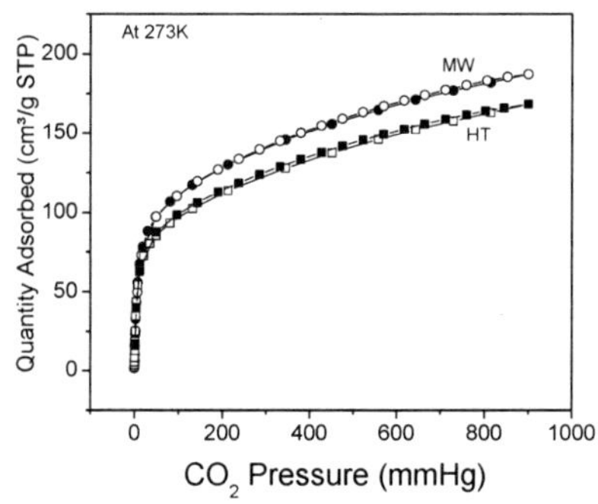

Figure 2. CO_2 sorption isotherms at 273 K for microwave (MW) and thermal (HT) Mg-MOF-74 samples.

By incorporating Pt nanoparticles in the pore structure or even depositing onto crystal surfaces of Mg-MOF-74, we can develop a bi-functional catalyst and sorbent material that could potentially be used to simultaneously to capture CO_2 using the Mg-MOF-74 component and then catalytically convert it into useful products (in presence of another gas reactant) using the Pt nanoparticles. Four different Pt-loaded Mg-MOF-74 samples with 0.5, 2.0, 5.0 and 10.0 wt% Pt were prepared by microwave irradiation (in absence of any reducing agent). Figure 3 display the TEM images for 2.0 and 10.0 wt% Pt/Mg-MOFF-74. It is clear that Pt nanoparticles are homogenously dispersed on Mg-MOF-74 crystals specially at low levels of loadings [4] (Figure 3a), while at higher loading Pt nanoparticles start to aggregate on the surfaces (Figure 3b) but the primary particle sizes are still below 10 nm. At 2 wt% the Pt particle size is less than 2-3 nm while increases to 8-10 nm at 10wt % Pt loading. We believe that the crystal surfaces especially at the edges act as a controlled reducing agent for Pt ions, where the terminal un-bound ligands are functionalized with carboxylate and hydroxide chemical groups. These groups are strong Lewis base sites that are capable of providing electrons to reduce Pt^{2+} into Pt^{0}. Figure 3d is a zoom-in picture from the edge of the crystal. It is clear that the concentration of Pt nanoparticles is greater at the surface than at the bulk of the crystal (Figure 3c) as confirmed from the EDX spectra that shows high Pt concentrations at the surfaces of the Mg-MOF-74 (Table 1).

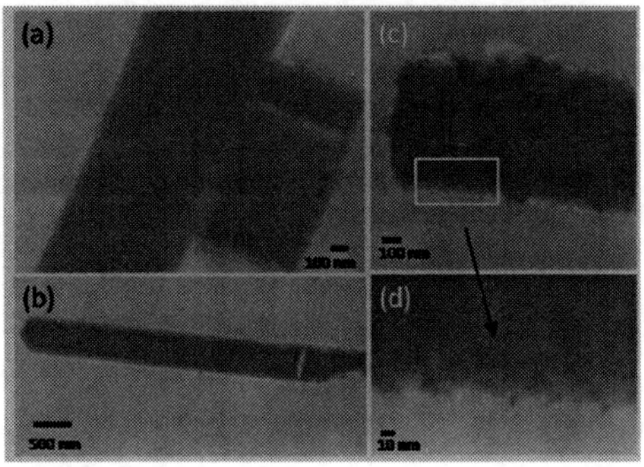

Figure 3. TEM images for 2wt% Pt (a,b) and 10wt% Pt (c,d) nanoparticles supported on microwave Mg-MOF-74 samples.

The crystal nature of the prepared Mg-MOF-74 was examined by X-ray diffraction. Figure 4 shows the XRD patterns of Mg-MOF-74 samples prepared by both microwave and by solvothermal methods, the two patterns have the same diffraction peaks compared to the corresponding simulated XRD obtained from CrystalMaker software. The crystalline nature of microwave sample is more pronounced as evident from the diffraction peak intensities which also matches the SEM results shown above. For the Pt loaded sample, a very broad diffraction peaks that can be indexed to the Pt crystal diffraction at 2θ equals 39.96°, 46.60° and 67.92° corresponding to the diffraction at 111, 200 and 220 lattice planes. The Pt average particle size was calculated using Scherrer equations from the XRD results and is shown in Table 1 along with the surface Pt concentration obtained from the EDX analysis. The structure of Mg-MOF-74 with an ordered array of 1D channels of 10-11Å width is also shown in the insert of Figure 4.

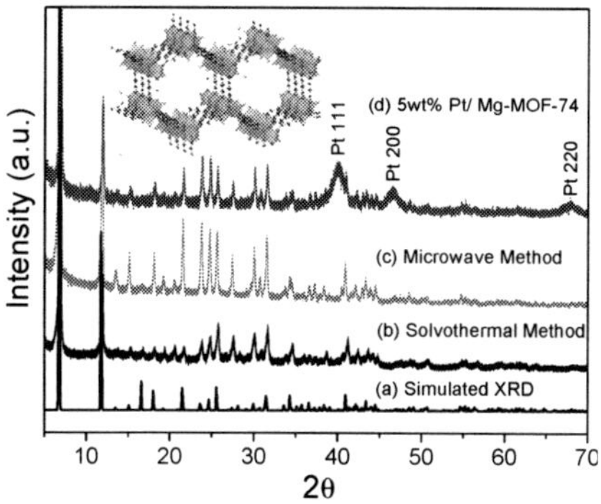

Figure 4. X-ray diffraction for (a) simulated pattern using CrystalMaker software, (b-c) Mg-MOF-74 samples prepared by solvothermal and microwave methods, respectively, and (d) Mg-MOF-74 microwave sample loaded with 5 wt% Pt nanoparticles. Also shown is the 1D channel crystal structure of Mg-MOF-74.

The BET surface area and room temperature CO_2 uptake were calculated from the nitrogen and CO_2 sorption isotherms at 77 and 298 K, respectively. The results shows that the surface area decrease from 1096 to 674 m^2/g as the Pt loading increases from 0 to 10 wt% as expected due to the solid Pt substitution for porous Mg-MOF-74 (table 1). For the same reason the CO_2 uptake followed the same trend (Figure 5) where it decreases from 28 to 16 wt% at 1 atm. The high uptake of CO_2 is due to the open Mg sites which will provide a strong electrostatic interaction between the Mg ion and the quadruple moment of CO_2 gas molecule. Also, Mg an alkaline earth metals with basic character offers a strong binding energy to CO_2 as reported in the literature [1,3].

Table 1. Pt concentration, particle size and effect on BET and CO_2 uptake of Mg-MOF-74.

Sample	Pt loading (wt.%)	Surface Pt[1] (wt.%)	Avg. Pt particle size[2] (nm)	BET surface area (m^2/g)	CO_2 uptake[3] (wt.%)
Mg-MOF-74	0.0	0.0		1096	28.74
0.5Pt-Mg-MOF-74	0.5	1.9	< 5	837[#]	22.35
2Pt-Mg-MOF-74	2.0	4.2	6	937	21.49
5Pt-Mg-MOF-74	5.0	7.89	10	898	16.28
10Pt-Mg-MOF-74	10.0	8.34	10	674	16.19

[1] *EDS analysis was used to determine surface Pt concentration.*
[2] *Scherrer equation was used to determine the average particle size of Pt from the XRD.*
[3] *CO_2 uptake was calculated at atmospheric pressure and room temperature.*
[#] *Very small Pt nanoparticles could block the pores and channels of Mg-MOF-74*

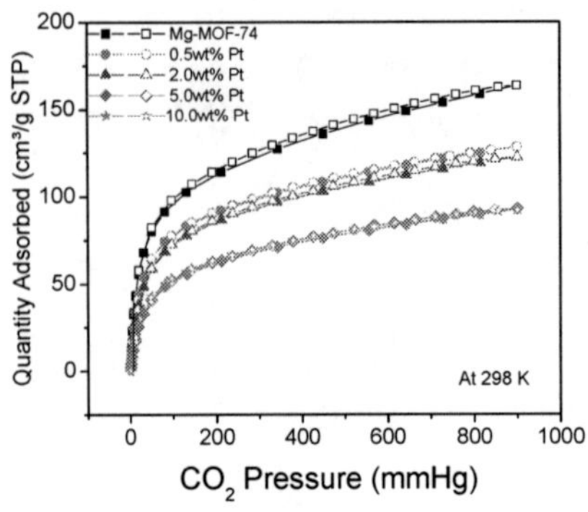

Figure 5. CO_2 sorption isotherms at room temperature for free and Pt-loaded Mg-MOF-74.

CONCLUSIONS

We have demonstrated the design of bi-functional catalyst based on Pt nanoparticles incorporated in Mg-MOF-74 that could be used simultaneous in CO_2 capture and its catalytic conversion. The microwave samples are well crystalline, with higher surface area and CO_2 uptake compared to solvothermal samples. The reduction of Pt ions in presence of Mg-MOF-74 suggests that the functional groups at the terminal linkers act as Lewis bases and reduce the Pt ions.

ACKNOWLEDGMENTS

We thank Dr. Lihua Zhang from the Center for Functional Nanomaterials at Brookhaven National Laboratory and Mr. James A. Poston from the National Energy Technology Laboratory for their support in the TEM and FE-SEM measurements.

REFERENCES

1. Caskey, S.R.; Wong-Foy, A.G.; Matzger, A.J. *J. Am. Chem. Soc.* **2008**, *130*, 10870.
2. Tranchemontagne, D,J.; Hunt, J.R.; Yaghi, O.M. *Tetrahedron* **2008**, *64*, 8553.
3. Britt, D.; Furukawa, H.; Wang, B.; Glover, T.G.; Yaghi, O.M. *PNAS.* **2009**, *106*, 20637.
4. El-Shall, M.S.; Abdelsayed, V.; Khder, A.S.; Hassan, H.M.A.; El-Kaderi, H.M.; Reich, T.E. *J. Mater. Chem.* **2009**, *19*, 7625.

Mater. Res. Soc. Symp. Proc. Vol. 1312 © 2011 Materials Research Society
DOI: 10.1557/opl.2011.123

Synthesis of luminescent nanoparticle embedded polymer nanocomposites for scintillation applications

Thomas Rogers, Chenlu Han, Brent Wagner, Jason Nadler and Zhitao Kang*
Georgia Tech Research Institute, Georgia Institute of Technology,
925 Dalney Street, Atlanta GA 30332-0826, U.S.A.

ABSTRACT

Detecting gamma-ray emission from radionuclides hidden within containers is a significant concern to national security and can be accomplished with scintillating materials such as NaI:Tl, $LaBr_3$:Ce crystals. However, the use of these high quality crystals limits the functionality of the detectors due to their high cost and scalability issues. Therefore the development of more durable, more easily manufactured, and more cost effective scintillating materials is desired. The incorporation of nanophosphors or Quantum Dots (QDs) into a polymer matrix to produce a transparent nanocomposite could potentially provide an alternative method to fabricate scintillating detectors. Embedded in a suitable polymer matrix, nanocomposite detectors may be easily made suitably large for portal monitors. Also, preparation of suitable particle sizes and/or compositions permits selection of a photon wavelength that optimally matches the photodetector response curve to increase the number of photons collected per pulse. In this paper a series of LaF_3:Ce nanophosphors with varying doping concentrations (1-30mol%Ce) were synthesized using a chemical precipitation method. Photoluminescence and photoluminescence excitation characterizations indicated that the highest luminescent intensity was obtained from the 20%Ce doped sample with a peak emission at 325 nm. The refractive indices of the nanoparticles were identified by index matching measurements. Then an index matched epoxy was selected for incorporation of these nanoparticles to prepare transparent nanocomposite scintillators. In addition, colloidal solutions of CdTe QDs with various emitting colors were synthesized and incorporated into a Polymethyl-methacrylate (PMMA) matrix to make transparent nanocomposites. An initial evaluation of the scintillation behavior of these nanocomposites was evaluated by exposure to gamma rays.

INTRODUCTION

A specific threat to national security is the detonation of a Radiological Dispersion Device (RDD) in an urban area. These devices can contaminate a large area by emitting high energy radiation via a nonfissile process. Because the resulting radiation from such an explosion can cause extensive and lasting destruction, prevention methods must remain capable of detecting threats quickly and reliably. In order to prevent the entry these types of devices into vulnerable population areas, detection and intervention are critical. The success of the prevention process relies heavily on the material properties of the detectors used. Current radionuclide detectors contain single-crystal scintillators such as $LaBr_3$:Ce and NaI:Tl.[1] However, due to their high cost and scalability issues the functional use of these high quality single crystals is limited.[2] Scintillation detectors made from nanocomposites encapsulated with luminescent nanoparticles can be of great advantage for gamma ray detection.[1-4] Embedded in a suitable polymer matrix, nanocomposite detectors can be easily made suitably large for various

*Correspondence author. Email: zhitao.kang@gtri.gatech.edu.

applications. The objective of this research is to study the scintillation properties of LaF_3:Ce nanophosphors and CdTe QDs embedded in transparent polymer matrices.

EXPERIMENT

LaF_3:Ce nanophosphors with various molar ratios of cerium (1-30%) were synthesized with a chemical precipitation method. In a typical synthesis of LaF_3:20%Ce, 3.2 g of NaF was dissolved in 200 ml of de-ionized water and mixed with 7.2 g of oleic acid dispersed in 200 ml of ethanol.[5] The resulting mixture was stirred and heated to 78°C. Then a solution of 7.0 g of $LaCl_3 \cdot 7H_2O$ and 2.0 g of $Ce(NO_3)_3 \cdot 6H_2O$ in 160 ml of water was added drop-wise into the NaF solution to form the white LaF_3:Ce precipitate. After stirring at 78°C for another hour, the mixture was cooled to room temperature. To remove the residual oleic acid and inorganic salt, the white precipitate was washed three times with ethanol and de-ionized water in turn by ultrasonication, followed by centrifugation at a speed of 3000 rpm for 15 min and then dried in vacuum. The refractive index of the LaF_3:Ce nanophosphor was experimentally determined with index-matching fluids of known refractive indices. After identifying the refractive index, a index-matched commercial epoxy product was selected to use as a matrix to incorporate the nanophosphors. Epo-TEK 301-2FL epoxy was used to encapsulate LaF_3:20%Ce nanoparticles with loading levels of 1, 5, 10 wt% by directly mixing the dry powders with one of the epoxy components. The dimensions of these samples are 32 mm in diameter and 5-15 mm thick. After curing at room temperature, the disks were polished to minimize surface scattering.

CdTe QDs with controlled particle size and emission wavelength were prepared using a colloidal synthesis technique as reported by Zhang et al.[6] A series of colloidal CdTe solutions were synthesized by adding freshly prepared NaHTe solution to N_2-saturated $Cd(ClO_4)_2$ solutions in water and in the presence of Mercaptopropionic Acid (MPA), which acted as stabilizing agents. Highly transparent CdTe-PMMA nanocomposite samples were prepared using a bulk polymerization method.[6] To avoid agglomeration and maintain transparency, a polymerizable surfactant, octadecyl-*p*-vinylbenzyl dimethylammonium chloride (OVDAC), was used to link to the surface ligands of the nanoparticles to the polymer matrix, and to immobilize the nanoparticles during the polymerization process. In this process, OVDAC was first used to extract the QD from water solution into a methyl-methacrylate (MMA) monomer. Then the monomer solution was polymerized in an oil bath at 65 °C for ~20 h using 0.1wt% 2,2'-Azobisisobutyronitrile (AIBN) as the initiator.

Photoluminescence (PL) and photoluminescence excitation (PLE) measurements were performed on the samples dispersed in ethanol with a Cary Eclipse Fluorescence Spectrometer and absorption measurements were performed using a Cary 5000 UV-Vis-NIR Spectrophotometer. X-Ray Diffraction (XRD) of packed nanophosphor powders was conducted with an X'Pert PRO α-1 diffractometer equipped with a Cu Ka X-ray tube emitting at 1.54 Å. Particle imaging was accomplished using a JEOL 100CX II Transmission Electron Microscope (TEM) operating at 100 kV. Gamma-ray measurements were conducted using a radionuclide source placed on top of the nanocomposite samples. The sample and source were placed on a Photomultiplier Tube (PMT) detector and the entire assembly was shielded to block external light. Photons generated by the scintillator were recorded by the PMT, and their pulse height spectra distributed over 1,024 channels.

DISCUSSIONS

From the wet-chemical synthesis process, nanophosphor particles with good dispensability in organic matrices were produced. For the 20%Ce doped LaF_3 sample, a broad and intense UV emission at ~325nm with width of 70nm at half maximum was observed at room temperature as shown in Figure 1. By monitoring at 325nm, the excitation spectrum of the sample was recorded with a main excitation peak centered at around ~250 nm. Another small peak at 220nm and a shoulder at lower energy side at ~280nm were also observed, which is in consistent with the absorption peaks as shown in the inset of Figure 1 due to the transitions in the 4f-5d band of Ce^{3+} ions. A Stokes shift of 90nm was detected which indicated relatively weak self-absorption.

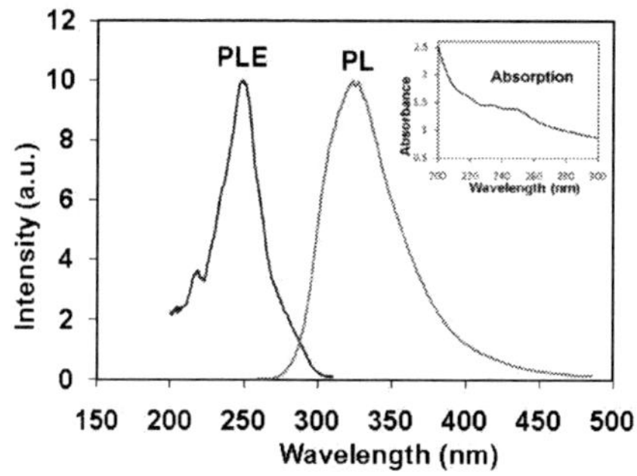

Figure 1: Photoluminescence and photoluminescence excitation spectra of LaF_3:20%Ce nanophosphor with oleic acid capping. The inset shows the absorption spectrum between 200-300nm.

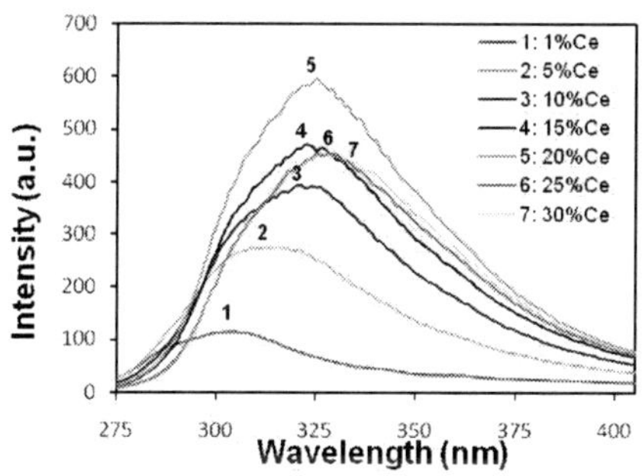

Figure 2: Photoluminescence spectra of LaF_3:Ce nanophosphors with various Ce doping ratio (1-30mol%).

In order to improve the luminescent efficiency of LaF_3:Ce nanophosphors, the concentration of Ce doping was optimized in a range of 1-30mol%. It was found that the LaF_3:20%Ce nanophosphor sample showed the greatest PL intensity as shown in Figure 2. With increasing doping concentration from 1-20%, the PL intensity increased continuously and the peak shifted to a relatively longer wavelength. However, when the Ce doping concentration was above 20%, concentration quenching was observed.

The structural properties of the nanophosphors were characterized by XRD and TEM. Figure 3 shows the XRD pattern of the LaF_3:20%Ce nanophosphor. The peak positions and intensities agree well with the data reported in the JCPDS standard card (32-0483) of pure hexagonal LaF_3 crystals. Broadening of the diffraction peaks was observed due to particle size reduction. A mean particle size of 9 nm was calculated based on the Debye-Scherrer formula. By TEM observation as shown in the inset of Figure 3, it was concluded that the sample was composed of individual rod-like nanoparticles although they were agglomerated in an ethanol solution. The particle size was about 10nm, which is in agreement with the XRD results.

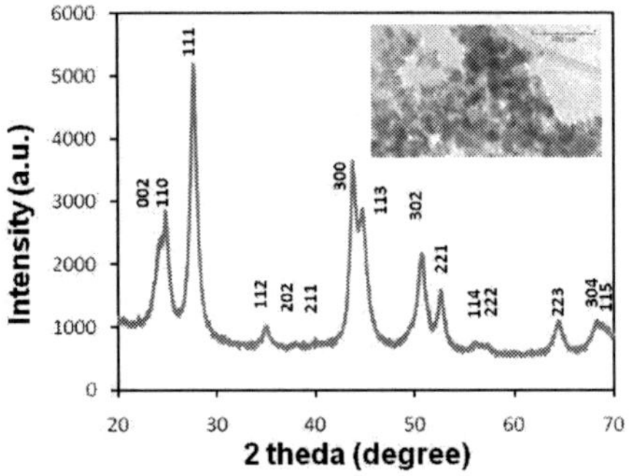

Figure 3: XRD pattern of LaF_3: 20%Ce nanophosphor capped with oleic acid. The inset shows the TEM image of the sample (scale bar 100 nm).

With index matching fluids, the refractive index of LaF_3:Ce nanophosphor was roughly determined. The LaF_3:20%Ce nanoparticle agglomerations almost "disappeared" under white halogen light when mixed into a fluid of n = 1.540, indicating a desirable index match. With an increase or decrease of the indices of the fluids, nanoparticle agglomerations were clearly observed. The encapsulating polymer chosen for this system, Epoxy Epo-Tek 301-2FL, was selected due to its reported refractive index value of. 1.512 at 589 nm. The epoxy's dispersion behavior will affect the extent of index matching with the nanophosphor as a function of wavelength. In the LaF_3:Ce/301-2FL composite system, reasonable index matching appears to have been achieved, with the composite exhibiting transparency under 365nm UV light illumination as shown in the inset of Figure 4. Figure 4 shows the gamma-ray responses of nanocomposite samples with 1-10wt% loadings of LaF_3:20%Ce nanoparticles using a ^{241}Am radionuclide source. Peaks with channel number from 9 to 14 were observed and attributed to the light emission from LaF_3: 20%Ce nanophosphors. The decrease of channel number with increase of LaF_3: 20%Ce content in the composite may be due to the increased scattering in the nanocomposite sample.

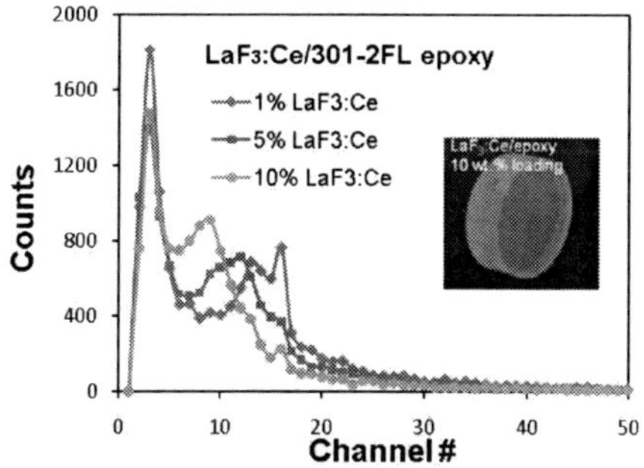

Figure 4: Responses obtained from LaF_3:20%Ce/epoxy composite samples with different loading using a ^{241}Am radionuclide source. The inset shows a transparent LaF_3:20%Ce/epoxy composite sample with 10wt% loading in a 301-2FL epoxy matrix under 365nm UV light illumination.

Highly transparent CdTe QD-PMMA nanocomposites were also prepared for gamma ray detection. As shown in the inset of Figure 5, composite samples with different sizes and various emission colors of 520~650 nm were synthesized. The QD loading concentrations in these samples were between 0.5-10wt%. The samples are transparent to light with wavelength equivalent or longer than the emission color of the embedded QDs. Figure 5 shows the gamma ray responses of CT36 QD-PMMA samples using a ^{241}Am radionuclide source for a counting time of 50000 s. The concentration of the ~520 nm emitting CT36 QD in all these samples were ~0.5wt% and the composite sizes were ranged from 10~34 mm thick in various orientations with volumes of 15~100 cm^3 as shown in the inset of Figure 5 (top). Light emission was observed from all these samples under the gamma ray excitation. A counting rate of about only 8~10 counts/min was obtained from these samples. The low concentration of QDs in the composites, self-absorption of QDs due to small Stokes shift, as well as the relatively low quantum efficiency of the PMT at this wavelength might be the reasons for the low counting rates.

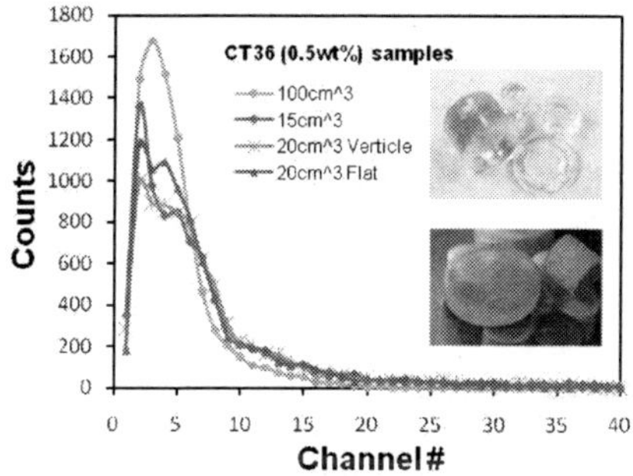

Figure 5: Responses obtained from CdTe QD-PMMA nanocomposite samples with different thicknesses/volumes using a ^{241}Am radionuclide source. The inset shows the 520 nm emitting CdTe/PMMA samples used for the gamma ray measurements (top) and various light emitting nanocomposites under a 365 nm UV lamp (bottom).

CONCLUSIONS

LaF_3:Ce nanophosphors were synthesized with a chemical precipitation method and epoxy encapsulated nanocomposites were prepared by matching the refractive index of the nanophosphor with the epoxy matrix. The prepared LaF_3:20%Ce/301-2FL epoxy nanocomposite samples were fairly transparent due to reasonable index matching between the nanophosphor and the matrix, leading to relatively low light scattering. Transparent CdTe QD-PMMA composites were also prepared for gamma ray detection and light emission was observed under gamma ray excitation. The feasibility of gamma-ray detection using nanoparticles incorporated in a transparent matrix was demonstrated, although clear photopeak spectral responses corresponding to gamma rays of various energies have not yet been realized. Further optimizations in terms of optical design and loading density are needed in order to achieve a system with the desired spectral response.

ACKNOWLEDGEMENTS

This material is based upon work supported by the U.S. Department of Homeland Security under Grant Award Number 2008-DN-077-ARI015-04. Rogers would like to thank the support of the President Undergraduate Research Assistantship from Georgia Tech.

DISCLAIMER

The views and conclusions contained in this document are those of the authors and should not be interpreted as necessarily representing the official policies of the, either expressed or implied, of the U.S, Department of Homeland Security.

REFERENCES

[1] B. D. Milbrath, A. J. Peurrung, M. Bliss, W. J. Weber: Radiation detector materials: An overview, *Journal of Materials Research*, 23, 2561, (2008)

[2] S. Letant and T. Wang: Semiconductor quantum dot scintillation under gamma-ray irradiation, *Nano Lett.*, 6, 2877, (2006).

[3] E. A. McKigney, R. E. Del Sesto, L. G. Jacobsohn, et.al.: Nanocomposite scintillators for radiation detection and nuclear spectroscopy, *Nuclear Instruments and Methods in Physcis Research*, A (579) 15, (2007)

[4] I. H. Campbell and B. K. Crone: Quantum-dot/organic semiconductor composites for radiation detection, *Adv. Mater.*, 18, 77, (2006)

[5] A. K. Burrell, K. C. Ott, J. C. Gordon, R. E. Del Sesto and T. M. McCleskey, US patent 20080011956A1

[6] H. Zhang, C. Wang, M. Li, J. Zhang, G. Lu, and B. Yang: Fluorescent nanocrystal-polymer complexes with flexible processability. *Adv. Mater.* 17, 853, (2005)

Mater. Res. Soc. Symp. Proc. Vol. 1312 © 2011 Materials Research Society
DOI: 10.1557/opl.2011.124

Thermal and viscoelastic behaviors of nanotube-reinforced polyethylene composite

Ananta Raj Adhikari[1], Mircea Chipara[2], and Karen Lozano[3]

[1]Department of Natural Sciences, Assumption College, Natick, MA-01609
[2]Department of Physics and Geology, University of Texas-Pan American, Edinburg, TX-78539
[3]Department of Mechanical Engineering, University of Texas-Pan American, Edinburg, TX-78539

ABSTRACT

The effect of processing (shear) time on the mechanical behavior and thermal stability of multiwalled nanotube reinforced polyethylene was investigated. It was observed that the mechanical property (storage modulus, loss modulus) of the composites is process dependant whereas the thermal stability does not. The increase in mechanical behavior is attributed to a stronger interface between the nanotube and the polymer matrix.

INTRODUCTION

Polymer composites have received increased interest given their light weight, increased strength, higher toughness and increased multifunctionality. The reinforcing of polymers with various fillers especially nano-sized structures has been described in numerous publications [1-4]. Among the many available nano-sized structures, carbon nanotubes (CNTs) have become a model system because of their exceptional properties [5-7]. The reinforcing fillers are distinct, physically separable and thermodynamically immiscible. The physical properties of the developed composites depend on the properties of the individual components, the degree of interaction and the meso-structural arrangement of these components in the matrix material.

It has been reported that processing effects play a significant role in the ultimate macro-properties. Lozano et al.[2] reported enhancements in strain and toughness of 16,000% and 20,000% respectively for carbon nanofiber reinforced polyethylene composites. These enhancements were obtained by subjecting the composite to extra processing steps after achieving initial dispersion. It is well known that optimum filler dispersion and stress transfer between fillers and polymers are crucial issues. In this regard, various approaches have been reported such as melt processing and extrusion techniques [8,9]. Independent of the processing method, processing parameters such as temperature, processing speed, and processing time are important. This study focuses on the effect of processing time on the thermo-mechanical behavior of multiwalled carbon nanotube (MWCNT) reinforced polyethylene (PE) composites prepared using a twin screw extruder.

EXPERIMENTS

Multiwalled carbon nanotubes (with reported purity of >95 % with outer diameter between 20 to 30 nm and 10 to 20 μm in length) were obtained from Cheap Tubes Inc. Polyethylene (density of 948 kg/m^3 and molecular weight of 125,000 g/mol) was supplied by

Phillips Chemical Company. Two sets of blend composites were prepared via high shear mixing using HAAKE minilab at 180 °C for 11min and 45min. The concentration of MWNT in PE was 0, 1, 4 and 8 % by weight.

Thermogravimetric analysis (TGA) was used to study the effects of nanotubes on the thermal degradation of PE using TGA Q500. For all experiments, samples of ~10 mg were heated from 40 °C to 700 °C at a ramping rate of 10 °C/min under nitrogen environment.

The extruded samples were pressed and cut into rectangular shapes required for dynamical mechanical analysis (DMA). DMA equipment (TA instrument DMA Q800) was used to study the mechanical behavior. Experiments were performed in tension mode over a temperature range of -125 °C to 100 °C at a rate of 5 °C/min and at a frequency of 1 Hz. The strain amplitude was 0.1 % with a preload force of 0.01 N. The data were analyzed for storage modulus and loss modulus.

RESULTS AND DISCUSSIONS

TGA was used to study the thermo-oxidative behavior for both sets of samples (11 min and 45 min). For comparison the temperature at which oxidation is maximum (T_{max}) was plotted (see Figure 1a). For the neat polymer, higher processing times (45 min) changes T_{max} by about 3 °C. T_{max} increased with the incorporation of MWNT into the polymer matrix for all samples. The fraction of mass located in the interface (m_{Ix}) was calculated from the TGA data simply by using the equation [10]:

$$m_{Ix} = \frac{m(x)_O - m_O}{m_O} \times 100(\%) \qquad (1)$$

where, m_O is the residual mass of the pristine processed polymer (this has no interface) at the temperature of maximum weight loss (T_{max}) and $m(x)_O$ is the mass of polymer composite with x wt% of MWNTs at T_{max}. Figure 1b shows the development of interface mass as a function of MWNT content. It is clearly observed that the interface mass is higher for 45min processed samples and increases with nanotube content. Increasing processing (shear) time increases the nanotube-polymer interactions therefore resulting in greater opportunities for bonding and/or wrapping of nanotubes by polymeric chains. Thus more and more polymer observed at T_{max} for all composites at longer processing (45 min) compared to shorter processing (11 min).

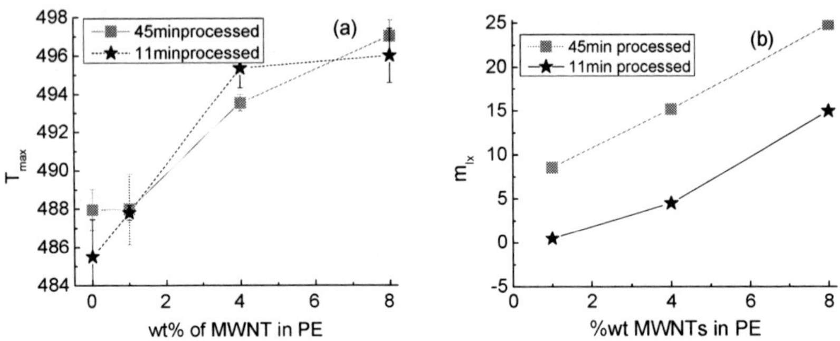

Fig. 1. (a) Variation T_{max} and (b) interface mass (m_{ix}) as a function of MWNT concentration in PE.

DMA was used to study the viscoelastic behavior of polymer materials over a wide range of temperatures. Material characteristics parameters such as storage modulus, loss modulus and mechanical damping (tanδ) can be extracted from the DMA graphs, providing useful information of the polymer-nanotube interactions.

The variation of storage modulus as a function of temperature is presented in Figure 2a and 2b for PE and MWNT/PE composites processed 11 min and 45 min respectively. These curves show that storage modulus of the composites are higher than those of neat polymer especially at lower temperature, demonstrating the reinforcing effects of nanofillers. This trend of higher modulus continues even at high temperatures (see inset Fig.1). A comparative account of variation of storage modulus at 25 °C is summarized in Table 1. It is evident that there is an increase in the storage modulus of PE with MWCNT. Increasing processing time results in a small increase in modulus.

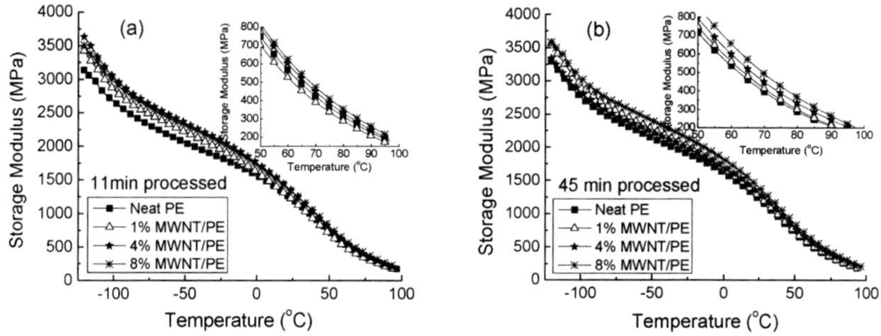

Figure 2. Variation of storage modulus with temperature of neat PE and MWNT-PE composites processed for (a) 11 min and (b) 45 min.

Table. 1. Storage modulus (E') in MPa of PE and MWNT/PE at 25 °C for different processings.

	Storage modulus (11 min)	% change in E'	Storage modulus (45 min)	% change in E'
Neat PE	1211		1215	
1% MWNT/PE	1201	-0.01	1269	4.4
4% MWNT/PE	1324	9.3	1310	9.1
8% MWNT/PE	1314	8.5	1398	14.0

Note: values of % change in E' are measured with respect to value of neat PE.

Loss modulus (E") was also evaluated. Figure 3 presents the loss modulus curves of neat PE and MWNT/PE composites for the 11min and 45min processed composites. High density polyethylene shows α and γ relaxations at about 59 °C and -110 °C respectively [11]. The α relaxation is associated with chain segment mobility in the crystalline phase due to reorientation of defects in crystals. It is clearly observed that the inclusion of nanotubes in polymer substantially increases the loss modulus (α relaxation) in both types of composites when compared to neat PE (see table 2). When compared the processing effects, the 45 min processed samples have higher viscous flow. Given the higher processing times, a better dispersion of nanotubes is expected.

Table.2. Loss modulus (MPa) corresponding to α (at 49 °C) transition.

Samples	11min processed		45min processed	
	E"(α)	% Change	E"(α)	% Change
Neat PE	116		114	
1% MWNT/PE	114	-1.7	119	4.4
4% MWNT/PE	124	6.5	118	5.6
8% MWNT/PE	127	9.5	134	17.2

Note: values of % change in E" are measured with respect to value of neat PE.

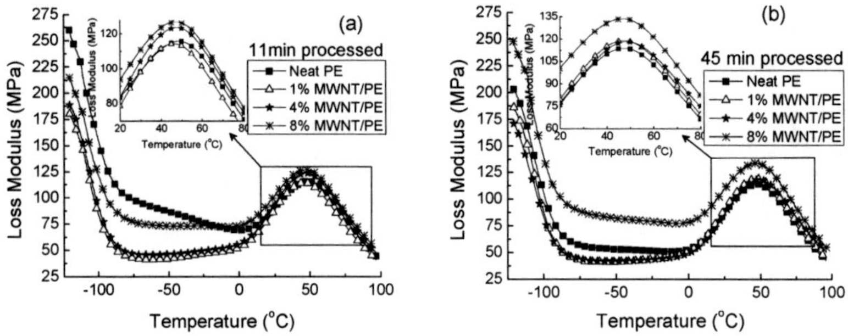

Fig. 3. Variation of loss modulus with temperature of neat PE and MWNT-PE composites after (a) 11 min and (b) 45 min processings.

CONCLUSION

Thermo-mechanical behavior of neat PE and MWNT/PE composites were studied. Thermal stability increased with the addition of nanotubes irrespective of the processing used. Mechanical behavior found process dependant. The 45min processed composites provided better mechanical properties when compared to the 11min processed composites. Better mechanical behavior could be explained due to the development of stronger interface between the nanotube and the polymer matrix with increasing processing times.

ACKNOWLEDGEMENTS

Financial support from National Science Foundation (NSF) under grant number DMR 0606224 is greatly acknowledged.

REFERENCES

1. D. Blond, V. Barron, M. Ruether, K. P. Ryan, V. Nicolosi, W. J. Blau, and J. N. Coleman, Enhancement of Modulus, Strength, and Toughness in Poly(methyl methacrylate)-Based

Composites by the Incorporation of Poly(methyl methacrylate)-Functionalized Nanotubes, Adv. Adv. Funct. Mater. **16**, 1608-1614 (2006).

2. K. Lozano, S. Yang and R. E. Jones, Nanofiber toughened polyethylene composites, Carbon **42**, 2329-2331 (2004).
3. B. Yang, J. Shi, K. P. Pramoda and S. H. Goh, Enhancement of stiffness, strength, ductility and toughness of poly(ethylene oxide) using phenoxy-grafted multiwalled carbon nanotubes, Nanotechnology **18**, 125606-125612 (2007).
4. R. B. Thompson, K. Ø. Rasmussen, and T. Lookman, Origins of Elastic Properties in Ordered Block Copolymer/Nanoparticle Composites, Nano Lett., **4**, 2455-2459 (2004).
5. M. S. Dressaulhaus and H. Dai, Carbon nanotubes: continued innovations and challenges, MRSbulletin/April 2004, 237-244.
6. M. M. J. Treacy, T. W. Ebbesen, and J. M. Gibson. Exceptionally high Young's modulus observed for individual carbon nanotubes. Nature **381**, 678-680 (1996).
7. J. N. Coleman, U. Khan, W. J. Blau and Y. K. Gun'ko, Small but strong: A review of the mechanical properties of carbon nanotube–polymer composites, carbon **44**, 1624-1652 (2006).
8. R. E. Gorga and R. E. Cohen, Toughness enhancements in poly(methyl methacrylate) by addition of oriented multiwall carbon nanotubes, J. Polym. Sci., Part B: Polym. Phys. **42**, 2690-2702 (2004).
9. R. Haggenmueller, H. H. Gommans, A. G. Rinzler, and J. E. Fischer, Aligned single-wall carbon nanotubes in composites by melt processing methods, K. I.Winey, Chem. Phys. Lett. **330**, 219-225 (2000).
10. M. D. Chipara, K. Lozano, A. Hernandez, and M. Chipara. TGA analysis of polypropylene-carbon nanofibers composites. Polym. Degrad. Stab. **93**, 871-876 (2008).
11. B. John, K. T. Varughese, Z. Oommen, P. Potschke, and S. Thomas, Dynamic Mechanical Behavior of High-Density Polyethylene/Ethylene Vinyl Acetate Copolymer Blends: The Effects of the Blend Ratio, Reactive Compatibilization, and Dynamic Vulcanization, J. Appl. Polym. Sci. **87**, 2083-2099 (2003).

Mater. Res. Soc. Symp. Proc. Vol. 1312 © 2011 Materials Research Society
DOI: 10.1557/opl.2011.333

PS-TiO_2 Nanocomposites: Thermal Investigations

Rafael Villegas[1], Yun Zhai[2], Hailan Xu[2], Dorina Magdalena Chipara[3], David Hui[2], Karen Lozano[1], and Mircea Chipara[3]
[1]Mechanical Engineering, The University of Texas Pan American, Edinburg, TX, United States.
[2]Mechanical Engineering, University of New Orleans, New Orleans, MO, United States.
[3]Physics and Geology, The University of Texas Pan American, Edinburg, TX, United States.

ABSTRACT

Nanocomposites of polystyrene loaded with various amounts of anatase (ranging between 0 % wt. and 20 % wt.) have been synthesized and investigated by thermal analysis. The research was focused on the simulation of Thermogravimetric Analysis data, aiming to a refined understanding of the interactions between of polystyrene and TiO_2 nanoparticles. The dependence of the first derivative of residual mass on temperature has been used to determine more accurately the temperature at which the mass loss is maxim. Several functions have been used to simulate the dependence of the first derivative of mass loss on the temperature of the sample. The highest correlation coefficient was obtained for the asymmetric Gaussian combination, which connects two halves of a Gaussian line with different linewidth. An increase of the thermal stability of the polymeric matrix upon loading with TiO_2 is reported.

INTRODUCTION

Thermogravimetric Analysis (TGA) examines the dependence of the mass of the sample on temperature, in a controlled atmosphere. TGA spectra of polystyrene (PS) and polystyrene-TiO_2 (PS-TiO_2) nanocomposites in nitrogen atmosphere are single sigmoids, characterized by a single inflection point. The mass of the sample starts to decrease slowly; as the temperature is increased the mass loss rate becomes larger and larger until the inflection temperature (T_i) is reached. At the inflection point, the mass loss rate (m_i) is highest. As the temperature exceeds T_i, the mass loss starts to decrease and eventually becomes zero (when all the mass of the sample is volatilized). Such dependencies have a sigmoid like character; Avrami like equation [1] can be used to describe the time evolution of the mass of a polymer or polymer-based nanocomposites.

$$m(T) = m(T_0) + Ae^{-C(T-T_0)^n} \qquad (1)$$

where m(T) is the residual mass at a temperature T, $m(T_0)$ is a fitting constant associated to the position of the zero line, A is the amplitude of the degradation process (amount of vaporized polymer), T_0 represents the temperature at which degradation starts, C is a constant connected to the volatilization rate, and n a parameter. Avrami equation has been frequently used to investigate crystallization processes [1]. In such cases, n was ranging between 0.5 and 4and C was positive and smaller than 1.

Typically, the filling of polymeric matrices by nanoparticles results in the formation of a polymer-nanoparticle interface. Due to the huge surface area of nanoparticles, the fraction of polymer belonging to this interface can be appreciable and consequently the physical properties of the polymeric matrix modified [2]. In most cases, this interface increases the stability of polymeric matrices shifting the inflection temperature towards higher values. The increase of the concentration of nanoparticles enhances the total surface of nanoparticles, confining more and more polymer within the interface. A saturation point is expected, where (almost) all polymer is interacting with the nanofiller (i.e. is confined within the interface).

TGA allows also the recording of the mass derivative as a function of temperature [2]. Such dependence presents a single maxim at the inflection point of the mass versus temperature dependence. A detailed analysis reveals that this derivative is not perfectly symmetric, while the Avrami equation (1) describes symmetrical sigmoids. This justifies the effort to find better functions to simulate the dependence of the mass derivative on temperature.

EXPERIMENT

TiO_2 nanoparticles with a dominant anatase structure, spherical morphology, and an average particle size of about 15 nm have been purchased from Nanostructured and Amorphous Materials Inc. and used as received. The polymeric matrix (atactic polystyrene) was purchased from Sigma-Aldrich. PS-TiO_2 nanocomposites loaded with various amounts of anatase ranging between 0 % wt. and 11 % wt., have been obtained by dissolving PS in cyclohexane (CH), who is a theta solvent at room temperature. Consequently, the interactions between the polymeric segments are balanced by the interactions with solvent's molecules. TiO_2 nanoparticles have been added to the PS-CH solution and the as obtained mixture was sonicated at 1 kW for 100 minutes by using a Hielscher sonicator. This process aimed to a homogeneous distribution of nanoparticles within PS. As no significant shifts in the glass transition temperature of polystyrene was noticed after the sonication of PS-CH solution for 100 minutes, it was concluded that PS is not degraded by sonication. The as obtained homogeneous solution was cast on microscope glass slides. It was noticed (from TGA data) that the solvent was not fully removed after storing the samples in air, at room temperature for a week. Consequently, an additional treatment of PS-TiO_2 films in a vacuum oven at 150 °C, for 3 hours, has been performed. TGA data confirmed the complete removal of the solvent after this additional thermal treatment, within an ± 1% accuracy. TGA measurements have been performed in nitrogen atmosphere by using a TGA Q 500 instrument. The as obtained thermograms (residual mass versus temperature and mass derivative versus temperature) were fitted by using Origin 8.1 Pro (both built-in functions as well as user defined functions) [3].

DISCUSSION

The thermal degradation (depolymerization) of PS is well known [4]. The effect of nanometer-sized filler on the thermal stability of PS can be easily assessed. In order to keep the system as simple as possible, the research was focused on cubic TiO_2 nanoparticles (anatase) with an edge of about 15 nm dispersed within a fully amorphous polymer (atactic PS). The evolution of the mass of PS-TiO_2 composites, as a function of temperature is shown in Fig. 1. As noticed from Fig. 2, the actual thermograms can be fitted with a good accuracy by Avrami-like expressions (see eq. 1). However, after fitting all experimental data shown in Fig. 1, it was concluded that the best fit is obtained for parameters ($m(T_0)$, T_0, a, c, and n) that do not have physical significance. An additional complication was originating from the fact that the hypersurface of the Correlation Coefficient as a function of the above mentioned parameters had a complex landscape with multiple local minima. A Monte Carlo approach is more appropriate for such situations. Local minima around T_0=50 °C (room temperature) and near T_i have been investigated in detail. For T_0<120 °C, no agreement between theoretical predictions and experimental data was observed. Systematically, the fit forces T_0 values that are above 250 °C and smaller than T_i. The estimated values for n and C are consistent with unreasonably low T_i values (ranging from 64 to 123 °C). As noticed from Fig. 1, T_i appears to be higher than 350 °C (for all samples) increasing as the concentration of nanoparticles is increased (at variance with the parameters corresponding to the best fit by using eq. 1). The observed shift of T_i is relatively large (over 20 °C for 11 % wt. TiO_2.), demonstrating a thick PS-TiO_2 interface.

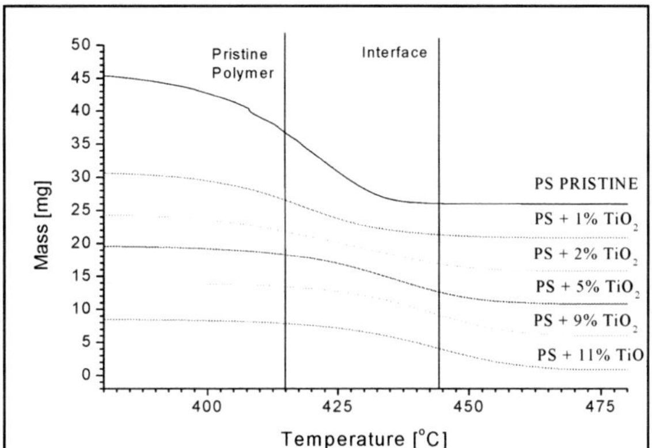

Figure 1. TGA thermograms of PS loaded with various amounts of TiO_2 [% wt.]

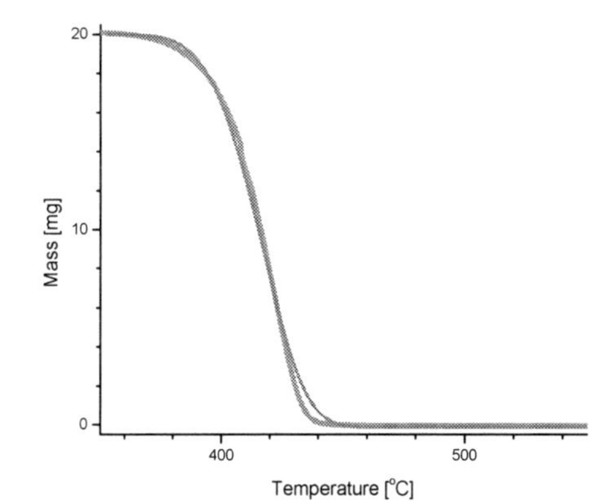

Figure 2. TGA thermograms of pristine PS. Light line represents the experimental data and the dark line the best fit obtained using eq. 1.

The temperature dependence of the mass derivative (which is the first derivative of the residual mass versus temperature) can be described with an acceptable accuracy by using symmetrical equations such as simple Gaussian (eq.2) and Lorentzian (eq. 3) line shapes. The inflection point of the mass versus temperature dependence is converted into an extreme point in the mass derivative versus temperature dependence.

$$m_d(T) = m_{d0G} + A_G e^{-\frac{(T-T_{iG})^2}{W_G^2}} \quad (2) \qquad m_d(T) = m_{d0L} + \frac{A_L W_L}{W_L^2 + (T-T_{iL})^2} \quad (3) \qquad m_d(T) = \frac{dm(T)}{dT}$$

where m_{d0G} adjust the zero, A_G and A_L are the amplitudes of mass loss rates, T_{iG} and T_{iL} are the temperature at which the mass loss is maxim, W_G and W_L describe the width of the dependence mass derivative m_d versus temperature. The subscripts G and L identifies the Gaussian and Lorentzian lineshapes. While this approach results in reasonable values for the extreme (inflection) temperatures, it fails to explain the asymmetry of experimental data.

Based on the fact that either Lorentzian or Gaussian lineshapes describe with an acceptable accuracy the dependence of m_d on temperature, a convolution like line shape (Voigt-see eq.4) was considered to fit the asymmetrical data.

$$m_d(T) = m_{d0V} + A_V(2\alpha/\pi)\frac{W_L}{W_L^2 + 4(T-T_{iV})^2} + (1-\alpha)\sqrt{\frac{4\ln 2}{\pi W_G}}e^{-\frac{4\ln 2}{W_G^2}(T-T_{iV})^2} \quad (4)$$

where α is a fitting parameter and the subscript V identifies the Voigt line shape.

Asymmetrical lineshapes have been simulated in spectroscopy by using various functions. The Breit-Wigner-Fano (BFW - see eq. 5) lineshape has been frequently used to simulate asymmetrical Raman spectra:

$$m_d(T) = m_{doBWF} + A_{BWF}\frac{1 + p(\frac{T-T_{iBWF}}{W_{BWF}})}{[1 + (\frac{T-T_{iBWF}}{W_{BWF}})^2]} \quad (5)$$

Where the fitting parameter p describes the deviation from the Lorentzian lineshape. An asymmetric lineshape is typically associated with a non constant width; i.e. the width of the left side is not identical to the width of the right side of the extreme (inflection) point. Accordingly, the dependence of the mass derivative on temperature has been fitted by asymmetrical Gaussians and BWF lineshapes (see eq. 6 and 7):

$$m_d(T) = \begin{cases} m_{d0G2} + A_{GL}\exp-\frac{(T-T_{iG})}{w_{GL}^2} & T < T_{iG} \\ m_{d0G2} + A_{GR}\exp-\frac{(T-T_{iG})}{w_{GR}^2} & T \geq T_{iG} \end{cases} \quad (6) \qquad m_d(T) = \begin{cases} m_{doBWF2} + A_{BWFL}\frac{1 + p_L(\frac{T-T_{iBWF}}{W_{BWFL}})}{[1 + (\frac{T-T_{iBWF}}{W_{BWFL}})^2]} & T < T_{iBWF} \\ m_{doBWF2} + A_{BWFR}\frac{1 + p_R(\frac{T-T_{iBWF}}{W_{BWFR}})}{[1 + (\frac{T-T_{iBWF}}{W_{BWFR}})^2]} & T > T_{iBWF} \end{cases} \quad (7)$$

where L and R identifies the left and right sides of the lineshape. Tentatively, a continuous and monotonous modification of the linewidth across the lineshape was considered and embedded within a extended BWF equation (see eq. 8).

$$m_d(T) = m_{d0BWFE} + A_{BWFE}\frac{[2W_{BWFE}e^{-\alpha T} + p_0(T-T_{iBWFE})]}{[W_{BWFE}^2 e^{-2\alpha T} + 4(T-T_{iBWFE})^2]} \quad (8)$$

Where α describes the dependence of the line width on T. For p_o=0 and α=0 eq. 8 degenerates into a symmetrical Lorentzian line. Fig. 3 shows the best fits obtained by using the above

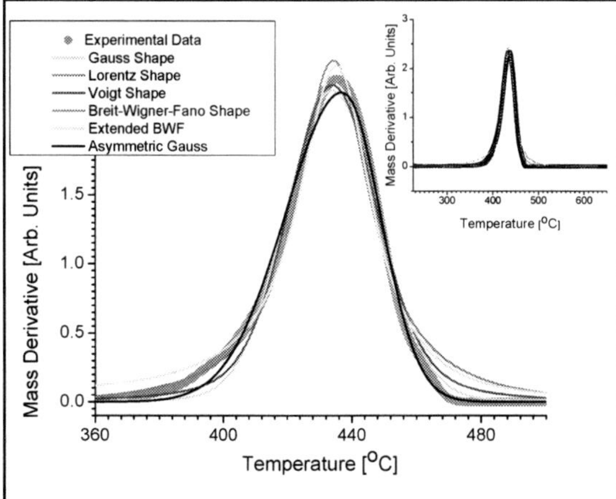

Figure 3. Mass derivative versus temperature for pristine PS. Experimental data (thick lines) and best fits (thin lines)

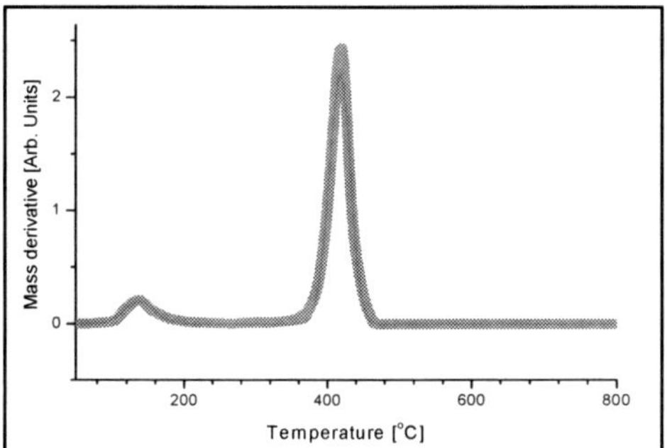

Figure 4. Thermogram of pristine PS containing solvent (CH) residues

mentioned equations for pristine PS. Similar fits have been obtained for each PS-TiO_2 sample. For the fit of Gaussian, Lorentzian, and Voigt line shapes the built-in functions of Origin 8.1 PRO have been used [3]. All other lineshapes were fitted by using the user-defined function capability, in C.

The analysis of the resonance lineshape revealed that the overall degradation process can be described by several parameters such as the temperature at which the mass loss rate is maxim. the maxim value of the mass loss, the width of the mass derivative dependence on temperature and the asymmetry parameter, defined as the ratio between the left and the right widths. Unexpectedly, the continuous variation of the linewidth (as exp-αT) did not result in an improved agreement between experiment and fitting function.

Samples of PS-TiO_2 containing residual amounts of solvent (CH) have been measured by TGA. Figure 4 depicts the temperature dependence of the mass derivatives (thick line) and the best fit obtained by connecting two Gaussians. The fit is excellent, demonstrating that equation (6) can be used even for two competing processes. The fitting reveals that the peak related to solvent evaporation is completely symmetric whereas the proper thermal degradation peak has an asymmetry (i.e. ratio between the left width and right width equal to 1.18±.05 °C) significantly larger than experimental errors. Previous TGA data on isotactic polypropylene filled with vapor grown carbon nanofibers demonstrated that the width of mass derivative versus temperature reflects the changes in the thermal conductivity of the sample. Based on TGA data, we suggest that the width of mass derivative versus temperature is affected by diffusion processes, thermal conductivity, and energy barriers. Hence, a complete analysis of TGA data should include the extreme (inflection) temperature, the mass loss at the extreme (inflection) temperature, the asymmetry ratio, and the width of the recorded line shape.

CONCLUSIONS

The best fit of the mass derivative dependence on temperature has been obtained for the double Gaussian dependence (see eq.6). This demonstrates that the shape asymmetry implies different linewidth on the left and on the right of the extremum temperature T_i. While thermal degradation processes that involve complex chemical reactions were characterized by asymmetric lineshape, the thermal evaporation of residual solvent was described by a symmetric lineshape. This suggests that the possibility to determine the width and the asymmetry of thermograms may add additional information about the overall thermal degradation process. It was speculated that the liewidth includes information about the thermal conductivity of the sample and diffusion constants.

ACKNOWLEDGMENTS

This research has been supported by DARPA under grant HR0011-08-1-0084 to AMRI-University of New Orleans. The research done at UTPA was supported by NSF PREM (DMR 0934157).

REFERENCES

1. Arnaldo T. Lorenzo, Marıa Luisa Arnal, Julio Albuerne, Alejandro J. Muller, *Polymer Testing* **26**, 222–231 (2007).
2. M. D. Chipara, K. Lozano, A. Hernandez, M. Chipara, *Polymer Degradation and Stability*, **93**, 871-876 (2008).
3. *http://www.originlab.com/*
4. M. Chipara, D. Barb, M. Cojocaru, M. Rogalski, M. D. Chipara, D. Tudor, *Polymer Degradation and Stability* **37**, 61-76 (1992).

Mater. Res. Soc. Symp. Proc. Vol. 1312 © 2011 Materials Research Society
DOI: 10.1557/opl.2011.161

Image Analysis Optimization for Quantifying Nanoparticle Dispersions in Polymer-based Nanocomposites Using Transmission Electron Microscopy (TEM)

Anand S. Badami, Mark W. Beach, Stewart P. Wood, Steven J. Rozeveld, and William A. Heeschen

The Dow Chemical Company, 1897 Building,
Midland, MI 48667

ABSTRACT

When comparing large numbers of TEM micrographs of insoluble additives in polymer-based nanocomposite systems, the ability to determine or estimate the dispersion quality (i.e. uniformity of size and/or spatial distribution) is often difficult. The objective of this study was to develop a method to quantify dispersions observed in TEM micrographs that enables both a numerical "ranking" to be assigned to individual dispersions as well as tabulation a multitude of images acquired over time. Several methods were reviewed and applied to a set of TEM dispersion images of an insoluble additive in polystyrene. Projected area diameter, particle area, and Euclidean distance between particle centroids were chosen from all the particle size distribution and spatial distribution parameters present in the literature, but none successfully yielded a quantitative indicator of dispersion quality for the micrographs. In contrast, generating cumulative volume percent curves for each sample appeared to be a preferred method of quantifying and comparing dispersions in TEM micrographs. The volume diameter values obtained by this method can be used for "ranking" and tabulation of dispersion quality and account for both "good" additive dispersions (i.e. those with small domains of a narrow size range around 1 μm or less) and "bad" additive dispersions (i.e. those with non-uniform domains ranging in size by several microns or more). As a result, the numerical values generated by this method can be used to quantitatively determine correlations between the dispersion quality of nanoparticles in polymer-based nanocomposite materials and various macroscale physical and/or performance properties of such materials. This method's precision was statistically determined to decrease with increasing particle size and be heavily dependent on representative sampling.

INTRODUCTION

Chemical additives such as polymer-based nanoparticles are often added to extend and modify the performance of many polymers. TEM images provide an analyst with a general visual interpretation of nanoparticle dispersion but lack a quantifiable value that could be used to compare to other TEM images to rate dispersion quality. This is especially true when reviewing multiple images collected over an extended period of time, as it becomes very difficult to interpret multiple images based solely on visual observations. It is also an issue when the analyst is trying to establish correlations between dispersion and other parameters, as the need for a specific quantitative parameter that relates to dispersion is required.

Very few studies focused on size distribution quantification of TEM micrograph dispersions exist in the literature. Most quantification is focused on optical micrographs or scanning electron micrographs. The methods of particle sizing and particle size distribution for these microscopy methods are largely applicable to TEM, however, and will be discussed here.

The first step in dispersion quantification first is to accurately measure particle size within the dispersion. An infinite number of linear dimensions, called statistical diameters, can be measured for any given particle in a dispersion when a chord length is randomly drawn across it. The preferred statistical diameter for use in particle sizing is projected area diameter, or equivalent circular diameter, because its measurement considers two dimensions of a particle [1,2]. It is defined as the diameter of a circle having the same projected area as the particle.

The second step in quantifying dispersion is generating a particle size distribution. A particle size distribution can be based on one dimension of the particles (e.g. a distribution by number based on diameter), two dimensions of the particles (e.g. a distribution based on area), or three dimensions (e.g. a distribution by mass or volume). The distribution is created by dividing the size range into classes and calculating the frequency of particles in each class. Number distribution, the easiest method by which to analyze size distribution via microscopy, is used commonly in the literature to quantify particle size [3]. Another quantifiable parameter is degree of clustering, or spatial distribution, which can be quantified by several methods including the random Voronoi polygon (RVP) model, also known as Dirichlet tessellation [4,5], average interparticle spacing (mean free path) [6,7], nearest and near neighbor distance distribution [6], quadrat method [6,8], radial distribution function [6,9], and Euclidean distance between centroids [10]. Although not as accurate as some of the other particle spatial distribution parameters, Euclidean distance between centroids is both computationally more simple and also the most inclusive method by which to measure interparticle spacing [5].

EXPERIMENT

Samples were blends of polystyrene with 3% (pph) of an insoluble nanoparticle additive sectioned into thin membrane slices approximately 50 nm thick by ambient ultramicrotomy and collected onto 400-mesh copper grids. Micrographs were obtained using a JEOL 1230 TEM.

Image analysis was conducted using ImageJ software, version 1.39q. Micrographs were analyzed as 8-bit images and threshold values were set to only highlight particles in their entirety and no two particles were grouped together as one. Particle area measurements included voids within particles and excluded any particles contacting an image edge. A macro was written within ImageJ to calculate interparticle distances and generate the related histogram data.

Particle area was used to calculate equivalent circular diameter (the diameter of a circle having the same area as the particle) which was used to calculate equivalent spherical volume (the volume of a sphere possessing the same diameter). Particle volumes were used to calculate a histogram of volume percent for each sample. A curve of cumulative volume percent versus equivalent diameter was obtained by sequentially summing the volume percent values with increasing diameter and used to determine D_v50 (the volume median diameter) and $D_v86.5$ (the diameter at which there is 86.5 percent of the total particle volume less than that size).

For precision calculations, data from multiple images of two samples were grouped together into series containing five sets each of 100, 500, 1000, and 5000 particles (i.e. five sets of 100 particles, five sets of 500 particles, etc.) for both samples. Image data was grouped to ensure that no particle was counted more than once within any of the four series. D_v50 and $D_v86.5$ values for each set were averaged with the same values from the other sets in the same particle count group (e.g. the five D_v50 values for all five 100 particle count sets were averaged) and standard deviation (SD) and relative SD (RSD) were calculated (RSD = SD/mean * 100%). Microsoft Excel was used for all cumulative volume percent related calculations.

DISCUSSION

Five TEM micrographs (figure 1) were selected for the purpose of evaluating potential dispersion quantification methods specifically because their dispersion qualities are easily distinguishable visually. Without much effort, one can describe the first and third images (1652, 1658) as "good" dispersions, the fifth image (1663) as a "bad" dispersion, and the second and fourth images (1653, 1659) as somewhere between good and bad.

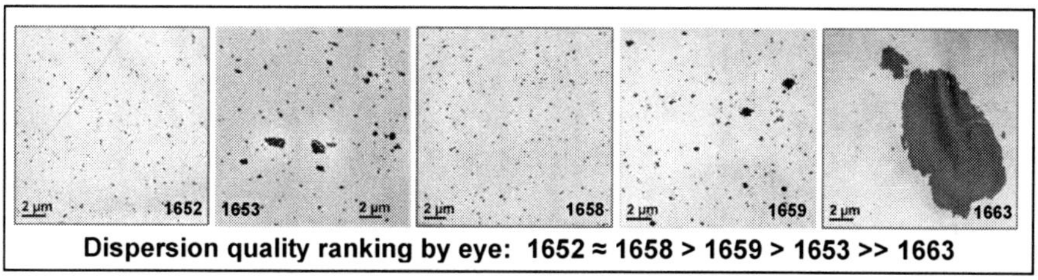

Figure 1. TEM micrographs of polymer-based nanoparticle dispersions with visible variations in dispersion quality.

The micrographs in figure 1 were quantified using projected area diameter (equivalent circular diameter), particle area, and Euclidean distance between particle centroids. Histograms of number of particles versus equivalent circular diameter (ECD; figure 2a), particle area versus ECD (figure 2b), and Euclidean distances between the centroids of all pairs of particles (figure 3) were generated for each of the five micrographs. The distributions of these linear and areal parameters largely overlapped, offering no single clear quantitative indicator of dispersion quality or any numeric value that would conclusively rank the samples as in figure 1.

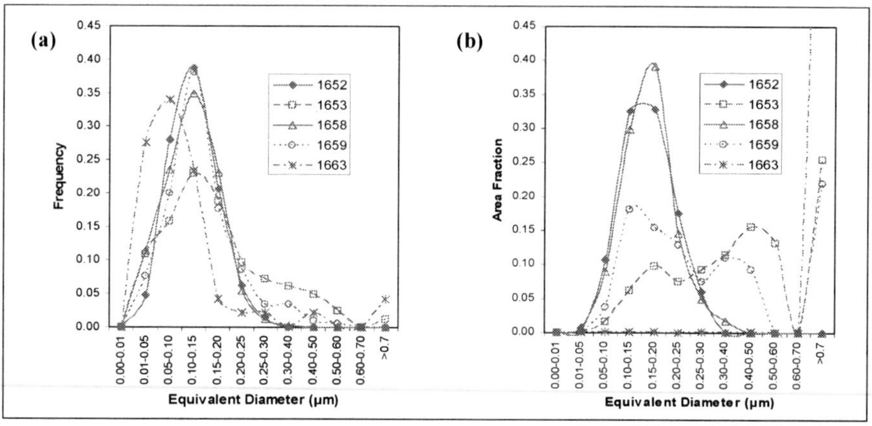

Figure 2. Histograms of (a) particle number versus equivalent circular diameter, and (b) particle area versus equivalent diameter for micrographs in figure 1.

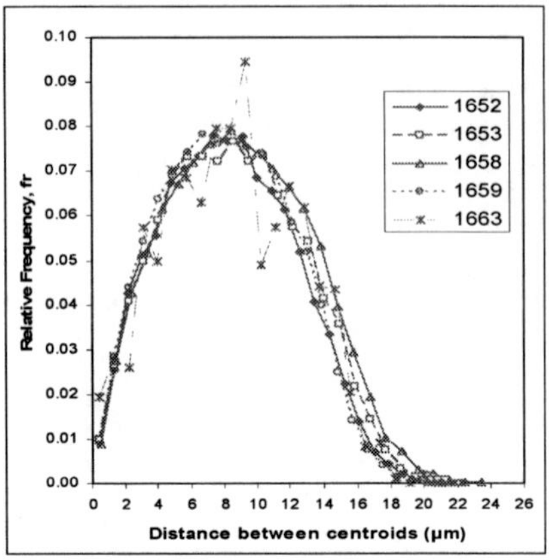

Figure 3. Histograms of Euclidean distances between particle centroids for figure 1 micrographs.

A quantitative indicator of dispersion quality was found by calculating cumulative volume percent curves [1] for each micrograph. These calculations are performed as follows. The first step is to generate a histogram of particle count versus equivalent diameter (Figure 4a).

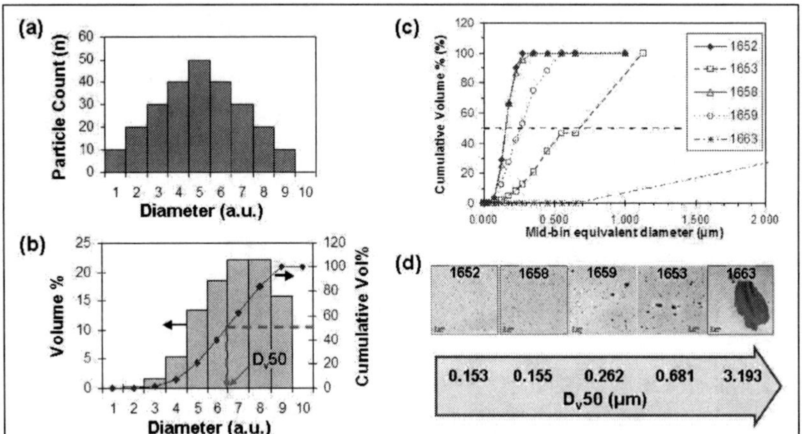

Figure 4. Cumulative volume percent method of providing a "ranking" of dispersions: (a) a histogram of particle number vs. particle diameter is generated, (b) the histogram in (a) is tranformed into a histogram of volume % vs. diameter from which a cumulative volume percent curve is generated. A volume diameter corresponding to 50 cumulative volume percent can be read from the curve, (c) cumulative volume percent curves are generated for the dispersions in figure 1, and (d) the D_v values obtained for each dispersion can be used to rank them by quality.

The second step is to calculate the volumes occupied by spheres having the same equivalent diameters as the particles. The volumes of all of these spheres is summed and used to generate a second histogram of volume percent versus equivalent diameter (figure 4b). The third step is to plot a curve of total particle volume percent versus equivalent diameter (figure 4b) by sequentially summing the volume percent values with increasing diameter. In the fourth and final step, a diameter corresponding to any cumulative volume percent can be read from this curve and used as a quantitative value for dispersion quality (e.g. D_v50, the diameter at 50 volume percent, is the diameter at which 50% of the total particle volume is below that diameter; figure 4b). Figure 4c displays the cumulative volume percent curves generated for each of the micrographs in figure 1. When the micrographs are placed in ascending order of the D_v50 values interpolated from each curve (figure 4d), the "good" dispersions gravitate to the left and the "bad" dispersions gravitate to the right, indicating that this method provides a way to quantitatively rank dispersions in the same way they would be ranked visually (as in figure 1). Volume diameter (D_v) varies inversely with dispersion quality. An added benefit of cumulative volume percent compared to the other quantification methods evaluated is that cumulative volume percent should correlate with the amount of particle loading (wt%) because particle volume is proportional to particle weight to the degree that particle density remains constant.

In order to evaluate the precision of the cumulative volume percent method, two samples displaying distinctly different "fine" and "course" dispersion were chosen. Multiple images of each sample were grouped into sets based on total number of nanoparticles. The mean and relative standard deviation of D_v50 and $D_v86.5$ values were calculated for multiple sets of the same size (figure 5). The precision of the cumulative volume percent method was found to decrease with an increase in particle size and is heavily dependent on representative sampling; this is due to the fact that there are very few of the largest particles present in the sample. If the sample population depicted in TEM micrographs does not accurately reflect the true population present in the sample, the diameters (i.e. D_v values) obtained from the cumulative volume percent curves will be, at best, prone to large error and, at worst, meaningless. The issue of representative sampling highlighted by this study underscores the need for an expeditious particle size screening method to precede TEM imaging and quantification.

Fine Dispersion

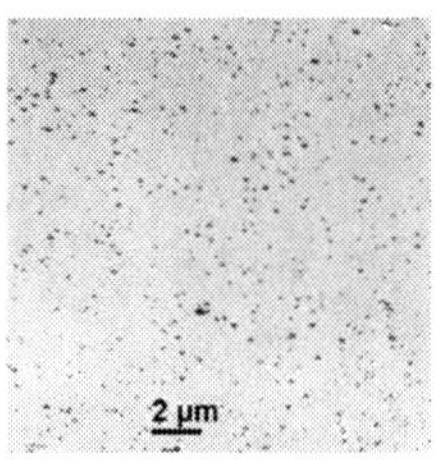

Sample #1787

Particles/set*	Dv50 mean (µm)	Dv50 Relative Std Dev. (%)	Dv86.5 mean (µm)	Dv86.5 Relative Std Dev. (%)
100	0.136	10.0	0.235	30.3
500	0.117	6.5	0.194	14.3
1000	0.121	3.4	0.196	7.6
5000	0.155	5.4	0.459	61.1

*Five sets of particles used for each set of calculations.

Coarse Dispersion

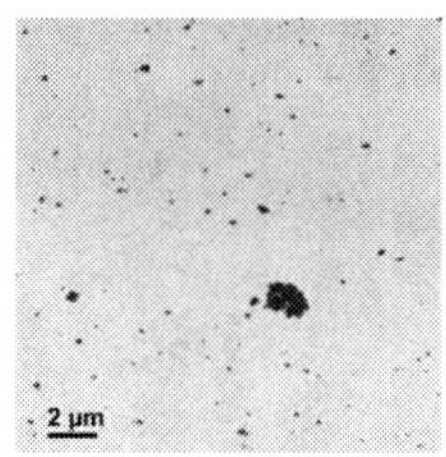

Sample #1788

Particles/set	Dv50 mean (µm)	Dv50 Relative Std Dev. (%)	Dv86.5 mean (µm)	Dv86.5 Relative Std Dev. (%)
100*	3.914	207.8	4.096	198.6
500	2.079	75.8	2.521	54.9
1000	6.909	105.2	24.853	186.6
5000	5.477	149.5	6.521	119.4

*Six sets used for these calculations. All others use five sets.

Figure 5. Results of precision calculations for cumulative volume percent method

CONCLUSIONS

Generating cumulative volume percent curves appears to be a preferred method of quantifying and comparing nanoparticle dispersions in TEM micrographs of polymer-based nanocomposites. The volume diameter values obtained by this method can be used for "ranking" and tabulation of dispersion quality. This method proved much more successful in quantifying dispersion quality than equivalent circular diameter, particle area, or Euclidean distance between particle centroids. The precision of the cumulative volume percent method is inversely related to particle size and is only as good as the degree to which the images analyzed represent the true sample because there are so few of the largest particles present in the sample.

REFERENCES

1. T. Allen, *Particle Size Measurement, Volume 1: Powder sampling and particle size measurement*, 5th ed. (Chapman and Hall, New York, 1997).
2. Ed Brewer and Arnold Ramsland, Journal of Pharmaceutical Sciences **84** (4), 499 (1995).
3. Nadine Gabas, Noël Hiquily, and Claude Laguérie, Particle and Particle Systems Characterization **11** (2), 121 (1994); Tatsu Nakano and Hiroshi Yuasa, International Journal of Pharmaceutics **215** (1-2), 3 (2001); Hiroshi Yuasa, Tatsu Nakano, and Yoshio Kanaya, International Journal of Pharmaceutics **158** (2), 195 (1997); Hiroshi Yuasa, Tatsu Nakano, and Yoshio Kanaya, International Journal of Pharmaceutics **178** (1), 1 (1999).
4. B. N. Boots, Metallography **15** (1), 53 (1982); P. Ganguly and W. J. Poole, Materials Science and Engineering A **332** (1-2), 301 (2002).
5. P. J. Wray, O. Richmond, and H. L. Morrison, Metallography **16** (1), 39 (1983).
6. P. A. Karnezis, G. Durrant, and B. Cantor, Materials Characterization **40** (2), 97 (1998).
7. E. E. Underwood, *Quantitative Stereology*. (Addison-Wesley, Reading, MA, 1970).
8. A. Rogers, *Statistical Analysis of Spatial Dispersions: The Quadrat Method*. (Pion, London, 1974).
9. B. D. Ripley, J. R. Stat. Soc. **B39**, 172 (1977).
10. N. Silva and A. Velhinho, Materials Science Forum **514-516**, 779 (2006).

Mater. Res. Soc. Symp. Proc. Vol. 1312 © 2011 Materials Research Society
DOI: 10.1557/opl.2011.125

Depth sensing indentation of nanoscale graphene platelets in nanocomposite thin films

Ardavan Zandiatashbar, Catalin R. Picu and Nikhil Koratkar
Department of Mechanical, Aeronautical and Nuclear Engineering, Rensselaer Polytechnic Institute, 110 8th Street, Troy, NY, 12180, U.S.A.

ABSTRACT

Significant improvement of mechanical properties was observed recently in graphene platelet-epoxy nanocomposites relative to unfilled epoxy, such as an increase of the fracture toughness by 50% and dramatic decrease of fatigue crack growth rate. In this work, thin films of 0.1 wt.% of graphene platelet (GPL) – epoxy nanocomposites were fabricated and the nanoscale mechanical properties of the nanocomposite were investigated by nanoindentation. This provides information about the presence of characteristic length scales induced by the microstructure and the strength of the filler-matrix interface.

INTRODUCTION

Carbon based nanoparticles attracted considerable attention recently as a result of their outstanding properties. Quasi-one dimensional carbon nanotubes and two dimensional graphene were used to enhance the mechanical properties of epoxy-based nanocomposites. In particular, significant toughening was observed in epoxy-based nanocomposites with graphene as nanoscale fillers [1-4]. The fracture toughness increases by approximately 20% and 50% upon the addition of only 0.1 wt% of carbon nanotubes and graphene platelets, respectively [5]. In cyclic loading the crack growth rate decreases by an order of magnitude relative to the unfilled epoxy [5,6].

The nanoindentation technique is widely used today to probe the local material behavior. As long as the film stiffness if smaller than that of the substrate and the indentation depth is kept smaller than a fraction of the film thickness (less than 10%), the measured response is considered to be characteristic for the properties of the film material [7-10].

In this article, results of nanoindentation tests performed on 0.1 wt. % GPL – epoxy nanocomposite and pure epoxy films are presented. The films are deposited on a silicon substrate by spin coating. Atomic force microscopy (AFM) is used to investigate the surface roughness of samples as well as to characterize the fillers located close to the upper surface.

EXPERIMENT

Samples preparation

Nanocomposite films with 0.1% weight fraction of nanofillers were prepared in a sequential manner. The graphene platelets used in this work were prepared by thermal reduction of graphite oxide [3]. Nanofillers were dissolved in 100-200 ml acetone per 0.1 gr of nanofiller and dispersed using an ultrasonic probe sonicator (Sonics Vibracell VC 750, Sonics and Materials Inc., USA) for 45 minutes, followed by adding epoxy resin (System 2000 Epoxy Resin, Fibreglast Inc., USA) to this suspension while continuing sonication. The solution was heated at 70°C and stirred using a Teflon-coated magnetic bar on a magnetic stir plate for 3 hours in order to evaporate the solvent. The procedure was followed by keeping the mixture in vacuum (-30mmHg) at 70°C for 12 hours. Hardening agent (2120 Epoxy Hardener, Fibreglast Inc., USA) was added with the ratio of 5.4:20 curing agent to resin after cooling down the mixture to room temperature (in order to prevent premature curing), and mixed using a high speed shear mixer (ARE-250, Thinky, Japan) for 2 minutes at 2000 rpm. To remove air bubbles formed during shear mixing, the mixture was degassed under vacuum for 20 minutes. The mixture was used immediately to coat a silicon substrate. This was achieved by spin coating at 8000 rpm for 40 seconds. The coated silicon samples were left at room temperature for 24 hours, followed by 4 hours post-cure at 90°C.

The cured samples were etched by immersion in acetone followed by immersion in isopropylene. The samples were then dried by blowing nitrogen to remove any solvent left on the surface.

Fillers characterization, surface roughness and thickness measurement

To characterize the nanofillers protruding from the surface as well as the surface roughness after etching, atomic force microscopy (AFM) was used. Figure 1 shows a cluster of graphene embedded in epoxy and emerging from the surface. Note that the roughness of epoxy away from the cluster is very small (few nm). The film thickness was measured by breaking the samples and observing their edge by SEM.

Nanoindentation

Nanoindentation tests were performed using a NanoTest® nanoindenter from Micro Materials Ltd. and utilizing a Berkovich diamond indenter [12].

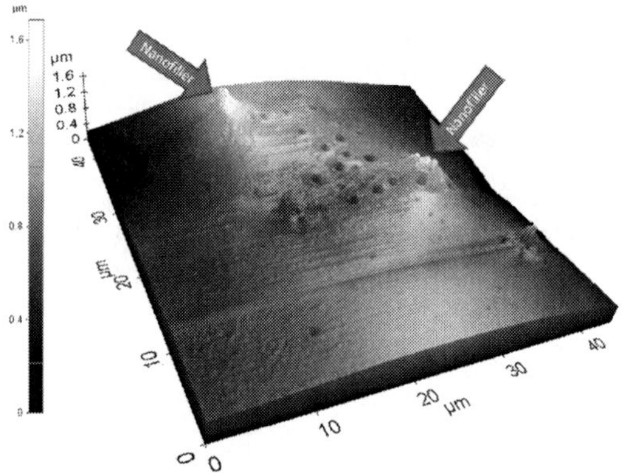

Figure 1. Atomic force microscopy (AFM) image of graphene platelets emerging from the film surface.

RESULTS AND DISCUSSION

Experiments were performed in load control with displacement limits, at 0.1 mN/sec loading and unloading rates. Two different load functions were used to capture possible time dependent behavior. In the single-cycle load function, the samples were loaded up to the depth of 1500 nm, followed by 10-second hold time and unloaded with a rate identical to the loading rate. In the multiple-cycle load function, multiple load-hold-unload cycles were used, with similar loading and unloading rates. At the end of each cycle samples were unloaded to 30% of the maximum load of the prior cycle. The indents were investigated by optical and atomic force microscopy.

As observed in the SEM of the sample cross-section, the film thickness is approximately 80 μm and hence is transparent even though opaque graphene clusters are present. This is opposed to bulk samples which are opaque. This makes possible the direct observation of nanofiller distribution in the film. The position of fillers in the through thickness direction can be estimated by adjusting the focus of the microscope. The accuracy of this measurement is defined by the microscope depth of focus, which was approximately 1 μm for the present instrument.

The surface roughness in regions where fillers do not protrude is very small (smaller than 10 nm). This roughness is similar to that measured for unfilled epoxy films prepared in similar conditions.

Nanoindentation was performed on both unfilled epoxy sample and 0.1 wt% GPL-epoxy samples. This filler volume fraction was selected since it was observed in previous work to lead to maximum toughness enhancement relative to the unfilled epoxy [1]. Figure 2a shows a set of 8 load – displacement curves for unfilled epoxy, both monotonic and cyclic. Each indent is performed at a different location, while an entire cyclic curve involving multiple loading, hold and unloading sequences is obtained by indentation at same location. The repeatability of the test on pure epoxy is excellent; the curves essentially overlap. This indicates that the heterogeneity length scale in unfilled epoxy is much smaller than the probing length scale, or the indentation size/depth. In fact, heterogeneity is expected to be observed on the scale of the molecular network (order of few nm). The maxima of the cyclic curve follow closely the monotonic curve.

In filled samples, indentation was performed close to and away from fillers that protrude from the surface of the sample. Figure 2b indicates that the material behavior in regions of the surface where fillers are not protruding is essentially identical to that of unfilled epoxy (Figure 2a). This conclusion is valid for indentation depths as small as 200 nm, which is the magnitude of the indentation depth of the first cycle in Figure 2b, and as large as 1.5 μm, which corresponds to the indentation depth of the last cycle. Clearly, the inherent heterogeneity of the material at higher depths does not seem to influence these results. Indenting very close to a protrusion (top curve in Figure 2b) leads to a significantly stiffer effective response. This is attributed primarily to the presence of a stiffer graphene cluster. These measurements do not provide evidence for the presence of a modified polymer layer (of interphase) whose existence was postulated for most filled polymers. In fact, in networks of high cross-link density, the presence of such modified layer is doubtful, except for the case when the crosslink density is affected by the presence of the filler. We conjecture that this is not the case in the present materials in which the graphene surface was not chemically modified to promote adhesion with epoxy. Surface functionalization of nanofillers (carbon nanotubes) was seen to lead to a dramatically different behavior of epoxy-based composites. [6]. The stiffer response obtained when probing close to fillers also indicates that the matrix-GPL interface is strong enough to transmit this level of traction.

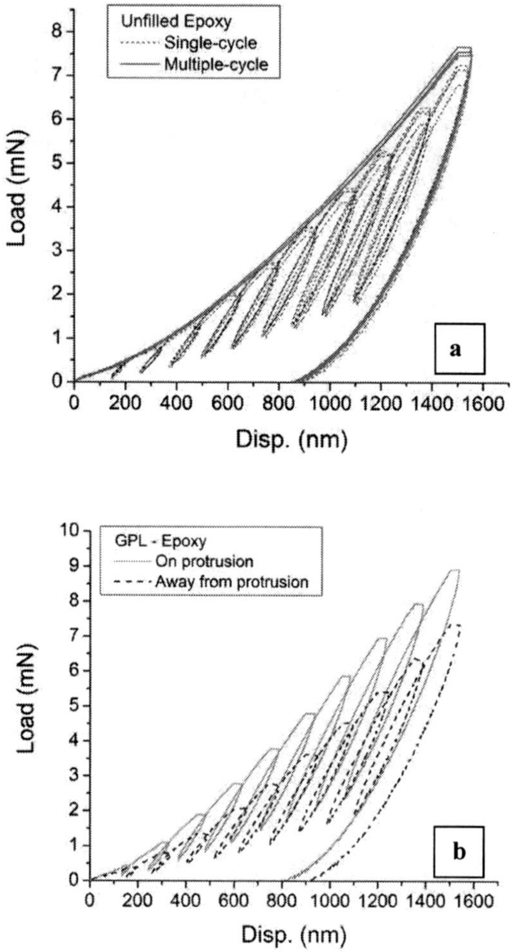

Figure 2. a) Load – displacement curves for unfilled epoxy and protrusion free regions on nanocomposite film, **b)** Load – displacement curves for indentation close to a protrusion in filled epoxy.

The cyclic curves indicate that both the unfilled and filled epoxy behave viscoplastically with similar dependence of the mechanical behavior on the deformation rate. In Figure 3 we plot the indenter displacement during hold periods of 10s each performed at the maximum load of each cycle (see Figure 2). The horizontal axis represents the load at which the hold was performed. Three sets of data are shown, corresponding to unfilled epoxy, and to nanocomposites probed at locations away and close to the graphene particle protruding from the surface. The creep response is identical in unfilled epoxy and nanocomposite probed away from the particle. Less creep is measured when indenting close to a filler.

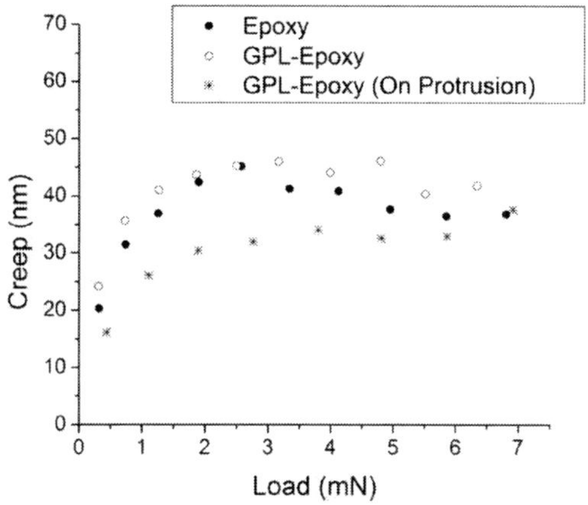

Figure 3. Indentation displacement during hold periods of 10s (creep) performed in unfilled epoxy and in the nanocomposite, away and close to a filler protruding from the film.

These results indicate that the length scale of heterogeneous materials properties in this nanocomposite is not different from that defined by the geometry/phase composition. The measurements do not reveal a modified polymer layer surrounding fillers, which, if present, would control the effective stiffness distribution in the material.

CONCLUSION

Nanoindentation tests were performed on samples of unfilled epoxy and GPL – epoxy nanocomposites. It is observed that there is no significant difference between the stiffness of unfilled epoxy and protrusion free regions; however, a stiffer response is observed when indenting at the location of a nanofiller cluster emerging from the film. A similar conclusion can be drawn regarding the effective creep rates measured during hold periods of a cyclic load function. Cyclic loading results indicate viscoplastic behavior of both unfilled epoxy and the nanocomposite. It can be also concluded that the fillers stick to the matrix well which results in smaller creep values.

ACKNOWLEDGMENTS

We gratefully acknowledge support from the NSF under grant CMMI 0900188 and from the ONR under grant N00014-09-1-0928.

REFERENCES

1. M.A. Rafiee, J. Rafiee, I. Srivastava, Z. Wang, H. Song, Z. -Z. Yu, N. Koratkar, Small 6, 179 (2010).

2. F.H. Gojny, M.H.G. Wichmann, U. Kopke, B. Fiedler, K. Schulte, Compos. Sci. Technol. 64, 2363 (2004).

3. M. A. Rafiee, J. Rafiee, Z. Wang, H. Song, Z.-Z. Yu and N. Koratkar, ACS Nano 3, 3884 (2009).

4. S. Stankovich, D.A. Dikin, G.H.B. Dommett, K.M. Kohlhaas, E.J. Zimney, E.A. Stach, R.D. Piner, S.T. Nguyen, R.S. Ruoff, Nature 442, 282 (2006).

5. W. Zhang, R.C. Picu, N. Koratkar, Appl. Phys. Lett. **91**, 193109 (2007); Nanotechnology **19**, 285709 (2008).

6. W. Zhang, I. Srivastava, Y.-F. Zhu, C.R. Picu, N. Koratkar, Small 5, 1403 (2009).

7. G.M. Pharr and W.C. Oliver, MRS. Bulletin **17** (7), 28-33 (1992).

8. R. Saha and W.D. Nix, Acta Materialia 50, 23 (2002).

9. M. Wang, K.M. Liechti, J.M. White, R.M. Winter, J. Mech. Phys. Sol. 52, 2329 (2004).

10. X. Li, and B. Bhushan, Mat. Char. 48, 11 (2002).

11. Micro Materials Ltd., Unit 3, The Byre, Wrexham Technology Park, Wrexham, United Kingdom.

Mater. Res. Soc. Symp. Proc. Vol. 1312
DOI: 10.1557/opl.2011.204

Supercapacitor Based on Graphene – Polyaniline Nanocomposite Electrode

Humberto Gomez[1, 4], Farah Alvi[2], Pedro Villalba[1, 4], Manoj. K. Ram[3], and Ashok Kumar[1, 3]
[1]Mechanical Engineering Department, University of South Florida, 4202 E. Fowler Ave, ENB 118 Tampa, FL, 33620 USA.
[2]Electrical Engineering Department, University of South Florida, 4202 E. Fowler Ave ,ENB 118 Tampa, FL, 33620 USA.
[3] Nanotechnology Research and Education Center (NREC), University of South Florida, Tampa, FL 33620 USA
[4] Departamento de Ingeniería Mecánica, Universidad del Norte, Barranquilla, Colombia.

ABSTRACT

This paper aims to investigate the concept of high power density and wider voltage window supercapacitors based on graphene-conducting polymer electrodes. The high conductivity, wider tunable potential window, stability of the electrode material in the doped form, faster charge transfer rate, and short charging times in graphene (G)-polyaniline (PANI) systems, contribute to the design and fabrication of high performance supercapacitors. Novel chemistries were used to synthesize a higher electronic conducting nanocomposite material based on graphene and polyaniline. The G-PANI nanocomposites were characterized, where an interesting structure was observed using different ratios of graphene with the aniline monomer. The G-PANI nanocomposite materials has shown the specific capacitance of 300 to 500 F/g at a current density of 0.1A/g. Based on the present experimental findings, it could be conclude that the G-PANI polymer nanocomposite supercapacitor technology could be transformative, and could surpass the existing technologies when the present synthesized novel material approach will be taken.

INTRODUCTION

Supercapacitors "electrochemical capacitors (ECs)" represent an attractive alternative energy source for portable electronics and automotive applications due to their high specific power and extended life. In fact, the growing demand of electronic portable systems and hybrid electric vehicles require high power in short-term pulses. In recent years, supercapacitors have been fabricated using activated carbon, ruthenium oxide, tantalum oxide, carbon nanotubes, graphene, and conducting polymers (polypyrroles, polythiophenes and polyanilines) due to their environmental stability and electrochemical properties **[1]**. Success has been achieved using nanocrystalline ruthenium and ruthenium oxides, but the high cost of ruthenium has resulted in a search for other metal oxides and nitrides because the supply of ruthenium is limited in the world. High power supercapacitors are a challenge to build from simply using conducting polymers because the exhibition of poor stabilities during the charge/discharge process, whereas the activated carbon and carbon nanotubes capacitance is limited due to their microstructures. The carbon nanotubes-polyaniline composite have been tested as supercapacitor electrodes with high capacitances and improved stability. Graphene, which is a two-dimensional honeycomb lattice structure of graphite, has exhibited unusual and intriguing physical, chemical and mechanical properties, and recently, have been used in supercapacitor devices **[2]**.

With the knowledge that graphene-conducting polymer have similar active surface compared with carbon, it is worth to investigate this nanocomposite as a potential candidate for capacitor technology, particularly with the additional benefits deriving from their oxidation and reduction at the molecular level. Such characteristics contribute to the capacitance of the device, which is typically due to the double layer only in carbon capacitors. Another important issue at the core of capacitor technology is the electrode material processability. Graphene-conducting polymer can be easily and efficiently obtained in the form of thin films or coatings directly onto the current collector. This results in lowering the current collector/electrode interfacial resistance. In this work, it was possible to dissolve the G-PANI nanocomposie at two different graphene – monomer (aniline) molar ratios, in N-Methyl 2- pyrrolidinone to make the film on graphite surfaces. In this work, the novel graphene (G)-polyaniline (PANI) nanocomposite material have been synthesized for the use in supercapacitor applications. A specific capacitance of 300 to 500 F/g at a current density of 0.1A/g is observed for G-PANI nanocomposite electrodes.

SYNTHESIS AND CHARACTERIZATION

The synthesis of graphene – polyaniline (G-PANI) was conducted according to Gomez et al **[3]**, kept at different aniline to graphene ratios, and added in a 1M HCl solution. The conducting polymer nanocomposites were characterized using Fourier Transform Infrared (FTIR), UV-Vis, and Raman spectrophotometer. Figure 1 shows the FTIR of Graphene-PANI (undoped) deposited in a silicon substrate where stretching bands from N-H (3410 cm^{-1}), C=C (1596 and 1514 cm^{-1}), emeraldine base (1319 cm^{-1}), and N=Q=N where Q represents the quinoid ring (1171 cm^{-1}) are depicted **[4]**. The characteristic bands of graphene for C=O and C-C appears to be shifted (1668 and 1008 cm^{-1}) possibly due to its incorporation in the polyaniline as a composite.

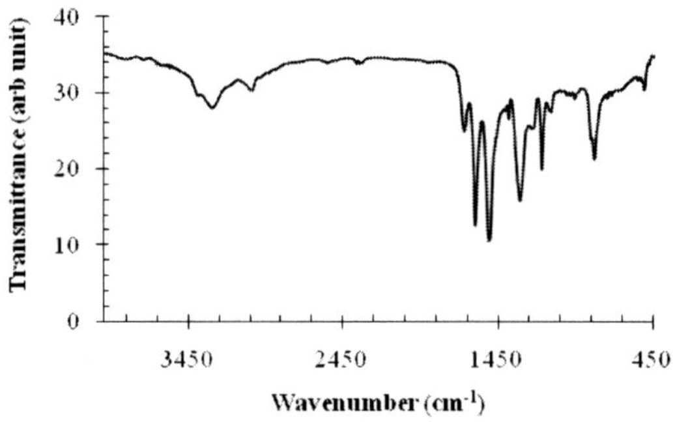

Figure 1. FTIR spectra of G-PANI (undoped)

Raman studies on G-PANI were conducted to derive information regarding the Raman Shift and absorption bands of inter- and/or intra-gap states. Figure 2 shows the Raman spectroscopy of the G-PANI-1 (1:1 ratio) and G-PANI-2 (1:2) nanocomposite systems, and polyaniline films deposited on silicon substrates. Raman spectroscopy shows typical bands for polyaniline systems which are shifted in G-PANI-1 and G-PANI-2 due to the incorporation of the graphene in the polymer structure. The graphene does not only make the nano-composite material but also could be acting as a doping for polymeric film as supported by the data shown in the following studies.

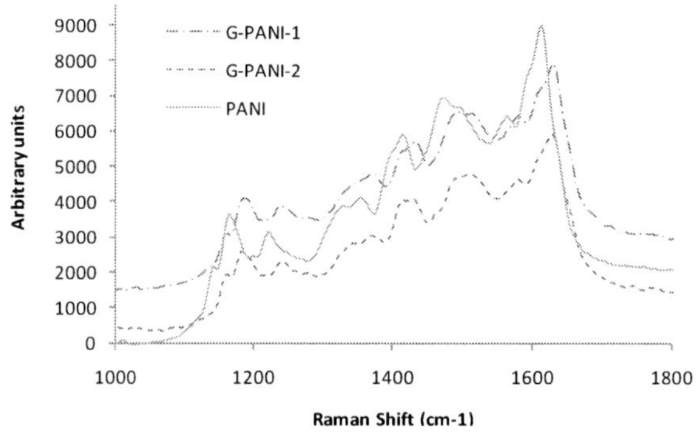

Figure 2. Raman spectra of polyaniline (PANI) and Gaphene-PANI nanocomposites at two aniline to graphene ratios.

The structure of the resulting graphene-polymer nanocomposite and the possible interconnection between the graphene platelets and the polymer matrix is shown in Figure 3. Due to the presence of graphene in the conducting polymer, the final conductivity of the polymer increases by > 2 orders of magnitude. High conductivity electrodes are beneficial to power densities of the material. The high electrical conductivity gives graphene consistently good performance over a wide range of voltage scan rates.

DISCUSSION

The conducting polymer with graphene nanocomposite material has been found to be stable at higher potential. Energy density is proportional to the square of voltage ($E \sim V^2$), so any increase in the operating potential (voltage) will certainly have an impact in the energy density. Figure 4 depicts the wider potential window of G-PANI based supercapacitor electrode showing the stability between -0.5 to 2.0V (curve 1 and 2), also revealing that the G-PANI nanocomposite goes from completely oxidized to reduced state in full potential window. This nanocomposite material will not be able to ultimately fulfill all the desired expectations; however, it will provide scientists with challenges for many years to come. In a way, what makes graphene special is that it is "all surfaces and no bulk" and therefore it can be used for its surface dependent .

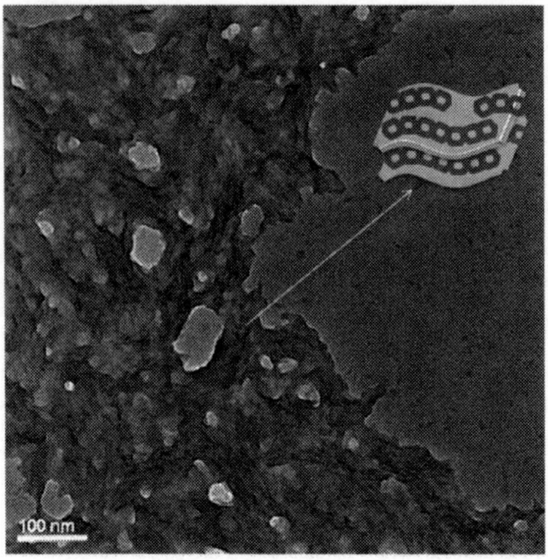

Figure 3. TEM micrograph of the graphene-polyaniline nanocomposite material and the possible interconnection between the polymer chains and the graphene platelets; scheme adapted from [2]

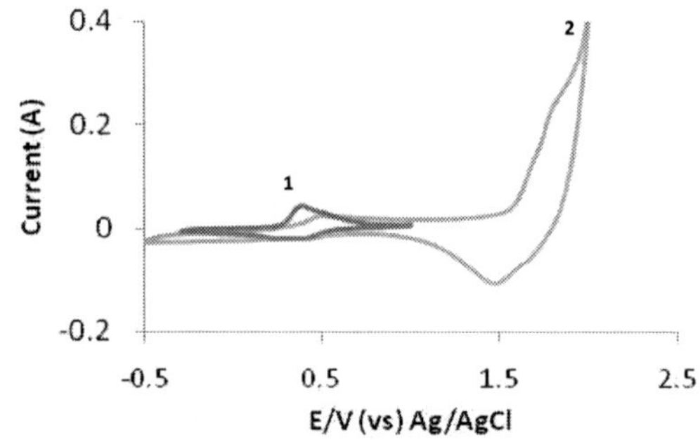

Figure. 4. Wider potential window of G-PANI based supercapacitor showing the stability between -0.5 to 2.0V (curve 2).

Cyclic voltammetry (CV) of graphene deposited on ITO plate was performed in 0.1 M camphor sulfonic acid at scan rate (5, 10, 20, 50, 100 mV/s) as shown in Figure 5. A distinct peak centered at -0.6V corresponds to the oxidation of graphene platelets. The graphene can also exhibit interesting electrochemical behavior. The CV plots shown in Figure 5 confirm the diffusion controlled system where the conductivity is induced by a charge transfer doping mechanism. Figure 6 shows the schematic of charging and discharging using G-PANI electrode in 2M H_2SO_4 electrolyte in the electrochemical cell. The specific capacitance is obtained by using the equation $C = -1/2\pi fZ''$ where Z" is the imaginary part of total complex impedance **[5]**. The single electrode specific capacitance values were evaluated by multiplying the overall capacitance by a factor of two and divided by the mass of a single electrode material. Capacitance values have been also obtained by using the equation C=i/s where "i" is the current and "s" the scan rate. The charge–discharge characteristics of the capacitor cells are evaluated at constant current.

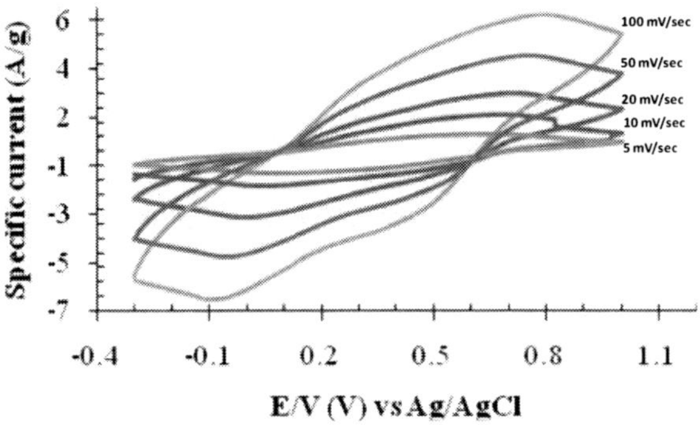

Figure 5. CV plots of G-PANI-1 in nafion-in 2M HCl

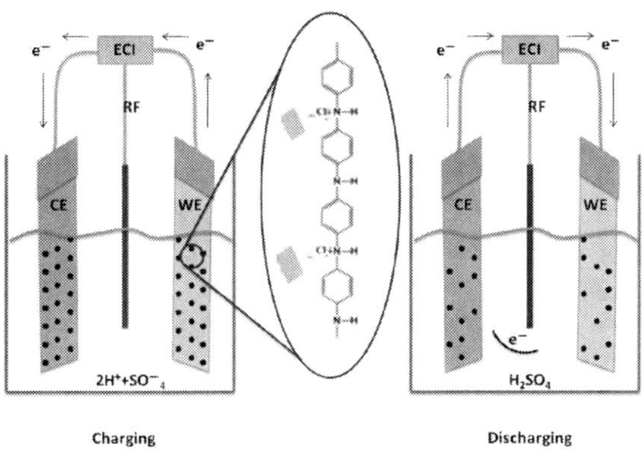

Figure 6 Charging and discharging of G-PANI-1 in 0.2M TBATFB.

The discharge capacitance "Cd" is evaluated from the linear part of the discharge curves using the relation. $Cd=i\Delta t/\Delta V$ where "i" is the constant current and 'Δt' is the time interval for the voltage change of ΔV. The columbic efficiency 'η' was nearly 1.2 calculated using (η

=charging time/discharging time x 100%). The G-PANI measured in the Figure 7as charging time in TBATFB it shows the better response but the quite similar charging and discharging is shown for the film studied in the 2M H_2SO_4 systems.

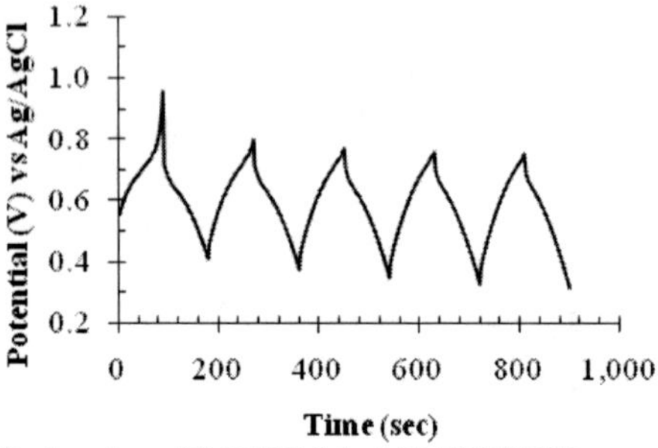

Figure 7 Charging and discharging of G-PANI-1 in 0.2M TBATFB.

CONCLUSIONS

G-PANI nanocomposite material was synthesized using oxidative polymerization technique. The FTIR and Raman studies of G-PANI films have shown that graphene does not only make the nano-composite material but also acting as a dopant to the polyaniline system. The G-PANI current charging and discharging plot have been used to estimate specific capacitance of 300 to 500 F/g at a current density of 0.1A/g. This manuscript provides a new material and technique for the design and fabrication of graphene-conducting polymer nanocomposite electrodes to be used for supercapacitor.

REFERENCES

[1] V.D. Patake, C.D. Lokhande, and Oh Shim Joo. "*Electrodeposited ruthenium oxide thin films for supercapacitor: Effect of surface treatments*". Applied Surface Science, 255, 7, 15, 2009, 4192-4196.

[2] Kai Zhang, Li Li Zhang, X. S. Zhao, and Jishan Wu. "*Graphene/Polyaniline Nanofiber Composites as Supercapacitor Electrodes*". Chem. Mater. 2010, 22, 1392–1401

[3] Humberto Gómez, Manoj K. Ram, Farah. Alvi, P. Villalba, Lee Stefanakos, and Ashok Kumar. "*Graphene-Conducting Polymer Nanocomposite as Novel Electrode for Supercapacitors*". Journal of Power Sources, In Print, 2010.

[4] Humberto Gomez, Manoj K. Ram, Farah Alvi, Elias Stefanakos, and Ashok Kumar. "*Novel Synthesis, Characterization, and Corrosion Inhibition Properties of Nanodiamond−Polyaniline Films*". J. Phys. Chem. C, 2010, 114 (44), pp 18797–18804.

[5] S. R. C. Vivechand, C. S. Rout, K. S. Subrahmanyam, A. Govindaraj, C. N. R. Rao. "*Graphene-based electrochemical supercapacitors*". J. Chem. Sci. 2008, 120, 1.

Mater. Res. Soc. Symp. Proc. Vol. 1312 © 2011 Materials Research Society
DOI: 10.1557/opl.2011.126

Facile synthesis of polypyrrole/graphene nanosheet-based nanocomposites as catalyst support for fuel cells

Burcu Saner, Selmiye Alkan Gürsel and Yuda Yürüm
Faculty of Engineering and Natural Sciences, Sabanci University Orhanli, Tuzla, Istanbul 34956, Turkey

ABSTRACT

The integration of catalyst metals into the graphene-based composites can be a new way to ensure thermal and electronic conductivities of the catalyst support materials in polymer electrolyte membrane fuel cells. In this work, graphene nanosheets were synthesized via a mild and safer chemical route including three major steps: graphite oxidation, ultrasonic treatment and chemical reduction. Then, polypyrrole was coated on graphene nanosheets by *in-situ* polymerization to fabricate polypyrrole/graphene nanosheet-based nanocomposites as the catalyst supports. Pt nanoparticles were uniformly dispersed on the surface of nanocomposites by sonication technique.

INTRODUCTION

Polymer nanocomposites incorporating carbon black, carbon nanotubes, and carbon nanofibers have been utilized for improved mechanical, thermal, electrical and gas barrier properties of polymers especially in the field of energy storage and electronics [1]. Recently, graphene sheets are promising materials to be used as nanofillers in the polymer matrices due to their high electrical conductivity, excellent mechanical strength, high surface area and high chemical stability [2]. Graphene is a single flat monolayer of sp^2-carbon atoms in 2D crystal structure. With the proper surface modifications, single graphene sheets can be separated from the graphite material and the layer-to-layer distance can be extended [3]. Chemical exfoliation is one of the extensively used methods to break van der Waals forces between graphene sheets in graphite and to receive monolayer graphene sheet.

In fuel cells, the catalyst layer must be very effective at breaking molecules into protons and electrons have high surface area, adequate porosity, high electronic conductivity, and it should be chemically and mechanically stable [4]. Catalyst has a great effect on both the cost and durability of polymer electrolyte membrane fuel cells (PEMFCs). In PEMFCs, platinum (Pt) can be considered as the best electrocatalyst in acidic media since it serves the lowest overpotentials and the highest activity [5]. Conducting polymers, especially polypyrrole (PPy), are widely preferred for fuel cell operations because of good electronic and proton conductivity, and dispensability [6]. PPy-modification can increase the electrochemical surface area and enhance the electrocatalysis ability of Pt/carbon catalyst [7].

In present work, graphene nanosheets as nanofillers were reinforced in polypyrrole by *in situ* polymerization to be used as the catalyst support material. In these nanocomposites, graphene sheets act as electron acceptors while PPy serves as an electron donor. Graphene nanosheets were exfoliated from graphite flakes in large amounts by applying a mild and safer method. The distribution and the size of Pt nanoparticles on the surface of composites were investigated to understand the importance of deposition technique.

EXPERIMENT

Chemical exfoliation of graphene nanosheets

Graphene nanosheets were produced by a safer and mild chemical route consists of oxidation, ultrasonic treatment and chemical reduction in large quantities [8]. Potassium permanganate ($KMnO_4$) is mostly used as an oxidizing agent in oxidation process but this chemical is very severe and leads to explosions when it is used together with H_2SO_4. Hence, potassium dichromate ($K_2Cr_2O_7$) was used as a milder oxidant in order to reduce the risks. In graphite oxidation, graphite flakes (1 g) was added into chromic acid prepared by using potassium dichromate, sulfuric acid and distilled water and then acetic anhydride (1 g) was poured dropwise into the mixture and the mixture was stirred at 45 °C for 10 days. The amounts of $K_2Cr_2O_7$ and H_2SO_4 were adjusted in the weight ratio 2.1:55. Filtration and neutralization of samples were conducted by using 0.1 M NaOH and distilled water until washings were neutralized. After oxidation process, ultrasonic treatment was performed for the homogenous dispersion of graphite oxide (GO) sheets. At last, sonicated GO sheets were chemically reduced through refluxing with hydroquinone in water to produce graphene nanosheets.

Synthesis of PPy/graphene nanosheet-based nanocomposites

PPy was coated on graphene nanosheets by *in situ* polymerization of pyrrole (Py) using $FeCl_3$ as the oxidant in the mixture of H_2O and ethanol in 1:1 (v/v) at room temperature for 24 hr [9]. The Fe^{3+}/Py molar ratio was approximately 2.4 [10]. The precipitated sample was filtered and rinsed several times by ethanol and distilled water to remove excess Py, catalyst and side products. The black powder was dried overnight in a vacuum oven at 60°C. The feeding mass ratio of Py and graphene nanosheets was adjusted as 1:1.

Pt nanoparticle deposition on nanocomposites

Pt deposition on the surface of composites was conducted by applying sonication technique. At first, graphene nanosheets were exposed to ultrasonic vibration for the homogeneous dispersion of sheets, and 8 % H_2PtCl_6 solution was added into mixture, and then 1 M $NaBH_4$ as a reducing agent was poured into mixture, respectively. The deposition time was adjusted to 2 hr.

All samples were characterized by Scanning Electron Microscopy (SEM), X-Ray Diffraction (XRD), Thermal Gravimetric Analyzer (TGA), Atomic Force Microscope (AFM), Raman spectroscopy and surface area analyzer. The electrical conductivities of pellet electrodes were measured by a conventional four-probe method at room temperature.

DISCUSSION

Graphene nanosheets

Pristine graphite contained smooth and sharp layers as shown in Figure 1 a. The oxidation process led to the formation of expanded and ripple structure as shown in Figure 1 b. During the oxidation, acetic anhydride diffused into layers and layers started to extend.

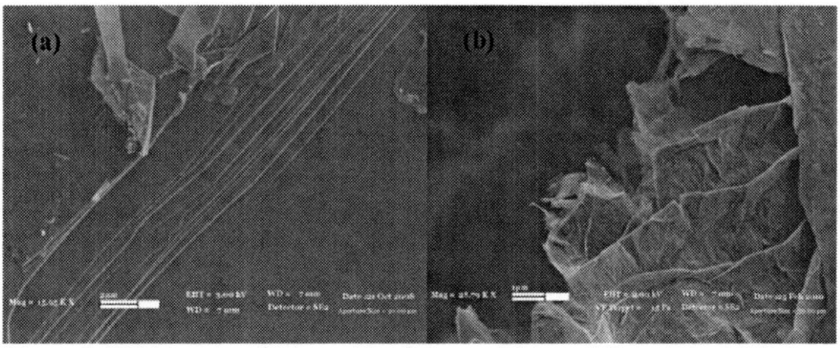

Figure 1. SEM images of **(a)** graphite flakes and **(b)** partially oxidized GO sheets

The XRD pattern of pristine graphite had a very sharp and high intense 002 peak but the intensity of this peak decreased comparably due to the destruction of structure (Figure 2 a and 2 b). Furthermore, the change of crystal structure can be estimated by using XRD data by applying Debye Scherrer equations [11]:

$$L_a = 0.89\lambda / \beta_{002} \cos\theta_{002} \tag{1}$$

$$n = L_a / d_{002} \tag{2}$$

(**L_a** : stacking height, **β** : full width half maxima (FWHM), **n:** average number of graphene layers, **d_{002}** : interlayer spacing). Average number of graphene layers for pristine graphite and GO sheets were estimated as 86 and 21, respectively. This also revealed how oxidation process destruct the graphitic structure.

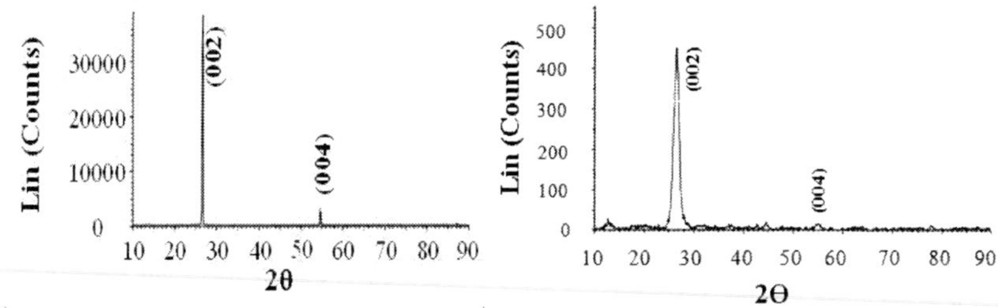

Figure 2. XRD patterns of **(a)** pristine graphite and **(b)** partially oxidized GO sheets

After ultrasonic treatment of GO sheets, samples were reduced by using hydroquinone in order to receive graphene nanosheets, Figure 2. There are several attempts in literature to separate graphene nanosheets from graphite. Herein, there are two distinct differences which are a milder oxidant and a simplified and cost effective procedure. Graphene nanosheets can be produced in large quantities to be used as the nanofillers in polymer composites.

Figure 2. SEM image of graphene nanosheets after chemical reduction

PPy/Graphene nanosheet-based nanocomposites

Py as a monomer diffused into graphene nanosheets during *in situ* polymerization and polymerized on the surface of graphene nanosheets layer by layer. Figure 3 depicted the successful layer by layer coating and sphere-like morphology of PPy particles. In this composite, graphene sheets act as electron acceptors while PPy serves as an electron donor.

Figure 3. SEM image of PPy/graphene nanosheet-based nanocomposite

Raman spectroscopy is an efficient technique to estimate the layer numbers and observe the change of crystal structure. All measurements were conducted using a 514 nm argon ion laser. There are three distinct peaks in Raman spectrum of graphene nanosheets which are the D band around 1364 cm^{-1}, the G band around 1582 cm^{-1} and the 2D band around 2709 cm^{-1}, Figure 4. G band intensity is directly proportional to crystallite size, La [12]. After coating PPy on the surface of graphene nanosheets, G band was broadened and its intensity was increased. This showed that crystallite size of structure was increased due to coating.

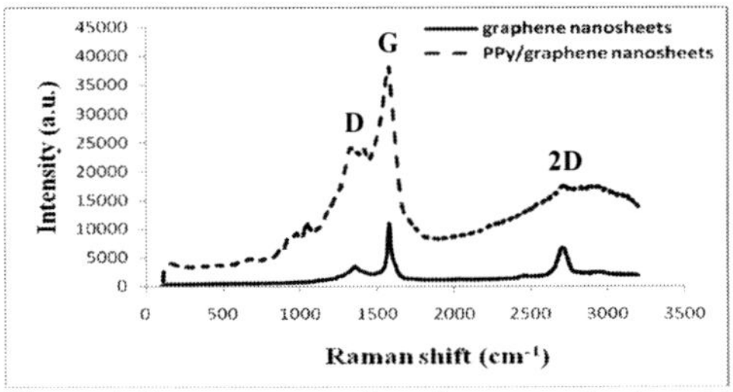

Figure 4. Raman spectra of graphene nanosheets and PPy/graphene nanosheets

The electrical conductivity of samples in the pellet form was measured by a conventional four-probe method. Conductivity of pristine PPy was relatively poor because of weak compactness and weak orientation of PPy particles [13] but conductivity was enhanced by using graphene sheets as filler in polymer matrix (Table I).

Table I. Electrical conductivities of PPy, graphene nanosheets and PPy/graphene nanosheet composites

Electrodes	Conductivity (S/cm)
Pristine PPy	0.00076
Graphene nanosheets	1.5
PPy/graphene nanosheet composites	0.13

High surface area has a significant influence on the dispersion and distribution of catalytic metals on catalyst supports. Nitrogen adsorption isotherms exhibited that BET surface area of reduced GO sheets was 507 m^2/g and after coating BET surface area decreased down to 290 m^2/g.

Pt/PPy/Graphene nanosheet composites

The particle size of Pt might be correlated with the degree oxidation of graphite which means that the efficient deposition of Pt nanoparticles is due to the strong interaction between the metal salt precursor and the functional group of GO sheets [14]. Herein, functionalized graphene nanosheets after 10 days of oxidation acting as the metal-anchoring sites provide metal nuclei formation and electrocatalysts deposition. During Pt deposition on graphene nanosheets, some Pt particles were aggregrated and some of them were dispersed uniformly on the surface. Figure 5 a exhibited that Pt particles were aggregated during deposition of Pt on the surface of graphene nanosheets. However, uniform dispersion of Pt particles was observed when graphene sheets were coated by PPy, Figure 5 b. This indicated a strong interaction between support and catalyst particles. Size distribution of Pt particles on the surface of the nanocomposites changed from 20 nm to 30 nm. Furthermore, sonication process provided the best exfoliation and dispersion of both Pt nanoparticles and graphene nanosheets.

Figure 5. SEM images of **(a)** Pt/graphene nanosheets via inlens detector and **(b)** Pt/PPy/graphene nanosheets via secondary electron detector

CONCLUSIONS

Graphene nanosheets were exfoliated from graphite by applying a mild and cost effective chemical route including oxidation, sonication and reduction. Graphene nanosheets based nanocomposites were prepared by a layer-by-layer PPy coating by *in situ* polymerization of Py monomer. The incorporation of graphene sheets in polymer matrix provided uniform dispersion of Pt nanoparticles on the surface with the help of sonication process. The support material has great importance to achieve high catalytic activity of fuel cell catalyst by lowering the catalyst deposition. Therefore, PPy/Graphene nanosheet-based nanocomposites having enhanced thermal stability, and electrical conductivity, and high surface can be a good alternative to carbon black and utilized as a catalyst support for Pt in PEMFCs.

REFERENCES

1. H. Kim, A. A. Abdala, and C. W. Macosko, *Macromolecules* **43**, 6515–6530 (2010).
2. S. Stankovich, D. A. Dikin, G. H. B. Dommett, et al., *Nature* **442**, 282-286 (2006).
3. M. J. McAllister, et al., *Chem. Mater.* **19**, 4396 (2007).
4. S.Natarajan and J.Hamelin, *Journal of The Electrochemical Society* **156**, B210-B215 (2009).
5. H. A. Gasteiger, S. S. Kocha, B. Sompalli, F. T. Wagner, *Appl. Catal. B.* **56**, 9 (2005).
6. R. Dersch, M. Steinhart, U. Boudriot, A. Greiner, J.H.Wendorff, *Polym. Adv. Technol.* **16**, 276 (2005).
7. H. Zhao, L. Li, J. Yang, Y. Zhang, *Journal of Power Sources* **184**, 375 (2008).
8. B. Saner, F. Okyay, Y. Yürüm, *Fuel* **89**, 1903–1910 (2010).
9. T. V. Vernitskaya, O. N. Efimov, *Russian Chemical Reviews* **66**, 443-457 (1997).
10. S. P. Armes, *Synth. Met.* **20**, 365-371 (1987).
11. B. Sakintuna, S. Cetinkaya, Y. Yürüm, *Energy Fuels* **18**, 883–888 (2004).
12. K. Sato, R. Saito, et al., *Chemical Physics Letters* **427**, 117–121 (2006).
13. N. G. Sahoo, Y. C. Jung, H. H. So, J. W. Cho, *Synth. Met.* **157**, 374-379 (2007).
14. N. Rajalakshmi, H. Ryu, M. M. Shaijumon, S. Ramaprabhu, *J. Power Sources* **140**, 250–257 (2005).

Mater. Res. Soc. Symp. Proc. Vol. 1312 © 2011 Materials Research Society
DOI: 10.1557/opl.2011.482

Permeability of Polymer/Clay Nanocomposites

Georgios A. Choudalakis and Alexandros D. Gotsis
Department of Sciences, Technical University of Crete, GR-73100, Chania, Greece

ABSTRACT

Because of their large interfacial area, the presence of nanoplatelets in the polymeric matrix decelerates the process of diffusion of gases through the material. The particles are impermeable barriers to the diffusing molecules, forcing them to follow complicated paths, increasing, thus, the diffusion length. The barrier properties of the nanocomposites depend on the properties of the polymeric matrix, the volume fraction of the nanoplatelets, their aspect ratio, their orientation, and their interactions with the matrix. The mobility of the molecules is hindered by the crystallinity but it is facilitated by the free volume within the material. The size and shape of the free volume holes in the polymer affect, thus, the rate of diffusion. Interactions between the nanoparticles and the matrix may lower the barrier properties because they may increase the free volume in the material. The estimation of free volume in the nanocomposite is important for the proper choice of components and the manufacturing of nanocomposite coatings with optimum barrier properties. Detailed information about the diffusion mechanisms at atomic and molecular levels can be obtained using the approach of free volume.

INTRODUCTION

Polymer/clay nanocomposites have attracted much attention in recent years due to their potential performance advantages on mechanical and barrier properties. Nanocomposites as barrier materials are used primarily in food, pharmaceuticals and electronics packaging, in protective coatings and in gas storage tanks. Oxygen and water are the most unwanted penetrants for the food industry, as they affect the quality and shelf-life of the products. Water can also cause cracking of silica or corrosion on metal surfaces. The carbon dioxide influences the quality of several drinks, while atmospheric pollution gasses, such as sulfur dioxide, causes serious degradation of exposed marble surfaces. The permeation of light gases, e.g. hydrogen, should be prevented in order to keep them in a tank for a long period. Although some of the above requirements are fulfilled by multilayer structures of conventional materials, high cost and, often, environmental factors, such as, e.g., organic solvents, create the need of alternative solutions.

Nanocomposite materials are two-phase systems that consist of a polymeric matrix and dispersed inorganic particles of nanometer scale. The most common inorganic particles belong to the family of 2:1 phyllosilicates. The layer thickness of each platelet is around 1 nm, and their lateral dimensions may vary from 30 nm to several microns. The layers are located on top of each other like the pages of a book. Van der Waals gaps exist between the layers, called galleries. Two particular characteristics of layered silicates play an important role in the creation of nanocomposites: The first is the ability of silicate sheets to disperse into individual layers, and the second is the possibility to modify their surface chemistry through ion exchange reactions with organic and inorganic cations.

The simple mixing of polymer and layered silicates does not always result in the generation of a nanocomposite, as this usually leads to dispersion of stacked sheets. This failure is due to the weak interactions between the polymer and the inorganic component. If these interactions become stronger, then the inorganic phase can be dispersed (exfoliate) in the organic matrix in nanometer scale. The phyllosilicate particles present a very high aspect ratio of width/thickness, in the order of 10 – 1000. For very low concentrations of particles, the total interface between polymer and layered silicates is much greater than that in conventional composites.

THEORY

Permeation process through nanocomposites

The mass transport mechanism of gasses permeating a nanoplatelet reinforced polymer is described schematically in Figure 1. In most theoretical treatises the nanocomposite is considered to consist of a permeable phase (polymer matrix) in which non-permeable nanoplatelets are dispersed. Figure 1 presents an ideal case in which the nanoparticles are evenly dispersed and considered to be rectangular platelets with finite width, L, and thickness, W. Their orientation is perpendicular to the diffusion direction. It is generally accepted that the transport mechanism within the polymer matrix follows Fick's law, and that the matrix maintains the same properties and characteristics as the neat polymer. Therefore, a decrease of the solubility of the gas is expected in the nanocomposite due to the reduced polymer matrix volume

$$S_{comp} = S_{polym}(1-\phi), \tag{1}$$

where S_{polym} is the solubility coefficient of the neat polymer, and ϕ is the volume fraction of the

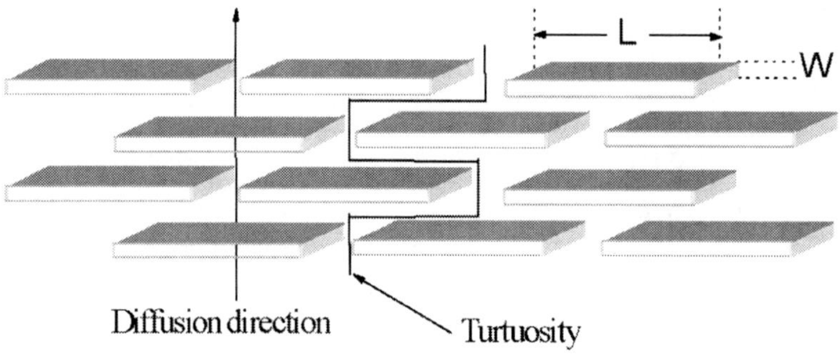

Figure 1: Regular arrangement of orthogonally shaped platelets in a parallel array with their main direction perpendicular to the diffusion direction.

nanoplatelets that are dispersed in the matrix. Further, a decrease in diffusion is expected due to a more tortuous path for the diffusing molecules. The platelets act as impermeable barriers to the diffusing molecules, forcing them to follow longer and more tortuous paths in order to diffuse through the nanocomposite. The major factor, then, is the tortuosity, which is connected directly to the shape and degree of dispersion of the nanoplatelets. The tortuosity factor τ is defined as

$$\tau = \frac{\ell'}{\ell}, \quad (2)$$

where ℓ' is the distance that a solute must travel to diffuse through the membrane when nanoplatelets are present and ℓ is the membrane thickness. The diffusion coefficient, D, is influenced by the tortuosity, τ:

$$D_{comp} = \frac{D_{polym}}{\tau}, \quad (3)$$

where D_{polym} is the diffusion coefficient of the matrix. It is generally accepted that the permeability coefficient is defined as the product of the diffusion and solubility coefficients $P = D \cdot S$ thus the combination of the above equations leads to the following expression for the relative permeability:

$$\frac{P_{comp}}{P_{polym}} = \frac{1-\phi}{\tau}. \quad (4)$$

Permeability models

Several models have been developed in order to describe the mass transfer within the nanocomposites. Most models assume that the platelets have a regular and uniform shape (rectangular or circular) and form a regular array in space. They are either parallel to each other or have a distribution of orientations, with the average at some angle relative to the main diffusion direction of the gas molecules.

A simple geometrical permeability model for a regular arrangement of platelets has been proposed by Nielsen [1] and is presented in Figure 1. After some geometrical considerations the permeability ratio that the model predicts becomes

$$\frac{P_{comp}}{P_{polym}} = \frac{1-\phi}{1+\frac{\alpha}{2}\phi}, \quad (5)$$

where $\alpha = L/W$ is the aspect ratio of the nanoplatelets.

This equation shows that the permeability of the nanocomposite decreases with the increase of ϕ and α. In practice, however, the limit for its validity is $\phi \leq 10\%$, as the particles have a tendency to aggregate, which increases with ϕ.

The simple Nielsen model describes qualitatively the physical process of the permeation through a nanocomposite system. Various other models for the permeability coefficient have been reported. In fact, all models agree in that the dependence of the relative permeability is determined mainly by the aspect ratio and the volume fraction of the inorganic phase.

The model proposed by Cussler [2] differs from that of Nielsen on the effectiveness of the barrier properties, as the exponents of the parameters α and ϕ are higher here:

$$\frac{P_{comp}}{P_{polym}} = \left(1 + \frac{\alpha^2 \phi^2}{1-\phi}\right)^{-1}. \tag{6}$$

The experimental results show that this model usually overestimates the barrier properties of the polymer nanocomposites. Fredrickson and Bicerano [3] developed a model using multiple scattering theory for oriented disk-shaped platelets. Their results are in accordance with Nielsen's formula in the dilute regime, $\alpha\phi << 1$, and with Cussler's formula in the semi-dilute regime, $\alpha\phi >> 1$.

In a different approach, Gusev and Lusti [4] developed a 3-D computational model for a random array of parallel circular disks. This model is based on the solution of the Laplace equation for the local chemical potential. The proposed expression for the global permeability is:

$$\frac{P_{comp}}{P_{polym}} = \exp\left[-\left(\frac{\alpha\phi}{x_0}\right)^{\beta}\right], \tag{7}$$

where the parameters β and x_0 are usually evaluated by fitting experimental data.

It is obvious that the orientation of the platelets to the diffusion direction influences the barrier properties through the tortuosity factor, τ. The tortuosity is the highest when the nanoplatelets are aligned perpendicular to the diffusion direction. For other orientation angles Nielsen's equation should be modified accordingly. For non-uniform orientation of the platelets an order parameter, S', is introduced to quantify the degree of their orientation around the diffusion direction [5]:

$$S' = \frac{1}{2}\langle 3\cos^2\theta - 1\rangle , \tag{8}$$

where θ is the angle between the diffusion direction and the unit vector normal to the surface of a platelet, and the average is taken over all platelets with all possible orientations. For non-uniform orientation Nielsen's equation takes the following form [5].

$$\frac{P_{comp}}{P_{polym}} = \frac{1-\phi}{1 + \frac{\alpha}{3}\left(S' + \frac{1}{2}\right)\phi} \tag{9}$$

Measurements of the transport coefficients

The permeation of gases through an organic membrane, such as a polymer film, is a complex phenomenon that consists of four processes: (i) the sorption of gas molecules on the surface of the membrane; (ii) the dissolution of the gas inside the membrane; (iii) the diffusion through it; and, finally, (iv) the desorption of gas from the other surface of the membrane.

For the direct measurement of the permeability, one can use an arrangement such as shown in Figure 2. A membrane separates the measuring cell into two compartments. The gas is introduced in the compartment V_0 (left hand side in Figure 2) under constant pressure, P_0. The gas permeates progressively the membrane and its concentration or partial pressure can be recorded as a function of time. In a flat membrane of thickness d, the solution of Fick's first law at steady state is [6]

$$c_d(t) = \frac{ADSP_0}{Vd}(t - t_0) \tag{10}$$

where $t_0 = d^2/6D$.

The characteristic time, t_0, is a measure of the time that is required to establish a constant flow. If we plot $c_d(t)$ vs. t for sufficiently long times to reach a linear response, then we can determine the time lag, t_0, as shown in Figure 3, and the diffusion coefficient D from eq. 10. The slope of the line gives the product $D \cdot S$, i.e. the permeability coefficient.

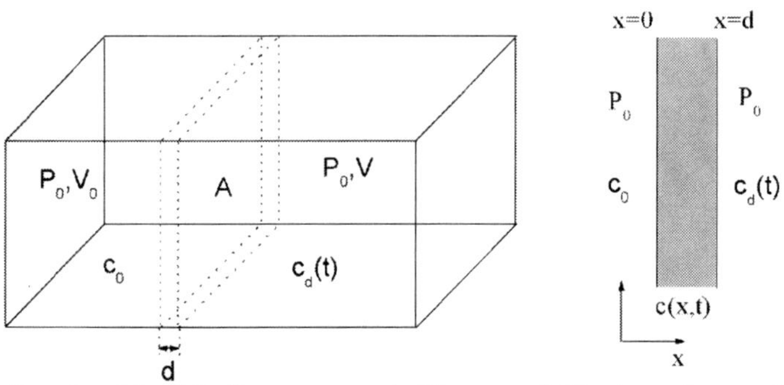

Figure 2: Experimental cell for the measurement of the permeability. A membrane, with thickness d and area A, divides the cell in two chambers. The gas is introduced in volume V_0 under constant pressure, P_0. The concentration of the test gas is c_0 on the left surface of the membrane and $c_d(t)$ on the right surface. The concentration of the gas inside the membrane is $c(x,t)$.

Alternatively, the diffusion coefficient can be determined by sorption/desorption experiments via an in situ gravimetric method. At initial times the mass gain/loss of the sample is described by:

$$\frac{M_t}{M_\infty} = \frac{8}{\pi^{1/2}}\left(\frac{Dt}{d^2}\right)^{1/2} . \tag{11}$$

By plotting M_t / M_∞ against $\sqrt{t}$ one obtains a straight line, at least up to the half-life time, $t_{1/2}$, where $M_t / M_\infty = 1/2$. The diffusion coefficient is calculated from the slope of this line.

The solubility coefficient is found by measuring the concentration of the soluble gas as a function of pressure. In general, the total concentration, C, of the absorbed gas molecules is

$$C = k_d P + \frac{C_H bP}{1+bP} , \tag{12}$$

with k_d being Henry's law constant, P the partial pressure of the gas, and C_H and b the hole saturation and hole affinity constants. Evaluating the coefficients, k_d, C_H and b by an appropriate fitting procedure, the solubility is then calculated by:

$$S = k_d + \frac{C_H b}{1+bP} . \tag{13}$$

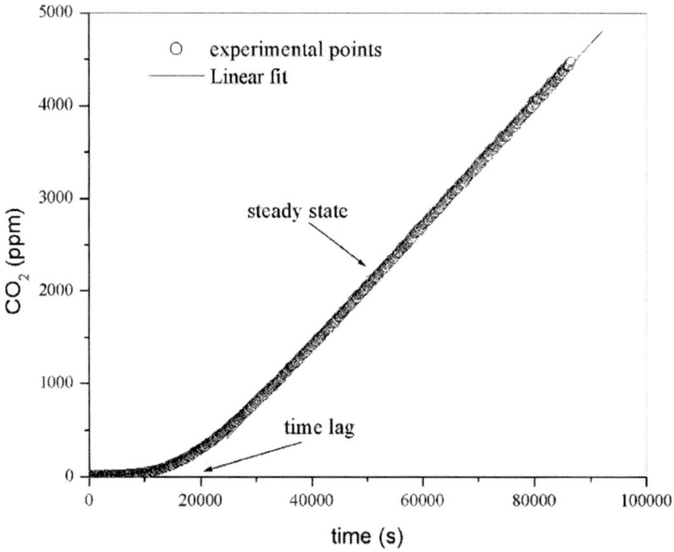

Figure 3: Experimental curve of the concentration of the permeating gas as a function of time. The time lag and steady state regime are depicted.

DISCUSSION

The decrease of the relative permeability is attributed in all models mainly to the inorganic phase. If their aspect ratio and volume fraction are the same, the particles will induce the same relative reduction of the permeability in any matrix they are dispersed, provided that the same degree of exfoliation is obtained. Further, if there are no special interactions between the solute gas and the particles, the relative enhancement of the barrier properties is not affected by the kind of the gas that is used (Figure 4).

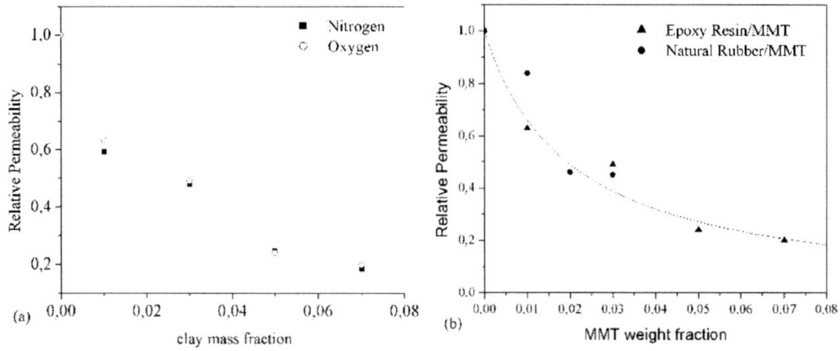

Figure 4: (a) Relative permeability as a function of clay mass fraction for an Epoxy Resin/MMT system for two different gases. (b) Oxygen relative permeability as a function of clay mass fraction for the same particles dispersed in different matrices.

In other words, the improvement of the barrier properties of the material is determined solely by the changes in the dispersed inorganic phase. This feature has been verified experimentally for various systems, such as Epoxy Resin/MMT [7] for oxygen and nitrogen, Polyethylene/MMT [8], PP/EPDM/MMT [9] for oxygen and carbon dioxide, Butyl Rubber/Vermiculite [10] for hydrogen, helium, oxygen, carbon dioxide and nitrogen.

Turning now to the results of the models, the simple model of Nielsen seems to predict adequately the improvement of the barrier properties of some of the experimental data like these shown in (Figure 5). The values of the particle aspect ratio that were used for the model fit are close to what is expected for the montmorillonite particles. On the other hand, Cussler's model requires significant lower values of the aspect ratio in order to fit to the experimental points. This seems to lead to the underestimation of the aspect ratio, especially for samples that present relatively high degree of exfoliation.

Despite the apparent success of the above results on the improvement of the polymer barrier properties there are also some disappointing experimental results, which show that this desirable barrier enhancement is not always achievable, even when particle exfoliation is accomplished. This may be due to the influence of the nanoplatelets on the free volume characteristics of the polymer matrix. The study of the polymer free volume variations caused by the inorganic particles can be performed using the PALS (positron annihilation life time) method, which is non destructive and adequately sensitive. Several investigators have reported an increase in the measured free volume of the nanocomposite as a function of the particle loading. Some of these results concern systems like PVA/MMT [11], Cyanate Ester/Bentonite [12], PA6/MMT [13].

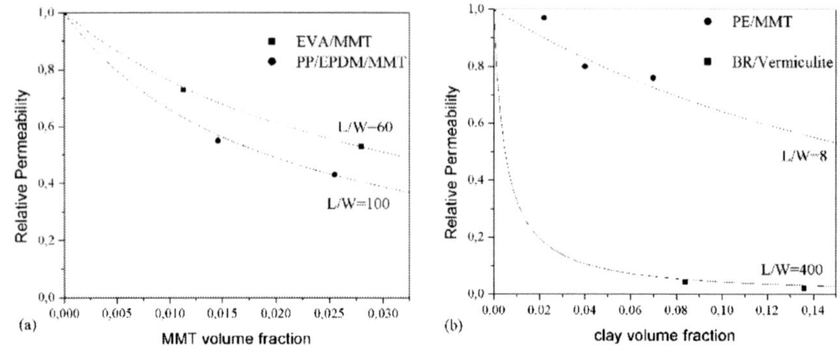

Figure 5: (a) Oxygen relative permeability as a function of clay volume fraction for an Ethylene Vinyl Acetate/MMT and Polypropylene/EPDM/MMT nanocomposites and the fitted model of Nielsen. (b) Carbon dioxide relative permeability as a function of clay volume fraction for a Polyethylene/MMT and Butyl Rubber/Vermiculite nanocomposite. The curves are the Nielsen equation fittings for the selected values of aspect ratio.

The effect of the particles on the free volume characteristics has three origins: The first is the creation of interfacial regions between the two phases; the second is the immobilization of the side-chain segmental motions of the polymer molecule; and the third is the effect on crystallinity or cross-linking density of the polymer. In a PVA/MMT system an increase in local free volume at the clay-PVA interface has been observed of the order of 25% for 10% w/w of MMT, while in Cyanate Ester/Bentonite the free volume enhancement was attributed to the reduction of the cross-link density. Kim et al [14] reported an increase in carbon dioxide permeability of the PEBAX/Silica nanocomposite due to the increase of the solubility coefficient, which was attributed to the crystallinity reduction in the polymer matrix. Similar results have been reported for the oxygen permeability of the Polyimide/TiO2 nanocomposite [15].

The diffusion coefficient depends on the fractional free volume, f_V, via the relation

$$D = A\exp\left(-\frac{B}{f_V}\right), \tag{14}$$

where parameter A is related to the size and the shape of the diffusion molecule, while parameter B is related to the minimal hole size available for diffusional jumping. On the other hand, the solubility coefficient is proportional to the fractional free volume: $S = f_V / \rho_g$. Thus the permeability coefficient of the nanocomposite should be $P \sim \frac{f_V}{\tau}\exp\left(-\frac{B}{f_V}\right)$, where τ is the tortuosity factor. If the free volume variations overwhelm the tortuosity effect, then the barrier properties of the nanocomposite will be lower than that of the neat polymer.

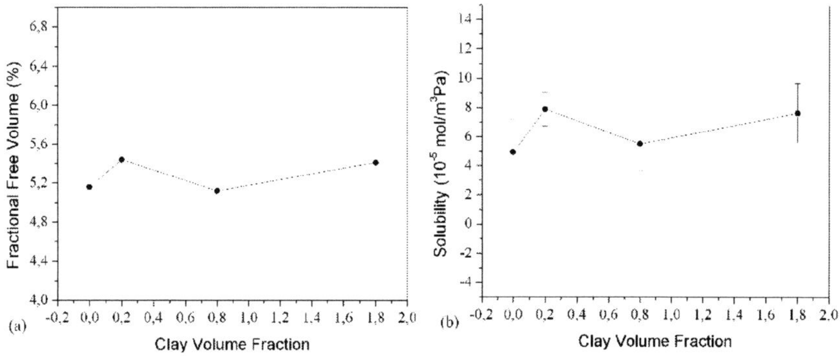

Figure 6: (a) Fractional free volume as a function of laponite volume fraction. (b) Solubility as a function of laponite volume fraction.

Important features of the polymer free volume properties are not only the total free volume variation but also its distribution over the hole sizes. The nano-particles may cause redistribution of the free volume hole sizes, while leaving the total amount of the free volume unaffected. This fact can strongly influence the barrier properties of the nanocomposite, something that was observed in an acrylic resin/laponite system [16]. The fractional free volume was studied by PALS, while the transport coefficients were extracted from curves like those in Figure 3. A remarkable agreement between the solubility coefficient and fractional free volume as a function of laponite volume fraction was observed (Figure 6). The permeability coefficient, however, increased with the laponite content in this case, mainly due to the increase of the diffusion coefficient (Figure 7).

This feature may be explained on the basis of the free volume hole size distribution. Since the sizes of the holes are in the same order as the size of the diffusing gas molecules, there is a hole size threshold, below which the gas molecules cannot penetrate the hole. The small holes, therefore, cannot contribute to the diffusion of the gas molecules and the diffusion is biased towards the larger holes. The broader the free volume hole size distribution, especially at the higher tail, the larger the gas diffusion coefficient for the same average hole size. On the other hand, this redistribution may act positively on the permeselectivity properties of the material.

CONCLUSIONS

The barrier properties of polymer/clay nanocomposites are determined mainly by the clay loading and the aspect ratio of the platelets. However, this is true only if there are no significant effects on the physical properties of the polymer as consequences of the incorporation of clay. In fact, in many cases the previous condition is not valid and the barrier properties deviate from the expected or predicted results. The key parameter to improve the barrier properties is not only the high exfoliation degree of the nano-clays but also the specific interactions that they develop in the matrix and their effect on the average size and size distribution of the free volume holes in the system.

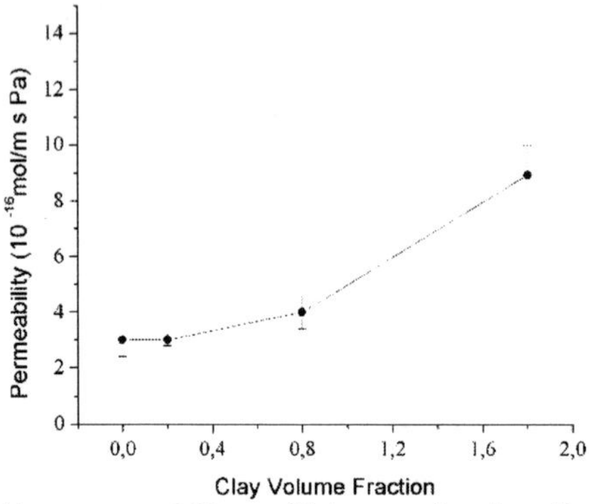

Figure 7: Carbon dioxide gas permeability coefficient as a function of laponite volume fraction.

ACKNOWLEDGMENTS

Part of this research project has been supported by the European Commission under the 6th Framework Programme through the Key Action: Strengthening the European Research Area, Research Infrastructures, Contract Nr: RII3-CT-2003-505925. Mr. Choudalakis was partly supported by ELKE (the Research Funds Financial Administration) of the Technical University of Crete. We thank Dr. Louise Nobel for providing us with acrylic resin/laponite samples.

REFERENCES

1. L.E. Nielsen, J. Macromol. Sci. A, **1**, 929-942 (1967).
2. A.A. Gusev, H.R. Lusti, Adv. Mater. **13**, 21 (2001)
3. E.L. Cussler, S.E. Hughes, W.J. Ward, R. Aris, J. Membr. Sci. **38**, 161-174 (1988)
4. G.H. Fredrickson, J. Bicerano, J. Chem. Phys. **110**, 2181-2188 (1999)
5. R.K. Bharadwaj, Macromolecules, **34**, 9189-9192 (2001)
6. G. Choudalakis, A.D. Gotsis, European Polymer Journal, **45**, 967-984 (2009)
7. J.M. Yeh, H.Y. Huang, C.L. Chen, W.F. Su, Y.H. Yu, Surface and Coatings Technology, **200**, 2753-2763 (2006)
8. E. Jacquelot, E. Espuche, J.F. Gerard, J. Duchet, P. Mazabraud, J. Polym. Sci. Part B:Polym. Phys., **44**, 431-440 (2006)
9. M. Frounchi, S. Dadbin, Z. Salehpour, M. Noferesti, J. Membr. Sci. **282**, 142-148 (2006)
10. S. Takashi, H.A. Goldberg, C.A. Feeney, D.P. Karim, M. Farrel, K. OLeary, D.R. Paul, Polymer, **47**, 3083-3093 (2006)

11. C.M. Paranhos, B.G. Soares, J.C. Machado, D. Windmoller, L.A. Pessan, European Polymer Journal, **43**, 4882-4890 (2007)
12. D.H. Yu, B. Wang, Y. Feng, Z.P. Fang, Journal of Applied Polymer Science, **102**, 1509-1515 (2006)
13. P. Winberg, M. Eldrup, N. Pedersen, M. Van Es, F. Maurer, Polymer, **46**, 8239-8249 (2005)
14. J.H. Kim, Y.M. Young, J. Membr. Sci. **193**, 209-225 (2001).
15. Y. Kong, H.D. Du, J. Yang, D. Shi, Y. Wang, Y. Zhang, W. Xin, Desalination, **146**, 49-55 (2002)
16. G.A. Choudalakis, A.D. Gotsis, H. Schut, S.J. Picken, European Polymer Journal, corrected proof (2011) doi: 10.1016/j.europolymj. 2010.12.008

Mater. Res. Soc. Symp. Proc. Vol. 1312 © 2011 Materials Research Society
DOI: 10.1557/opl.2011.127

Morphological Investigations of Organic/Inorganic Nanocomposites Fabricated to Achieve Controlled Dispersion at High Loadings

Andrew J. Duncan, Andrew B. Schoch, Christopher S. Gold,
Joseph L. Lenhart, Frederick L. Beyer

U.S. Army Research Laboratory, Aberdeen Proving Ground, MD 21005-5069, U.S.A.

ABSTRACT

Realization of property enhancements inherent to the presence of nanoparticles continues to be a challenge for the production of bulk nanocomposite materials with commercially available techniques. This study combines twin-screw compounding with surface modification of SiO_2 nanoparticles to enable targeted dispersion in a SEBS block copolymer. Production of these composites with high levels of well-dispersed particulates aims to leverage aggregation for production of hierarchical structure. The aggregation state of the particles as well as the level of order in the block copolymer morphology was determined through USAXS and TEM. Particles coated with ligands miscible with the end-blocks of the BCP (minority component) increased dispersion at all loading levels observed up to 10 vol%. Ligands employed to increase miscibility of the nanoparticle with the mid-block (majority component) resulted in large aggregates for all loadings without disturbance of the BCP morphology.

INTRODUCTION

Most research efforts to realize the multifunctional benefits of polymeric nancomposites have focused primarily on low volume percent loadings and precise control of functional nanoparticle placement in polymer hosts.[1-4] Our goal is to maximize loading of the functional particulate in the polymer template with minimal or controlled variation of the morphological structure. Coincident dispersion of the nanoparticle and stability of the polymer morphology provide hierarchically structured functional materials. Local aggregation extends the hierarchical structure and influences the composite's properties similar to microscopic fillers.[5] Bulk processing techniques were employed to create scalable functional materials that retain the benefits of nanofillers within the polymer nanostructure.

Controlled incorporation of multi-functional particulates into polymer matrices requires a complex balance of thermodynamic and entropic contributions to the free energy of the composite system. Prior efforts to create well-dispersed systems that realize the multi-functional benefits of incorporating high surface area to volume ratio species are quite extensive and have often employed the microphase separated morphologies associated with block copolymers.[6-7] With respect to the thermodynamic contribution, minimization of the Flory-Huggins parameter (χ) between species tethered to the surface of the nanoparticle and the matrix polymers has demonstrated success in organization of variously sized particles in block copolymers.[8]

The approach detailed here tailors non-specific interactions between the particulate and polymeric host. Aliphatic or aromatic ligands were grafted onto the surface of nano-SiO_2 to tailor the miscibility of the nanoparticle to a targeted domain of poly(styrene-*b*-ethylene-*co*-butylene-*b*-styrene) (SEBS). The composites were processed using a twin-screw compounder/extruder. The primary methods employed to characterize the polymer morphology and particle distribution of the nanocomposites were synchrotron-based ultra small-angle X-ray scattering (USAXS) and transmission electron microscopy (TEM). Variation in loading levels and ligand composition when combined with the block copolymer allows for identification of critical parameters to achieve controlled dispersion at high loadings without extended disruption of the polymer morphology.

EXPERIMENTAL MATERIALS & METHODS

The poly(styrene-*b*-ethylene-*co*-butylene-*b*-styrene) (SEBS, Kraton G1652) block copolymer was generously provided by Kraton Polymers in flake form. The G1652 SEBS has a M_n of approximately 79 000 g/mol and a polydispersity index of nearly 1.1. The styrenic block content is stated as 30 wt %. 10-15 nm SiO_2 (MEK-ST, CS) nanoparticles were purchased from Nissan Chemical as a 30 wt % dispersion in methyl ethyl ketone (MEK). N-octyltrimethoxysilane (Oct) and phenethyltrimethyoxysilane (Phen) coupling agents were purchased from Gelest Inc.

Surface functionalization of CS with Oct and Phen proceeded as follows. 150 mL of toluene was placed in a 250 mL roundbottom flask. 5-9 g of the coupling agent was added to the flask with a syringe. 15 g of CS was measured out in a 50 mL beaker and added to the flask. A further 100 mL of toluene was used to wash any remaining CS from the beaker into the flask. The contents of the flask were then heated to reflux and stirred for 24 h with a heating mantle and a reflux condenser open to air. MEK and toluene were then removed with a rotovap to provide a final dispersion of approximately 30 wt % particles. This dispersion was washed with a non-solvent (methanol/water for CS-Oct, hexanes for CS-Phen) to remove any unreacted coupling agent. After precipitation from the dispersion due to addition of the non-solvent, the dispersions were centrifuged and the liquid layer was decanted off. The functionalized CS was redispersed in a good solvent (hexanes for CS-Oct, toluene for CS-Phen). The wash process was repeated three times.

CS dispersions were added to neat SEBS to provide 3 mL batches with 1, 5, and 10 vol % of CS. The batch was fed into a DSM Xplore twin-screw extruder with a 3 mL recirculating path and a co-rotating screw configuration to compound at 180 °C for 10 min. Rotation of the screws was set to 50 rpm during the feed and reduced to 10 rpm during the compounding stage. Extrusion proceeded through a circular path (170 °C) and die with a 3 mm diameter at 1-10 rpm into air. All samples were then annealed at 130 °C in air for 48 h.

Ultra-small-angle X-ray scattering (USAXS) data were collected on the Bonse-Hart double-crystal USAXS instrument at the Advanced Photon Source, Argonne National Laboratory.[9] Extruded samples were mounted such that the X-ray beam impinged on the center of the extrudate, perpendicular to the flow direction. The slit was configured to 0.3 mm high and 2.0 mm wide with the width parallel to the flow direction. For each sample, usable slit-smeared one-dimensional data were collected over a scattering angular range of 10^{-4} – 0.3 $Å^{-1}$, at an incident photon energy of 16.85 keV (λ = 0.736 Å). All data correction, processing, and analysis were performed using IGOR Pro v6.12 (Wavemetrics, Inc.) and procedures available through Argonne

National Laboratory.[10] The raw data were corrected for background scatter and sample absorption. Desmearing was not performed on these data due to the difficulties encountered when desmearing well-defined features, such as Bragg diffraction maxima, in the scattering data.

Transmission electron microscopy (TEM) was used to collect real space images of each sample. Samples were prepared for TEM through cryo-ultramicrotomy using a Leica UCT microtome and a diamond knife (Micro Star Technologies) to prepare sections 70 nm or less in thickness. Microtomy was performed to provide sections representative of the plane orthogonal to the extrusion direction. Microtomed sections were stained using fresh RuO_4 (aq.) for a minimum of 30 min, making domains containing PS and CS-Phen appear dark in micrographs obtained using bright field imaging. Imaging was performed on a JEM-2100F field emission TEM (JEOL USA, Inc.) operated at 200 kV. Bright field images were collected using an 11 megapixel SC1000 Orius CCD camera (Gatan, Inc.). Electron energy-loss spectroscopy (EELS) and energy filtered imaging (EFTEM) were carried out with an 863 Gatan Imaging Filter (GIF) Tridiem with an UltraScan 1000 FT CCD camera. The K edge for carbon at 283.8 eV and the $L_{2,3}$ edge for silicon at 99.2 eV were employed for elemental mapping. A 20 eV slit was employed for two pre-edge (edge - 52 eV, edge - 12 eV) windows and a post-edge (edge + 40 eV) window for EFTEM imaging. Image analysis was performed using DigitalMicrograph (Gatan, Inc.).

RESULTS & DISCUSSION

This study attempts to disperse nanoparticles with comparable size to the domain spacing of the host BCP while modulating the thermodynamic miscibility of the particle surface with one of the copolymer blocks. Short ligands are used to minimize the entropic penalty associated with mixing high molecular weight surface ligands in a microphase separating triblock matrix. 10 nm SiO_2 nanoparticles were functionalized with n-octyl (Oct) and phenethyl (Phen) ligands with projected miscibility with the polystyrene (PS) end-blocks and poly(ethylene-*co*-butylene) (PEB) mid-block respectively. Post-functionalization the nano-SiO_2 was washed and dispersed in a good solvent for the ligand, however the CS-Oct (hexanes) did show greater tendency for aggregation in the dispersed state than the CS-Phen (toluene) over extended storage. After washing, target amounts of the nanoparticle dispersions were mixed with bulk SEBS and compounded in a twin-screw extruder to produce well-dispersed nanocomposites. The volume percent of functionalized CS in the composites was set at 1, 5, and 10 vol % to observe the applicability of this approach across a larger range of compositions than is typically addressed.

Ultra-small-angle X-ray scattering

The SEBS composites containing three different loadings of CS nanoparticles were examined with USAXS to simultaneously investigate the dispersed state of the particulates and the microphase separated morphology of the BCP. The scattering curves for the total sample set including neat SEBS are shown in Figure 1. SEBS without any CS provides scattering characteristic of hexagonally close packed PS cylinders in a matrix of PEB as indicated by the multiple Bragg reflections. Four of the five initial Bragg reflections were observed for this sample. The absence of the reflection at $q/q^* = 2.00$ is likely due to a minimum in the cylinder form factor. SEBS G1652 was previously reported to have an order-order transition (OOT) temperature of 240 °C and an order-disorder transition (ODT) temperature greater than 350 °C.[11] Both the extrusion and annealing temperatures are sufficiently below these values to limit morphological transitions while allowing for increased long-range order.

Scattering from the CS nanoparticles effectively obscures the characteristic scattering peaks originating from the BCP at loadings greater than 1 vol %. The shoulder in the scattering data for the composites at $q \approx 0.038$ Å^{-1} is relatively constant despite changes in loading or the ligand identity on CS and is consistent with scatter from particulates approximately 12 nm in diameter, indicating that this feature is from the individual nanoparticles. At smaller values of q, the effect of surface functionalization of the CS becomes apparent. Between $q = 10^{-2} - 10^{-3}$ Å^{-1} a broad, shallow peak appears in the scattering data. This feature is typically referred to as a Guinier knee and indicates aggregation of primary particles.[12] Overall, the CS-Phen particles appear to form aggregates ~60 nm in size while the CS-Oct form much larger aggregates, ~220 nm in size. Composites of CS-Phen demonstrate very little dependence of aggregate size on the percent loading of particles. For CS-Oct, increased volume percent loading also appears to result in increasingly large NP aggregates as indicated by the shift of the Guinier knee to lower q values.

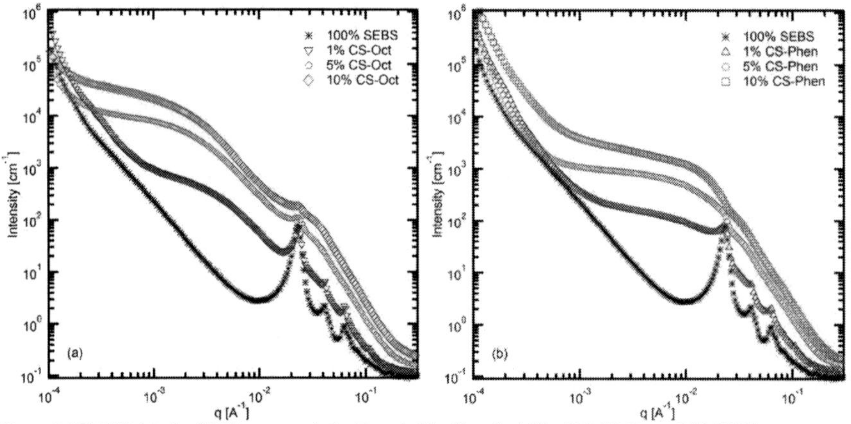

Figure 1. USAXS data for SEBS compounded with varied loadings (vol %) of (a) CS-Oct and (b) CS-Phen

Preliminary scattering data were also taken after rotation of the extrudate 90° clockwise with respect to the beam slit. Comparison of scattering in the two configurations indicated that extrusion imparted significant anisotropy to the alignment of the BCP morphology while demonstrating very little effect upon the aggregate state of the nanoparticles.

Transmission electron microscopy

Observation of the morphologies indicated through scattering experiments was performed in real-space with bright field TEM. The presence of either CS-Oct or CS-Phen does not appear to significantly alter the ability of the BCP to microphase separate. Furthermore, the cylindrical morphology with PS in a matrix of PEB expected from this composition of SEBS is apparent in all cases with a domain spacing of ~25 nm.

Both nanocomposites lacked sufficient contrast for at least one composition to simultaneously observe both the CS dispersion and the BCP morphology. To attain this information the composites were imaged with EFTEM to allow specific observation of the elemental components within a target area. EFTEM was carried out for all CS-Oct and CS-Phen composites with constant experimental parameters for mapping Si and C. Elemental maps for Si content are shown in Figure 2. This technique provided spatial resolution of the CS particles in the CS-Oct composites and confirmed that significant aggregation occurred for all loadings of the CS-Oct not visible with BFTEM. The dispersed state of the CS-Phen was further confirmed with EFTEM. For CS-Phen SEBS composites the functionalized nanoparticles appear well dispersed for all loadings investigated. At 1 vol % small aggregates of CS-Phen composed of 3-4 particles are observed in comparison to aggregates greater than 100 nm for CS-Oct. The most striking comparison is between the two composites at 10 vol % loading. The presence of CS-Oct results in large aggregates several hundred nanometers in size while the CS-Phen composites retain homogeneous dispersion.

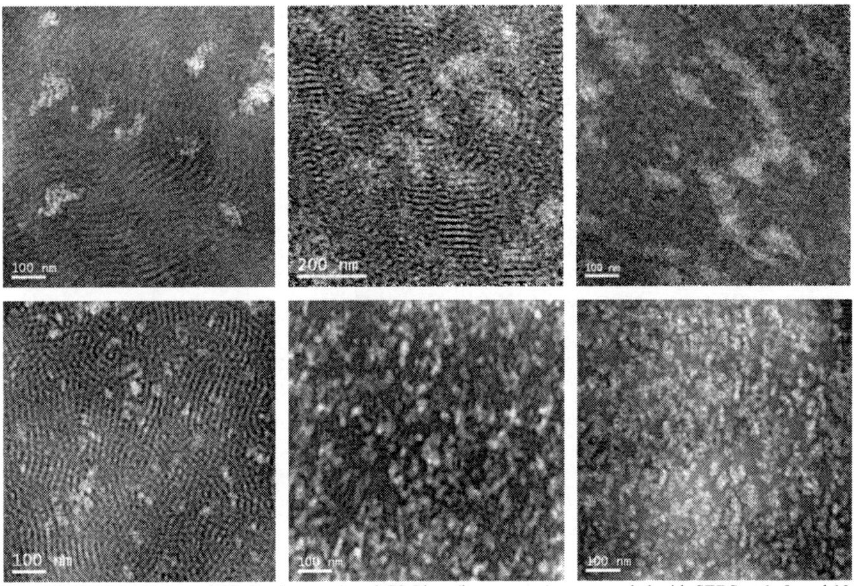

Figure 2. EFTEM Si map of CS-Oct (top row) and CS-Phen (bottom row) compounded with SEBS at 1, 5, and 10 vol % (left to right, respectively)

CONCLUSIONS

Nanocomposites composed of surface functionalized nano-SiO_2 and SEBS block copolymer were successfully produced in bulk across a range of filler loadings through twin-screw extrusion. Characterization with USAXS revealed that varied loadings resulted in minimal changes in aggregation compared to miscibility controlled by particulate functionalization. Aromatic ligand functionalization led to better dispersion of the particulates at high loadings than particulates modified with the aliphatic ligands as indicated by both USAXS and TEM. Quantitative analysis of the particle dispersion and agglomerate distributions from USAXS is in progress based on a unified fitting model. EFTEM maps of Si and C content in the nanocomposites allowed for imaging of the nano-SiO_2 within the block copolymer morphology and correlation with scattering experiments. This demonstration of synergistic control over nano-filler dispersion in BCP where the domain size is comparable to the particle diameter may enable further optimization of highly-filled nanocomposites in the future.

ACKNOWLEDGEMENTS

This research was supported in part by the Postgraduate Research Participation Program at the U.S. Army Research Laboratory, administered by the Oak Ridge Institute of Science and Education through an interagency agreement between the U.S. Department of Energy and Army Research Laboratory (contract number ORISE-1120-1120-99). Use of the Advanced Photon Source was supported by the U. S. Department of Energy, Office of Science, Office of Basic Energy Sciences, under Contract No. DE-AC02-06CH11357.

REFERENCES

1. R. Shenhar, T. B. Norsten, V. M. Rotello, *Adv. Mater.* **17,** 657-669 (2005).
2. H. Kang, F. A. Detcheverry, A. N. Mangham, M. P. Stoykovich, K. C. Daoulas, R. J. Hamers, M. Muller, J. J. de Pablo, P. F. Nealey, *Phys. Rev. Lett.* **100,** 4 (2008).
3. S. J. Jeon, S. M. Yang, B. J. Kim, J. D. Petrie, S. G. Jang, E. J. Kramer, D. J. Pine, G. R. Yi, *Chem. Mater.* **21,** 3739-3741 (2009).
4. J. Listak, I. F. Hakem, H. J. Ryu, S. Rangou, N. Politakos, K. Misichronis, A. Avgeropoulos, M. R. Bockstaller, *Macromolecules* **42,** 5766-5773 (2009).
5. P. Akcora, H. Liu, S. K. Kumar, J. Moll, Y. Li, B. C. Benicewicz, L. S. Schadler, D. Acehan, A. Z. Panagiotopoulos, V. Pryamitsyn, V. Ganesan, J. Ilavsky, P. Thiyagarajan, R. H. Colby, J. F. Douglas, *Nat. Mater.* **8,** 354-359 (2009).
6. A. Haryono, W. H. Binder, *Small* **2,** 600-611 (2006).
7. R. A. Vaia, J. F. Maguire, *Chem. Mater.* **19,** 2736-2751 (2007).
8. M. R. Bockstaller, Y. Lapetnikov, S. Margel, E. L. Thomas, *J. Amer. Chem. Soc.* **125,** 5276-5277 (2003).
9. J. Ilavsky, P. R. Jemian, A. J. Allen, F. Zhang, L. E. Levine, G. G. Long, *J. Appl. Crystallogr.* **42,** 469-479 (2009).
10. J. Ilavsky, P. R. Jemian, *J. Appl. Crystallogr.* **42,** 347-353 (2009).
11. K. A. Mauritz, R. I. Blackwell, F. L. Beyer, *Polymer* **45,** 3001-3016 (2004).
12. R.-J. Roe, Methods of X-ray and Neutron Scattering in Polymer Science, Oxford University Press, New York, 2000.

Mater. Res. Soc. Symp. Proc. Vol. 1312 © 2011 Materials Research Society
DOI: 10.1557/opl.2011.170

Long Cycle Life Nanocellulose Polypyrrole Electrodes

Gustav Nyström[1], Henrik Olsson[1], Martin Sjödin[1], Daniel O. Carlsson[1], Albert Mihranyan[1], Leif Nyholm[2] and Maria Strømme[1]

[1]Division of Nanotechnology and Functional Materials, The Ångström Laboratory, Uppsala University, Box 534, 751 21 Uppsala, Sweden

[2]Department of Materials Chemistry, The Ångström Laboratory, Uppsala University, Box 538, 751 21 Uppsala, Sweden

ABSTRACT

A polypyrrole (PPy) nanocellulose composite was shown to cycle well over 3000 cycles in 2.0 M NaCl electrolyte when used as the active material for both electrodes in an energy storage device. SEM micrographs show a highly porous nature of the conductive paper material and electrochemical charge-discharge measurements, as well as external electrode potential monitoring, confirm the good cycling behavior of the material.

INTRODUCTION

Limited cycling stability has long been considered one of the main drawbacks of conducting polymer based electrodes [1, 2]. Recent results, however, show progress towards acceptable cycle life [3-5]. In some cases, an improved cycle life was seen as a result of depositing the conducting polymer on a substrate, for instance PPy on Carbon nanotubes (CNTs) [6] or in a Nafion-PPy composite [7]. It was argued that the rigid CNT network may have a stabilizing effect on the PPy during cycling and that Nafion, besides having a mechanical stabilizing effect, also stabilizes the radical cations formed during the electrochemical oxidation of PPy. In both these studies, the CNT and Nafion composites showed a drastically improved cycle life compared to that of the pristine PPy polymer.

We have recently reported on an electrode material consisting of cellulose nanofibers extracted from the green algae Cladophora subsequently coated with a thin layer of PPy [8-11]. This composite conductive paper material has a nitrogen BET surface area of 80 m^2/g and was found to function as an electrode material in aqueous electrolyte all-polymer batteries [12]. In the present study, it is shown that long cycle life can be obtained with the present nanocellulose PPy electrodes in aqueous electrolytes.

EXPERIMENT

Cellulose from the Cladophora algae was extracted as previously described [13]. Preparation of the polypyrrole (PPy) nanocellulose composite has been described elsewhere [8]. In short, the nanocellulose PPy composite was prepared by thorough mixing of 300 mg Cladophora cellulose, well suspended in 100 mL water, with a 50 mL 0.43 M pyrrole solution followed by chemical polymerization of pyrrole on the cellulose fibers by adding 50 mL 0.3 M $FeCl_3$ while stirring.

The electrochemical characterizations were carried out using an Autolab PGSTAT302N potentiostat (Ecochemie, the Netherlands) and 2.0 M NaCl electrolyte solutions. A Pt auxiliary electrode (placed in a separate compartment) and a 3 M NaCl Ag/AgCl reference electrode were used in the 3-electrode measurements. For the 2-electrode measurements, small sandwich type battery cells were assembled with a symmetric setup of two nanocellulose-PPy electrodes separated by a filter paper and held together using two glass slides. Pt foils were used as current collectors and the weight of each electrode was kept around 10 mg. For illustration and photographs of the setup see ref [12]. The testing of the battery cells was done in a crystallization dish filled with the electrolyte, 2.0 M NaCl. Two multimeters (34401A Agilent technologies, USA) were used to individually monitor the potential of both electrodes versus a common 3 M NaCl Ag/AgCl reference electrode.

DISCUSSION

Chemical polymerization of polypyrrole onto cellulose nanofibers in water suspension yields a black conductive paper-like material composed of nanofibers of cellulose coated by a thin layer of polypyrrole as is seen in Figure 1. In the SEM micrographs, a porous fiber network is seen on the surface of the PPy-cellulose composite, cf. Fig. 1A. The porosity was maintained throughout the bulk of the ~1 mm thick material, as is seen in the lower magnification cross-section SEM image in Fig. 1B.

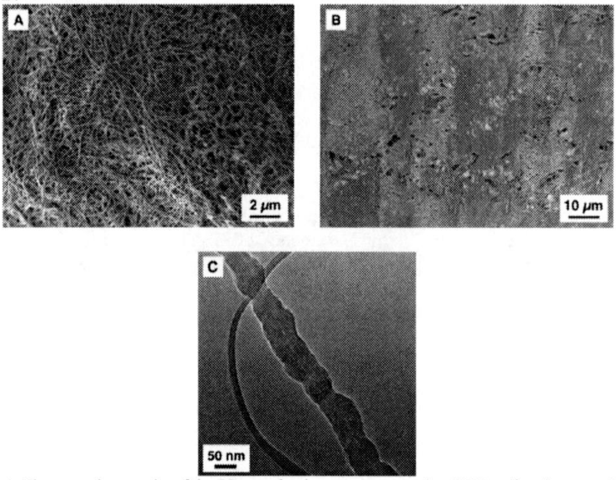

Figure 1. Electron micrographs of the PPy conductive paper composite, SEM surface image at 20 000 times magnification (A) and SEM cross section image at 1000 times magnification (B) as well as a TEM image of an individual nanocellulose-PPy fiber (C). Individual cellulose fibers coated with PPy as well as bundles of fibers are seen in (A). In (B) it is seen that the porosity is maintained throughout the bulk of the composite material. Wedge shaped regions are artifacts due to the Ar gas cross section polishing procedure (IB-09010CP Cross Section Polisher, JEOL, Japan). In (C) the true nanometer dimension of the cellulose-PPy fiber is clearly revealed.

The PPy conductive paper material can be used as the active electrode material in a symmetric aqueous (2.0 M NaCl) energy storage device, as was reported previously [12]. Five typical charge-discharge curves between 0 V and 1.2 V with 10 mA charging current are seen in Fig. 2. The typical capacitor-like charge and discharge curves allow for an easy determination of the capacitance from the slope of the discharge curve using the following equation:

$$C_{cell} = \frac{i}{\left(-\left(\frac{\Delta V}{\Delta t}\right) \cdot m\right)}$$

where i is the charging current and m is the total dry electrode weight of both electrodes. For a symmetric capacitor, the electrode specific capacitance is 2 times the cell capacitance. Normalizing with respect to the mass of one electrode adds another factor 2 to the cell capacitance.

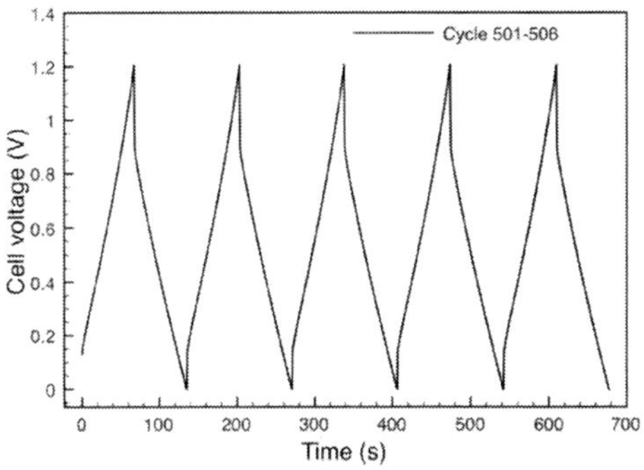

Figure 2. Five charge and discharge cycles at 10 mA. The total dry weight of both electrodes was 21.2 mg.

The cell capacitance and single electrode specific capacitance during extended cycling between 0 V and 1 V and 0 V and 1.2 V are seen in Fig. 3. It is seen that the initial capacitance is well maintained even after 3000 cycles in the aqueous 2.0 M NaCl electrolyte.

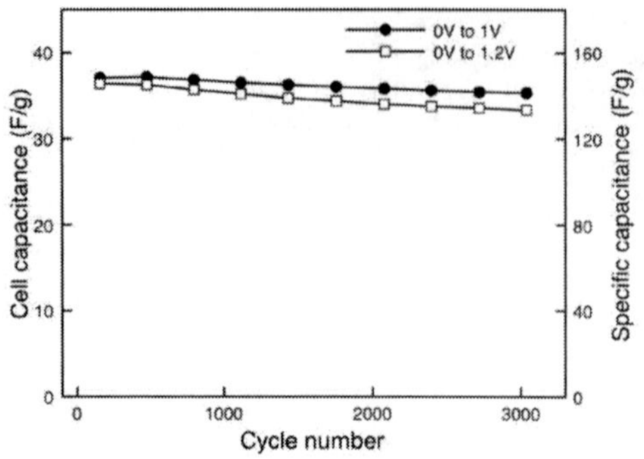

Figure 3. Cell capacitance and specific electrode capacitance as a function of cycle number when charging the cell to 1 V (closed circles) and 1.2 V (open squares) using a 10 mA charging current.

When assembling the energy storage device, both electrodes were in their as-synthesized, and slightly oxidized, state with an open circuit voltage of 0.2 V vs. 3 M NaCl Ag/AgCl. During the first cycles, the positive electrode will therefore reach potentials where degradation or overoxidation of PPy is expected [14], cf. Fig. 4. A minor degradation of the polymer is expected which would lead to a decrease in the pH. Such a bulk electrolyte pH decrease was indeed also found. It is believed that the local pH in the cell remains low and that this protects the PPy material from further degradation. This was confirmed by 3-electrode galvanostatic measurements at neutral and acidic pH. In the acidic electrolyte the electrode cycles without any observed degradation for more than 3000 cycles whereas in the neutral electrolyte the capacitance drops to zero after 600 cycles. It is also likely that the cellulose nanofibers are contributing positively to the good cycle life of the PPy-cellulose composite in a similar way as has been proposed for CNTs [6] and Nafion [7] composites. In addition, the adhesion between the nano thin polypyrrole layer and the cellulose substrate is likely to be higher than for corresponding thicker layers due to a higher degree of hydrogen bonding between cellulose and polypyrrole in the former. The thin PPy layers are also expected to tolerate volume expansions better than thicker layers. Both these effects, resulting from the nano thin PPy layers on the cellulose fibers, are expected to positively influence the cycle life of the composite material.

After the first few hours of cycling, the potential of the electrodes reached a common 0 V turning point and then cycled in a symmetric potential window for a total operation time of 120 hours.

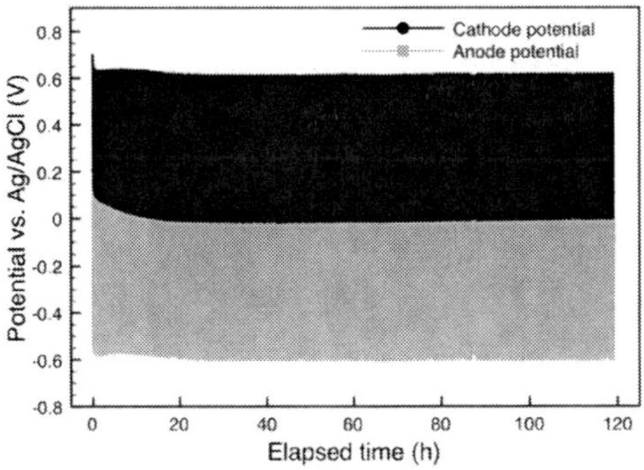

Figure 4. Individual potential of the cathode (black field) and the anode (grey field) versus an Ag/AgCl reference electrode during a total charge-discharge cycling time of 120 hours.

Encouraged by the stable cycling behavior of the device in the – for PPy-based devices in aqueous electrolytes – rather large potential regime of 0 V to 1.2 V, the potential window was further increased up to 1.8 V (not shown here). To our great surprise, it was found that the device cycled without degradation also in this potential interval. The underlying reason for this is a subject of further investigation in our laboratory.

CONCLUSIONS

A long cycle life over 3000 cycles has been demonstrated for a PPy conductive paper composite electrode material in aqueous 2.0 M NaCl electrolyte. It is believed that a local acidic pH in the energy storage device is protecting the material against degradation. The highly reversible cycle life shows that it is possible to use the presented composite material in cost-efficient and environmentally friendly water based energy storage devices.

ACKNOWLEDGMENTS

Kjell Jansson at Stockholm University is gratefully acknowledged for performing the cross section SEM imaging. The Swedish Foundation for Strategic Research (SSF, grant RMA08-0025), the Knut and Alice Wallenberg foundation, the Swedish Science Council (VR) and the Nordic Innovation Centre (contract number 10014) are acknowledged for financial support of this work.

REFERENCES

1. P. Novák, K. Müller, K. S. V. Santhanam, O. Haas, *Chem. Rev.* **97**, 207 (1997).
2. K. Naoi, M. Morita, *Electrochem. Soc. Interface* **17**, 44 (2008).
3. Y. G. Wang, H. Q. Li, Y. Y. Xia, *Adv. Mater.* **18**, 2619 (2006).

4. V. Gupta, N. Miura, *Electrochim. Acta* **52**, 1721 (2006).
5. W.-C. Chen, T.-C. Wen, *J. Power Sources* **117**, 273 (2003).
6. C. Peng, J. Jin, G. Z. Chen, *Electrochim. Acta* **53**, 525 (2007).
7. B. C. Kim, J. M. Ko, G. G. Wallace, *J. Power Sources* **177**, 665 (2008).
8. A. Mihranyan, L. Nyholm, A. E. Garcia Bennett, M. Strømme, *J. Phys. Chem. B* **112**, 12249 (2008).
9. K. Gelin, A. Mihranyan, A. Razaq, L. Nyholm, M. Strømme, *Electrochim. Acta* **54**, 3394 (2009).
10. A. Razaq, A. Mihranyan, K. Welch, L. Nyholm, M. Strømme, *J. Phys.Chem B* **113**, 426 (2009).
11. M. Strømme, G. Frenning, A. Razaq, K. Gelin, L. Nyholm and A. Mihranyan, *J. Phys. Chem. B* **113**, 4582 (2009).
12. G. Nyström, A. Razaq, M. Strømme, L. Nyholm, A. Mihranyan, *Nano Lett.* **9**, 3635 (2009).
13. A. Mihranyan, A. P. Llagostera, R. Karmhag, M. Strømme, R. Ek, *Int. J. Pharm.* **269**, 433 (2004).
14. T. W. Lewis, G. G. Wallace, C. Y. Kim, D. Y. Kim, "Studies of the overoxidation of polypyrrole", presented at *International Conference on the Science and Technology of Synthetic Metals*, Snowbird, Ut, Jul 28-Aug 02, 1996.

Nanostructured Polymeric Materials—Synthesis and Assembly

Mater. Res. Soc. Symp. Proc. Vol. 1312 © 2011 Materials Research Society
DOI: 10.1557/opl.2011.128

Inorganic-Organic Hybrid Copolymers derived from Silsesquioxanes or Carborane Building Blocks

Gunjan Gadodia,[1] Gregoire Cardoen,[1] Yoan Simon,[1] Hiroshi Abe,[2] E. Bryan Coughlin[1]
[1]Department of Polymer Science and Engineering, University of Massachusetts Amherst, Amherst MA 01002 U.S.A.

[2]Sekisui Integrated Research Inc. 2-2, Kamichoshi-cho, Kamitoba Minami-ku, Kyoto 601-8105 JAPAN

ABSTRACT

To explore competitive or cooperative effects novel organic-inorganic hybrid copolymers are being prepared and studied. The use of polyhedral oligomeric silsesquioxanes (POSS), a molecularly precise isotropic comonomer, is being utilized to take advantage of the inherent size scale of these particles, average diameters of 1-2 nm. The organic component selected for study in these hybrid systems are either semi-crystalline or amorphous polymers. The architectures of the hybrid copolymers range from random, to precise block copolymers, as well as telechelic and hemi-telechelic end-functionalized model compounds. The degree of POSS aggregation that occurs is found to be a function of thermal history, and processing conditions. Templating, or arresting, aggregation can be achieved using either crystalline organic polymer scaffolds in the bulk. The second inorganic comonomers for study has been constructed from icosahedral carboranes. Dicarbo-*closo*-decaboranes have been widely investigated for their thermal stability, chemical resistance, unique geometry, and the high cross-section for the capture of thermal neutrons. While carboranes have been widely incorporated into small molecules, metal complexes, and on a limited basis in polymer systems relatively little work exists relating their unique properties to systems with extended π-conjugation. Details of the syntheses, characterization and performance properties of both sets of hybrid systems will be discussed.

INTRODUCTION

POSS molecules with short peripheral groups act as spheres and pack in a hexagonal arrangement.[1] However, when POSS is attached to the backbone, the connecting linkage restricts the mobility of the cage and imposes considerable spatial constraints on the crystal shape.[2] Crystalline polymers have properties much different than amorphous polymers and the crystalline content must be considered for a given polymer application. In semi-crystalline block copolymers, one block can crystallize while the other block remains amorphous. The crystalline copolymer can effect the mechanical properties,[3] morphology[4-7] domain spacing,[8] and the kinetics of microphase separation.[9-11] When POSS is incorporated into the backbone of the polymer there is a competition of microphase separation between minimizing the interfacial energy or the thermodynamic driving force of crystallization.[12] As a result interesting morphologies can be obtained with the current system. A bottom-up approach was used to combine the newly emerging class of Polyhedral Oligomeric Silsesquioxanes (POSS) with conventional polymers to produce semi-crystalline conformationally asymmetric hybrid materials which can be used for novel applications. POSS have a generic formula of $R_n(SiO_{1.5})_n$ where R are alkyl groups used for polymerization, grafting or solubility and n is an even integer with $n \geq 4$.[13] POSS molecules have an inorganic silicon-oxygen core and are surrounded by an organic alkyl periphery. The periphery of the POSS occupies 70% volume of the molecule and plays a very important role in determining the physical properties of the molecule

The synthesis of a novel silyl-protected oxanorbornene imide carborane (SONIC) monomer as well as its copolymerization by ring-opening metathesis polymerization (ROMP) to obtain amphiphilic diblock copolymers is described. The subsequent functionalization and labeling of the polymers, as well as their solution behavior have been investigated. Initial biological studies have shown the incorporation of the carborane-containing polymers in carcinoma cells. These findings pave the way for future investigations of these polymeric architectures as delivery agents for boron neutron capture therapy of cancer.

EXPERIMENT

Synthesis of poly(ethylene-butylene) macroinitiator

Poly(ethylene-butylene) homopolymer (3 g, 0.51 mmol) was dissolved in THF (10 mL) containing triethylamine (2.22 mL, 1.61 g, 16 mmol), the temperature was reduced to 0 °C and E B-bromoisobutyryl bromide (0.61 mL, 1.14 g, 5.0 mmol) was added to the flask dropwise over 15 min. After stirring for 24 h, the reaction solution was dissolved in 20 mL methylene chloride. The solution was washed with water (2 x 20 mL), 2% HCl solution (1 x 20 mL) and water (1 x 20 mL) and then filtered over a plug of silica with dichloromethane as eluent. Column chromatography was carried out with hexane : ethyl acetate (20 : 1) as eluent. The solution was precipitated in cold methanol, filtered and vacuum dried. Yield 1.00 g of a viscous clear liquid (33%). ^1H NMR ($CDCl_3$, 400 MHz): δ 0.8 ($CH_2CHCH_2C\underline{H}_3$) 1.1 – 2.0 (all methylene protons), 1.29 ($(CH_3)_2C$), 1.64 ($CH_2C\underline{H}CH_2CH_3$), 4.2 ($CH_2C\underline{H}_2OCO$).

Synthesis of poly(EB-*b*-MA-POSS (isobutyl)) diblock copolymers

To an oven dried 10 mL round bottom flask was added poly(ethylene-butylene) macroinitiator (0.14 g, 0.0241 mmol), MA-POSS(isobutyl) (0.91 g, 0.96 mmol) (Ratio of PEB macroinitiator : MA-POSS(isobutyl) 1 : 40), PMDETA (5.55 μL, 0.0265 mmol), THF (1.0 mL). Three freeze-pump-thaw cycles were performed. CuCl (2.38 mg, 0.024 mmol), $CuBr_2$ (0.539 mg, 0.00241 mmol, 10 mol% relative to CuCl) were added to the mixture and again three freeze-pump thaw cycles were performed. Polymerization was carried out for 24 h at 50 °C. The reaction solution was then diluted with 10 mL THF and passed through a column of neutral alumina to remove catalyst. The colorless transparent solution was concentrated by evaporation. The polymer was then precipitated in methanol. The polymer was vacuum dried. Yield 0.21 g of a white powder (19.0%). ^1H NMR ($CDCl_3$, 400 MHz): δ 0.54 ($SiC\underline{H}_2CH(CH_3)_2$), 0.58 ($SiC\underline{H}_2CH_2CH_2OC(O)$-), δ 0.8 ($CH_2CHCH_2C\underline{H}_3$), 1.01 ($SiCH_2CH(C\underline{H}_3)_2$), 1.1 – 2.0 (all PEB methylene protons), 1.29 ($(CH_3)_2C$), 1.34 ($CH_2C(C\underline{H}_3)$), 1.6 ($SiCH_2C\underline{H}_2CH_2OC(O)$-), 1.64 ($CH_2C\underline{H}CH_2CH_3$), 1.8 ($SiCH_2C\underline{H}(CH_3)_2$), 1.91 ($C\underline{H}_2C(CH_3)$), 3.9 ($SiCH_2CH_2C\underline{H}_2OC(O)$), 4.2 ($C\underline{H}_2CH_2OC(O)$) ppm.

Polymer Characterization

Gel permeation chromatography (GPC) measurements were performed in tetrahydrofuran (THF) at 1.0 mL/min using a Knauer K-501 pump with a K-2301 refractive index detector and K-2600 UV detector, and a column bank consisting of two Polymer Labs PLGel Mixed D columns at 40 °C. Molecular weights are reported relative to polystyrene standards (Polymer Labs, Inc.). Nuclear magnetic resonance (NMR) spectra were collected on Bruker 400 MHz instrument and obtained from $CDCl_3$ solution. ^1H NMR spectra were referenced to residual $CHCl_3$ at 7.26 ppm and ^{13}C NMR spectra were referenced to residual $CHCl_3$ at 77.16 ppm. Infrared (IR) spectra were obtained on a Perkin-Elmer Spectrum One FTIR spectrometer equipped with an ATR accessory. The spectra were obtained on vacuum dried bulk samples. Thermogravimetric analysis (TGA) was carried out using a TA Instruments TGA 2950 thermogravimetric analyzer with a heating rate of 10° C/min from room temperature to 700° C under air. (40 mL/min). Samples for Differential Scanning Calorimetry (DSC) were thermally annealed at 150 °C and allowed to cool to room temperature over a period of 3 h. DSC was performed on TA Instruments DSC Q1000 and the heating and cooling rates of the sample were 10 °C / min.

DISCUSSION

POSS-Diblock Copolymers

The hydroxyl terminated PEB was reacted with α-bromoisobutyryl bromide in the presence of triethylamine to obtain PEB macroinitiators for atom transfer radical polymerization (ATRP) (Scheme 2). The macroinitiator synthesis was monitored by ^1H NMR, the resonances corresponding to methylene protons α to the hydroxyl group in PEB shift downfield from $\delta = 3.4$ ppm to $\delta = 4.2$ ppm indicating the formation of an ester bond and no residual resonance is observed at δ 3.4 ppm after the reaction indicating complete consumption of the starting material. Methacrylate functionalized POSS monomer was polymerized by ATRP using standard conditions and the PEB macroinitiator (Scheme 2). Unreacted monomer was removed either by precipitation or by Soxhlet extraction in methanol for 2 – 7 days. The GPC traces of PEB of molecular weight M_n 19,000 g/mol and p(EB-*b*-MAPOSS (isobutyl)) of molecular weight M_n 35,000 g/mol are shown in Fig. 1. A clear shift in the peak to higher molecular weight and very little overlap with the peak corresponding to the macroinitiator clearly shows the incorporation of the POSS block. The PDI of the p(EB-*b*-MA-POSS (isobutyl)) diblock copolymers was fairly narrow (PDI < 1.14). The number of units of POSS attached to PEB backbone were calculated by comparing the methyl peak integrations of the butylene units at $\delta = 0.8$ to the integration of methyl peaks of the isobutyl periphery of POSS at $\delta = 1.01$. The maximum number of units of POSS attached to PEB backbone was 31. Various compositions of the diblocks synthesized are listed in Table 1. The nomenclature of the diblock copolymers is on the basis of number of repeat units of EB and POSS and % ethylene content in PEB block, copolymer **1** with 107 units of EB, 9 units of POSS and a ethylene content of 44 is thus named $E(44)B_{107}POSS_9$.

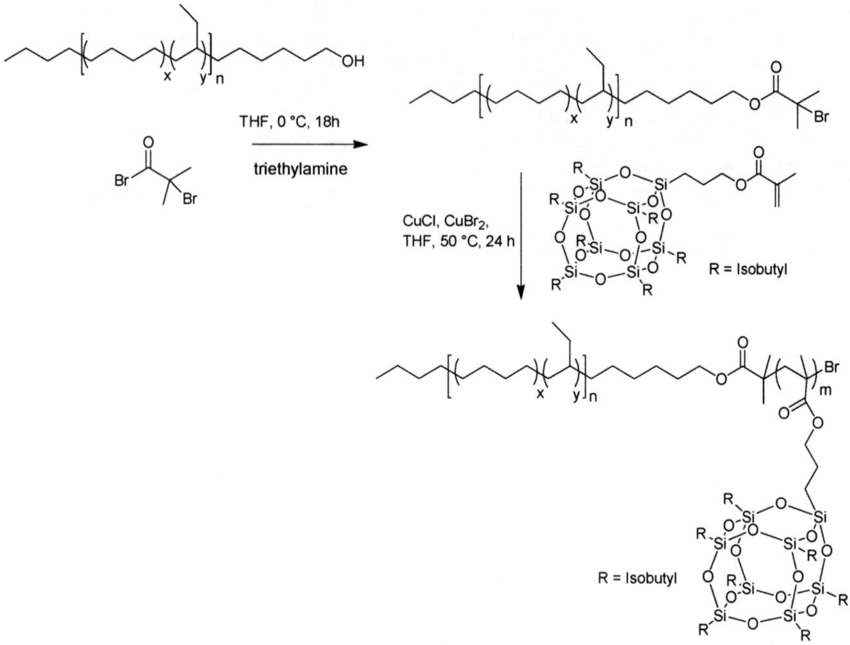

Figure 1 . Synthesis of PEB macroinitiator and p(EB-*b*-MA-POSS (isobutyl)) diblock copolymers.

Table 1. Molecular weights of poly(ethylenebutylene) and p(EB-*b*-MA-POSS(isobutyl)) diblock copolymers.

No.	Copolymer	M_n PEB g/mol (PDI)[a]	% 1,4 content	M_n p(EB-*b*-MA-POSS (isobutyl)) g/mol (PDI)[a]	$N_{(ethylene-butylene)}$	$N_{(POSS)}$	$f_{(EB)}$
1	$E(44)B_{107}POSS_9$	5,800 (1.04)	44	14,200 (1.05)	107	9	0.46
2	$E(65)B_{350}POSS_{17}$	19,000 (1.14)	65	35,000 (1.1)	350	17	0.60
3	$E(63)B_{632}POSS_{28}$	34,200 (1.09)	63	60,600 (1.08)	632	28	0.62
4	$E(72)B_{1072}POSS_{15}$	56,000 (1.05)	72	70,100 (1.06)	1072	15	0.84

a - Molecular weight determined by GPC against PS standard, b - Molecular weight determined by ^1H NMR, N - number of repeat units, *f* - volume fraction, assuming density of poly(ethylene-butylene) 0.91 g/cm^3 and density of poly(MA-POSS (isobutyl)) 1.15 g/cm^3 (by ASTM D792-00).

Carborane Diblock Copolymers

Ring-opening metathesis polymerization was utilized to control the architecture of amphiphilic diblock copolymers of carboranes in order to utilize them as efficient delivery agents for BNCT. The novelty of this concept lies in the combination of BNCT with the latest findings in terms of macromolecular drug delivery. Post-polymerization PEGylation was successfully undertaken as a way to prevent opsonization and prolong the half-time blood circulation. The monomers and polymers were analyzed by ^1H NMR, ^{13}C NMR and the molecular weights of the polymers were obtained by gel permeation chromatography (GPC). The copolymers were also labeled with rhodamine-B for *in vitro* studies. The copolymers were not observed to have a cytoxic effect on laryngeal carcinoma cells, and were furthermore, shown to be internalized by confocal microscopy. The formation of micelles was confirmed by dynamic light scattering and the micelles are currently under investigation using transmission electron microscope and scanning-force microscopy.

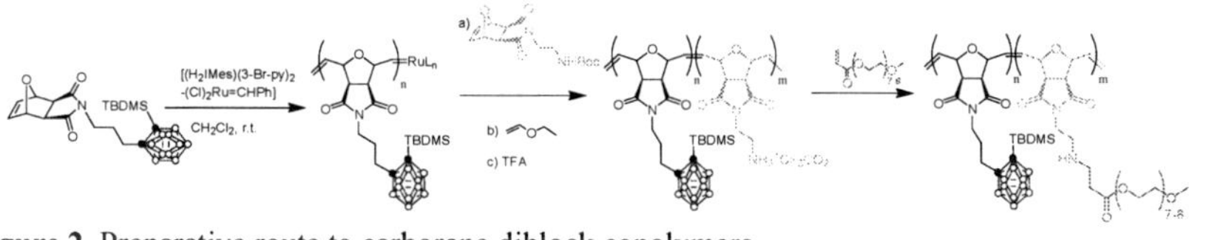

Figure 2. Preparative route to carborane diblock copolymers

CONCLUSIONS

Low PDI poly(ethylene-butylene) homopolymers and p(EB-*b*-MA-POSS(isobutyl)) diblock copolymers were synthesized by a combination of anionic polymerization and ATRP. Both phases of MA-POSS and EB crystallize depending upon the number of repeat units of MAPOSS and 1,4-content in the EB chains. Three different morphologies i.e. cylinders, lamellae and spheres of the semicrystalline diblock copolymer were obtained by changing the relative volume fractions of EB and POSS phases. Inverse cylindrical morphology was observed, with majority 60 volume % EB block forming the cylindrical phase while the minor MAPOSS phase occupies the periphery around the cylinder. This is presumably was due to conformational asymmetry or relative flexibility of the phases. Crystalline lamellar morphology was obtained due to the crystallization of the POSS cubes or due to high χ values. Further investigations are ongoing to study the morphology of the block copolymers under different annealing conditions, different crystalline content and lowering EB volume fractions.

The synthesis of a novel silyl-protected oxanorbornene imide carborane (SONIC) monomer as well as its copolymerization by ring-opening metathesis polymerization (ROMP) to obtain amphiphilic diblock carborane copolymers was described. The subsequent functionalization and labeling of the polymers, as well as their solution behavior have been investigated. Initial biological studies have shown the incorporation of the carborane-containing polymers in carcinoma cells. These findings pave the way for future investigations of these polymeric architectures as delivery agents for boron neutron capture therapy of cancer.

REFERENCES

1. Waddon, A. J.; Coughlin, E. B. *Chemistry of Materials* **2003**, *15*, 4555-4561.
2. Pyun, J.; Matyjaszewski, K.; Wu, J.; Kim, G. M.; Chun, S. B.; Mather, P. T. *Polymer* **2003**, *44*, 2739-2750.
3. Mohajer, Y., Wilkes, G. L., Wang, I. C., McGrath, J. E. *Polymer* **1982,** *23,* 1523-1535.
4. Nojima, S. Kato, K., Yamamoto, S. Ashida, T. *Macromolecules* **1992,** *25,* 2237-2242.
5. Ryan, A. J., Hamley, I. W., Bras, W., Bates, F. S. *Macromolecules* **1996,** *28,* 3860-3868.
6. Hamley, I. W., Fairclough, J. P. A., Terril, N. J., Ryan, A. J., Lipic, P. M., Bates, F. S., Towns-Andrews, E. *Macromolecules* **1996,** *29,* 8835-8843.
7. Castillo, R. V., Muller, A. J., Lin, M., Chen, H., Jeng, U. Hillmyer *Macromolecules* **2008,** *41,* 6154-6164.
8. Rangarajan, P., Register, R. A., Fetters, L. J., Bras, W., Naylor, S., Ryan, A. J. *Macromolecules* **1995,** *28,* 4932-4938.
9. Shiomi, T., Takeshita, H., Kawaguchi, H., Nagai, M., Takenaka, K., Miya, M. *Macromolecules* **2002,** *35,* 8056-8065.
10. Chen, H-J., Wu, J-C., Lin T-L., Lin, J. S. *Macromolecules* **2001,** *34,* 6936-6944.

11. Shiomi, T., Tsukada, H., Takeshita, H., Takenaka, K., Tezuka, Y. *Polymer* **2001,** *42,* 4991-5004.
12. Quiram, D. J., Register, R. A., Marchand, G. R. *Macromolecules* **1997,** *30,* 4551-4558.
13. Li, G. Z.; Wang, L. C.; Ni, H. L.; Pittman, C. U. *Journal of Inorganic and Organometallic Polymers* **2001**, *11*, 123-154.

Mater. Res. Soc. Symp. Proc. Vol. 1312 © 2011 Materials Research Society
DOI: 10.1557/opl.2011.205

Assembly of Nanoparticles from Bioactive Peptides and Chitosan

B. Hu,[1,2] Q. R. Huang, [1] and X. X. Zeng, [2]
[1]Department of Food Science, Rutgers University, 65 Dudley Road, New Brunswick, New Jersey 08901, USA;
[2]College of Food Science and Technology, Nanjing Agricultural University, Nanjing 210095, People's Republic of China;

ABSTRACT

Assembly of nanoparticles from bioactive peptides, caseinophosphopeptides (CPPs) and chitosan (CS) at physiological conditions and various CS/CPPs mass ratios have been systematically studied using a combination of turbidimetric titration, dynamic light scattering (DLS), electrophoretic mobility (zeta-potential) and transmission electron microscopy (TEM). Peptides, incorporated with CS forming nanoparticles, have already been prepared and identified using liquid chromatography-tandem mass spectrometry (LC-MS-MS). They are characteristic with different amount of clusters of phosphorylated seryl residues. At low salt concentration, an increase of CS/CPP mass ratio shifted the critical $pH_{\varphi 1}$, which designated the formation of CS/CPP nanocomplexes, as well as pH_{max}, representing the neutralization of positive and negative charge to higher pH values. The peptide-polymer binding mechanism was analyzed according to the results of DLS, electrophoretic mobility, and TEM. First, negatively charged CPPs absorbed to positively charged CS molecular chain to form intrapolymer nanocomplexes saturated with CPPs (CPPNP). Then, the negatively charged CPPNP was bridged by added positively charged CS. Finally, novel nano-scaled spherical brushes were formed as additional CS molecule absorbed back to and bound the CPPNP. Phosphorylated groups in the CPPs might be the dominant sites for interaction with $-NH^{3+}$ on the CS molecular chain.

INTRODUCTION

Recently, significant efforts have been devoted to investigate the assembly between polyelectrolyte and protein forming different hierarchical supramolecular structures on the nm–μm length, such as coacervates, multilayers and brushes[1-3]. A number of non-covalent forces can contribute to complex formation between polymers and proteins, among which, the predominance of electrostatic interaction is widely accepted. Furthermore, the Voorn–Overbeek, Veis–Aranyi theories as well as their modifications have already been brought forward focusing on the mechanism of the interaction between polyelectrolyte and protein[3,4]. However, as the constitution elements of protein, the detailed information about the interaction between natural peptides and polyelectrolyte as well as the corresponding interaction model has been scarcely reported.

Food is a fascinating and important material, making up some of the most complex examples of soft condensed matter (SCM) with which we interact daily[5]. It is a natural resource of biopolymer soft materials, not only biodegradable and biocompatible, but also biofunctional[6,7]. Tryptic digestion of the casein proteins yields caseinophosphopeptides (CPPs) from the N terminus polar region, which contains clusters of phosphorylated seryl residues[8]. CPPs reveal multifunctional bioactivities[9,10,11,12,13]. Chitosan (CS) is composed of glucosamine, known as 2-amino-2-deoxy-(1→4)-β-D-glucopyranan. CS is a non-toxic, biodegradable and biocompatible polyelectrolyte. Moreover, it possesses the antimicrobial and antitumor activities[14,15].

Herein, the interaction between CS and CPPs was comprehensively characterized. The interaction between CS and CPP as a function pH in different mass ratios was monitored by turbidity titration. DLS was used to have access to the size of the peptide/polysaccharide complexes. Measurements of zeta-potential were used to study the CPPNP induced by CS as well as the adsorption of CS on the CPPNP surface. Direct evidence for the mechanism of the interaction between the peptides and polysaccharide was gained from the image of TEM.

EXPERIMENT

CPPs were prepared and identified with HPLC-MS-MS as described in our previous study[16]. Chitosan (CS) of the molecular weights 100 kDa, derived from crab shell, was obtained from Golden-Shell Biochemical Co. Ltd (Hangzhou, China). The degree of deacetylation was 90%.

The pH-dependent turbidity measurements, reported as 100 – T%, were carried out at 420 nm using a Brinkmann PC910 colorimeter equipped with a 1 cm path length optical probe. The colorimeter was calibrated to read 100% transmittance with Milli-Q water. NaOH solution (suitable concentration) was used to adjust the pH of the mixture containing equal volume of 0.5 mg/mL CPP and CS with varying concentrations (0, 0.0625, 0.125, 0.25, 0.5 and 1.0 mg/mL) under gentle stirring. All the measurements were conducted at 25 °C.

The mean particle size and size distribution were determined using dynamic light scattering – based BIC 90 plus particle size analyzer equipped with a Brookhaven BI-9000AT digital correlator (Brookhaven Instrument Corporation, New York, NY, USA) at a fixed scattering angle of 90° at 25 ± 1 °C. The change of the particle size with different peptide/polysaccharide mass ratios was studied. All these measurements were made in triplicate at 25 ± 1 °C. Electrophoretic mobility (UE) (a particle's velocity in an electric field) for homogenous and mixed CPP-CS systems was investigated as a function of CPP-CS mass ratios using a Zetasizer Nano-ZS90 (Malvern Instruments, Westborough, MA, USA). All measurements were made in triplicate. The morphological characteristics of nanocomplex with the increase of CS/CPP mass ratio were examined by a high performance digital imaging TEM machine (JEOL H-7650, Hitachi High-Technologies Corporation, Japan). One drop of the suspension was stained with 2 % (w/v) phosphotungstic acid, placed on a copper grid and allowed to evaporate in air. Once evaporated the samples were placed in a TEM for imaging. The accelerating voltage used was 100 kV and the images were taken on a Gatan electron energy loss spectrometry system using 6 eV energy slit.

DISCUSSION

Figure 1 showed turbidimetric titration curves of mixtures of CPP/CS as a function of pH at 0, 0.0625, 0.125, 0.25, 0.5, and 1mg/mL CS concentrations (C_{CS}). It was generally accepted that the abrupt change of turbidity arose mainly from the change of either the mass or the size of complex in the solution[17,18]. The abrupt increase of turbidity at $pH_{\varphi 1}$ was supposed to denote the point of global phase transition, where the cross-linking of CS with CPP resulted in the formation of insoluble peptide/ polysaccharide complexes. It was well-known that electrostatic interactions played an important role in the formation of protein-polysaccharides complexation[2,19]. Increasing the negative net charge of CPP via increase in pH would enhance its binding to CS, which leaded to the increase in complex size. The turbidity maximum might be caused by the simultaneous effects of increasing complex size and solubility. Before pH_{max}, neutralizing of the positive and negative net charge leaded initially to a loss of solubility causing sharply increase in turbidity. After the CS was saturated with binding CPP, the solubility of the complex increased with further increase of pH because the surface peptides became negatively charged again under higher pH value. The competition of these size and solubility effects caused the maxima in turbidity. Generally, $pH_{\varphi 2}$ was used to denote the dissociation of protein/polysaccharide complex. At higher CS concentrations, $0.25 mg/mL \leq C_{chitosan} \leq 1 mg/mL$, upon increase in the polymer concentration, the intrapolymer complex saturated with binding CPP aggregated to interpolymer species trapped by CS molecular chain. As a weak base polysaccharide, CS was insoluble at neutral and alkaline pH solutions. The excess CS and the CPP/CS interpolymer complex lost the positive charge and precipitated as the pH increased up to 7.

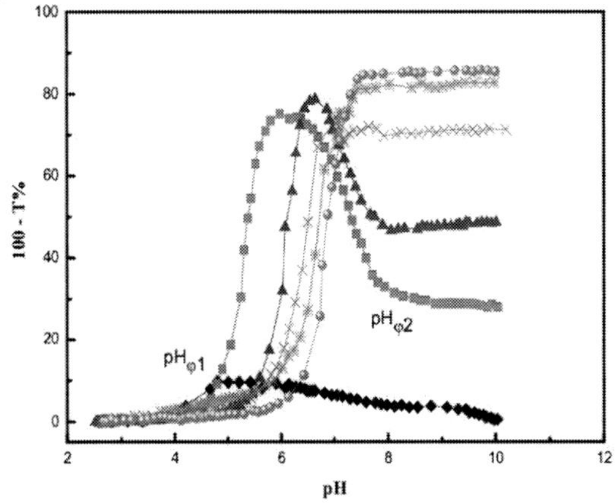

Figure 1. Turbidity (100 - T%) as a function of pH of the mixture of 0.5 mg/mL casein phosphopeptide with different concentrations of chitosan (◆, 0 mg/mL chitosan; ■, 0.0625 mg/mL chitosan; ▲, 0.125 mg/mL chitosan; ×, 0.25 mg/mL chitosan; □, 0.5 mg/mL chitosan; ● 1.0 mg/mL chitosan) in 0.01M NaCl.

Table1 showed the apparent diameters and surface charge for the CPP-CS nanoparticles formed as a function of CS/CPP mass ratio at pH 6.0, which has been proximity to physiological pH. Clearly, particle size increased with CS/CPP mass ratio at low Ccs, and reached the maximum. In this process, the surface charge (Zeta potential) of CPP-CS nanoparticles increased linearly from -17.1 to -2.3. Interestingly, further increase of CS/CPP mass ratio caused the particle size sharply decreased back to a plane around 115 nm again. The PDI ranged from 0.1 to 0.15, indicating that a homogeneous dispersion of nanoparticles was obtained. Meanwhile, surface charge of the nanoparticles (Zeta potential) increased sharply from negative to positive then reached a plateau with smooth increase to +29.4 mV. Figure 2 shows the TEM images. Firstly, irregular shaped CPPNP is initiated at the CS/CPP mass ratio 0.02 (Figure 2A). Then, as the CS/CPP mass ratio increased to 0.125 (Figure 2B), highly bridging of the CPPNPs with CS molecular chains could be seen clearly. Eventually, superfluous CS molecular adsorbed back to the surface of the nanoparticle forming a so-called "hairy layer", which can be seen clearly in Figure 2C. Until now, the CS-CPP nano-scaled spherical brushes were formed.

Table1: Change in mean particle size and electrical charge (ζ-potential) of CS/CPP nanoparticles as the CS/CPP mass ratio increased (both the pH of CPP and CS are pH6).

CS/CPP mass ratio	0.02	0.05	0.08	0.125	0.2	0.5	1.0
Particle Size (nm)	118±1.3	122±1.8	608±1.1	877±3.2	174±1.7	115±1.4	117±1.3
Zeta Potential(mV)	-17.1±0.3	-11.2±0.7	-7.1±0.2	-2.3±0.3	19.3±0.6	24.1±0.9	29.4±0.7

Combining the results of DLS□Zeta potential as well as the TEM image, we supposed the interaction model between CPP and CS as following. At low CS/CPP mass ratio, negatively charged CPPs easily bound to and cross-linked the positive charged CS molecular, partially neutralized the positive charge, forming intrapolymer nanocomplex saturated with CPPs (CPPNP)[20]. Following, the sharp increase in the particle size with increasing CS/CPP mass ratio might be contributed to the bridging of the negatively charged CPPNP with added positive charged CS. At high CS concentration, superfluous CS molecular adsorbed back to the surface of CPPNP to form the regular positive charged CS-CPP nano-scaled spherical brushes (CS-brushes). CPPNP was saturated with and packed within chitosan. Then, there was essentially only the CS-brushes and free CS. The electronic repulse between the positively charged CS-brushes and free CS hinders the further largely adsorption of free CS to the surface of the nanoparticles, keeping the particle size and surface charge steady with small change. In addition, the absorption of CS back to the CPPNP as well as the electronic repulse between the CS-brush and free CS made the nanoparticles more compact. This observation was consistent with the results found from the addition of cationic chitosan to the Blg/pectin anionic secondary emulsion[21]. This interaction model was similar to the one based on micelle–polyelectrolyte coacervation at low salt concentration. The binding between the micelle and polyelectrolyte increased at low salt concentration. The excess of macroion in stoichiometric produced complexes with net charges all of the same sign, which by mutual repulsion could not coacervate[22].

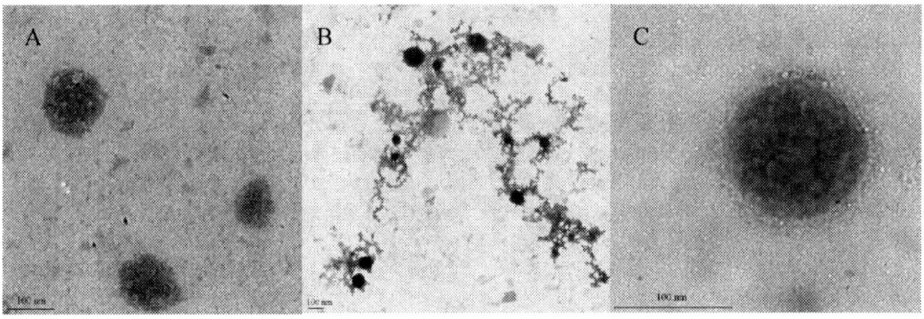

Figure 2. TEM images of the nano-suspension as the CS-CPP mass ratio are: 0.02 (A), 0.125 (B) and 1 (C) at pH6.

CONCLUSIONS

In low salt concentration, bioactive peptides, CPPs and CS assembled forming novel polysaccharide nano-scaled spherical brushes encapsulating peptides. First, negatively charged CPPs bound to and cross-linked the positive charged CS molecular forming intrapolymer nanocomplex saturated with CPPs (CPPNP). Following, the negatively charged CPPNP was bridged by added positively charged CS. Finally, novel nano-scaled spherical brushes were formed as additional CS molecule absorbed back to and bound the CPPNP. This study represents a new approach for preparing CS nanoparticles with bioactivity. Further work focusing on the bioactivities of these nanoparticles will be done in the future.

ACKNOWLEDGMENTS

This work was supported by a grantin-aid from 863 Program, Ministry of Science and Technology, People's Republic of China (2007AA10Z351 and 2007AA100403). United States Department of Agriculture National Research Initiative (#2009-35603-05071, QH).

REFERENCES

1. F. Caruso, H. Mohwald, Protein multilayer formation on colloids through a stepwise self-assembly technique. *J. Am. Chem. Soc.* **121,** 6039 (1999).
2. C. L. Cooper, P. L. Dubin, A. B. Kayitmazer, S. Turksen, Polyelectrolyte–protein complexes. *Curr. Opin. Colloid Interface Sci.* **10,** 52 (2005).
3. C. G. de Kruif, F. Weinbreck, R. de Vries, Complex coacervation of proteins and anionic polysaccharides, *Curr. Opin. Colloid Interface Sci.* **9,** 340 (2004)
4. D. J. Burgess, "Complex coacervation: Microcapsule formation," Macromolecular Complexes in Chemistry and Biology, Eds. P. L. Dubin, J. Bock, R. M. Davies, D. N. Schulz, C. Thies, (Springer–Verlag, 1994) pp. 285–300.
5. M. S. Shoichet, Polymer Scaffolds for Biomaterials Applications. *Macromolecules* **43,** 581 (2010).

6. R. Mezzenga, P. Schurtenberger, A. Burbidge, M Michel, Understanding foods as soft materials. *Nat. Mater.* **4,** 729 (2005).
7. J. Ubbink, A. Burbidge, R. Mezzenga, Food structure and functionality: a soft matter perspective. *Soft Mater.* **4,** 1569 (2008).
8. R. X. Su, W. Qi, Z. M. He, S. X. Yuan, Y. B. Zhang, Pancreatic hydrolysis of bovine casein: Identification and release kinetics of phosphopeptides. *Food Chem.* **104,** 276 (2007).
9. D. Erba, S. Ciappellano, G. Testolin, Effect of the ratio of casein phosphopeptides to calcium (w/w) on passive calcium transport in the distal small intestine of rats. *Nutr.* **18,** 743 (2002).
10. K. J. Cross, N. L. Huq, J. E. Palamara, J. W. Perich, E. C. Reynolds, Physicochemical characterization of casein phosphopeptide-amorphous calcium phosphate nanocomplexes. *J. Biol. Chem.* **280,** 15362 (2005).
11. J. M. Laparra, A. Alegria, R. Barbera, R Farre, Antioxidant effect of casein phosphopeptides compared with fruit beverages supplemented with skimmed milk against H_2O_2-induced oxidative stress in Caco-2 cells. *Food Res. Int.* **41,** 773 (2008).
12. M. Kampa, S. Loukas, A. Hatzoglou, A. Daminaki, P. M. Martin, E. Castanas, Opioid alkaloids and casomorphin peptides decrease the proliferation of prostatic cancer cell lines (LNCaP, PC3 and DU145) through a partial interaction with opioid receptors. *Eur. J Pharmacol.* **335,** 255 (1997).
13. I. Hata, S. Higashiyama, H. Otani, Identification of a phosphopeptide in bovine α_{s1}-casein digest as a factor influencing proliferation and immunoglobulin production in lymphocyte cultures. *J. Dairy Res.* **65,** 569 (1998).
14. L. F. Qi, Z. R. Xu, X. Jiang, C. H. Hu, X. F. Zou, Preparation and antibacterial activity of chitosan nanoparticles. *Carbohydr. Res.* **339,** 2693 (2004).
15. L. F. Qi, Z. R. Xu, In vivo antitumor activity of chitosan nanoparticles. *Bioorg. Med. Chem. Lett.* **16,** 4243 (2006).
16. B. Hu, S.S. Wang, J. Li, L.M. Meng, X.X. Zeng, Q.R. Huang, Assembly of Nanoparticles from Bioactive Peptides and Chitosan, ACS NANO (Submitted).
17. J. L. Xia, P. L. Dubin, H. Dautzenberg, Light scattering, electrophoresis, and turbidimetry studies of bovine serum albuminpoly(dimethyldiallylammonium chloride) complex. *Langmuir* **9,** 2015 (1993).
18. H. L. Jiang, K. J. Zhu, Polyanion/gelatin complexes as pH sensitive gels for controlled protein release. *J. Appl. Polym. Sci.* **80,** 1416 (2001).
19. k. Laos, G. J. Brownsey, S. G. Ring, Interactions between furcellaran and the globular proteins bovine serum albumin and β-lactoglobulin. *Carbohy. Polym.* **67,** 116 (2007).
20. J. L. Xia, H. W. Zhang, D. R. Rigsbee, P. L. Dubin, T Shaikh, Structural elucidation of soluble polyelectrolyte-micelle complexes: intra-vs interpolymer association. *Macromolecules* **26,** 2759 (1993).
21. D. Guzey, D. J. McClements, Formation, stability and properties of multilayer emulsions for application in the food industry. *Adv. Colloid Interface Sci.* **128–130,** 227 (2006).
22. Y. Wang, K. Kimura, Q. R. Huang, P. L. Dubin, W. Jaeger, Effects of salt on polyelectrolyte– micelle coacervation. *Macromolecules* **32,** 7128 (1999).

Mater. Res. Soc. Symp. Proc. Vol. 1312 © 2011 Materials Research Society
DOI: 10.1557/opl.2011.129

Self-Assembling Functionalized Amino Acids into Unusual Shapes

Justin R. Barone, Naresh K. Budhavaram, and Katherine J. Harvey
Biological Systems Engineering Dept., Virginia Tech, 303 Seitz Hall (0303), Blacksburg, VA 24061, U.S.A.

ABSTRACT

Multi-component small molecule systems that are amphiphilic or that can hydrogen bond end-to-end or side-to-side have been shown to self-assemble into a variety of shapes including fibers, rods, sheets, plates, spheres, and tubes. Recently, we have identified a simple route to self-assemble the same shapes from one-component systems. The structures form by attaching ethyl vinyl sulfone (EVS) to amino acids in water at room temperature. Choice of amino acid, amount of EVS substitution, and solvent conditions determine the final shape. Functionalized amino acids spontaneously form structures like fibers, spheres, tubes, and donuts when dried from solution. Here we focus on fibers and tubes.

INTRODUCTION

Self-assembly is an interesting processing concept because it relies on minimizing the free energy of interaction embodied in the molecules to create structures rather than the application of external energy. Nature does this by very precisely forming molecules that will spontaneously interact in aqueous solution and presumably is its way to conserve energy. Self-assembly proceeds from the molecular level by first constructing an elementary building block. For example, prions are composed of elementary nanometer-sized β-sheets.[1] In nature, the β-sheet is a fairly common protein building block. A typical β-sheet forms when two peptide molecules form a β-strand through amine-carbonyl hydrogen bonding on the peptide main chain. Self-assembly continues by many β-strands forming a β-sheet and then the sheets stack into larger structures until a nanometer to micrometer sized object results such as a fiber.[2] β-structures in general perform a variety of functions in nature including bearing load, stabilizing enzymes, protecting bacteria, and acting as recognition motifs for enzymes.[3,4] Recently, it has been shown that molecules other than the carbonyl can form β-structures. Sulfonyls have been used in β-strand mimics for therapeutics.[5,6] Essentially, the sulfonyl replaces the carbonyl in a β-strand hydrogen bond. One origin of many common diseases is protease cleavage of proteins required for biochemical function. The protease attacks the protein rendering it unable to do its job and initiating disease. A lot of these protease enzymes are known to bind to a β-strand motif. Therefore, β-strand mimics introduced into the body could act as enzyme inhibitors by forming similar structures recognized by the enzyme but not be made of natural protein.

Two important structures self-assembled in nature are micron-sized fibers and tubes. The fiber is the material of choice in nature for structural applications. Micron-sized fibers of (a) silk capture insects, (b) keratin provide protection and support to animals, and (c) collagen impart rigidity to cartilage.[4] Tubes are a common structure in nature, for instance in the form of veins or viruses like tobacco mosaic virus.[7] Compared to spheres or fibers, tubes are thermodynamically more complex to assemble. These self-assembled structures rely on some elementary building block formation such as the β-sheet.

Synthetic self-assembling systems usually rely on organic solvents and multiple components and not aqueous solutions at standard conditions. However, the structures observed: ribbons, tubes, and sheets along with other shapes, are formed by noncovalent interactions between the molecules of the various components, which is very similar to a natural motif.[8-17] Two component systems in particular have been used in the past because of the ease of having different charged ends, geometries, and amphiphilic properties that can interact to form a shape. Using smaller biobased molecules is a rather interesting motif because of a potentially easier synthesis and less complicated assembly. Derivatives of lysine[8], alanine[18], glycine,[19] aspartic acid,[19] and phenylalanine[20] have been reported in the synthesis of structures. The largest drawback to most of these approaches is the deviation from the aqueous standard conditions at which nature operates.[8] Hirst et al. reported fibrous and flat platelet structures from two component gel systems containing lysine based dendrites and diaminododecane.[15] Fibers from derivatives of glycine and aspartic acid were developed by Kuang et al.[19] while Battacharya et al.[20] used bipolar urethane amides of L-phenylalanine. Jang et al. synthesized fibers, wires, ribbon-like structures, tubules and plate-like structures using a three component system of terephthaloylbisalanine, dodecylamine, and a metal complex of zinc, cadmium, or copper.[18] The materials described in this were formed from one component systems of substituted amino acids in aqueous solution.

EXPERIMENT

Materials. L-lysine (Lys or K, mol. wt. 146.19 g/mol) and alanine (Ala, A, mol. wt. 89.09) were purchased from Sigma Aldrich and ethyl vinyl sulfone (EVS, mol. wt. 120.17 g/mol) was purchased from VWR.

Amino Acid Substitution. 0.5 g of amino acid was added to 50 ml of pure water and the pH was adjusted to 9. After the amino acid completely dissolved, the desired amount of EVS was added to the solution and reacted for 24 hr at 30°C. After about 2 hrs, the pH dropped to 6. Following reaction, the solution was dried in Teflon-coated aluminum boats for 2 days at ambient conditions. The molar ratio of amino acid to EVS was based on the number of potential reactive sites (amine groups) on lysine or alanine. Lysine had 2 NH_2's and so the total number of reactive sites was counted as 4, since both primary and secondary amines could participate in the reaction.[21]

Fourier Transform-Infrared (FT-IR) Spectroscopy. All samples were analyzed with a Thermo-Nicolet 6700 FT-IR spectrometer. Dried samples were collected on a SmartOrbit diamond ATR cell with a total of 64 scans and a resolution of 4 cm^{-1}. Solution samples were collected on a ZnSe trough also in ATR mode over 256 scans and a resolution of 4 cm^{-1}. Spectra were baseline corrected and analyzed using OMNIC v. 7.3 software.

X-ray powder diffraction (XRD). XRD patterns were recorded on a PANalytical X'Pert PRO X-ray diffractometer (Westborough, MA) using Co radiation generated at 40 kV and 40 mA. Scanning was done with a Theta/Theta goniometer from 2-70° 2θ with a step size of 0.0668545° 2θ at a time of 600 s. The incident wavelength was λ=0.154 nm.

Scanning Electron Microscopy (SEM). A Zeiss LEO 1550 Field Emission Scanning Electron Microscope was used for measuring the modified lysine samples. Dried samples were mounted on stubs, Au coated, and imaged at 5 kV and 5 mm working distance.

RESULTS AND DISCUSSION

We have previously characterized the substitution reactions of EVS on amino acids using Raman spectroscopy.[22] It was found that lysine preferably substitutes at the α-carbon at pH 9, which would be expected from the pK_a of the α- and ε-amine. Using FT-IR spectroscopy, substitution reactions can be followed in solution through the change in the amine on the amino acid (Fig. 1a) and increase in CH_2 absorbance on EVS (Fig. 1b). The sulfonyl absorbance increase around 1100 cm^{-1} could also be used. All spectra were normalized at $\nu_s(COO^-)$~1400 cm^{-1}, which showed no shifting with reaction conditions, suggesting the carboxylate was not affected by EVS substitution or subsequent aggregation.

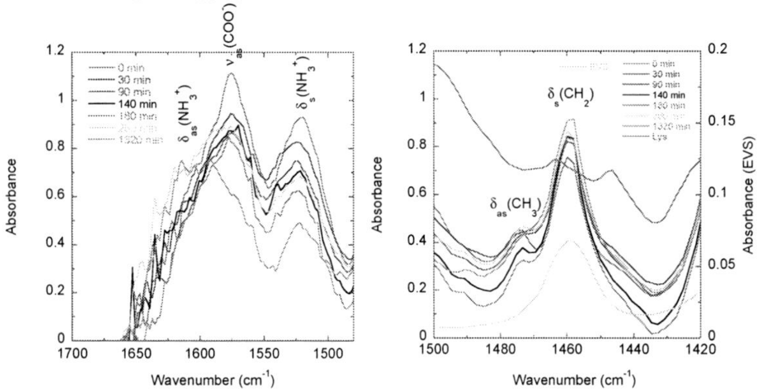

(a) (b)

Figure 1. FT-IR spectroscopy results of reacting lysine solutions showing (a) shifting at $\delta_s(NH_3)$~1500 cm^{-1} suggesting that the reaction is that the α-NH_2 and (b) increase in EVS CH_2 absorbance.

Observations of dried EVS-substituted alanine and lysine samples showed unusual microstructures forming under certain conditions. Specifically, at moderate molar substitutions of EVS on alanine and lysine, fibers and tubes were observed, respectively, as shown in Fig. 2.

(a) (b) (c)

Figure 2. (a) Ala:EVS 1:0.2 (42 Kx), (b) Lys:EVS 1:0.5 (15Kx), and (c) Lys:EVS 1:1 (5Kx) form fibers, tubes, and rosettes, respectively.

FT-IR spectra of dried materials suggested that aggregation of EVS-substituted lysine was through interaction of an EVS sulfonyl (S=O) and secondary α-NH on lysine. Fig. 3 shows spectra of dried materials as a function of various preparation and drying conditions. No shifting was observed at $\nu_{as}(COO^-)$ at 1580 cm^{-1} and $\nu_s(COO^-)$~1400 cm^{-1} suggesting the carboxylate did not react with EVS or participate in the aggregation process. All spectra were therefore again normalized at $\nu_s(COO^-)$~1400 cm^{-1}. Large shifts were observed in $\delta_s(NH_3^+)$ at 1500 cm^{-1} on lysine. Shifting and peak shape changes were observed on $\nu_s(S{=}O)$ at 1306 cm^{-1} and $\nu_{as}(S{=}O)$ at 1125 cm^{-1} on EVS (Fig. 3a,b). Part of the absorbance at 1275 cm^{-1} originated in S=O of EVS and part was the C-N reaction product of lysine and EVS, which was why this absorbance increased in intensity over $\nu_s(S{=}O)$ at 1306 cm^{-1} as EVS substitution increased.

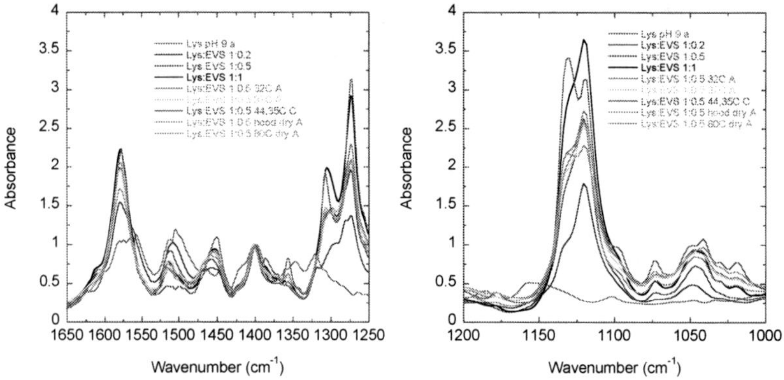

(a) (b)

Figure 3. Changes in (a) amine on lysine and (b) sulfonyl on EVS of dried materials showed potential interaction sites between molecules.

The splitting of the sulfonyl absorbance around 1125 cm^{-1} showed that one of the EVS sulfonyls was favored as the hydrogen bond acceptor in a hydrogen bonding interaction with an amine on another molecule, which was observed as the shift at 1500 cm^{-1}. Experiments showing absorbance changes of $\delta_s(NH_3^+)$ at 1500 cm^{-1} on lysine as a function of pH delineated the contributions from the α- and ε-amines. Shifts in $\delta_s(NH_3^+)$ appeared consistent with hydrogen bonding on the remaining secondary amine at the α-carbon. The fact that both lysine and alanine only formed structures at intermediate substitutions, i.e., below full substitution of the α-amines, also corroborates this.

EVS-substituted alanine showed a similar trend in the amine peak around 1500 cm^{-1} (Fig. 4a). However, there did not appear to be a similar shape change in the sulfonyl absorbance around 1125 cm^{-1} (Fig. 4b). In the alanine FT-IR analysis, alanine had coincident NH_3 and C-C absorbances in this region, making interpretation difficult.[23]

XRD showed new peaks around 0.45-0.46 nm for both EVS-substituted lysine and alanine as shown in Fig. 5. This length scale was consistent with the carbonyl-amine distance in a peptide β-sheet. It is possible that sulfonyls and amines form β-sheet mimics.[5, 6] For EVS-substituted alanine, fibrillation may simply be dependent on β-sheet like formation and growth, similar to an amyloid. However, the tube formation in EVS-substituted lysine appeared more complex. The rosettes in Fig. 2c were thicker than the tube wall thickness in Fig. 2b. Our hypothesis is that EVS-substituted lysine aggregates side by side into β-sheet mimics that then assemble into larger sheets. If the sheets are too thick, they cannot roll and end up as rosettes. If the sheets are thin enough, they can roll left under the natural chirality of the amino acid into a tube.

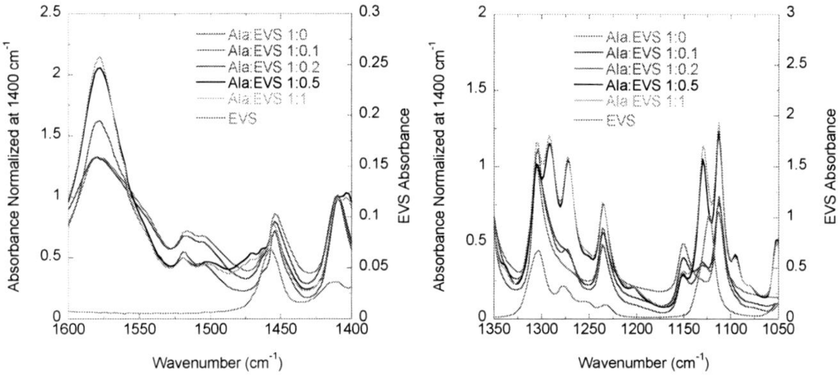

(a) (b)

Figure 4. FT-IR spectra of (a) amine and (b) sulfonyl regions in EVS-substituted alanine.

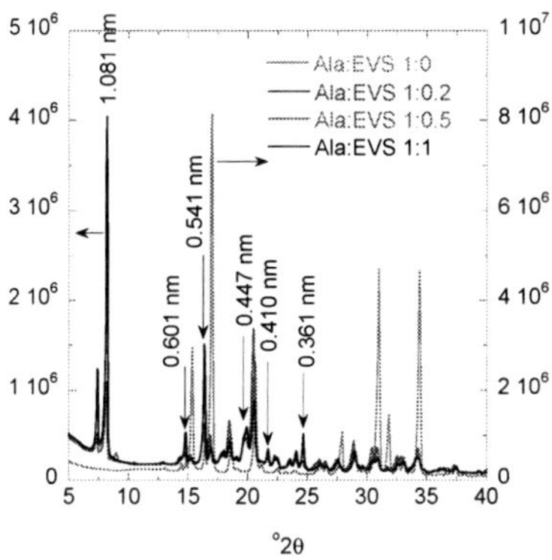

Figure 5. XRD of dried EVS-substituted alanine material.

ACKNOWLEDGMENTS

Parts of this work were carried out using instruments in the Nanoscale Characterization and Fabrication Laboratory, a Virginia Tech facility operated by the Institute for Critical Technology and Applied Science.

REFERENCES

1. C. M. Dobson, Nature **426**, 884-890 (2003).
2. C. E. MacPhee and C. M. Dobson, Journal of the American Chemical Society **122**, 12707-12713 (2000).
3. D. M. Fowler, A. V. Koulov, W. E. Balch and J. W. Kelly, TRENDS in Biochemical Sciences **32** (5), 217-224 (2007).
4. J. Vincent, *Structural Biomaterials*. (Princeton University Press, Princeton, 1990).
5. W. A. Loughlin, J. D. A. Tyndall, M. P. Glenn and D. P. Fairlie, Chemical Reviews **104** (12), 6085-6117 (2004).
6. S. Baskar Raj, V. Sethuraman, S. Francis, M. Hemamalini, P. T. Muthiah, G. Bocelli, A. Cantoni, U. Rychlewska and B. Warzajtis, CrystEngComm **5** (15), 70-76 (2003).
7. D. Philp and J. F. Stoddart, Angewandte Chemie-International Edition **35** (11), 1155-1196 (1996).
8. D. K. Smith, Chem. Commun. (1), 34-44 (2006).
9. H. Basit, A. Pal, S. Sen and S. Bhattacharya, Chemistry - A European Journal **14** (21), 6534-6545 (2008).
10. A. Brizard, D. Berthier, C. Aimé, T. Buffeteau, D. Cavagnat, L. Ducasse, I. Huc and R. Oda, Chirality **21** (1E), E153-E162 (2009).
11. A. Brizard, C. Dolain, I. Huc and R. Oda, Langmuir **22** (8), 3591-3600 (2006).
12. G. M. Dykes and D. K. Smith, Tetrahedron **59**, 3999-4009 (2003).
13. Andrew R. Hirst, Juan F. Miravet, B. Escuder, L. Noirez, V. Castelletto, Ian W. Hamley and David K. Smith, Chemistry - A European Journal **15** (2), 372-379 (2009).
14. A. R. Hirst and D. K. Smith, Chemistry - A European Journal **11** (19), 5496-5508 (2005).
15. A. R. Hirst, D. K. Smith, M. C. Feiters, H. P. M. Geurts and A. C. Wright, J. Am. Chem. Soc. **125** (30), 9010-9011 (2003).
16. T. Sagawa, S. Chowdhury, M. Takafuji and H. Ihara, Macromolecular Symposia **237** (1), 28-38 (2006).
17. K. T. Nam, S. A. Shelby, P. H. Choi, A. B. Marciel, R. Chen, L. Tan, T. K. Chu, R. A. Mesch, B. C. Lee, M. D. Connolly, C. Kisielowski and R. N. Zuckermann, Nature Materials **9** (5), 454-460 (2010).
18. D. Jang, Ho Y. Lee, M. Park, Seong R. Nam and J.-I. Hong, Chemistry - A European Journal **16** (16), 4836-4842 (2010).
19. G. C. Kuang, Y. Ji, X. R. Jia, Y. Li, E. Q. Chen and Y. Wei, Chem. Mater. **20** (13), 4173-4175 (2008).
20. S. Bhattacharya, S. N. G. Acharya and A. R. Raju, Chem. Commun. (17), 2101-2102 (1996).
21. B. D. Mather, K. Viswanathan, K. M. Miller and T. E. Long, Prog. Polym. Sci. **31**, 487-531 (2006).
22. N. K. Budhavaram and J. R. Barone, Journal of Raman Spectroscopy, n/a-n/a (2010).
23. B. Hernandez, F. Pfluger, M. Nsangou and M. Ghomi, Journal of Physical Chemistry B **113**, 3169-3178 (2009).

Mater. Res. Soc. Symp. Proc. Vol. 1312 © 2011 Materials Research Society
DOI: 10.1557/opl.2011.507

Polypyrrole Nanowire Array with High Aspect Ratio Fabricated by Block-Copolymer-Templated Electropolymerization

Hideaki Komiyama,[1] Tomokazu Iyoda,[1,*] and Kaori Kamata[1,2]
[1]Division of Integrated Molecular Engineering, Chemical Resources Laboratory, Tokyo Institute of Technology, 4259 Nagatsuta-cho, Midori-ku, Yokohama, Kanagawa 226-8503, Japan
[2]JST-PRESTO, 4-1-8 Honcho, Kawaguchi, Saitama 332-0012, Japan

ABSTRACT

Conducting polypyrrole (PPy) nanowire array was electrochemically fabricated by using microphase-separated block copolymer (PEO_m-*b*-$PMA(Az)_n$) thin film as an electrode template. Electropolymerization proceeds selectively through the perpendiculary oriented PEO nanocylindrical microdomains on an ITO electrode coated by the above polymer film. The length of the PPy nanowires was controllable in range of 10-120 nm by the thickness of the coated template film and the amount of passed charge. A 10 nm of diameter, 27 nm of periodicity, and 12 of aspect ratio of the crystalline PPy nanowires were successfully achieved. Selective removal of the template is also achieved simply by rinsing with solubility-controlled mixed solvent.

INTRODUCTION

Conducting polymer nanowires are of great advantage for applications in electronics, energy conversion and storage [1], and photovoltaic cell [2] because of their high electrical conductivity, large surface area, and mechanical flexibility. At present, conducting polymer nanowires have been commonly fabricated by chemical or electrochemical polymerization inside nano-scaled cylindrical pores as a template such as track-etched membranes and anodic porous alumina membranes [3-5]. Recently, the porous block copolymer membranes, of which one of the microdomains can be removed, has drawn much attention due to simple fabrication of high surface density of nanowires. However, the aspect ratio of conducting polymer nanowires fabricated by using block copolymer template has been limited to less than 4, much smaller than that of nanowires obtained by other porous templates [6,7]. The reason is the difficulty in perpendicular alignment of cylindrical microdomain with large aspect ratio in block copolymer templates.

We have developed a series of liquid crystalline block copolymers consisting of poly(ethylene oxide) and poly(methacrylate) bearing azobenzene mesogens in the side chain, here denoted by PEO_m-*b*-$PMA(Az)_n$, which form perpendicularly aligned and highly ordered PEO nanocylinder array with very high aspect ratio in thin film [8,9]. Since the PEO of the nanocylinder is in under-cooled liquid state at room temperature and a well-known medium of ion-conducting polymers, the thin film coated on an electrode could be worked as an electrode template without removal of PEO cylindrical microdomains. The PEO nanocylinders in the template is expected as ethereal solvent and electrochemical channels. Here, we report the fabrication of nanowire array of conducting polypyrrole (PPy) with high aspect ratio through the

PEO_m-*b*-$PMA(Az)_n$ template. Furthermore, the nanowire length and aspect ratio could be controlled by film thickness of the template and the amount pf passed charge (*Q*) in electropolymerization.

EXPERIMENT

The synthesis of PEO_{114}-*b*-$PMA(Az)_{45}$, which chemical structure was shown in Figure 1, has been reported previously [8]. PEO_{114}-*b*-$PMA(Az)_{45}$ template was prepared by spin-coating on an ITO electrode and annealing at 140 °C. Electropolymerization of pyrrole (Py) was carried out with conventional three-electrode system. The ITO electorde coated with the PEO_{114}-*b*-$PMA(Az)_{45}$ template was used as a working electrode for electropolymerization of Py (10 mM) in 1 mM $LiClO_4$ propylene carbonate solution. Pt plate and Ag/AgCl were employed as a counter and reference electrodes, respectively. The electropolymerization was performed under constant potential of 1.3 V (vs Ag/AgCl).

Atomic force microscope (AFM) observations were performed using a Nanoscope IIIa (Veeco) in the tapping mode. The cantilever had a 5 Nm^{-1} spring constant, 119-154 kHz resonant frequencies, and an approximately 10 nm tip radius. Field emission scanning electron microscopic (FESEM) images were taken using an S-5200 (Hitachi High-Technologies Co.).

DISCUSSION

Figure 1 shows a schematic illustration of the fabrication process for PPy nanowire array by using PEO_{114}-*b*-$PMA(Az)_{45}$ block copolymer template. After coating and thermal annealing, PEO_{114}-*b*-$PMA(Az)_{45}$ template gave a perpendicularly aligned and hexagonally arranged PEO nanocylindrical microdomains. The thickness of the template was about 150 nm. The PPy grew from the ITO interface along the PEO nanocylinders selectively. We encountered the collapse of the perpendicular orientation of PPy nanowires when the template was washed out with toluene as good solvent for the template. In order to maintain the perpendicular orientation of the PPy nanowires, the dissolution process of the template was controlled by using a mixed solvent of toluene with acetone as poor solvent (70 vol% acetone) under spinning condition, called chemical etching here. We succeeded in gradually dissolving the template out from the top

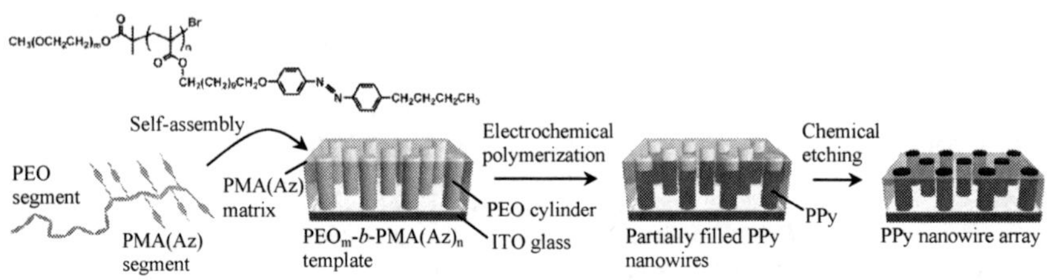

Figure 1. Chemical structure of PEO_m-*b*-$PMA(Az)_n$ diblock copolymer and schematic illustration of fabrication of polypyrrole (PPy) nanowire array through block copolymer template.

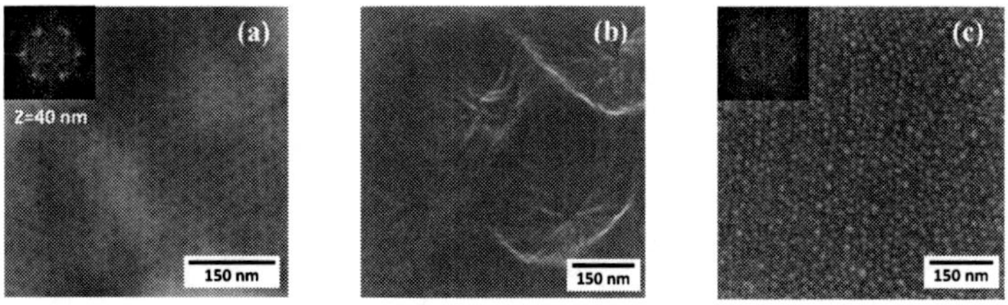

Figure 2. (a) Tapping mode AFM height images of the PEO_{114}-*b*-PMA$(Az)_{45}$ template after electropolymerization of pyrrole. FESEM images of PPy nanowires structures prepared by different etching process for the PEO_{114}-*b*-PMA$(Az)_{45}$ template. The etching processes were (b) rinse with toluene and (c) chemical etching with mixed solvent of toluene/acetone (3/7 (v/v)) at spinning-condition, respectively. Image of (b) is 25° tilted view. FFT images are shown in insets of (a) and (c).

surface with minimal damage of the perpendicular its orientation of the PPy nanowire array.

Figure 2(a) shows AFM height image of the PEO_{114}-*b*-PMA$(Az)_{45}$ template on ITO electrode after electropolymerization of Py. After electropolymerization with 14.2 C/cm^2 of passed charge, no change in the surface morphology of the template was observed. Dark dots and bright surroundings in the images can be assigned to cylindrical PEO microdomains and PMA(Az) matrix, respectively. After dissolution of the template with toluene, the nanowire structure appeared (Figure 2(b)). The observed nanowires are assigned to PPy grown through the PEO nanocylinders. However, the remaining PPy nanowires would be coaggregated during solvent evaporation and lay down on the ITO electrode. Finally, the PPy nanowires formed a hair whorl like structure in Figure 2(b). Although it was difficult to measure the diameter and length of the PPy nanowires from the image of Figure 2(b), the other experiments, which were mentioned later,

revealed that the diameter and length of the PPy nanowires were 10 nm and 120 nm, respectively. Accordingly, we succeeded in the high aspect ratio of 12 for PPy nanowires through the present block copolymer templating process. To the best of our knowledge, the aspect ratio of 12 is higher than any other block copolymer templating process for conducting polymer nanowires. Kim and co-workers reported the aspect ratio of up to 4 for conducting polymer nanorods through block copolymer template [6].

In order to keep the orientation of the PPy nanowires with high-aspect ratio, we applied the chemical etching method that the template is removed from the surface in a stepwise fashion by using mixed solvent of toluene and acetone (3/7 (v/v)). Through this chemical etching, the template partially remains to support the orientation of PPy nanowires. Here, we examined quantitative analysis of the chemical etching for the template without PPy nanowires. The mixed solution of toluene and acetone (3/7 (v/v)) was dropped onto the spinning sample on a spin-coater with rotating speed of 2000 rpm. The very small amount of the template at surface could be dissolved and be blown away through this chemical etching process. The etched thickness was controlled by the number of casting of droplet of the mixed solution. The chemical etching was performed to as-prepared PEO_{114}-*b*-PMA$(Az)_{64}$ template (without electropolymerization). Here, we casted mixed solvent droplet of the amount of 8 $\mu l/cm^2$ onto the spinning template

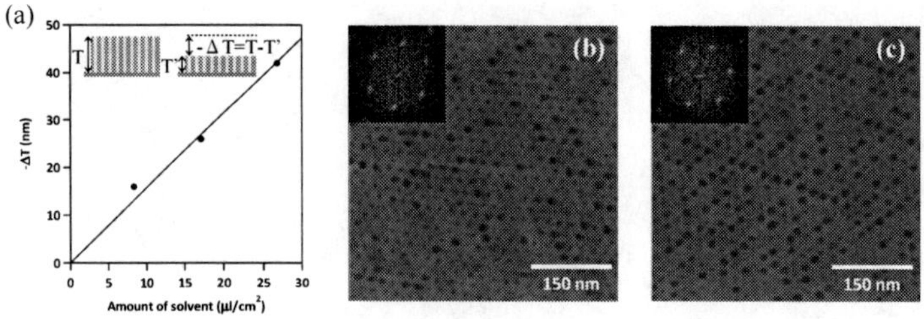

Figure 3. (a) The etched thickness of as-prepared PEO_{114}-*b*-$PMA(Az)_{64}$ template by chemical etching. AFM phase images of (b) as-prepared PEO_{114}-*b*-$PMA(Az)_{64}$ template and (c) the template after two times of chemical etching. FFT images are shown in insets.

every one process. The result is summarized in Figure 3(a). The thickness became around 12-nm thinner through the one chemical etching process. Interestingly, after chemical etching, the hexagonal dots appeared in AFM images (Figures 3(b, c)). The hexagonal arrangement of PEO nanocylinder was maintained throughout the chemical etching process.

The chemical etching was carried out three times to the template after electropolymerization. The resulting FESEM image is Figure 2(c). The surface morphology changed from hexagonal dot pattern to pit pattern after the chemical etching. It is illustrated that the chemical etching process would chip away topmost surface of the template to selectively leave the PPy nanowires. Based on the chemical etching efficiency (Figure 3(a)) and 150 nm of initial thickness of the template, the thickness of PPy nanowire array in Figure 2(c) is calculated to be 114 nm. The actual thickness was measured by AFM. The bump between the template surface and the ITO surface was made with a knife. The depth of the bump was 120 nm, corresponding to the thickness obtained by chemical etching result. Section profile of AFM height image revealed that the around 10-nm of tip of PPy nanowires exposed from the template surface. It should be noted that PPy nanowires never lay down and gathered each other unlike the removal of all template as shown in Figure 2(b). The PPy nanowires were surrounded by the remaining PEO_{114}-*b*-$PMA(Az)_{45}$ template. Herein, we describe this structure as PPy nanowire array. The inset of FFT image in Figure 2(c) also showed a perfect hexagonal pattern. The diameter and the periodicity of PPy nanowires were 10 nm and 27 nm, respectively, in good agreement with those of PEO nanocylinders. To the best of our knowledge, the highest density conducting polymer nanowire array was reported in Kuila and Stamm's polyaniline nanorods [7]. In comparison with this report, the size of diameter and periodicity for our PPy nanowires is quite similar to that of reported polyaniline nanorods. As a consequence, PPy nanowire array with the high aspect ratio of 12 and the high density of 1×10^{12} pit/inch^2 was successfully fabricated by PEO-domain selective electropolymerization with chemical etching.

The amount of deposition of conducting polymer is controlled by the amount of passed charge (Q). The PPy nanowires with different length were obtained as a function of Q. Figure 4 shows FESEM images of the PPy nanostructures fabricated by different Q. Note that the PEO_{114}-*b*-$PMA(Az)_{45}$ template was completely removed by rinse with toluene to observe the

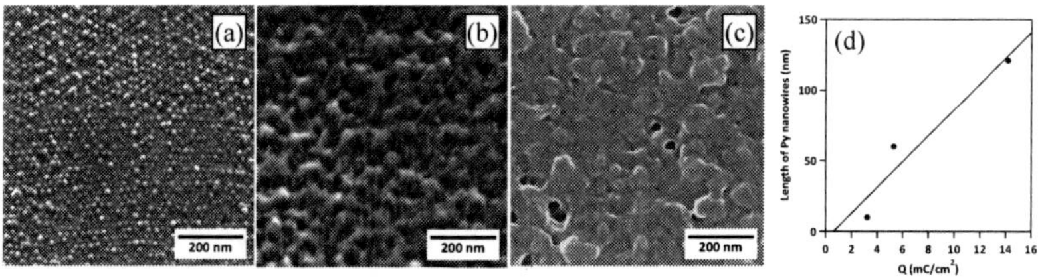

Figure 4. FESEM images of PPy nanostructures prepared by electropolymerization at the amount of passed charge (Q) of (a) 3.2 mC/cm^2, (b) 5.3 mC/cm^2, and (a) 79.3 mC/cm^2. The block copolymer template was removed by washing with toluene. These images are 30° tilted view. (d) The relationship between the length and the amount of passed charge (Q).

nanostructures more easily, here. In the case of small Q of 3.2 mC/cm^2 (defined as the area of electrode), PPy nanodot array was observed in Figure 4(a). The size parameters of PPy nanodot array were the diameter of 10 nm and the periodicity of the nanodots of 27 nm. In addition, the height of the nanodots was about 10 nm from section profile of AFM image. A lager Q of 5.3 mC/cm^2 gave PPy nanowires with the length of 60 nm as shown in Figure 4(b). In smaller aspect ratio, the tip of PPy nanowires gathered each other without the collapse of perpendicular alignment. PPy nanowires with length of 120 nm shown in Figure 2(b) were obtained by the Q of 14.2 mC/cm^2. Up to Q of 79.3 mC/cm^2, the PPy nanosutructure would be covered by the continuously grown PPy film over the emplate (Figure 4(c)). After the tips of the PPy nanowire reached at the top surface of PEO_{114}-*b*-$PMA(Az)_{45}$ template, the PPy formed the covered film on the surface of template. The length of the PPy nanowires are proportional to the amount of passed charge, as is shown in Figure 4(d). We can expect to achieve higher aspect ratio because the PEO_m-*b*-$PMA(Az)_n$ template has a potential to present longer length and smaller diameter of nanocylinders by control of the thickness of the template and molecular weight of PEO and PMA(Az) segments, respectively.

CONCLUSIONS

We have demonstrated to fabricate the high-aspect-ratio PPy nanowire array using PEO_m-*b*-$PMA(Az)_n$ block copolymer as a template. The aspect ratio of 12 is much higher than conducting polymer nanowires obtained by any other porous block copolymer templates. The PPy selectively grew thorough the PEO nanocylindrical microdomain during electropolymerization. The PPy nanowires buried under the template were exposed by the chemical etching process without collapse of the orientation. The remaining template supported to stand the nanowires although the high-aspect-ratio nanowires tended to lose their fine alignment. The different length from PPy nanodots to nanowires was obtained by control of the amount of passed charge.

ACKNOWLEDGMENTS

K.K. thanks a Grant-in-Aid for Science Research in a Priority Area "Super-Hierarchical Structures" (No. 19022008) from the Ministry of Education, Culture, Sports, Science and Technology, Japan. H.K. thanks a Japan Society for the Promotion of Science for a research fellowship for young scientists.

REFERENCES

1. S. I. Cho and S. B. Lee, *Acc. Chem. Res.* **41**, 699 (2008)
2. J. S. Kim, Y. Park, D. Y. Lee, J. H. Lee, J. H. Park, J. K. Kim and K. Cho, *Adv. Funct. Mater.* **20**, 540 (2010)
3. C. R. Martin, *Science* **266**, 1961 (1994)
4. C. R. Martin, *Acc. Chem.* Res. **28**, 61 (1995)
5. R. Xiao, S. I. Cho, R. Liu and S. B. Lee, *J. Am. Chem. Soc.* **129**, 4483 (2007)
6. J. I. Lee, S. H. Cho, S. M. Park, J. K. Kim, J. K. Kim, J. W. Yu, Y. C. Kim and T. P. Russell, *Nano Lett.* **8**, 2315 (2008)
7. B. K. Kuila and M. Stamm, *J. Mater. Chem.* **20**, 6086 (2010)
8. Y. Tian, K. Watanabe, X. Kong, J. Abe and T. Iyoda, *Macromolecules* **35**, 3739 (2002)
9. M. Komura and T. Iyoda, *Macromolecules* **40**, 4106 (2007)

Mater. Res. Soc. Symp. Proc. Vol. 1312 © 2011 Materials Research Society
DOI: 10.1557/opl.2011.508

Nanostructure Dependent Surface Energy of Silica Nanorod Arrays through Block Copolymer Templating Processes

Yongbin Zhao, Aihua Chen, Tomokazu Iyoda*
Division of Integrated Molecular Engineering, Chemical Resources Laboratory, Tokyo Institute of Technology, Yokohama, Kanagawa 226-8503, Japan.

ABSTRACT

The self-assembled block copolymer films with poly (ethylene oxide) (PEO) and hydrophobic polymetharylate (PMA) with azobenzene mesogen in the side chain, denoted as PEO_m-b-$PMA(Az)_n$, were used as the template to prepare hexagonally ordered silica nanorod arrays by immersing the template films in the silicate precursor containing tetraetoxysilane (TEOS). The diameter and the center-to-center distance of the SiO_2 nanorod arrays were controlled by selecting the block copolymer with different PEO volume fraction. In addition, the contact angles of different kinds' solvents for the SiO_2 nanorod arrays were characterized. We further found, the diameter and the period distance of silica nanorods are very important factors for controlling the contact angle of different kind's solvents on the surface of the SiO_2 nanorod arrays.

INTRODUCTION

Self-organization, as a powerful route to the 'bottom-up' fabrication of nanostructures for use in optical, optoelectronic, and magnetic storage devices, has been attracted in the world[1-2]. Recently, the self-assembly of block-copolymeric system into aligned, highly ordered arrays that cover the range between approximately 10 and 100nm, makes them ideal candidates as template and scaffolds for the fabrication of nanostructured materials. For example, by using perpendicular PS-b-PMMA cylinder nanopatterns as template, the ferromagnetic Co nanowires with high-density vertical arrays has been fabricated through subsequent direct current electrodeposition[4]. In addition, using the similar process, it also has been demonstrated the replication of nanoscale feature into a variety of materials [5-6], including the formation of metal nanorods, nanoporous metal films, nanoelectrode arrays, etc.

Recently, our group has developed a new series of amphiphilic liquid crystalline diblock copolymers consisting of poly(ethylene oxide) (PEO) and poly(methacrylate) bearing an azobenzene mesogen in the side chain as template to prepare high ordered nanoarrays, such as Ag nanodot arrays on the flexible and rigid substrated[7], mesoporous silica nanorods arrays with tunable aspect ratios[8], etc.

In this work, the self-assembled block copolymer, denoted as PEO_m-b-$PMA(Az)_n$ films with normal PEO cylindrical domains was used as the template to prepare hexagonally ordered silica nanorod array by immersing the template films in the silicate precursor containing tetraetoxysilane (TEOS). The diameter and the center-to-center distance of the SiO_2 nanorod arrays were further controlled by selecting the block copolymer thin films with different PEO volume fraction. In addition, the contact angles of different kinds' solvents for the SiO_2 nanorod arrays were characterized. We found, the diameter and the period distance of silica nanorods were very important factors for controlling the contact angle of different kind's solvents on the surface of the different SiO_2 nanorod arrays.

EXPERIMENTAL DETAILS

Preparation of block copolymer films with perpendicular PEO cylindrical orientation

The amphiphilic liquid crystalline diblock copolymers PEO_m-b-$PMA(Az)_n$, was synthesized by atom transfer radical polymerization (ATRP). The details in synthesis and characterization of the block copolymer were described in our previous publication [9]. The solutions f PEO_m-$PMA(Az)_n$ block copolymers in toluene or chloroform with 1~5 wt% were spin-coated on the Si wafer with the rotation rate of 2000 rpm and 30 sec, subsequently annealed at 140°C under vacuum for 12 h, the block copolymer films with perpendicular PEO cylindrical orientation were obtained. The thickness of block copolymer films between 100 nm to 500 nm could be controlled by adjusting the condition of spinning-coated and concentration of solution.

Preparation of perpendicular aligned silica nanoarrays using block copolymer films with perpendicular PEO cylindrical orientation as template

To form the silica nanoarrays on the surface of Si wafer, a silica sol was firstly prepared in the following typical procedure: a mixture of ethanol (5.5 g), TEOS (2.08 g), H_2O (0.5 g) and 0.4 g of HCl aqueous solution (0.1 M) was heated at 70 °C for 1 h, and cooling to room temperature to form the silica sol. And then, the PEO_m-b-$PMA(Az)_n$ films with perpendicular PEO cylindrical orientation as template, were immersed in the silica sol solution at room temperature for 2 h, and washed by de-ionized water, formed the silica/block copolymer nanocomposite films. After dried at 50 °C for overnight, the silica nanoarrays with perpendicular orientation were obtained by calcinated in air at rate of 1°C/min to 550 °C for 6h.

Measurement

The AFM images were taken with a Digital Instrument Dimension 3000 Scanning Force Microscope. The imaging was conducted in tapping mode using a silicon cantilever with a resonance frequency of 300 kHz [10]. Prior to the recording of FE-SEM images on a Hitachi S-5200 field-emission scanning electron microscope a Pt/Au layer (thickness ~ 2 nm) was deposited onto the surface of the samples using a HITACHI E-1010 ion sputter. Water contact angles were measured by the sessile drop method using a conventional drop shape analysis technique (Cruss DSA 100, Hamburg, Germany). Deionized regent grade water was used for contact angle measurements. Liquid droplets (3ul) were dropped carefully onto the sample's surface.

RESULTS AND DISCUSSION

As we all know, the simple process of solvent evaporation could produce highly orderly arrays of cylindrical microdomains oriented normal to the surface in block copolymers with long-range lateral order. In our lab, microphase-separated diblock copolymer films of amphiphilic PEOm-b-PMA(Az)n, consisting of ploy (ethylene oxide) and Poly(methacrylate) bearing an azobenzene segment in the side chain, could be prepared after annealing at 140°C for 12h in vacuum through self-assembling process. Fig. 1 shows the AFM top-images of different kinds of BCP films on the silicon wafer. It can be seen that the PEO cylindrical domains are perpendicular to the substrate and form highly ordered hexagonal structure. In addition, the diameter of the PEO cylindrical domains in the PEO_{114}-$PM(Az)_{45}$, PEO_{272}-$PM(Az)_{78}$ and PEO_{272}-$PM(Az)_{116}$ films is about 11 nm, 20nm, 20nm, respectively. In addition, the period distance between the two PEO cylinder is about 22nm, 34nm and 35nm, respectively.

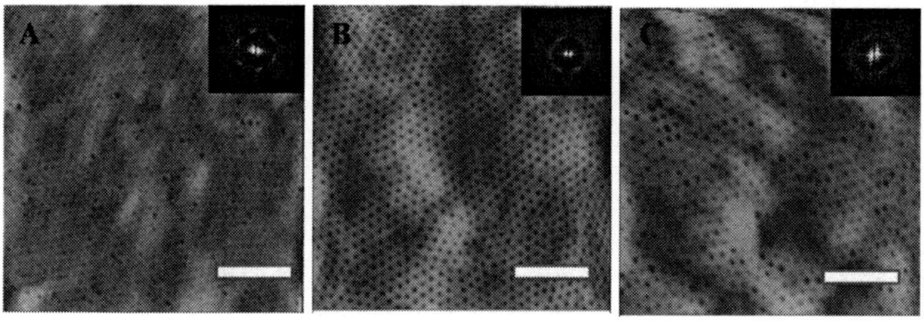

Fig.1 The AFM images of BCP films: (a) PEO_{114}-$PM(Az)_{45}$; (b) PEO_{272}-$PM(Az)_{78}$; (c) PEO_{272}-$PM(Az)_{116}$; (Insert: Relative FFT images, respectively; Scale: 250nm).

In addition, both out-of-plane orientations of the hexagonal cylinder phase were applied for various transcription processes. The three kinds of BCP films with different perpendicular PEO cylinders' diameters and periods as templates were immersed in a tetraetoxysilane (TEOS) precursor solution containing acidic solution for sol-gel reaction, the silicate sol would penetrate into the PEO cylindrical domains with the immersing time increasing. And then, washed by water in order to remove the overlayered sol on the surface of BCP films and dried in oven overnight. After removal of the templates by calcination, silica nanorod arrays with perfect perpendicular orientation were obtained. Fig. 2 shows the FE-SEM images of Silica nanorod arrays with different template. The nanorods' diameter of silica nanorods from the PEO_{114}-$PM(Az)_{45}$ (Fig.2a), PEO_{272}-$PM(Az)_{78}$ (Fig.2b) and PEO_{272}-$PM(Az)_{116}$ (Fig.2c) films is about 10 nm, 19nm and 20nm and the distance between the nanorods is about 21nm, 35 nm and 38 nm, respectively, which were corresponding to the PEO domain's diameter and the distance between the PEO cylinders, further indicating that the PEO structure can be easy to transcript to silica nanorod arrays by this method.

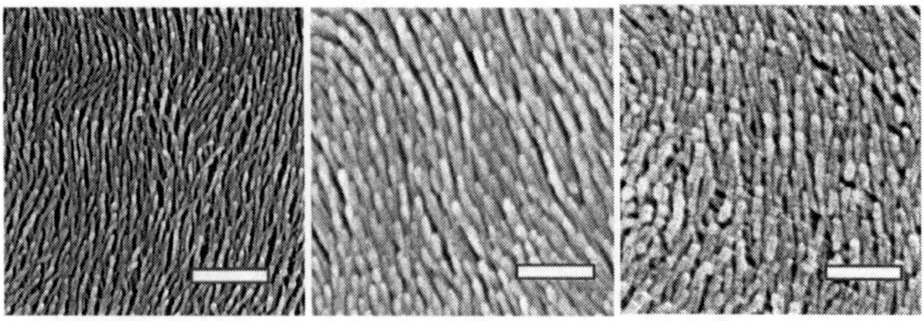

Fig.2 The FE-SEM images of Silica nanorod arrays with different tmeplate: (a) PEO_{114}-$PM(Az)_{45}$; (b) PEO_{272}-$PM(Az)_{78}$; (c) PEO_{272}-$PM(Az)_{116}$; (Scale: 250nm).

Fig. 3a shows that the diameter and periodicity of silica nanorod arrays increased slowly by using the different block copolymer as templates. Furthermore, it can be seen from Fig. 3b, all of the water contact angles of the perpendicular aligned silica nanorod arrays were higher than that of silica films (CA. 72.5°), showed hydrophobic properties. Duo to the nanosacle structure of one-dimensional arrays consists of silica nanorod and air, the proportion of water/air interfaces can greatly intensify the water repellency, as shown in the left inserted Cassie's model of Fig.4a. Based on this Cassie's equation [11], the theoretical value of the water contact angles were calculated as show in Fig. 3b. With the diameter and the period distance of silica nanorod arrays increasing, the theoretical water contact angles were also decreasing slowly from 134° to 109.3°, with corresponding in the regulation of experimental values. However, duo to the existence of the -OH functional groups on the surface of silica nanorods, the experimental water contact angles of silica nanorod arrays was lower than the theoretical values.

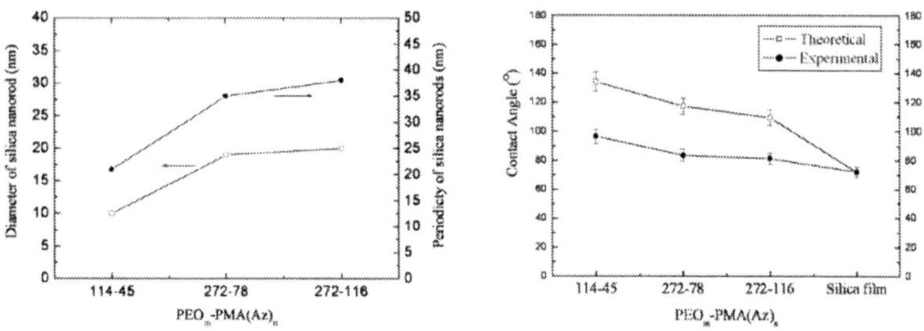

Fig. 3 (a) The diameter and periodicity of silica nanorod arrays by different block copolymer templates; (b) the theoretical and experimental water contact angles of silica nanorod arrays based on the cassie's equations.

Interestingly, Fig.4a shows that the contact angles of ethanol for different silica nanorod arrays were lower than that of silica films (CA. 20°), showed the superhydrophilicity. However, when non-polar solvents, such as toluene and hexane instead of water were used to measure the contact angles of the silica nanrod arrays, both of the contact angles of silica nanorod arrays obtained from different BCP films was lower than the silica films in Fig. 4a, showed superlipophilicity. Toluene and hexane, as the sedimentary solvents were easier to immerse into the gap in the silica nanrods than water, so the wettability could be further improve by the roughness of substrate [12]. Based on the above deduction, the wenzel model [13] was proposed to describe the toluene absorbance of silica nanorods, as shown in the right inserted model of Fig.4a. Moreover, base on the surface tensions and contact angles of four kinds of solvents on the surface of silica nanrod arrays, the surface free energy of the silica nanrod arrays from the different block copolymer films can be calculated by the Fowks's equation [14], as shown in Fig. 4b. With the diameter and the period distance of silica nanorod arrays from the PEO_{114}-$PM(Az)_{45}$, PEO_{272}-$PM(Az)_{78}$ and PEO_{272}-$PM(Az)_{116}$, respectively, increasing, the surface free energy of the silica nanorod arrays also grew up slowly, so the water contact angles decreased slowly, indicating that the diameter and the period distance of silica nanorods were very important factors for controlling the contact angle of different kinds solvents on the surface of the different materials.

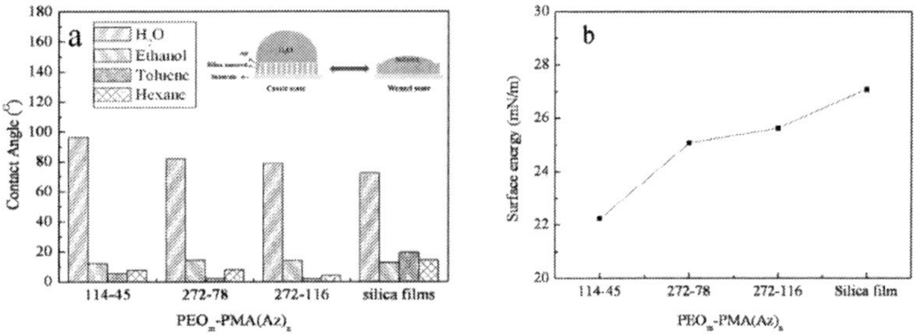

Fig. 4 (a) The contact angles of silica nanorod arrays from the different BCP films (Insert figure is the models of contact angles); (b) the surface free energy of the silica nanorods based on the Fowks's equation.

CONCLUSION

In this study, the amphiphilic liquid crystalline block copolymers films with perpendicular cylindrical domains were prepared by spin-coating toluene or chloroform solutions on the surface of silicon wafers, followed by annealing at 140°C for 2h in vacuum. In addition, the perpendicular aligned silica nanorod arrays were fabricated by applying the sol-gel method for the orientation-controlled template films. Furthermore, the diameter of silica nanorods and distance of inter-rods also could be controlled by adjusting the molecular weight of PEO and PMA (Az) segment in block copolymer. In addition, with the diameter and the center-to-center distance of the SiO_2 nanorod arrays were adjusted, the contact angles of different solvents also cloud be controlled, showed the hydrophobic properties.

ACKNOWLEDGMENTS

Y. Zhao (Ph.D) acknowledges the support of the postdoctoral fellowship for foreign researchers from the Japan Society for the Promotion of Science (JSPS).

REFERENCES

1. K.J. Vahalar, nature **424**, 839 (2003).
2. J. Y. Cheng, C. A. Ross, V. Z. H. Chan, E. L. Thomas, R. G. H. Lammertink, G. J. Vancso, Adv. Mater. **13**, 1174 (2001).
3. S. Krishnammoorthy, Y. Ferbig, C. Hibert, R. Pugin, C. Hinderling, J. Brugger, H. Heinzelmann, Nanotech. **19**, 285301 (2008).
4. T. Thurn-Albrecht, J. Schotter, G. A. Kästle, N. Emley, T. Shibauchi, L. Krusin-Elbaum, K. Guarini, C. T. Black, M. T. Tuominen, T. P. Russell, Science **290** , 2126 (2000).
5. H. C. Kim, X. Q. Jia, C. M. Stafford, D. H. Kim, T. J. McCarthy, M. Tuominen, C. J. Hawker, T. P. Russell, Adv. Mater. **13**, 795 (2001).
6. K. Shin, K. A. Leach, J. T. Goldbach, D. H. Kim, T. P. Russell, Nano Lett. **2**, 933(2002).
7. J. Li, K. Kamata, S. Watanabe, T. Iyoda, Adv. Mater. **19**, 1267 (2007).
8. A.Chen, K. Kamata, M. Komura, T. Iyoda, Adv. Mater. **20**, 763 (2008).
9. Y. Q. Tian, K. Watanabe, X. X. Kong, J. Abe, T. Iyoda, Macromolecules **35**, 373 (2002).
10. M. Komura, T. Iyoda, Macromolecules **40**, 4106 (2007).
11. A. B. D. Cassie, Baxter, Trans. Faraday Soc. **40**, 546 (1944).
12. R. Blossey. Nature Mater. **2**, 301 (2003).
13. R. N. Wenzel, Ind. Eng. Chem. **28**, 988 (1964).
14. F. M. Fowks, J. Phys. Chem. **66**, 382 (1962).

Mater. Res. Soc. Symp. Proc. Vol. 1312 © 2011 Materials Research Society
DOI: 10.1557/opl.2011.130

Synthesis of Thermo-Responsive Fluorinated Star-Shaped Polymers by Living Cationic Polymerization

Tomomi Irita[1], Katsuhiko Imoto[1], Takabumi Nagai[1] and Sadahito Aoshima[2]
e-mail address tomomi.irita@daikin.co.jp

[1]Chemical R&D Center, Daikin Industries, Ltd., 1-1 Nishi Hitotsuya, Settsu-shi, Osaka, 566-8585, Japan
[2]Department of Macromolecular Science, Graduate School of Science, Osaka University, 1-1 Machikaneyama, Toyonaka, Osaka 560-0043, Japan

ABSTRACT

Well-controlled fluorinated star-shaped polymers with oxyethylene units were designed and synthesized by living cationic polymerization. These fluorinated star-shaped polymers were prepared using successive fluorinated initiators to generate fluorinated, end-functionalized and hydrophilic segments with oxyethylene. Poly(vinyl ether)s with oxyethylene units are known to be thermo-responsive polymers that exhibit a lower critical solution temperature (LCST) in aqueous solutions. These novel fluorine-containing star-shaped polymers with oxyethylene were also thermo-responsive, as expected. In addition, the hetero-fluorinated star-shaped polymers with oxyethylene segments and different fluorine compositions were synthesized with various bifunctional initiators by living cationic polymerization having narrow molecular weight distributions (Mw/Mn=1.23 - 1.29) and contained very little, if any, presence of the linear homopolymer precursor, as evidenced by GPC. The LCST of this polymer depended on the end group fluorine composition and significant changes in cloud points were observed in water, as confirmed by static light scattering (SLS) measurements. As an example, aqueous solutions the star-shaped polymers without fluorine terminal groups showed no cloud point below up to 60 °C in aqueous solution. After the fluoro-functionality was incorporated into the terminal groups, the cloud point of the aqueous solutions decreased to almost 20 °C.

INTRODUCTION

Fluorine is a remarkable element that can impart outstanding properties to a macromolecule, including low surface energy, low refractive index, high thermal stability, enhanced chemical resistance and so on when compared to their non-fluorinated analogues [1,2].

Miktoarm star-shaped copolymers show interesting properties in the solid state, as well as in solution, due to their unique architectures. Star-shaped polymers, which are usually defined as branched macromolecules that consist of linear polymer arms joined together by a central core, have been of great interest in the past decade.[3]

We have demonstrated that cloud points of aqueous solutions of end-functionalized poly(vinyl ethers) with oxyethylene side groups synthesized by living cationic polymerization were governed by the properties of a terminal group[4]. Stimuli responsive polymers with various shapes were prepared. In the case of star-shaped polymers, the effect of end groups on the thermo sensitivity was controlled by the structure of the arms. The properties of star-shaped polymers and polymer micelles consisting of amphiphilic diblock copolymers were significantly affected by end groups.

In this paper, we describe the synthesis of well-defined hetero-fluorinated star-shaped polymers with oxyethylene segments and different fluorine compositions by living cationic polymerization. A star-shaped architecture combined with conjugated character in the arms would bring new electrical, optical, and morphological properties to the system. The properties of the star-shaped polymers and polymer micelles were more effected by the terminal groups than those of end-functionalized linear homopolymers.

EXPERIMENTAL

Tridecafluorononyloxy ethyl vinyl ether were supplied by Daikin Industries Co., Ltd. and distilled over calcium hydride under reduced pressure. [2-Methoxyethylvinyl ether were supplied by Maruzen and distilled. 1,4-Dioxane (Wako; >99.5%) was distilled over calcium hydride and then lithium aluminum hydride. Toluene (Wako; >99.5%) were dried by passage through solvent purification columns (Glass Counter; Solvent Dispensing System) and then distilled over metallic sodium and calcium hydride, respectively.

Polymerization was carried out under dry nitrogen in a baked glass tube equipped. Toluene (1.0 mL), 1,4-dioxane (0.20 mL; 2.34 mmol), tridecafluorononyloxy ethyl vinyl ether (0.42 mL; 1.62 mmol), and 40 mM IBEA solution in hexane (1/4, v/v; 0.20 mL; 8.0×10-3 mmol) were successively added into the tube using dry syringes. Polymerization was initiated by adding a prechilled 40 mM Et1.5AlCl1.5 solution in toluene at 0 °C. The reaction mixture was constantly stirred during polymerization using a magnetic stirring bar. The polymerization was terminated with prechilled methanol (0.5 mL) containing a small amount of aqueous ammonia solution (0.1 wt%). Methanol was removed by decantation, and the final colorless polymer was dried under vacuum.

The molecular weight distribution (MWD) was measured by size-exclusion chromatography (SEC) in chloroform by using three polystyrene gel columns [Tosoh; TSK gel G-4000HXL, G-3000HXL, and G-2000HXL; exclusion limit molecular weight = 4×105, 6×104, and 1×104, respectively; column size = 7.8 mm (internal diameter)×300 mm; flow rate = 1.0 mL/min] connected to a pump, column oven, ultraviolet detector, and a refractive-index detector (Tosoh; DP-, CO-, UV and RI-8020, respectively). The number-average molecular weight (Mn) and polydispersity ratio [weight-average molecular weight/number-average molecular weight (Mw/Mn)] were calculated from the chromatographs with respect to 16 poly(styrene) standards (Tosoh; Mn = 577–1.09×106, Mw/Mn < 1.1).

The turbidity of the polymer solutions was monitored using the transmittance at 500 nm, recorded on a UV/Vis spectrometer (Jasco V-550) with a Peltier-type thermostatic cell holder (Jasco ETC-505), on cooling at 1.0 °C/min.

DISCUSSION

The polymerization of fluorinated initiator were conducted in toluene at 0 °C under conventional conditions ([monomer]0 = 0.80 M, [IBEA]0 = 4 mM, [Et1.5AlCl1.5]0, [1,4-dioxane] = 1.2 M. Figure 1 shows time-conversion curves for the polymerization of a series of fluorine-containing vinyl ether monomers using the IBEA/Et1.5AlCl1.5 initiator system in

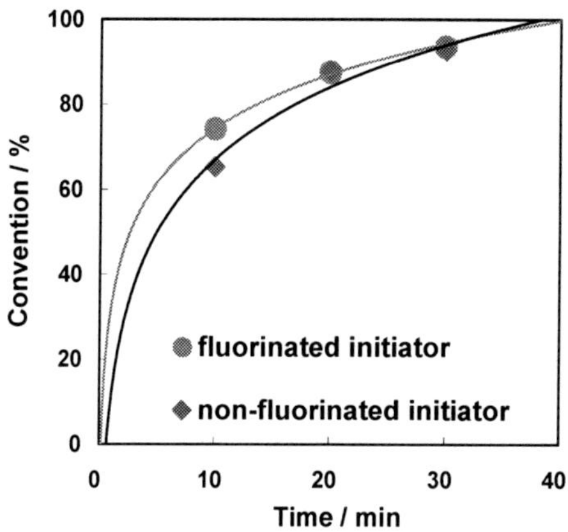

Figure 1. Comparison of time-conversion curves for the polymerization of fluorinated initiator ([monomer]0 = 0.80 M, [fluorinated initiator]0 = 4 mM, [Et1.5AlCl1.5]0 = 20 mM, [1,4-dioxane] = 1.2 M, in toluene at 0 °C).

toluene in the presence of 1,4-dioxane at 0 °C. For comparison, non-fluorinated initiator, such as IBEA were polymerized under the same conditions. In all cases, the polymerization proceeded homogeneously throughout the reaction, and the time-conversion curves were reproducible. In addition, polymerization was initiated without an induction period and reached almost 100% within 40min. Both the fluorinated initiator and non-fluorinated initiator have the similar reaction times. The polymerization results are summarized in Table 1, which show narrow PDI and high yield. Scheme 1 illustrates the synthetic routes for homo and star-shaped polymers based on our novel fluorinated initiator.

Table 1. Cationic polymerization of fluorinated-containing polymers by fluorinated initiator

Type	Polymer Structure	Mn x 10^4	Mw/Mn	Yield %
Homopolymer	F*-MOVE $_{200}$	24.4	1.11	99.9
Star-shaped copolymer	F*-MOVE $_{200}$-DVE star	200.0	1.29	99.0

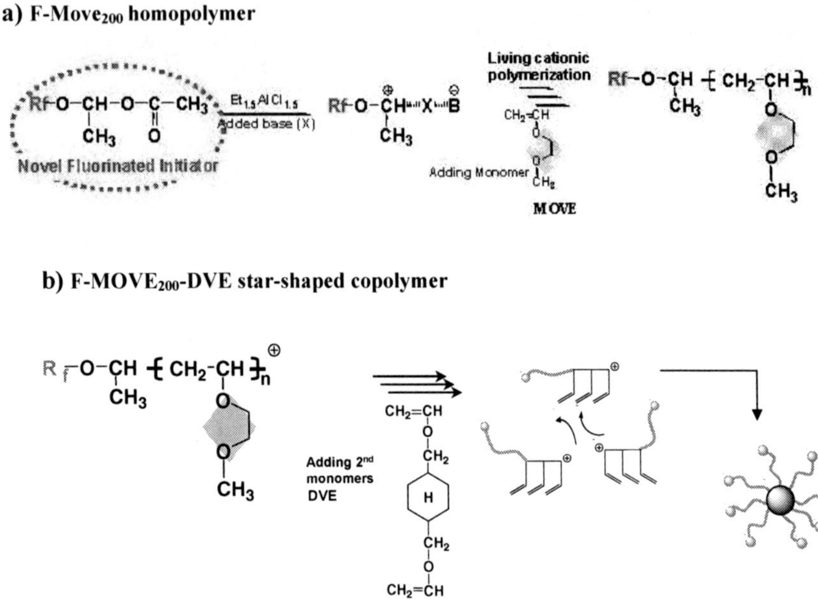

Scheme 1. Synthesis of a) homo polymer with fluorinated end-group by novel fluorinate initiator, and (b) star-shaped polymer with fluorinated end-group at arm segment.

The LCST of this polymer depended on the end group fluorine composition and significant changes in cloud points were observed in water, as confirmed by static light scattering (SLS) measurements. Figure 2 shows aqueous solutions the star-shaped polymers without fluorine terminal groups showed no cloud point below up to 60 ºC in aqueous solution. After the fluoro-functionality was incorporated into the terminal groups, the cloud point of the aqueous solutions decreased to almost 20 ºC.

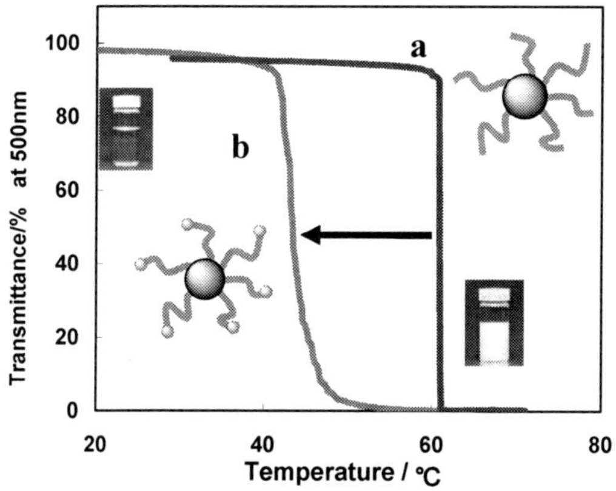

Figure 2. Temperature dependence of transmittance at 500nm of star-shaped polymers in 1 wt% H_2O, heating rate is 1°C/min. (a) Non-fluorinated functionalized star-shaped copolymer (b) Fluorinated end-functionalized star-shaped copolymer.

CONCLUSIONS

We synthesized well-defined fluorinated star-shaped polymers by using fluorinated initiators to generate fluorinated end-functionalized and hydrophilic segments with oxyethylene. These star-shaped polymers showed narrow PDI and different fluorine end-group compositions. Star-shaped polymers were achieved by living cationic polymerization. The properties of the star-shaped polymers and polymer micelles were more effected by terminal groups than those of end-functionalized linear homopolymers.

The aqueous solutions of the star-shaped polymers without fluorine terminal groups showed no cloud point below up to 60 °C in aqueous solution. After the fluoro-functionality was incorporated into the terminal groups, the cloud point of the aqueous solutions decreased to almost 20 °C.

REFERENCES

1. (a) D. Valade, C. Boyer, B. Ameduri, B. Boutevin, Macromolecules 39 (2006) 8639–8651, and references cited herein;
   (b) J.J. Reisinger, M.A. Hillmyer, Prog. Polym. Sci. 27 (2002) 971–1005.
2. (a) E. Kissa, Fluorinated Surfactants: Synthesis, Properties, Applications, Marcel Dekker, New York, 1994 (Chapter 1);
   (b) J. Scheirs, Modern Fluoropolymers, Wiley, New York, 1997.

3. Hadjichristidis, N.; Pitsikalis, M.; Pispas, S.; Iatrou, H. Chem. ReV. 2001, 101, 3747-3792.
4. Aoshima, S.; Kanaoka, S. Adv. Polym. Sci. 2008, 210, 169.

Mater. Res. Soc. Symp. Proc. Vol. 1312 © 2011 Materials Research Society
DOI: 10.1557/opl.2011.131

Layer-by-layer Stacking Method for 3-D Nano Structure Fabrication Using Block Copolymer Self Assembly

Shin'ichi Warisawa[1], Ryosuke Kanameda[1], Reo Kometani[1] and Sunao Ishihara[1]
[1]Department of Mechanical Engineering, School of Engineering, The University of Tokyo, 7-3-1 Hongo Bunkyo-ku, Tokyo 113-8656, Japan

ABSTRACT

In this paper, we present layer-by-layer stacking method to fabricate self-assembled structures of block copolymers (BCP) toward the out-of-plane direction. Layer-by-layer stacking is realized by transferring a BCP film on one substrate to another. Specifically, a water-soluble polymer film is coated on the former substrate, which is placed and fixed in contact with a target substrate. Consequently, the BCP film is released from the substrate and transferred to the target substrate when immersed in de-ionized water. In our experiment, PS-b-PMMA is used to form and transfer self-assembled structures, and polyvinyl alcohol is used as a water-soluble polymer. We prepared two kinds of target substrates; one has horizontal cylindrical structures by BCP self assembly, and the other has groove structures by EB lithography. In the case of BCP patterned substrate, BCP film with vertical cylindrical structures is transferred onto the line structures of BCP. In the case of EB lithography patterned substrate, BCP film with vertical cylindrical structures is transferred in a doubly suspended condition. Furthermore, vertical and horizontal cylindrical structures are also observed to align along the grooves.

INTRODUCTION

Recent block copolymer self-assembly research has demonstrated the potential to form aligned line patterns [1], periodical sphere structures [2], and directed line and sphere patterns using chemical and physical templates at the nanoscale. However, many of these have been examined mainly on two-dimensional flat substrates [3, 4]. On the contrary, BCP thin films with microdomains oriented normal to a substrate have paid much attention, and thus many methods such as shear force flow [5, 6], electric field [7], solvent evaporation [8] and so on are employed to obtain ordered thicker or larger film effectively. Since most of these apply a unidirectional field to the entire film, obtained structures have an orientation in the applied field direction, and then fabricated structures are limited. In this research, we offer the possibility of extending self-assembled structures in the out-of-plane direction based on a layer-by-layer stacking method using thin film transfer.

LAYER-BY-LAYER STACKING METHOD

Figure 1 shows a layer-by-layer stacking method that we propose. Layer-by-layer stacking is realized by transferring a BCP film of one substrate (source substrate) onto another (target substrate). The source substrate is coated by a water-soluble polymer, and then coated by a block copolymer. The source substrate may be either flat or topographically patterned. The target substrate is prepared in advance with its surface patterned by self-assembled block copolymer or topographically patterned by lithography. In the next step, the source substrate is placed and fixed in contact with the target substrate.

These two substrates are then immersed in de-ionized water. The block copolymer film on the source substrate is released and transferred onto the target substrate due to solution of the water-soluble polymer. Annealing of the block copolymer film on the source substrate can be done either before or after it is placed onto the target substrate. Figure 1 shows the case of annealing after transfer.

Advantages of using the proposed method are expected as follows: (i) The source substrate with topographically patterned surface can localize block copolymer films to the target substrate as is patterned. This case will realize complicated three-dimensional structures if each transfer is well aligned. (ii) Different block copolymer self-assembled structures can be stacked in different layers, and thus complicated 3-D structures can be fabricated as designed. (iii) When the target substrate is topographically patterned, self-assembled block copolymer films can be transferred onto the target substrate in a suspended condition.

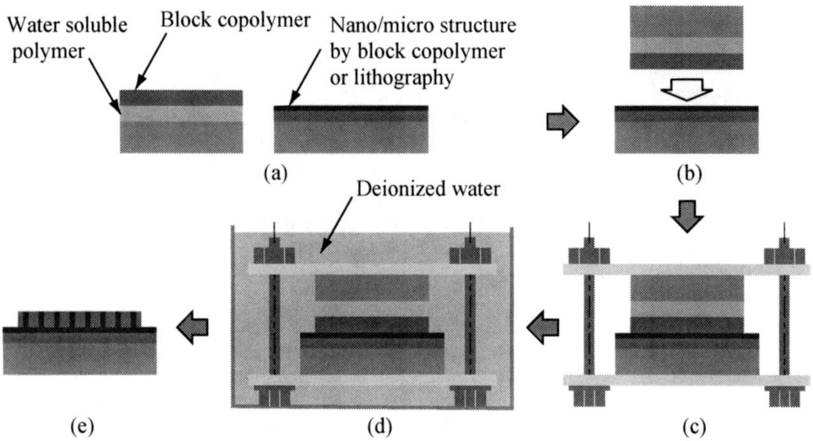

Figure 1. A layer-by-layer stacking method. (a) Two substrates are prepared. A water-soluble polymer is coated under a block copolymer film on one substrate, and nano/micro structure is patterned on the other substrate. (b, c) Two substrates are faced together, (d) and then immersed into de-ionized water. (e) The block copolymer film is released and transferred onto the latter substrate when immersed into de-ionized water.

EXPERIMENT

We prepared two kinds of target substrates; one has horizontal cylindrical structures formed by block copolymer self assembly, and the other had groove structures fabricated by EB lithography. For each of the target substrates, we prepared source substrates as follows. As a water-soluble polymer, we used Polyvinyl alcohol (PVA) as a water-soluble polymer, JMR-10HH of Japan Vam & Poval Co., Ltd. PVA is insoluble in toluene, a solvent for block copolymer. Therefore, it is expected to act as a release layer for our proposed method. The block copolymer we used for all experiments here is PS-b-PMMA of Polymer Source Inc., which has a molecular weight of 50.5 kg/mol for PS and 20.9 kg/mol for PMMA.

PVA was spin-coated on a Si substrate. Consecutively, a 3.0 % toluene solution of PS-b-PMMA was spin-coated in 100 nm-thick onto the PVA film. We tried to anneal the source substrate coated with PVA and PS-b-PMMA films before transferring them to the target substrate to promote a self-assembled structure. However, PVA that we used denatured into water-insoluble by annealing at 190 °C. Therefore, annealing of the block copolymer on the source substrate was done after transferring onto the target substrate.

Transfer onto block copolymer self-assembled structures

For preparation of a target substrate, a 1.5 % toluene solution of the block copolymer was spin-coated onto a flat Si substrate and annealed at 190 °C for 24 hours. The source substrate explained above and the target substrate were faced in contact with each other and fixed by screws as shown Fig. 1. The fixed substrates were immersed in a de-ionized water bath to dissolve the PVA film. At this stage, PS-b-PMMA film is detached from the source substrate and transferred onto the target substrate. The substrates were then taken out of the de-ionized water bath and separated from each other. Finally, the target substrate with two layered films was annealed at 190 °C for 24 hours.

Figure 2 (left) shows an optical microscope image of the annealed target substrate. The transferred film at center and the target film on both sides of the transferred film are observed. The transferred film exhibits many cracks and defects. The target film can be observed through the cracks. Figure 2 (right) shows an SEM image of the transferred film with a circular shaped defect. Vertical cylindrical structures and small cracks are observed on the transferred film, while horizontal cylindrical structures are observed on the target film.

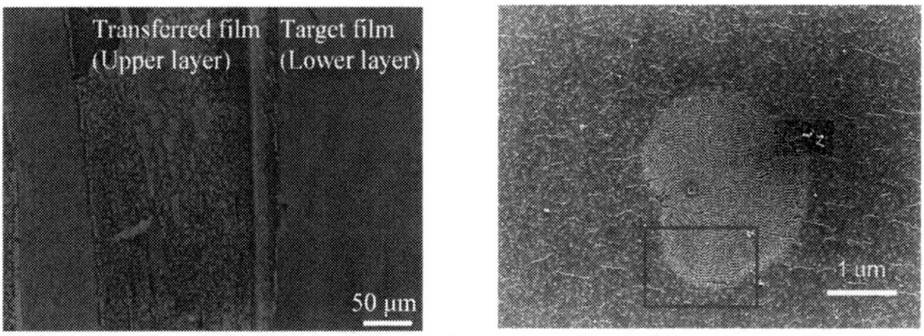

Figure 2. (left) Optical microscope image shows a transferred film and target film. The transferred film has many cracks and defects, through which the target film can be also observed. (right) SEM image of one of defect area shows that vertical cylindrical structures form on the transferred film and horizontal cylindrical structures on the target film.

Figure 3 shows the thickness of the transferred film by AFM. Figure 3 (left) shows the same morphology as that observed by SEM shown in Figure 2. Figure 3 (right) shows the height information around the boundary between the transferred and target films. This measurement exhibits the film thickness of the transferred film is about 14 nm. The boundary forms a very gentle slope. From these observations, the transferred and target films seem to melt together.

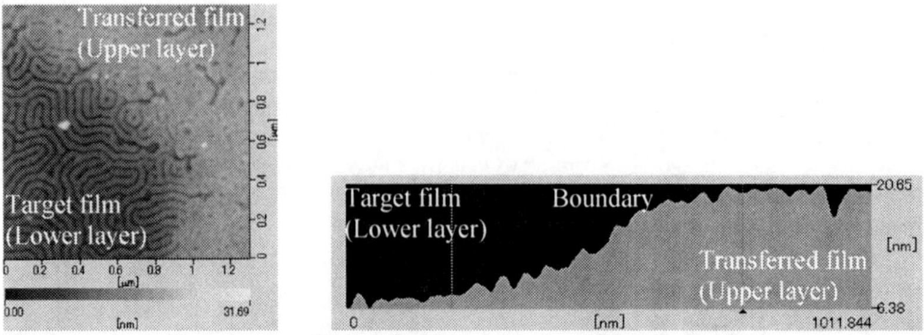

Figure 3. Amplitude image of the defect area taken by AFM. (left) Horizontal and vertical cylindrical structures are observed just as observed by SEM. (right) The transferred film has approximately 14 nm of thickness, which is thinner than as spin-coated.

Transfer onto line and space structures by EB lithography

We employed our proposed method to a topographically patterned Si target substrate as shown in Figure 4 left. The patterned substrate has grooves with 280 nm valleys and 120 nm mesas. The source film transferring procedure is the same as that described in the previous section.

One result after transferring and annealing the source film is shown in Figure 4 right. It is clearly observed that horizontal cylinders were formed on the mesas and vertical structures in the valleys. The horizontal cylindrical structures were aligned to some extent along with the mesas. This localized morphology difference between on the mesas and in the valleys results from the block copolymer confinement by the two substrates and the topographically patterned substrate causing film thickness variations.

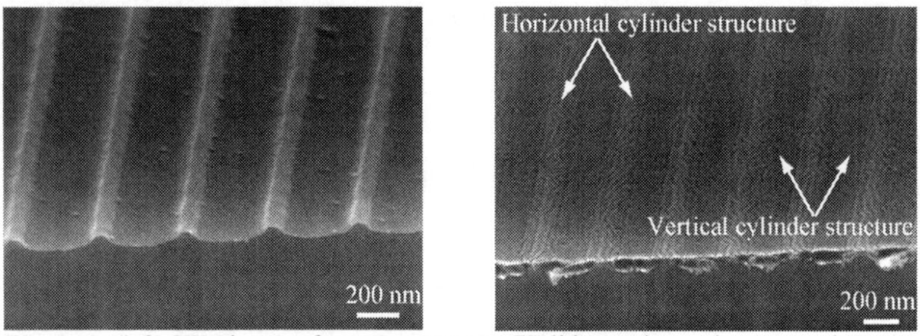

Figure 4. (left) SEM image of nanostructures by EB lithography on the target substrage. Widths of a groove and mesas are 280 nm and 120 nm, respectively. (right) SEM image of transferred and annealed of block copolymer film. It is clearly observed that horizontal cylinder structures are on the mesas and vertical structures are in the valleys.

Another result is shown in Figure 5. This result exhibits no specific localized or aligned morphology. What should be paid much attention is the center of the image shown in Figure 5 left. Figure 5 right shows that the block copolymer thin film was partially transferred in a suspended condition between the mesas of the grooves. It suggests that the transferred block copolymer film was not confined so much as the case of Figure 4, and thus that the film thickness variations were not large enough to induce a localized or aligned morphology. However, since the suspended portion was very limited, then it cannot be said the method was appropriate for such suspended films.

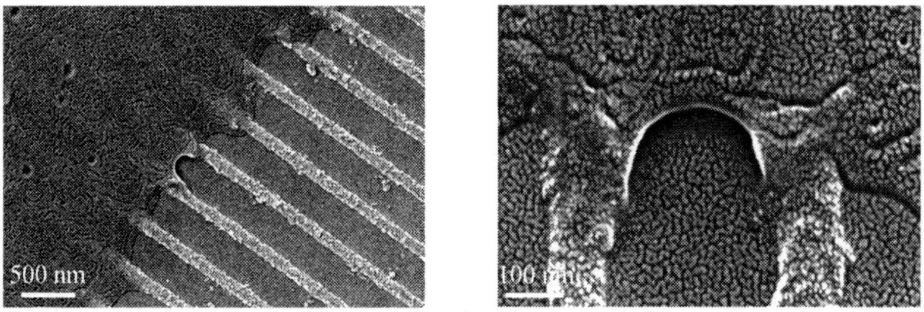

Figure 5. Both SEM images show transferred and annealed block copolymer film. (left) A part of the boundary of the transferred film observed at the center of the image is like a doubly suspended membrane. There are also observed horizontal and vertical cylindrical structures on the membrane. (right) A magnified image of the doubly suspended film shows the horizontal cylindrical structure in detail.

DISCUSSION

One of the problems of our proposed method is a material selection of the releasing layer between a substrate and block copolymer thin film. PVA used in our experiments is water soluble in nature, but denatures into water insoluble by annealing necessary for block copolymer self assembly. The decrease of the thickness of the transferred film and the immersion of the block copolymer film into the patterned grooves were cased partially because of melting during long-time annealing. If the annealing process can be done before transferring, those problems may be improved.

Another problem is the defect on the transferred film. It was observed in the case of transferring onto the flat surface coated by the other block copolymer film, not in the case of topographically patterned surface. The difference between the two cases is only the target substrate. Therefore, transferring a block copolymer film in de-ionized water using PVA is not main reason for it. The reason should be investigated in the next step.

The screw fixation was obviously more critical issue that caused poorly-reproducible results. Facing pressure between two substrates is also an important factor for better fabrication, but was not controlled by the screw fixation. Especially, in the case of facing flat surfaces, this problem is more important as shown in the experimental results.

The obtained pattern as shown in Figure 4 is almost the same as reported in the reference [9]. The difference is whether a pattern is transferred or a block copolymer film is transferred. The patterned mechanism is basically the same between the two.

CONCLUSIONS

We proposed a layer-by-layer stacking method in order to fabricate 3-D nano structures using block copolymer self assembly. Our proposed method describes how to transfer a block copolymer thin film onto a target substrate. The method defines in-plane structures by the block copolymer self assembly mechanism and out-of-plane structures by the staking technique. Experiments were conducted to show the capability of fabricating 3-D nano structures. The transferred block copolymer films are stacked on the target block copolymer films, and different microdomain morphologies were observed in each layer. Doubly suspended block copolymer films were partially observed on the topographically patterned substrate. The problems in the proposed method were discussed.

ACKNOWLEDGMENTS

This work was partly conducted at the Center for Nano Lithography & Analysis, the University of Tokyo, which is supported by the Ministry of Education, Culture, Sports, Science and Technology (MEXT) of Japan. This work was also supported by a Grant-in-Aid for Scientific Research (B) No. 21360059 from the Japan Society for the Promotion of Science.

REFERENCES

1. Stoykovich, M.P., et al., *Directed assembly of block copolymer blends into nonregular device-oriented structures.* Science (New York, N.Y.), 2005. **308**: pp. 1442-1446.
2. Segalman, R.A., H. Yokoyama, and E.J. Kramer, *Graphoepitaxy of Spherical Domain Block Copolymer Films.* Advanced Materials, 2001. **13**: pp. 1152-1155.
3. Yang, J.K.W., et al., *Complex self-assembled patterns using sparse commensurate templates with locally varying motifs.* Nature nanotechnology, 2010. **5**: pp. 256-260.
4. Bita, I., et al., *Graphoepitaxy of self-assembled block copolymers on two-dimensional periodic patterned templates.* Science (New York, N.Y.), 2008. **321**: pp. 939-943.
5. Chen, Z., *Pathways to Macroscale Order in Nanostructured Block Copolymers.* Science, 1997. **277**: pp. 1248-1253.
6. Angelescu, D.E., et al., *Shear-Induced Alignment in Thin Films of Spherical Nanodomains.* Advanced Materials, 2005. **17**: pp. 1878-1881.
7. Thurn-Albrecht, T., et al., *Ultrahigh-density nanowire arrays grown in self-assembled diblock copolymer templates.* Science (New York, N.Y.), 2000. **290**: pp. 2126-2129.
8. Kim, S.H., et al., *Highly Oriented and Ordered Arrays from Block Copolymers via Solvent Evaporation.* Advanced Materials, 2004. **16**: pp. 226-231.
9. Park, C., et al., *Double textured cylindrical block copolymer domains via directional solidification on a topographically patterned substrate.* Applied Physics Letters, 2001. **79**: pp. 848-850.

Mater. Res. Soc. Symp. Proc. Vol. 1312 © 2011 Materials Research Society
DOI: 10.1557/opl.2011.132

Process Optimization for Nanocrystalline Cellulose Production from Microcrystalline Cellulose

Christophe Danumah and Hicham Fenniri
National Institute for Nanotechnology and Department of Chemistry
University of Alberta, Edmonton, AB, T6G 2M9, Canada

ABSTRACT

A process optimization has been developed for obtaining nanocrystalline cellulose (NCC) by acid hydrolysis of commercially available microcrystalline cellulose (MCC) in high yield (~ 40-50%). This method was based on control of key parameters such as the rate of addition of sulfuric acid solution to the MCC/water suspension, the mixing speed, the volume of collected NCC suspensions and the volume ratio of NCC suspension to water during dialysis. The resulting NCC products were characterized by x-ray diffraction (XRD), thermogravimetric analysis (TGA), elemental analysis (EA), scanning electron microscopy (SEM) and atomic force microscopy (AFM). Electron microscopy results showed that the rod-shaped NCC had lengths and widths of about 40-400 nm and 5-40 nm, respectively.

INTRODUCTION

Cellulose is the world's most abundant natural, renewable, biodegradable polymer, and occurs as whisker like microfibrils. The latter are converted to nanocrystalline cellulose (NCC), a new family of nanoparticles, by careful acid hydrolysis of the widely available cellulose [1].

The growing interest in cellulose nanocrystals is due to their relatively low density, low cost, high aspect ratios (typically >30-100), high surface area, unique morphology and mechanical properties (rigid rods with 25%-30% strength of carbon nanotubes, stronger than steel) [1-4]. NCC also possesses a reactive surface covered with hydroxyl groups, providing the possibility for extensive chemical modification [1-4] and broad applicability. NCC can be used to develop an emerging class of nanomaterials, which is expected to capture a new market in transportation, medical and packaging applications, electronics and construction [5, 6]. Nevertheless, one of the biggest challenges involved is a reliable recipe for their large-scale production in high yield. The feasibility of the cost-effective commercial scale-up, and the manner in which the yield of NCC may be maximized are the two essential factors that would govern the success of any approach [7]. Previous attempts to optimize the hydrolysis conditions of NCC were based on hydrolysis parameters, such as temperature and reaction time, which resulted in a yield of 21-38 % [8, 9].

This work describes a straightforward method used to optimize the acid hydrolysis of cellulose and maximize the yield by carefully controlling factors, such as the rate of addition of sulfuric acid solution to the MCC/water suspension, the mixing speed, the volume of the collected NCC suspensions and the volume ratio of NCC suspension to water during dialysis.

EXPERIMENTAL DETAILS

NCC Preparation

The method described by Bondesson et al [9] was used as starting point to optimize the preparation and isolation of NCC using acid hydrolysis conditions (Figure 1 [10]). In a typical preparation MCC (5 g, Aldrich) was mixed with water (22.5 mL). The resulting suspension was placed in an ice bath and stirred for 10 min (500 rpm). Concentrated sulfuric acid (95-98%, 21.5 mL) was added dropwise to a final acid concentration of 64 wt% and the suspension was heated to 45°C while stirring at 500 rpm for 2 h. The reaction was then stopped by diluting the suspension with cold water (10×). The solid was isolated by centrifugation (7,500 rpm, 15 min, 20°C), re-suspended in deionized water (125 mL) and centrifuged down. This suspension/centrifugation step was repeated five times with sonication in an ice-bath (10 min) between each step. The supernatants (containing colloidal NCC) recovered between each centrifugation step were combined and subjected to dialysis against distilled water for 3 days at room temperature until a constant neutral pH was reached inside the dialysis membranes. Fisherbrand regenerated cellulose dialysis membranes (12,000-14,000 molecular weight cut-off) were used. The water bath was changed every 24 h and the ratio of NCC suspension to water was kept at 1:50 (v/v). The colloidal solution in dialysis bags was sonicated in an ice bath for 10 min to yield colloidal dispersions and freeze-dried for 48 h to obtain NCC. After dialysis, the dispersion may also be stored at 3°C prior to freeze-drying.

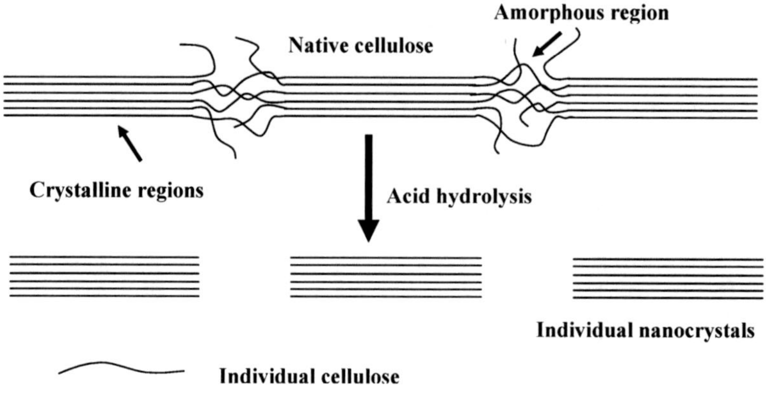

Figure 1. NCC formation by controlled acid hydrolysis [10].

Characterizations

x-ray diffraction (XRD) studies of NCC were recorded on a Bruker D8 x-ray diffractometer equipped with a CuKα (40 kV, 40 mA) radiation over a 2θ range of 1° to 40° at room temperature. Thermogravimetric analysis was carried out under normal atmosphere using a

TGA model Q500 thermogravimetric analyzer. SEM imaging was performed on a Hitachi model S-4800, operated at 15kV (30kV in STEM mode). Tapping mode atomic force microscopy (AFM) was performed using a Digital Instruments/Veeco Instruments MultiMode Nanoscope IV AFM equipped with an E scanner. Elemental analysis of the total sulfur content was performed with a Carlo Erba EA 1108 elemental analyzer.

DISCUSSION

Native cellulose is composed of amorphous and crystalline regions as shown by Figure 1. The harsh sulfuric acid hydrolysis removes amorphous cellulose to produce crystalline cellulose nanoparticles. The acid causes the scission of glycosidic linkages hydrolytically, thereby releasing the individual nanocrystallites. Sulfuric acid is commonly used because of the formation of sulfate ester groups during acid treatment, which leads to the formation of individual nanocrystallites with negative charges on their surface (Figure 2). Repulsion between the negatively charged NCC results in stable aqueous suspension [1, 11]. The final properties of NCC depend on different parameters, such as the cellulose used, acid concentration, time of reaction, temperature of reaction and the specific acid used; which need to be controlled.

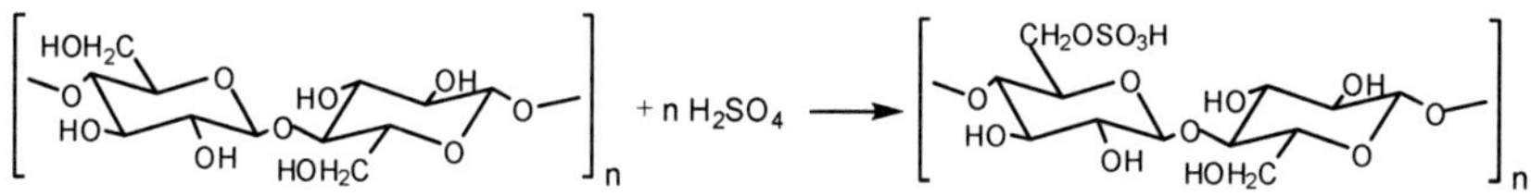

Figure 2. Esterification of hydroxyl groups by sulfate ions from sulfuric acid treatment of cellulose.

Table 1 shows that the yield of isolated NCC increases with the time of acid addition to the MCC/water suspension; while the total sulfur content remains relatively similar.

Table 1: Effect of addition rate of sulfuric acid to MCC/water suspension in an ice bath

NanoCrystalline Cellulose			
Sample ID	**Time of addition of acid (h)**	**Yield (wt %)**	**Total sulfur content (wt %)**
NCC Sample 1	0.5	20	0.76
NCC Sample 2	1.0	15	0.51
NCC Sample 3	2.0	30	0.66
NCC Sample 4	4.0	40	0.70
NCC Sample 5	4.0	50	0.65

Reaction time: 2h; reaction temperature: 45°C; mixing speed: 500 rpm

These results indicate that fast addition of acid solution to NCC/water suspension promotes the rapid hydrolysis of cellulose, which affects the relative amounts of cellulose nanocrystals formed.

x-ray powder diffraction (XRD) was used to monitor the structure of MCC and NCC samples and evaluate their crystallinity. Figure 3 shows that the solid-state structure of MCC and NCC are similar. Sharp peaks were observed at $2\theta = 22.5°$ and $34.5°$. In addition a shoulder was observed around the region $2\theta = 14°–17°$. These peaks are typical reflections of cellulose I structure [12]. Crystallinity is commonly measured as a ratio between the diffraction portion of the crystalline part of the sample and the total diffraction from the same sample. This ratio was found to be *ca.* 72 % and *ca.*75% for MCC and NCC samples, respectively, suggesting that the crystallinity of NCC was preserved even after acid hydrolysis.

Thermogravimetric analysis was used to study the degradation characteristic of the MCC and NCC obtained by acid hydrolysis. The samples were heated from room temperature to 600°C at a heating rate of 20°C/min under atmospheric conditions. Representative TGA results for MCC and NCC (Figure 3) illustrate that the thermal stability of NCC is lower than that of MCC. The onset decomposition temperatures were found to be *ca.* 325°C and *ca.* 178°C (sulfate ester groups) for MCC and NCC samples, respectively. The remaining 50% of NCC samples decomposed at *ca.* 260°C. This decrease in thermal stability may be attributed to the presence of sulfate groups on the NCC surface.

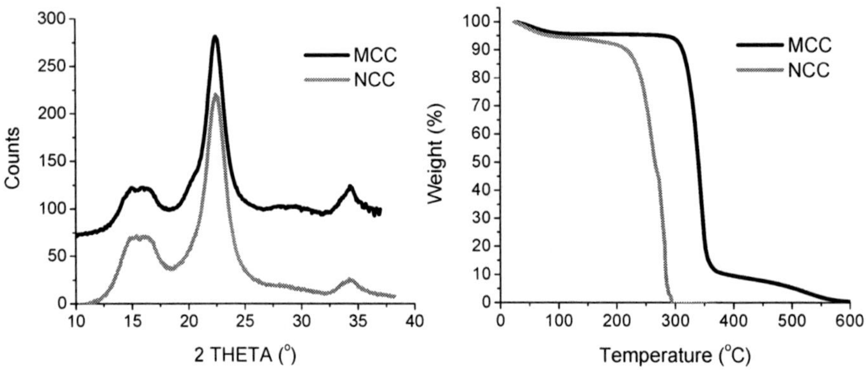

Figure 3. XRD pattern of MCC and NCCs (left) and TGA curves of MCC and NCCs (right).

The starting material (MCC) and NCC obtained after hydrolysis using different addition rates of acid in an ice bath were also investigated by SEM (Figure 4). The images show large aggregates/agglomerates of MCC particles formed by strong hydrogen bonding [13]. In contrast with this, NCC is composed of nanorod-like structures with a tendency for partial agglomeration by alignment of the NCC along their longest axis. Their widths and lengths were 5-40 nm and 50-400 nm, respectively.

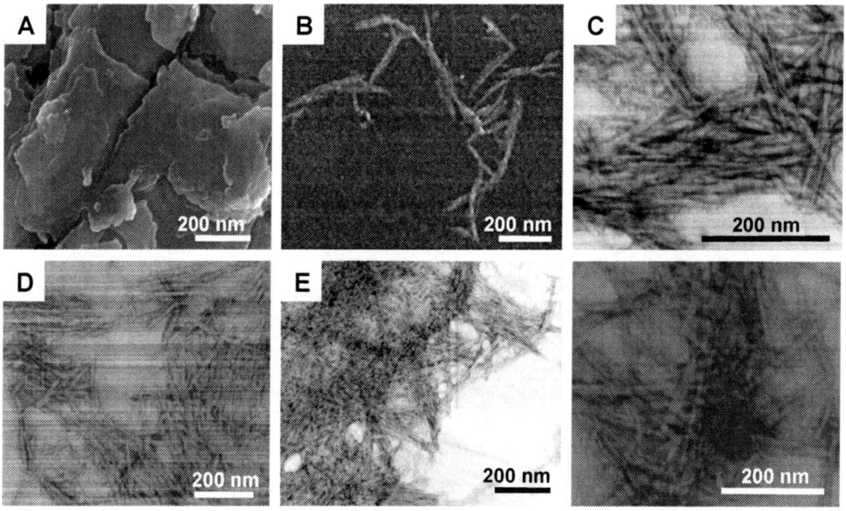

Figure 4. SEM images of MCC (A), NCC (B, sample 1) and STEM images of NCC (C–F, samples 2-5, respectively). All samples were prepared at 1 mg/mL in water (dried material dispersed in water).

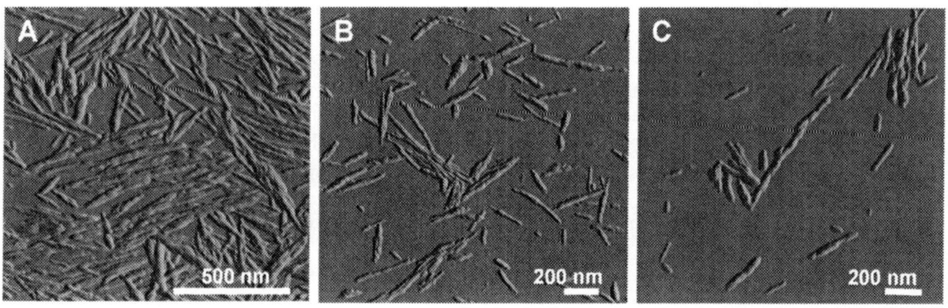

Figure 5. Representative AFM images of rod-shaped NCC dispersed in water: Original Suspension (A), 10 times dilution with water (B) and 100 times dilution with water (C).

The rod-shaped morphology for NCC materials is also supported by the representative AFM results (Figure 5). The latter also revealed the effect of dilution on the improvement of the rod-shaped NCC dispersion in water. It can be seen from SEM and the representative AFM results that the NCC samples are more heterogeneous in size. This is due to the non-selectivity of the acid hydrolysis process and the electrostatic repulsion between negatively charged NCC nanorods. AFM results from dialyzed NCC suspension in water demonstrate the possibility of producing a stable colloidal suspension of NCC with H_2SO_4 when the hydrogen bonding effect is minimized. The freeze-drying was found to promote aggregation of NCC, in agreement with earlier reports [14, 15].

CONCLUSIONS

In this report, we have optimized a procedure for the production of NCC in good yield. In particular, we found that the rate of addition of sulfuric acid to MCC/water suspension, the mixing speed and the work up procedure significantly impact the yield of the NCC produced. We anticipate that this approach could be applied to other cellulose sources such as cotton.

ACKNOWLEDGMENTS

We thank the National research Council and the University of Alberta for supporting this project, and Dr. Jay-Young Cho for AFM analyses.

REFERENCES

1. M.M. de Souza Lima and R. Borsali, Macromol. Rapid Commun. **25**, 771 (2004).
2. Y. Habibi, A.-L. Goffin, N. Schiltz, E. Duquesne, P. Dubois, A. Dufresne, J. Mater. Chem. **18**, 5002 (2008).
3. M.A.S. Azizi Samir, F. Alloin, A. Dufresne, Biomolecules **6**, 612 (2005).
4. S. Beck-Candanedo, M. Roman, D.G. Gray, Biomolecules **6**, 1048 (2005).
5. Polymer Nanocomposites and Their Applications, edited by S. Sinha Ray and M. Bousmina, (American Scientific Publishers, New York, 2006) 600p.
6. M.A. Hubbe, O.J. Rojas, L.A. Lucia, and M. Sain, BioRessources **3 (3)**, 929 (2008).
7. W. Hamad, Can J. Chem. Eng. **84**, 513 (2006)
8. W.Y. Hamad and T.Q. Hu, Can. J. Chem. Eng. **88**, 392 (2010).
9. D. Bondeson, A.P.Mathew, K. Oksman, Cellulose **13**, 171 (2006).
10. Figure1- Image from http://forestproducts.orst.edu/faculty/simonsen/Nanocomposites.pdf
11. R.H. Marchessault, F.F. Morehead and M.J. Koch, J. Colloid Sci. **16**, 327 (1961).
12. A. P. Mathew, K. Oksman, and M. Sain, J. Appl. Polym. Sci. **97**, 2014 (2005)
13. N.E. Marcovich, M.L. Auad, N.E. Bellesi, S.R. Nutt, M.I.J. Aranguren, J. Mater. Res. **21**, 870 (2002).
14. E. Dujardin, M. Blaseby and S. Mann, J. Mater. Chem. **13**, 696 (2003).
15. S. Dong and M. Roman, J. Am. Chem. Soc. **129**, 13810 (2007).

Mater. Res. Soc. Symp. Proc. Vol. 1312 © 2011 Materials Research Society
DOI: 10.1557/opl.2011.133

Graphitic Carbon Nanoparticles from Asphaltenes

Christophe Danumah, Andrew J. Myles and Hicham Fenniri
National Institute for Nanotechnology and Department of Chemistry, Alberta
Edmonton, Alberta T6G 2M9, Canada

ABSTRACT

We report preliminary studies on the preparation of carbon nanoparticles (CNPs) of ~ 20 nm in diameter derived from asphaltenes under mild conditions. This transformation occurred upon heating a thin film of asphaltenes cast on a carbon coated copper grid under both nitrogen and oxygen atmosphere. The resulting CNPs were characterized by thermogravimetric analysis (TGA), scanning/transmission electron microscopy (SEM/STEM), transmission electron microscopy (TEM), elemental analysis (EA) and Auger electron spectroscopy (AES). The findings point towards asphaltenes, a crude oil by-product, as a potential source for carbon nanomaterials.

INTRODUCTION

Carbon nanomaterials are of great industrial interest for applications in electronics [1], bio-imaging [2], catalysis, materials science and engineering [3] due to their high surface area, large pore volume, high conductivity [4], mechanical strength and thermal properties [1]. Several synthetic routes to carbon nanomaterials have been reported with the goal of reducing overall cost and avoiding harsh, non-scalable conditions [5, 6]. For example, inexpensive carbon sources such as candle soot [7], natural gas soot [8], sawdust [9] and biomass processed through hydrothermal carbonization [6] have been used for generating carbon nanomaterials. Asphaltenes have also been investigated in this regard and carbon nanoparticles (CNPs) were successfully generated using electron beam irradiation [9] and chemical vapor deposition [10].

The major aim in the area of carbon nanomaterial preparation is the cost and scalability, which are required for any industrial application. Thus, our ultimate goal is to develop a simple, energy efficient and cost-effective processing methodology for fabricating CNPs from asphaltenes, which is a class of oilsands by-product. Our preliminary investigations have shown that upon heating of a thin film of asphaltenes cast on a carbon coated copper grid under both nitrogen and oxygen atmosphere, graphitic CNPs are formed. The resulting CNPs are *ca.* 20 nm in size and have been characterized using TGA, SEM/STEM, TEM, EA, and AES.

EXPERIMENTAL DETAILS

General Experimental

C7-Asphaltenes samples used as carbon stock in this work were prepared in our laboratory from the Athabasca Crude Oil [11]. Reagent grade toluene was purchased from Fisher Scientific.

Preparation of colloidal solutions of Asphaltenes

A stock solution was prepared by dissolving asphaltenes (1 mg/mL) in toluene using a heat gun for few seconds followed by sonication with a Branson ultrasonic Cleaner (Model 1510R-MT) for 2 min. The heating/sonication process was repeated twice to ensure maximum solubilization of asphaltenes. The resulting stock solution was allowed to stand for 2-3 days. Aliquots from this solution were then diluted to the desired concentrations (0.0001–0.1 mg/mL) (Figure 1). The 0.0001 mg/mL solution was used as a starting material for the production of carbon nanoparticles.

Thin film preparation

One drop (10-15 μL) of the dilute colloidal solution (0.0001 mg/mL) was cast onto a 400-mesh carbon-coated copper grid and allowed to stand for 60s. Excess fluid was blotted using Whatman No. 1 filter paper. The resulting TEM grid was air-dried for 24 h before thermal treatment.

CNP preparation

The thin film samples were loaded onto an aluminum pan and heated under nitrogen atmosphere to 200°C (10°C/min) for 18 h using a TGA apparatus. The nitrogen gas was then switched to oxygen and the sample was treated or an additional 18h at 200°C. We have also investigated samples cast on carbon-coated grids and treated for 18 h at (a) room temperature, (b) 200°C under nitrogen atmosphere, and (c) 200°C under oxygen atmosphere.

Characterizations

TGA was carried out on a TGA model Q500 thermogravimetric analyzer. SEM imaging was performed on a field-emission gun Hitachi model S-4800, operated at 15kV (30kV in STEM mode). TEM images were obtained on a JEOL 2200 FS TEM Schottky field emission instrument equipped with an in-column omega filter working at 200 kV. AES measurements were carried out using JAMP-9500F Auger microprobe (JEOL) and Elemental analysis of the total sulphur content was performed with a Carlo Erba EA 1108 elemental analyzer.

DISCUSSION

TGA of C7-Asphaltenes was used to investigate the decomposition temperature of asphaltenes so that the temperature for processing the thin film could be optimized. Thus, the sample was heated up to 1000°C under nitrogen and oxygen at a heating rate of 20°C/min . The TGA profile under N_2 (Figure 1) shows two mass transitions. The first relatively small transition occurred between 200–250°C (0.9 wt%) and was likely due to evaporation of volatiles. The second large mass reduction occurred between 350–550°C (53 wt%) was due to a combination of evaporation and cracking [12]. The TGA profile under air showed two mass transitions, the first between 380–500°C (40 wt%) and was due to a combination of evaporation of cracking, whereas the second occurred between 500–700°C (59 wt%) was due to combustion of organic materials. The thermograms show stability up to 375°C, which indicates that there is no decomposition at the processing temperature of 200°C. Based on these results, the asphaltenes samples were subjected to thermal treatment under nitrogen atmosphere followed by oxygen atmosphere at 200°C for 18h.

SEM imaging of the asphaltenes samples (Figure 2) showed that heating of a thin film on a carbon-coated copper grid at 200°C under nitrogen atmosphere followed by oxygen atmosphere resulted in the formation of well-defined nanoparticles, *ca.* 20 nm in diameter, whereas an earlier method based on chemical vapor deposition of asphaletenes gave nanospheres of 300–400 nm in diameter [10]. The absence of decomposition of asphaltenes at 200°C as revealed by TGA (Figure 1) is in agreement with elemental analysis that showed only 0.8% loss of hydrogen content upon heating, therefore suggesting a direct transformation of the asphaltenes from a thin film to nanoparticles. The generation of CNPs using this process was also confirmed by TEM (Figure 3).

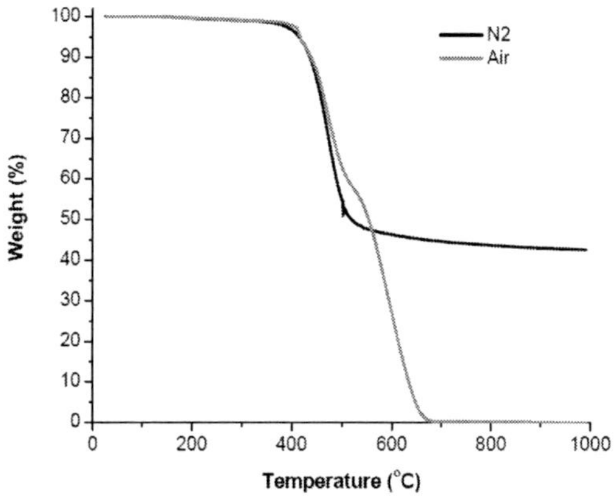

Figure 1. TGA profiles of C7-Asphaltenes powder under nitrogen and air.

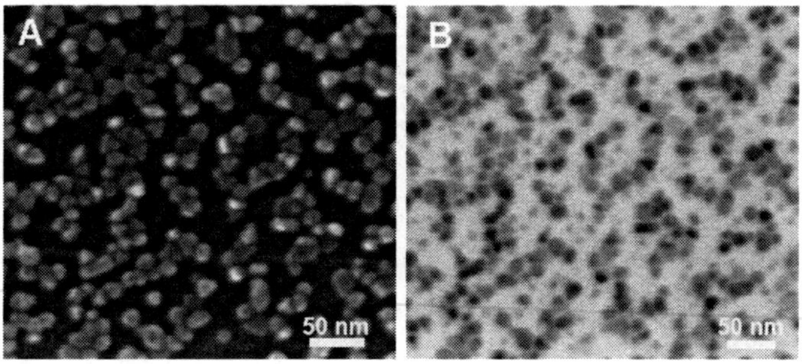

Figure 2. SEM (A) and TEM (B) images of CNP's derived from asphaltenes.

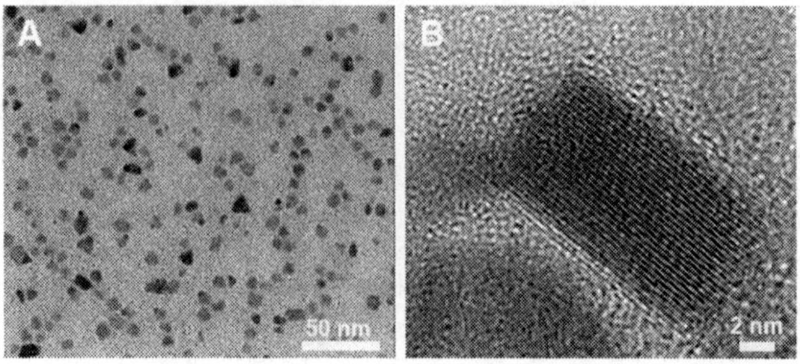

Figure 3. Low (A) and high-resolution (B) TEM images of CNP's derived from asphaltenes.

To investigate the chemical composition of these nanoparticles, Auger electron spectroscopy was performed on the samples after thermal treatment. The results (in % atomic concentration) revealed that the particles are mostly carbonaceous (94.7 %) with a small amount of oxygen (2.6 %) and copper (1.7 %). The presence of copper arises from the carbon coated copper TEM grid used as substrate. Therefore, these results are consistent with the parent sample's composition, which is composed mostly of carbon (82.03% carbon, 1.70 % oxygen). The small differences in carbon and oxygen content may arise from the underlying carbon film and the partial oxidation occurring during the heating process under oxygen.

Further examination using high-resolution transmission electron microscopy revealed highly ordered arrangement of atoms as evidenced by distinct lattice features (Figure 3B). The interlayer d-spacing of 0.35 nm is consistent with that of graphitic carbon [13], which suggests that the carbon nanoparticles obtained are graphitic in nature [10, 13].

Thin films of asphaltenes with (a) no thermal treatment, (b) thermal treatment at 200°C under nitrogen atmosphere for 18 h, and (c) thermal treatment at 200°C under oxygen atmosphere for 18 h, were analyzed by SEM. As shown in Figure 4, none of these samples showed CNPs similar to those in Figure 2. Figure 4 shows (A) *ca.* 0.5-1 μm, (B) *ca.* 50-100 nm, and (C) *ca.* 50 nm particles. Small nanoparticles (*ca.* 10 nm) were observed on all SEM images and come from the carbon coated copper grid.

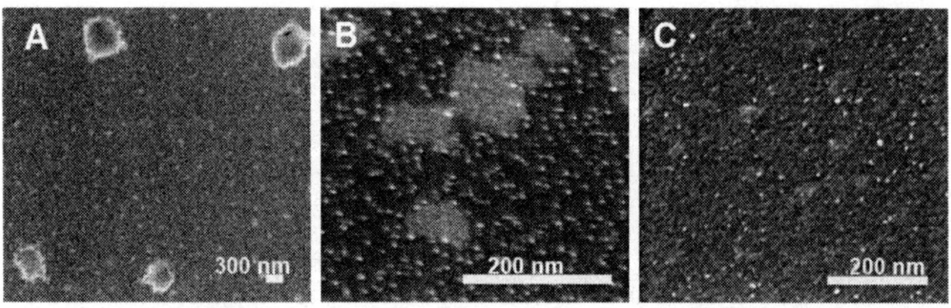

Figure 4. SEM images of thin films of asphaltenes from toluene solutions cast on carbon coated copper grids. (A) Non heated sample, (B) sample heated under nitrogen at 200°C for 18 h, (C) Sample heated under oxygen at 200°C for18 h.

CONCLUSIONS

In summary, preliminary investigations into the generation of carbon nanomaterials from C7-asphaltenes have shown that upon heating a thin film under specific conditions, well-defined graphitic CNPs of *ca.* 20 nm in diameter are produced. Thermal analysis showed that the CNPs are very stable; while Auger analysis revealed their carbonaceous content. High resolution TEM established the graphitic structure of these CNPs with an interlayer d-spacing of 0.35 nm. The results indicate that asphaltenes, a by-product of bitumen processing, is a potential candidate for the cheap production of a high-value added carbon nanomaterials.

ACKNOWLEDGMENTS

We thank the Centre for Oil Sands Innovation (COSI), the University of Alberta and the National Institute for Nanotechnology for supporting this project, and Dr Jae-Young Cho for the TEM analyses.

REFERENCES

1. H. Li, C. Xu, N. Srivastava, K. Banerjee, IEEE Trans. Elect. Dev. **56**, 1799 (2009).
2. S.C. Ray, A. Saha, N.R. Jana, R. Sarkar, J. Phys. Chem. C **113**, 18546 (2009).
3. M.-M. Titirici, M. Antonietti, Chem. Soc. Rev. **39**, 103 (2010).
4. K.E. Prasad, B. Das, U. Maitra, U. Ramamurty, C.N.R. Rao, Proc. Nat. Acad. Sci, **106**, 13186 (2009).
5. M. Sevilla, C. Sanchis, T. Valdes-Solis, E. Morallon, A.B. Fuertes, J. Phys. Chem. C, **111**, 9749 (2007).
6. M.-M. Titirici, M. Antonietti, Chem. Soc. Rev. **39,** 103 (2010).
7. H. Liu, T. Ye, C. Mao, Angew. Chem. Int. Ed. **46**, 6473 (2007).
8. L. Tian, D. Ghosh, W. Chen, S. Pradhan, X. Chang, S. Chen, Chem. Mater. **21**, 2803 (2009).
9. G.A. Camacho-Bragado, P. Santiago, M. Marin-Almazo, M. Espinosa, E.T. Romero, J. Murgich, V.R. Lugo, M. Lozada-Cassou, M. Jose-Yacaman, Carbon **40**, 2761 (2002).
10. X. Wang, J. Guo, X. Yang, B. Xu, Mater. Chem. Phys. **113**, 821 (2009).
11. H. Groenzin, O.C. Mullins, Energy Fuels. **14**, 677 (2000).
12. F. Rakotondradany, H. Fenniri, P. Rahimi, K.L. Gawrys, P.K. Kilpatrick and M.R. Gray, Energy Fuels, **20**, 2439 (2006).
13. R.E. Franklin, Acta Cryst. **4**, 253 (1951).

Mater. Res. Soc. Symp. Proc. Vol. 1312 © 2011 Materials Research Society
DOI: 10.1557/opl.2011.134

Crystallization Kinetics in Isotactic Polypropylene Films with Carbon Nanotubes

Georgi Georgiev[1,2], Scott Schoen[2], Devin Ivy[2], Lauren Wielgus[2], Yaniel Cabrera[2], and Peggy Cebe[2]

[1]Department of Natural Sciences, Assumption College, Worcester, MA 01609
[2]Department of Physics and Astronomy, Tufts University, Medford, MA 02155

ABSTRACT

Polymer nanocomposites are the largest commercial application for carbon nanotubes (CNTs) which determines the interest in their effect on crystallization processes of polymers. We chose Isotactic Polypropylene (iPP) as one of the most widely used polymers. Nanocomposites with multiwall carbon nanotubes (MWCNTs) 0-5% by weight were studied, using differential scanning Calorimetry to measure the crystal nucleation and kinetics effects of MWCNTs. Isothermal crystallization at 138°C was performed and the data were analyzed using Avrami analysis. We obtained results for the effect of MWCNTs on the crystallization kinetics. The Avrami analysis showed a dramatic increase in the crystallization rate constant and constancy of the Avrami exponent with increase of the CNTs concentration. The full width at half maximum (FWHM) of the heat flow exotherm and the peak time for crystallization (t_p) change dramatically. The crystallinity shows a slight variation with the CNTs concentration dipping at 2% CNTs which can be explored further at higher concentrations.

INTRODUCTION

Kinetics of crystallization and the mechanisms of nucleation in polymers are influenced by the presence of carbon nanotubes (CNTs) [1]. We are exploring the effects of different concentrations of multiwall carbon nanotubes on the crystallization kinetics of isotactic polypropylene (iPP) isothermally crystallized at 138°C. The largest effect on kinetics that we observe is for concentrations of CNTs at or below 1% by mass, which is different than what other groups observe [2,3,4]. With increase of the concentration of nanotubes, we observe a decrease of the crystallization time during isothermal annealing, and an increase of the crystallization temperature during the nonisothermal crystallization. We analyze the nucleation and crystallization kinetics by using Kolmogorov-Johnson-Mehl-Avrami (JMAK) theory [5]. The Avrami exponent is not significantly affected by the addition of CNTs, but the rate constant K, the FWHM of the heat flow exotherm, the t_p and the crystallization half-time $\tau_{1/2}$ are affected greatly, results consistent with other reports [6].

THEORY

The theory of phase transitions is extremely useful because it can be applied to phenomena from different disciplines exhibiting the same type of phase transition. Kolmogorov-Johnson-Mehl-Avrami (JMAK) theory describes the decay of a metastable system to a unique equilibrium phase. This decay is driven by the difference between the free energy densities of the metastable and the equilibrium phase. The nucleation processes occur randomly in the metastable phase and the regions grow freely. The degree of phase transformation and in our case the fraction of volume transforming from one phase to another, f(t), is described by:

$$f(t)=1-\exp(-Kt^{n}) \qquad [1]$$

where t is the time, K is the temperature dependent crystallization rate constant and n – the Avrami exponent - represents growth and nucleation behavior. This equation has been applied to almost all areas where first order transitions occur, including liquid crystals and polymers, but also in cosmology, biophysics, magnetism, and surface science [5, 7]. It has been used in Chemistry to calculate chemical crystallization rates and in Ecology [5]. The equation can be linearized as:

$$\ln[-\ln(1 - f(t))] = \ln K + n \ln t \qquad [2]$$

where, $\ln K$ is the y-intercept, and n is the slope of the fitted straight line. The theory has been extended from time evolution of the order parameter for first order phase transitions to two point correlation functions, enabling it to predict and explain results from small angle scattering [8]. Ramos applies this model to kinetic Ising lattice gas model [5]. In this model the elementary kinetic processes act on microscopic length and time scales, making it possible to explain nucleation and growth rates. Gough derives a more general equation from the logistic curve where the JMAK theory is only a special case [9].

MATERIALS AND METHODS

iPP and CNTs

Isotactic polypropylene was obtained from Scientific Polymer Products, Inc. in powder form as Catalog #130. Mutli-walled carbon nanotubes came from MER Corporation. MWCNT's were produced by chemical vapor deposition (CVD). Catalytic CVD MWCNTs had 140 ± 30 nm diameters and 7 ± 2 micron length, with purity of >90% (Catalog # MRCSD).

Purifying MW Carbon Nanotubes by Oxidation

One gram of the MWCNT's was suspended in a mixture of concentrated sulfuric acid and nitric acid (3:1 vol. ratio). This solution was sonicated in a Misonix water bath sonicator for 24 hrs at 50°C. The resultant suspension was diluted with 400 mL of deionized water and filtered through a 400 nm pore membrane (PTFE) until the water passing through the filter had a pH between 6 and 7. The dispersions were subsequently filtered or diluted to the desired concentration. The resulting MWCNT's had a pH of between 3 and 3.5 and were stable. Atomic Force Microscopy (AFM) analysis shows that the CNTs were shortened between 30% and 40% of their original length, as expected. Small amounts of acids may be trapped by the CNTs. In a following study performing Size Exclusion Chromatography (SEC) on the iPP will show whether there are any changes in their molecular weight distribution in nanocomposites.

Preparation of iPP/CNT Nanocomposites

Nanocomposites were prepared by sonicating MWCNT in Xylene (75ml) at 50°C in a flask for 30min. IPP was added to the Xylene containing the MWCNT to form the different percent concentrations of the nanocomposites for the study. The solution was heated and stirred by placing the flask in an oil bath at 130°C sitting on a hot plate. Once the iPP is completely dissolved, we poured the solution slowly into a non-solvent (a polar compound, for example alcohol) with about a 5x volume dilution. The resulting powder precipitate was dried and then pressed into a film using a compression molding hot press at temperature of 200°C and pressure of 3-6000 lbs/sq in.

Differential scanning calorimetry

DSC analysis was performed on the TA Instruments calorimeter, Model # Q100.

Thermal treatment
The iPP nanocomposites were melted for two minutes at 200°C to erase any thermal history, and then cooled at 10°C/min to 138°C where they were held for various times in order to complete the isothermal crystallization for the nanocomposites with various CNTs content.
Data processing
We used the widely accepted Avrami model to analyze the DSC data for isothermal procedures. By plotting the relative crystallinity, x, as ln[-ln(1 – x)] vs. ln$t$, from eq [2], the slope of the curve, n, was determined over the range where a linear relationship exists between the plotted parameters. We obtain the value of K by K=exp(b), where b is the y-intercept of the linear fit of the plot [10,11,12]

RESULTS AND DISCUSSION

Figure 1 shows the heat flow vs. time during isothermal crystallization at 138°C, for iPP/CNT nanocomposites, at the indicated compositions. Time to maximum heat flow is a strong function of the CNT content, indicating that the crystal growth is accelerated by the addition of CNTs.

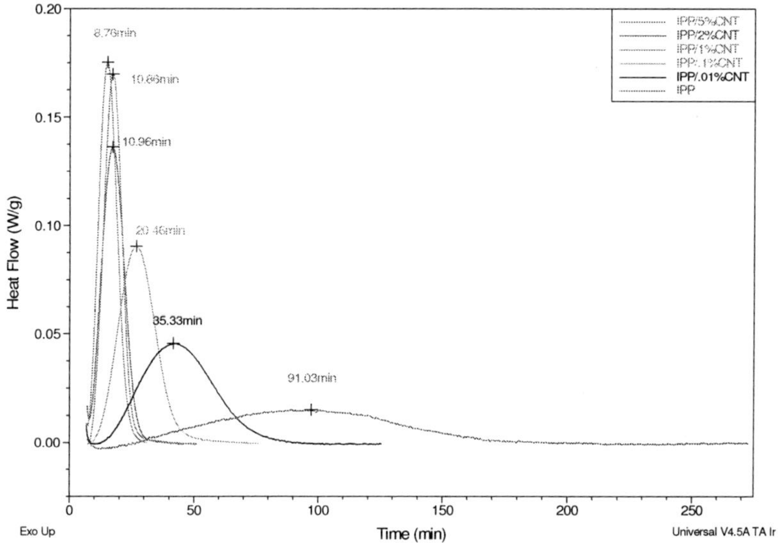

Figure 1. Heat flow vs. time for crystallization at 138°C for different iPP nanocomposites, as shown on the legend. The order of the curves for the decreasing CNT compositions is highest from left to right and top to bottom in the legend. The leftmost curve with t_p=8.76min corresponds to the 5% CNTs composition, and the rightmost curve with t_p=91.03min to the 0% CNTs composition. The time to maximum heat flow, t_p, decreases systematically with CNT addition, and these times are listed in Table 1. The only composition which reproducibly deviates slightly from the trend is the 2% CNTs which peaks at t_p=10.96min, while the 1% CNTs curve peaks at t_p=10.86min.

Figure 2 represents the relative crystallinity versus time from integrated heat flow data for the 138°C isothermal crystallization shown in Figure 1 for different % CNT nanocomposites. The dramatic increase of crystallization rate with increase of the CNT content is expressed in the decrease of the crystallization half time and it has most pronounced effect for the lowest concentrations of CNTs. With increase of the CNTs concentration the effect saturates. Figure 3 represents the Avrami plots of ln[-ln(1-X)] versus ln(t) for all of the concentrations from 0 to 5% CNTs. The slope of the lines does not depend significantly on the CNTs content.

As shown in Table 1, during the isothermal crystallization at 138°C the crystallinity is almost constant, but the crystallization time decreases dramatically, which is reflected in the increase of the value of K. The Avrami exponent n is a function of the dimensionality of growth: n between 1 and 2 indicates 1D growth; n between 2 and 3 represents a 2D growth; and n values between 3 and 4 indicates 3D growth [6] for heterogeneous nucleation processes. n also depends on the nucleation mechanism. The dimensionality reflected in the Avrami exponent n is also almost unchanged, and is around 3, which is typical for heterogeneous nucleation and three dimensional spherulitic growth. The crystal growth changes from spherulitic to fibrillar, originating on the surface of the nanotubes when CNTs are introduced [13, 14, 15]. The peak time dramatically deceases with adding nanotubes and the FWHM is also decreased greatly, pointing to a more uniform crystal population. The crystallization half time, for the crystallinity to reach 50% of the maximum crystallinity of the sample - $\tau_{1/2}$, is always longer than t_p, indicating the heat flow curves are asymmetric.

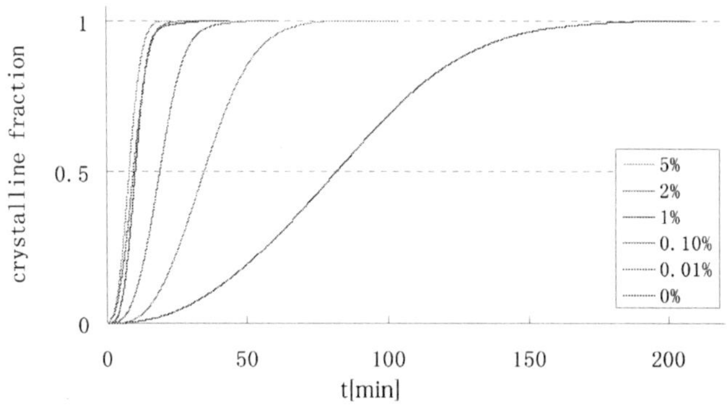

Figure 2. Relative crystallinity versus time from integrated heat flow data for the 138°C isothermal crystallization shown in Figure 1. The order of the curves is same as in figure 1. The concentration of the CNTs in the nanocomposites is highest on the left and lowest on the right and highest on the top and lowest at the bottom of the legend. The leftmost curve corresponds to the 5% CNTs composition, and the rightmost curve to the 0% CNTs composition. The crystallization halftime decreases systematically with CNT addition, and these times are listed in Table 1.

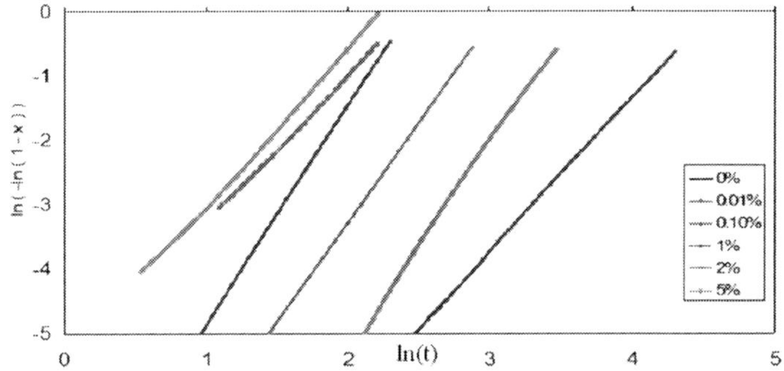

Figure 3. Avrami plots of ln[-ln(1-X)] versus ln(t). The y intercept is proportional to the rate of crystallization K and the slope depends on the nucleation and the dimensionality of crystal growth, which determines the Avrami exponent *n*. The order of the lines is same as in figures 1 and 2. The concentration of the CNTs in the nanocomposites is highest on the left and lowest on the right and highest on the top and lowest at the bottom of the legend. The leftmost line corresponds to the 5% CNTs composition and the rightmost line to the 0% CNTs composition. The data are shown in Table 1.

Table 1 – DSC Parameters# characterizing the rate of isothermal crystallization and the overall crystallinity of iPP/CNT nanocomposites.

CNTs content in Nano-composite	H_C (J/g)	t_p (min)	FWHM (min)	X_c	n	K (1/min)	$\tau_{1/2}$ (min)
0%	94.62	91.03	81	0.453	2.38	0.01	83
0.01%	96.68	35.33	29	0.465	3.14	0.03	32
0.10%	96.92	20.46	18	0.464	3.11	0.05	19
1%	92.26	10.86	8.5	0.441	3.41	0.09	9.1
2%	80.08	10.96	8.7	0.381	2.31	0.09	9.2
5%	88.39	8.76	7.0	0.445	2.45	0.11	7.5

The parameters are as follows: H_c is the integrated total exotherm area; t_p is the peak time, or the time to the maximum of the heat flow exotherm; FWHM is the Full Width at Half Maximum of the crystallization exotherm peak; X_c is the crystallinity mass fraction, determined by DSC as: $X_c=(H_c)/(H_f)$, where H_f is the heat of fusion for 100% crystalline iPP (H_f=209 J/g) [15]; n is the Avrami exponent, capturing the nucleation and dimensionality of crystallization; K is the crystallization rate constant with dimensions of 1/time, measures the rate of the crystallization process; n and K are determined from eq. 2; $\tau_{1/2}$ is the time for crystallization to reach 50% of the maximum crystallinity for the sample.

CONCLUSIONS

The crystallization rate constant K [min^{-1}] as determined from the Avrami analysis is much larger for the samples with nanotubes, which reflects the faster crystallization rates. The $\tau_{1/2}$ and the time to maximum heat flow in the crystallization exotherm, t_p, decrease dramatically. FWHM narrows with increase of CNT concentration, which suggests a more uniform distribution of the formed crystals, similar to observations from other authors [6]. The Avrami exponent *n* does not change with increasing of the concentration of CNTs. The crystallinity varies slightly with the CNTs concentration showing a slight dip at 2% CNTs. To determine whether there is a trend in crystallinity, higher concentrations of CNTs need to be explored.

ACKNOWLEDGMENTS

The authors thank: Assumption College for a Faculty Development Grant; The Natural Science Department at Assumption College for the summer student stipends; The Petroleum Research Fund, 44149-AC7; The National Science Foundation, Polymers Program of the Division of Materials Research, DMR-0602473, and MRI Program under DMR-0520655 for thermal analysis instrumentation.

REFERENCES

1. F. Ciardelli, S. Coiai, E. Passaglia, A. Pucci, and G. Ruggeri, Polymer International **57**, 805 (2008).
2. C.A. Avila-Orta, F.J. Medellýn-Rodrýguez, M.V. Davila-Rodrýguez, Y.A. Aguirre-Figueroa, K. Yoon, B.S. Hsiao, Journal Of Applied Polymer Science **106**, 2640 (2007).
3. D. Xu, Z. Wang, Polymer, **49**, 330 (2008).
4. L. Valentini, J. Biagiotti, M.A. López-Manchado, S. Santucci, J.M.Kenny, Polymer Engineering and Science, February **44**, 303 (2004).
5. R. Ramos, P.A. Rikvold, M.A. Novotny, Physical Review B, **59**, 9053 (1999).
6. B.P. Grady, F. Pompeo, R.L. Shambaugh, D.E. Resasco, J. Phys. Chem. B, **106**, 5852-5858 (2002).
7. K. Avramova, Cryst. Res. Technol. **37(5),** 491 (2002).
8. K. Sekimoto, Physics Letters 5 **105A**, 8 (1984).
9. T. Gough, R. Illner, VLSI Design **9**, No. 4, pp. 377-383 (1999).
10. J. Sandler, G. Boza, M. Nolte, K. Schulte, Y. Lam, M.S.P Shaffler, J. of Macromolecular Science **B42**, 479 (2003).
11. G. Xu, L. Du, H. Wang, R. Xia, X. Meng, Q. Zhu, Polymer International **57**, 1052–1066 (2008).
12. J. Young Kim, H.S. Park, S.H. Kim, Polymer **47**, 1379 (2006).
13. G. Georgiev, Y. Cabrera, M. Cronin, C. Rocheleau, B. Feinberg, P. Cebe, Bulletin of the American Physical Society, **53(1)** R1.00132 (2008).
14. E. Assouline, A. Lustiger, A. H. Barber, C. A. Cooper, E. Klein, E. Wachtel, H. D. Wagner, Journal of Polymer Science: Part B: Polymer Physics **41**, 520 (2003).
15. C. A. Avila-Orta Francisco, J. Medellýn-Rodrýguez, M. V. Davila-Rodrýguez,Y. A. Aguirre-Figueroa, K. Yoon, B. S. Hsiao, Journal of Applied Polymer Science, **106**, 2640 (2007).

Mater. Res. Soc. Symp. Proc. Vol. 1312 © 2011 Materials Research Society
DOI: 10.1557/opl.2011.135

Self-Assembly of a Water-Soluble Tricyclic Heterocycle into J-Type Rosette Nanotubes

Gabor Borzsonyi, Rachel L. Beingessner, Takeshi Yamazaki, Jae-Young Cho, Andrew J. Myles, Andriy Kovalenko and Hicham Fenniri*
National Institute for Nanotechnology, National Research Council (NINT-NRC) and Department of Chemistry, University of Alberta, Edmonton, Alberta, T6G 2G2, Canada.

ABSTRACT

The synthesis and self-assembly of a water-soluble, tricyclic, self-complementary heterocycle that features the hydrogen bond donor-acceptor arrays of both guanine (G) and cytosine (C) juxtaposed between a pyridine ring is presented. In solution, this tricycle, which has been termed xK1, self-assembles into Rosette Nanotubes (RNTs) that have an inner diameter of 1.4 nm. Unlike the RNTs formed from the bicyclic congener K1, we demonstrate that xK1 with its extended π system, forms a J-type RNT assembly determined through UV-Vis, CD and fluorescence spectroscopy experiments. This observation brings the possibility of developing electrically conducting RNTs for applications in the areas of photovoltaics and molecular wires.

INTRODUCTION

The G∧C motif (K1, Figure 1) is a self-complementary bicyclic hybrid molecule of the DNA bases guanine and cytosine [1]. In solution, it self-assembles into tubular architectures called Rosette Nanotubes (RNTs) [2] which have an internal diameter of ~1 nm. By covalently functionalizing the G∧C motif using standard synthetic strategies, the physical and chemical properties of the RNTs can be easily modified for specific applications, since these groups are expressed on the outer periphery.

For applications including drug delivery, we wanted to include in our repertoire, RNTs that have larger internal diameters. To this end, we have recently described the synthesis of a tricyclic version of the G∧C motif, called xG∧C, which has a pyridine ring juxtaposed between the guanine and cytosine faces and readily self-assembles in organic solvents (DMF and DMSO) into RNTs that have an extended internal diameter of 1.4 nm [3]. In this proceeding, we now describe the synthesis and self-assembly of the *water-soluble* congener called xK1 (Figure 1), which is functionalized with the amino acid lysine [4]. Unlike xG∧C or any of the functionalized bicyclic G∧C motifs we have previously studied, here we demonstrate that xK1 self-assembles into *J-type RNTs* that have an exceptionally high molar ellipticity value. As J-aggregate materials have been used as organic photoconductors [5], LED's [6], sensitizers [7] and photovoltaics [8], this work provides exciting application possibilities for this family of RNTs.

EXPERIMENT

The synthesis of xK1 was accomplished in 9 steps from trichloro compound **1** [9] with an average stepwise yield of 76% (Figure 2). All new compounds were characterized by ^{1}H and ^{13}C NMR and HRMS as well as by elemental analysis for xK1 [4].

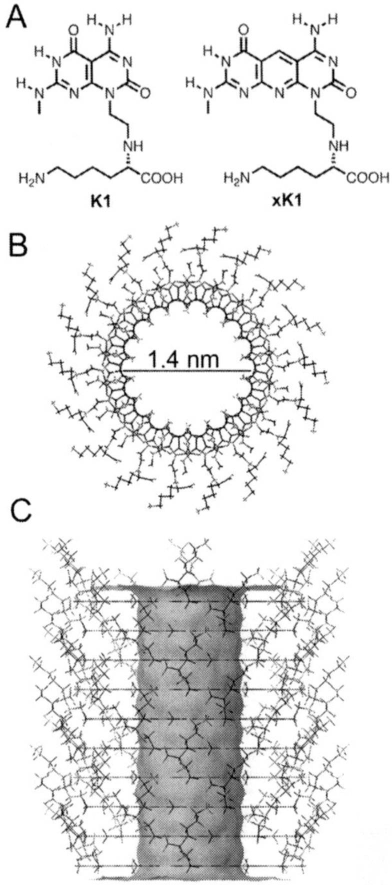

Figure 1. K1 and xK1 modules (A) and the RNTs self-assembled from xK1, top (B) and side (C) views.

SEM Imaging. SEM samples were prepared by floating a carbon-coated 400-mesh copper grid on a droplet of xK1-water solution (0.01 mg/mL, 1.3 x 10^{-5} M) for 30 s. The grid was then blotted, dried in air and on a vacuum hotplate for one day before imaging. SEM imaging was carried out on a Hitachi S4800, field emission high-resolution scanning electron microscope, using 10.0 kV accelerating voltage and 20 μA current.

TEM Imaging. TEM samples were prepared by floating the samples prepared for the SEM on one droplet of uranyl acetate (2% aqueous) for 40 s each. The grid was then blotted, dried in air and then by vacuum hotplate. TEM investigation was carried out on JEOL 2200 FS TEM – 200kV Schottky field emission instrument equipped with an in-column omega filter. Bright field TEM images were acquired using an energy-filtered zero loss beam (slit width 10eV).

AFM imaging. The sample surface was imaged using a Digital Instruments/Veeco Instruments MultiMode Nanoscope IV AFM equipped with an E scanner. In order to obtain optimized height profiles, silicon cantilevers (MikroMasch USA, Inc.) with low spring constants of 4.5 N/m were used in tapping mode (TM-AFM). Clear images of the surface were obtained using a low scan rate (0.5-1 Hz) and amplitude setpoint (1 V). Clean mica substrates (1 × 1 cm^2) were prepared and the samples were deposited by spin-coating a 0.1 mg/mL solution of xK1 at 2000 rpm for 20 s.

Modeling. Molecular models for xK1 RNTs were generated with Macromodel 8.5 using the OPLS-AA force field. In order to obtain the structure and thermodynamics of solvation in water, the three–dimensional reference interaction site model (3D–RISM) integral equation with the Kovalenko–Hirata (KH) closure approximation was employed [10].

DISCUSSION

The self-assembly of xK1 was studied in water at a pH_{final} of 2.8. Unlike the analogous bicyclic G∧C motif K1, it is not necessary to heat xK1 in order to drive the entropically driven self-assembly process due to the larger hydrophobic surface area. Characterization of the resulting xK1 RNTs by TEM (C) and AFM (A) images shown in Figure 3, established that their outer diameter is 4.4 $\pm$ 0.2 nm and 4.2 $\pm$ 0.2 nm respectively, which is in agreement with the theoretical value of 4.3 nm.

CD, UV-Vis and fluorescence spectroscopy revealed that xK1 RNTs form a J-type assembly that is not observed with K1. In the CD-spectra (2.1×10^{-5} M) shown in Figure 4A for example, two couplets were observed for xK1 that are centered at 257 and 380 nm, with positive maxima at 264 and 388 nm. During the 7 day period in which the growth of the RNTs were monitored, the CD intensity at 388 nm increased over 40x fold and reached an impressive molar ellipticity of 4×10^6 deg•M^{-1}•m^{-1} [11]. By definition, the intensity of the CD couplet is proportional to the fourth power of the dipole strength and the relative orientation of the chromophores and inversely proportional to the square of the interchromophoric distance [12]. Given that the interchromophoric distance is not expected to be much different between xK1 and K1, the high molar ellipticity indicates that xK1 has a significant increase in dipole strength and adopts a favorable arrangement of the chromophores for exciton coupling.

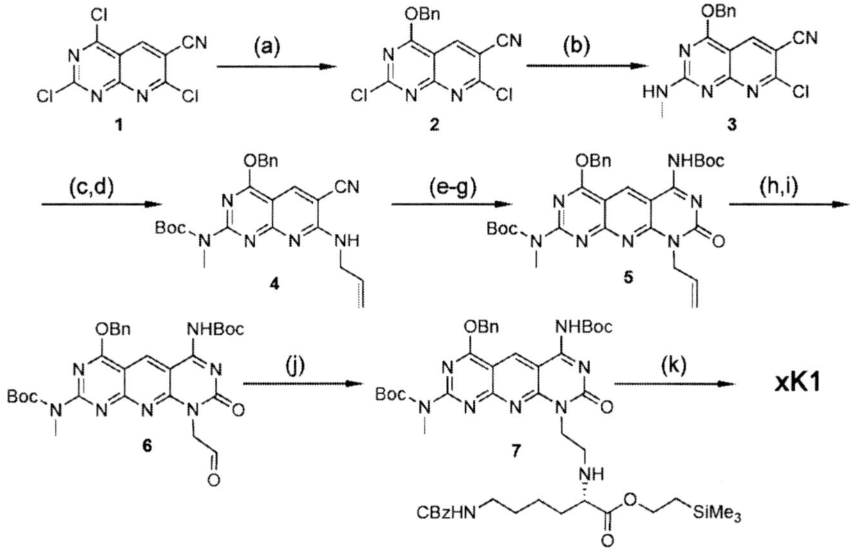

Figure 2. Synthesis of xK1. Reagents and conditions: (a) BnOH, Et_3N, CH_2Cl_2, -40°C, 70%; (b) CH_3NH_2, DIPEA, CH_2Cl_2, 0°C, 60%; (c) Boc_2O, DIPEA, DMAP, THF, 20°C; (d) Allylamine, DIPEA, CH_2Cl_2, 20°C, 50% (2 steps); (e) Cl_3CONCO, CH_2Cl_2, 20°C; (f) 7N NH_3/CH_3OH, 20°C, 74% (2 steps); (g) Boc_2O, DIPEA, DMAP, THF, 20°C, 96%; (h) OsO_4/t-BuOH, 50% NMO/THF, 20°C; (i) $NaIO_4/H_2O$, 20°C, 85% (2 steps); (j) *(S)*-2-trimethylsilyl)ethyl 2-amino-6-(benzyloxycarbonylamino)hexanoate, DIPEA, 1,2-DCE, $Na(OAc)_3BH$, 20°C, 82%; (k) 95% CF_3CO_2H/thioanisole, 20°C, 87%.

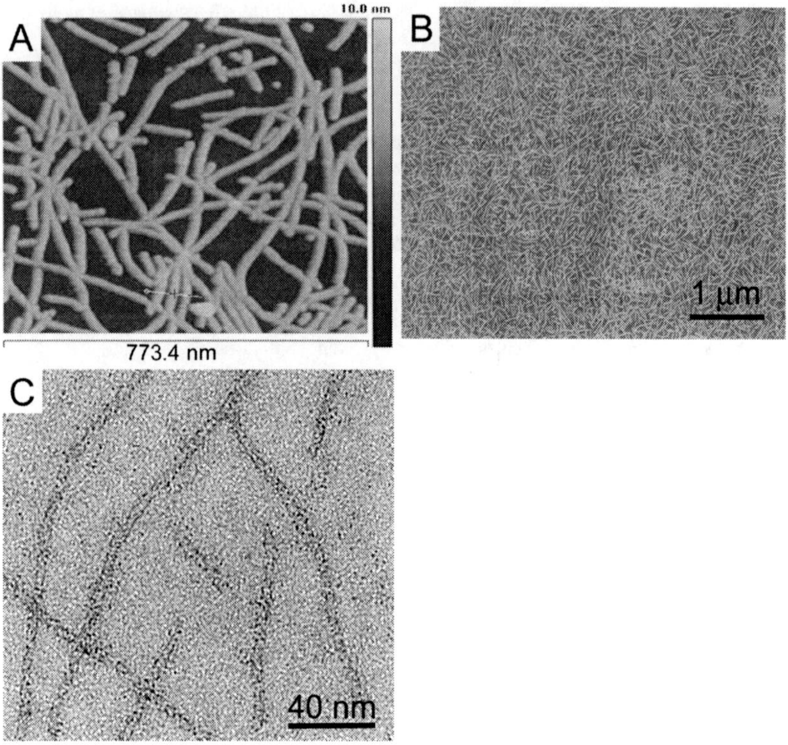

Figure 3. AFM, SEM and TEM image of xK1 in water.

UV-Vis experiments further established that the enhanced intermodular electronic communication observed in the CD spectra is the result of xK1 organizing into a J-type arrangement within the RNTs. This is evident by the growth of a sharp J-band (Figure 4B) over time at 388 nm, which corresponds to intense couplet observed in the CD spectra. Since this band decreases in intensity when the RNTs are disassembled by high temperatures or addition of TFA (data not shown), it is clear that the CD signal is associated with the supramolecular chirality of the RNTs rather than the intrinsic chirality of xK1. It is also important to note from the UV-Vis spectra that while a subtle hypochromic effect (*ca.* 8%) was observed at 246 and 363 nm (maxima) over 7 days, a more pronounced hyperchromic effect (up to *ca.* 50%) was noted for the thermal disassembly (data not shown). This implies that RNT formation has already progressed prior to our initial measurements. Moreover, since the J-band was weak at t=10 min, but grew between the 3h to 7 day observation period, this also suggests that the J-type RNTs are formed in two stages. The first stage is the rapid assembly of the RNTs and the second is the re-positioning of the chromophores into an optimal orientation for the exciton coupling.

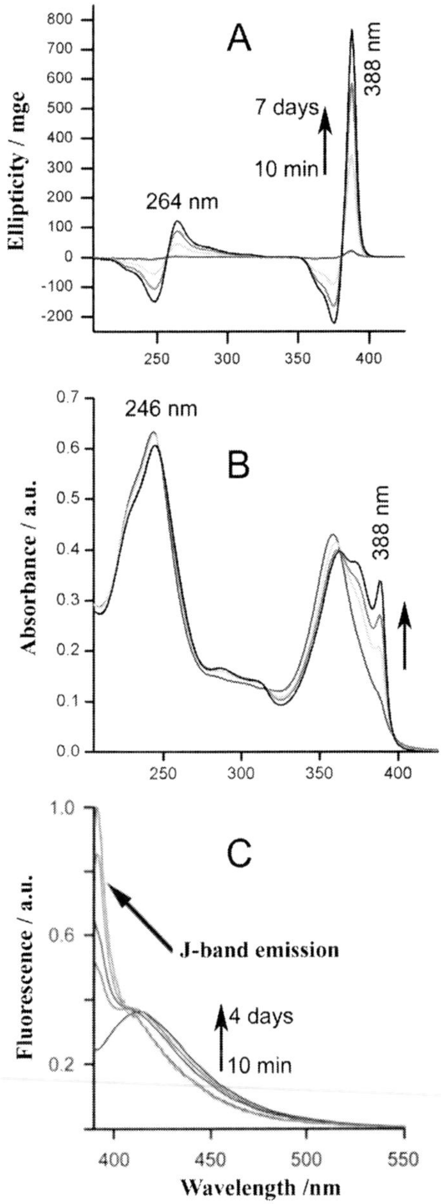

Figure 4. (A) CD; (B) UV-Vis; and (C) fluorescence spectra of xK1 RNTs (2×10^{-5} M in water).

In order to further characterize this transition from the first stage (λ_{max} 356 nm and isobestic point 369 nm) to the second stage (J-band, λ_{max} 388 nm), steady state fluorescence spectroscopy was performed. The spectra resulting from the excitation at 356 and 369 nm showed a decrease in fluorescence intensity and a change in band shape due to the growth of the narrow J-band fluorescence peak at 393 nm. While excitation at 388 nm (J-band) minimized the absorption of the first stage, there was a growth in the J-band fluorescence emission at 393 nm (Figure 4C). Typical of a J-type aggregate, the absorption and emission spectra were nearly mirror images of each other and the Stokes shift for the J-band was only 5 nm [13].

CONCLUSIONS

Here we have described the synthesis and self-assembly of a tricyclic G∧C motif called xK1 which self-assembles in water into J-type RNTs that have an impressive molar ellipticity. Evidence for this J-type assembly stems from the CD, UV-Vis and fluorescence spectroscopy data. Future efforts will be focused on carrying out additional photophysical studies of xK1 in order to have a more detailed understanding of the nature of this electronic communication.

ACKNOWLEDGMENTS

We thank the Natural Sciences and Engineering Research Council Canada, National Research Council and the University of Alberta for supporting this program.

REFERENCES

1. (a) M. Mascal, N. M. Hext, R. Warmuth, M. H. Moore, and J. P. Turkenburg, *Angew. Chem., Int. Ed. Engl.* **35**, 2204 (1996). (b) A. Marsh, M. Silvestri, and J.-M. Lehn, *Chem. Commun.* 1527 (1996). (c) H. Fenniri, P. Mathivanan, K. L. Vidale, D. M. Sherman, K. Hallenga, K. V. Wood, and J. G. Stowell, *J. Am. Chem. Soc.* **123**, 3854 (2001).
2. (a) H. Fenniri, B.-L. Deng, and A. E. Ribbe, *J. Am. Chem. Soc.* **124**, 11064 (2002). (b) H. Fenniri, B.-L. Deng, A. E. Ribbe, K. Hallenga, J. Jacob, and P. Thiyagarajan, *Proc. Natl. Acad. Sci. USA.* **99**, 6487 (2002). (c) J. G. Moralez, J. Raez, T. Yamazaki, R. K. Motkuri, A. Kovalenko, and H. Fenniri, *J. Am. Chem. Soc.* **127**, 8307 (2005). (d) R. S. Johnson, T. Yamazaki, A. Kovalenko, and H. Fenniri, *J. Am. Chem. Soc.* **129**, 5735 (2007). (e) R. L. Beingessner, B.-L. Deng, P. E. Fanwick, and H. Fenniri, *J. Org. Chem.* **73**, 931 (2008). (f) G. Tikhomirov, M. Oderinde, D. Makeiff, A. Mansouri, W. Lu, F. R. Heirtzler, D. Y. Kwok, and H. Fenniri, *J. Org. Chem.* **73**, 4248 (2008).
3. G. Borzsonyi, R. S. Johnson, A. J. Myles, J.-Y. Cho, T. Yamazaki, R. L. Beingessner, A. Kovalenko, and H. Fenniri, *Chem. Commun.* **46**, 6527 (2010).
4. G. Borzsonyi, R. L. Beingessner, T. Yamazaki, J.-Y. Cho, A. J. Myles, M. Malac, R. Egerton, M. Kawasaki, K. Ishizuka, A. Kovalenko, and H. Fenniri, *J. Am. Chem. Soc.* **132**, 15136 (2010).
5. (a) H. Lin, R. Camacho, Y. Tian, T. E. Kaiser, F. Würthner, and I. G. Scheblykin, *Nano Lett.* **10**, 620 (2010). (b) W. Jin, Y. Yamamoto, T. Fukushima, N. Ishii, J. Kim, K. Kato, M. Takata, T. Aida, *J. Am. Chem. Soc.* **130**, 9434 (2008). (c) K.-Y. Law, *Chem. Rev.* **93**, 449 (1993).
6. (a) J. R. Tischler, M. S. Bradley, Q. Zhang, T. Atay, A. Nurmikko, and V. Bulovic, *Org. Electron.* **8**, 94 (2007). (b) B. J. Walker, G. P. Nair, L. F. Marshall, V. Bulovic, and M. G. Bawendi, *J. Am. Chem. Soc.* **131**, 9624 (2009).
7. T. H. James, Ed. "The Theory of the Photographic Process" (Macmillan, 1977).
8. (a) K. Sayama, S. Tsukagoshi, T. Mori, K. Hara, Y. Ohga, A. Shinpou, Y. Abe, S. Suga, and H. Arakawa, *Sol Energy Mater. Sol. Cells* **80**, 47 (2003). (b) M. Kawasaki, and S. Aoyama, *Chem. Commun.* 988 (2004). (c) A. R. Tameev, A. Vannikov, and H. F. M. Schoo, *Thin Solid Films* **451/452**, 109 (2004). (d) X.-F. Wang, O. Kitao, H. Zhou, H. Tamiaki, and S. Sasaki, *J. Phys. Chem. C* **113**, 7954 (2009).
9. G. Beck, *Ger. Offen.* 3643456 (1988).
10. (a) A. Kovalenko, "Three-Dimensional RISM Theory for Molecular Liquids and Solid-Liquid Interfaces," *Molecular Theory of Solvation*, ed. F. Hirata (Kluwer Academic Publishers, 2003) pp. 169-276. (b) A. Kovalenko, and F. Hirata, *J. Chem. Phys.* **112**, 10391-10402 and 10403-10417 (2000).
11. (a) O.-K. Kim, J. Je, G. Jernigan, L. Buckley, and D. Whitten, *J. Am. Chem. Soc.* **128**, 510 (2006). (b) C. Nuckolls, T. J. Katz, L. Castellanos, *J. Am. Chem. Soc.* **118**, 3767 (1996). (c) H. von Berlepsch, S. Kirstein, and C. Böttcher, *J. Phys. Chem. B.* **107**, 9646 (2003).
12. N. Berova, L. Di Bari, and G. Pescitelli, *Chem. Soc. Rev.* **36**, 914 (2007).
13. J. M. W. Chan, J. R. Tischler, S. E. Kooi, V. Bulovic, and T. M. Swager, *J. Am. Chem. Soc.* **131**, 5659 (2009).

Mater. Res. Soc. Symp. Proc. Vol. 1312 © 2011 Materials Research Society
DOI: 10.1557/opl.2011.1

Highly-ordered Porous Coordination Polymer Nanofilms Grown by Layer-by-Layer Deposition Technique

Rie Makiura[1,2] and Hiroshi Kitagawa[2,3]

[1]Nanoscience and Nanotechnology Research Center, Osaka Prefecture University, Osaka 599-8570, Japan
[2] CREST, Japan Science and Technology Agency, Tokyo 102-0075, Japan
[3] Division of Chemistry, Graduate School of Science, Kyoto University, Kyoto 606-8502, Japan

ABSTRACT

The facile bottom-up fabrication of a perfect preferentially-oriented MOF nanofilm, NAFS-1 on a solid surface, which is endowed with highly crystalline order both in the out-of-plane and in-plane orientations to the substrate, as determined by synchrotron X-ray surface crystallography, was achieved by the unique combination of a layer-by-layer growth technique coupled with the Langmuir-Blodgett (LB) method.

INTRODUCTION

The assembly of molecular building blocks with metal ions and organic linkers, ranging from so-called metal-organic frameworks (MOFs) to extended organic-inorganic hybrids, has been attracting intense interest [1-2]. MOFs are highly ordered crystalline coordination polymers with well-defined porous networks. Because of their potential applications associated with the presence of pores and cavities, such materials have been examined for storage, separations, sensors, size- and shape-selective catalysis, molecular recognition, and nanoscale reactors. However, the assembly of MOF films whose size and growth direction are completely controlled has proven an extremely challenging task [3,4]. Although there have been experimental reports that showed promising results in obtaining partially ordered system, the structural details, especially an in-plane direction have remained completely unknown [3,4]. Here, we present how the layer-by-layer growth and the LB methods can be integrated to allow fabrication of a perfectly crystalline MOF thin film on a solid surface comprising metalloporphyrin building units and metal ion joints.

EXPERIMENTS

5,10,15,20-tetrakis-(4-carboxyphenyl)-porphyrin-Co-(II) (CoTCPP) was purchased from Porphyrin Systems. Silicon single crystal, Si(100) were used as a substrate in XRD measurements. Amorphous quartz was used as a substrate for UV-vis absorption spectroscopic measurements. Before the film fabrication, the substrates were immersed successively into ultrasonic baths of chloroform, acetone, and ethanol, each for 30 minutes.

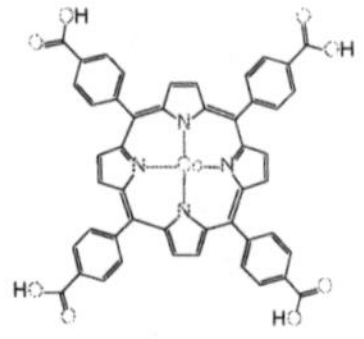

CoTCPP

CoTCPP and pyridine (py) were dissolved together into mixed chloroform/methanol solvent (3:1, v/v, 0.2 mM for CoTCPP). 0.1 M of $CuCl_2{\cdot}2H_2O$ aqueous solution was prepared as a subphase and poured carefully into a Langmuir trough. CoTCPP/pyridine solution was spread onto the $CuCl_2$ subphase. The surface was compressed by two barriers and the formation of the ordered array on the surface was followed by measuring surface pressure (π) with KSV minitrough system and the molecular area in the array was evaluated. The CoTCPP-py-Cu array formed at the surface was transferred onto the substrate at room temperature. The substrate with the CoTCPP-py-Cu array was then rinsed with distilled water and immersed into distilled water. The number of the CoTCPP-pu-Cu 2D layers is controlled by the number of cycles of sheet deposition. The same spread solution of CoTCPP /py solution was also applied onto a pure water subphase for comparison.

UV-visible absorption spectra of the CoTCPP- py-Cu sheets on the quartz substrate were measured with a Hitachi U-3310 spectrophotometer at room temperature. IR absorption spectra were collected on a JASCO FT-IR 620 spectrometer under vacuum condition at room temperature. Synchrotron X-ray diffraction (XRD) data were collected at room temperature with a multiaxis diffractometer at beamline BL13XU (λ = 1.554 Å) in SPring-8. Out-of-plane scans were performed in a θ-2θ scattering geometry. Grazing incident X-ray diffraction (GIXRD) measurements were performed at α = 0.1°. Diffraction from the sample surface was detected in the in-plane direction with a detection angle to the surface of 0.1°. The structural model for NAFS-1 was constructed using the Materials Studio (Accelrys Inc.) software.

RESULTS AND DISCUSSION

<u>π-a isotherms</u>

Figure 1 shows π-a isotherms of CoTCPP-py LB films. The larger molecular area observed for the CoTCPP-py array assembled on the $CuCl_2$ solution indicates that they lie flat on the solution surface (i.e. the CoTCPP 2D plane is parallel to the surface). On the other hand, in the case of pure water subphase, the smaller molecular area derived implies that the CoTCPP molecules either stand vertically at some angle to the solution surface or remain in the horizontal orientation but pack more closely.

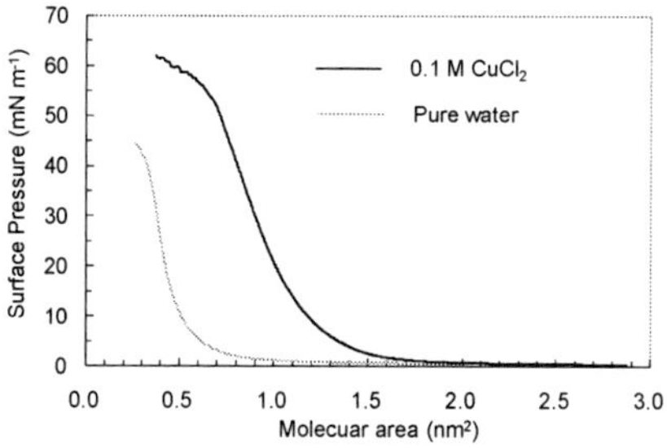

Figure 1. Surface-pressure–area (π–a) isotherms for CoTCPP-py monolayers on two different subphases.

UV-vis spectroscopy

The arrays on the surface were then transferred to a quartz substrate and UV-vis absorption spectroscopic measurements were performed (Fig. 2). The more intense absorbance of the Soret band of CoTCPP observed for the monolayer assembled on the pure water subphase is attributed to more CoTCPP molecules existing in the same area than in the case of the $CuCl_2$ solution subphase (Fig. 2a). This is consistent with the proposed molecular arrangements from π-a isotherm results. In addition, the measured absorbance of the CoTCPP-py-Cu monolayer is in good agreement with the calculated value considering the molecular orientation (the CoTCPP 2D plane is parallel to the surface). This indicates that the substrate surface is well covered with the CoTCPP-py-Cu array – the surface coverage is nearly 100%. The successive film stacking of CoTCPP-py-Cu monolayers was also monitored by UV-vis spectroscopy as shown in Fig. 2b. The linear increase in absorbance of the CoTCPP Soret band represents successful layer-by-layer growth resulting in assembling MOF nanofilm (NAFS-1; nanofilm of metal-organic frameworks on surface no.1) [5, 6].

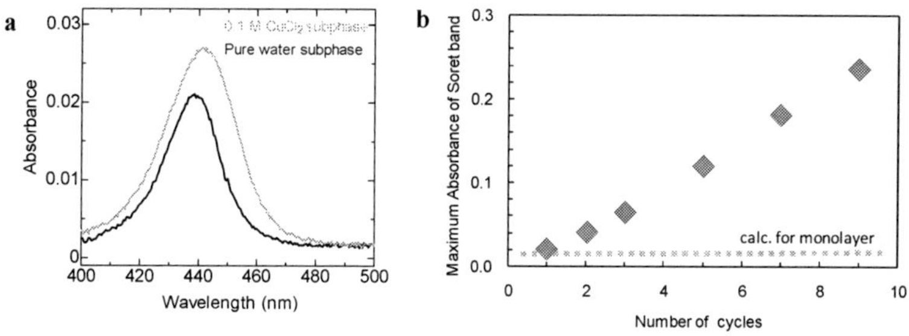

Figure 2. **a.** UV-Vis absorption spectra of CoTCPP-py monolayers fabricated on $CuCl_2$ aqueous solution and on pure water. **b.** Maximum absorbance versus the number of the deposition cycles of NAFS-1.

IR spectroscopy

Linkage motif of CoTCPP molecules and copper ions in NAFS-1 was investigated by IR spectroscopy (Fig. 3). A strong absorption band is observed around 1400 cm^{-1} in NAFS-1 as also seen in copper benzoate. Copper benzoate is well known to have the dinuclear 'paddlewheel' structure. When the dinuclear paddlewheel structure is formed, the COO stretching mode from the carboxylate ion (COO^-) becomes IR active and appears around 1400 cm^{-1}. On the other hand, the absorption around 1400 cm^{-1} was present neither in the CoTCPP-py LB film nor in benzoic acid.

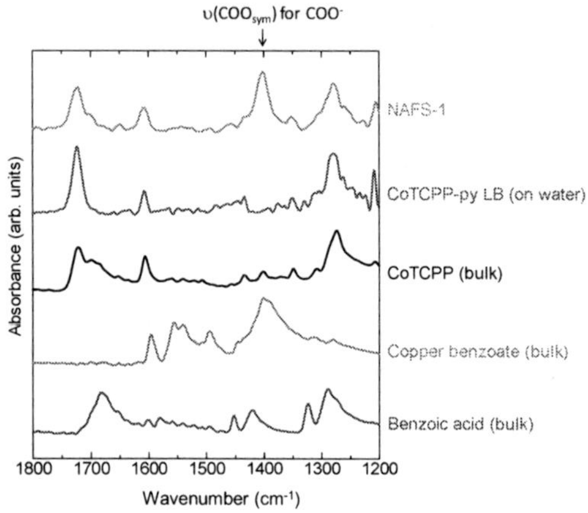

Figure 3. IR spectrum of the NAFS-1 film together with hose of key reference materials.

XRD and structural model construction

In order to investigate the crystallinity of NAFS-1, synchrotron X-ray ($\lambda = 1.554$ Å) diffraction measurements were performed at room temperature in two scattering geometries, out-of-plane mode and in-plane (grazing incidence) mode (Fig. 4). Figure 4a shows the XRD profiles measured in out-of-plane scans by using θ-2θ scattering geometry for NAFS-1 with 20 stacking cycles fabricated on a Si(100) substrate. Three diffraction peaks were observed and indexed as (001), (002) and (003), leading to a value of 0.9380(3) nm for the distance of the interlayer spacing. The total thickness of NAFS-1 is expected to be approximately 19 nm from the interlayer spacing and the number of stacking layers. This value is in good agreement with the crystalline domain size (20 nm) calculated from the full width at half maximum (FWHM) of the observed peaks.

The grazing incidence XRD (GIXRD) measurements were carried out for NAFS-1. Observation of a number of diffraction peaks as shown in Fig. 4b implies that NAFS-1 has a highly ordered characteristic in the parallel direction to the substrate surface. All Bragg reflections can be indexed as (*hk*0), namely, (110), (200), (320), (400), (330), (440), and (550)/(710) by using a pseudo-2D tetragonal unit cell with basal plane dimensions, a = b =

1.6460(3) nm. In addition, none of the Bragg reflections observed in grazing incidence geometry coincides with any of the peaks in the out-of-plane XRD pattern. This represents that the combination of the LbL protocol coupled with the LB method can lead to the preparation of nanofilms which grow in a perfect preferentially-oriented directionally-controlled manner on surfaces.

The observation of Bragg reflections in both out-of-plane and in-plane modes allowed us to develop a crystalline structural model for NAFS-1 – an in-plane structural motif of NAFS-1 comprising CoTCPP units linked by binuclear $Cu_2(COO)_4$ paddle wheels and resulting in a similar checkerboard pattern to that encountered in the bulk porphyrin paddle wheel frameworks (PPF) [7] as shown in Fig. 4c. The comparable lattice size of NAFS-1 (1.65 nm) to those of the bulk systems (~1.66 nm) [7] together with the IR results (Fig. 3) confirm the formation of the paddle-wheel structure. Regarding the interlayer packing manner, interdigitated stacking motif was considered which can be driven by π-π interaction between pyridine molecules projecting from the 2D planes (Fig. 4d and 4e). The in-plane XRD pattern calculated from the structural model with consideration of the metal-coordinated pyridine molecules and the interlayer stacking order is in excellent agreement with the obtained XRD profile (Fig. 4b) [5, 6].

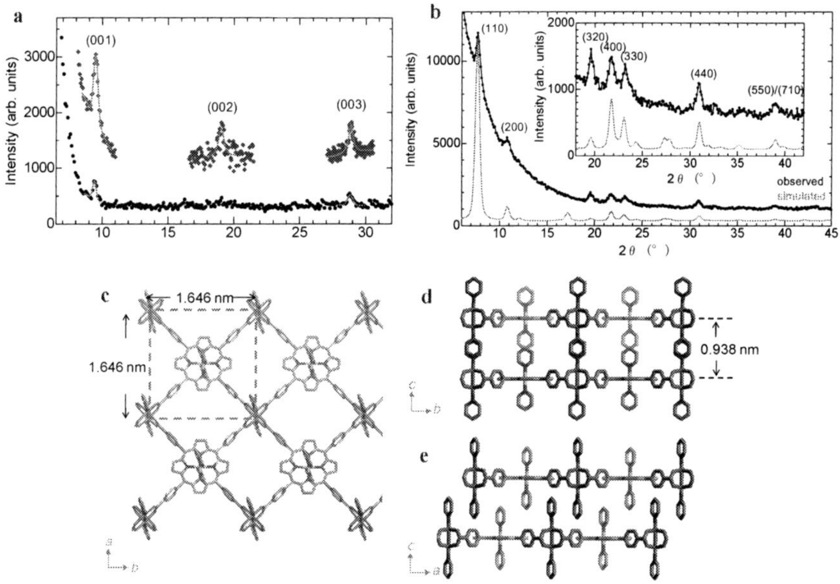

Figure 4. **a.** Out-of-plane XRD profiles of NAFS-1. **b.** In-plane XRD (GIXRD) profiles of NAFS-1 shown together with a simulated XRD pattern of the structural model. **c-e.** Schematic diagram of the proposed crystalline structure for NAFS-1 shown in different projection along the three unit cell axes.

CONCLUSIONS

We succeeded in the facile bottom-up fabrication of a perfectly crystalline, preferentially oriented MOF nanofilm, NAFS-1 on a solid surface, namely "a nanofilm of a metal-organic framework on a surface consisting of metalloporphyrin building units", and the ordered structure was demonstrated by surface synchrotron X-ray crystallography. As structural design has been the main focus of current attempts towards fabrication of surface assemblies, the future direction will surely be targeted towards various physical properties and functionalities, including structural dynamics associated with the frameworks and the responses of their cavities to external stimuli that may differ from comparable well-studied bulk porous systems.

ACKNOWLEDGMENTS

We acknowledge financial support by JST(CREST), NEDO, and JSPS under Grants-in-Aid for Scientific Research (No. 20350030), Challenging Exploratory Research (No. 20655030), Scientific Research on Innovative Areas (No. 22108524), and Young Scientists (B) (No.22750056). We thank Super Photon ring-8 GeV (SPring-8) for access to the synchrotron X-ray facilities.

REFERENCES

1. J. R. Long, O. M. Yaghi, *Chem. Soc. Rev.* **38**, 1213 (2009). (and references cited therein).
2. G. Férey, *Chem. Soc. Rev.* **37**, 191 (2008).
3. D. Zacher, O. Shekhah, C. Wöll and R. A. Fischer, *Chem. Soc. Rev.* **38**, 1418 (2009).
4. R. Haruki, O.Sakata, T. Yamada, K. Kanaizuka, R. Makiura, Y. Akita, M. Yoshimoto, H. Kitagawa, *Trans. MRS-J.* **33**, 629 (2008).
5. R. Makiura, S. Motoyama, Y. Umemura, H. Yamanaka, O. Sakata, H. Kitagawa, *Nature Mater.* **9**, 565 (2010).
6. R. Makiura, H. Kitagawa, *Eur. J. Inorg. Chem*, **24**, 3715 (2010).
7. E. Y. Choi, P. M. Barron, R. W. Novotny, H. T. Son, C. Hu, W. Choe, *Inorg. Chem.* **48**, 426 (2009).

Mater. Res. Soc. Symp. Proc. Vol. 1312 © 2011 Materials Research Society
DOI: 10.1557/opl.2011.2

Study of Selectively Permeable Coatings to Textiles

Joseph G. Sargent[1], Jun S. Lee[2], Emmanuelle Reynaud[1], Michael D. Gilbert[3], and James M. Sloan[4]
Department of [1]Mechanical and [2]Plastics Engineering, University of Massachusetts of Lowell, One University Avenue, Lowell, MA 01854, U.S.A.
[3]EIC Laboratories Inc., Norwood, MA 02062
[4]U.S. Army Research Laboratory, Aberdeen Proving Ground, MD 21005

ABSTRACT

Breathable barrier textiles for both chemical agent and moisture are being actively developed for military and industrial applications. An ideal approach is to coat textiles with a semi-permeable film that allows the transport of water while still serving as a barrier for chemical agents. Sulfonated poly (styrene-*block*-isobutylene-*block*-styrene) (SIBS) copolymer spontaneously phase separates upon drying from solution to produce a nanostructured film with the controlled barrier functionality for water permeation and repelling of chemical agents.

The objective of this research is to investigate coating uniformity and phase morphology of SIBS coating materials fabricated by novel solvent combinations. Scanning electron microscopy analysis is used for the assessment of the coating uniformity as well as the level of adhesion between the polymer coating and fabric substrate. Transmission electron microscopy is used to characterize the phase separation morphology of the SIBS copolymer coating. The mechanical behavior of the coated fabric is determined through tensile and shear tests and is compared to the bare fabric behavior. The goal of this study is to relate the processing conditions of the final nanostructured block copolymer coated fabric produced on industrial scale.

INTRODUCTION

The need for a breathable as well as durable polymer coated fabric for use as protective material in clothing has resulted in the development of a block copolymer film possessing both the desired membrane [1] and mechanical properties [2]. The block copolymer system developed consists of sulfonated polystyrene-*block*-polyisobutylene-*block*-polystyrene. It is the phase separated structure of the polymer that makes it an effective barrier to chemical agents while still allowing for the ionic diffusion of water molecules. The copolymer is prepared for processing by being dissolved in a combination of solvents. Studies have shown that the solvents used [3], as well as the evaporation rate [4], can have an effect on how copolymer phases separate. Work is currently being done to identify an industrial process which will produce a coating with the morphology necessary for sufficient semi-permeable membrane performance as well as the desired mechanical properties.

The work presented here is a study on the microstructure and mechanical properties of a woven cotton/nylon blend fabric coated with sulfonated polystyrene-isobutylene-styrene block copolymer. A mixture of solvents and SIBS copolymer has been applied on the fabric through a blade process, followed by heating for solvent evaporation. The microstructure of the coating is characterized through the transmission electron microscopy. The quality and uniformity of the coating is determined through scanning electron microscopy. The tensile strength and shear behavior of the coated and uncoated materials are determined through mechanical testing

EXPERIMENT

Poly (styrene-*block*-isobutylene-*block*-styrene) copolymer (SIBSTAR 103T-F, Kaneka Texas Corp., Mw 100K) was sulfonated using a variation of the standard procedure, using lauryl sulfonate in cyclohexane as the reaction medium. Sulfonation was carried out to a level of 97-98% styrene as determined by titration. The coating solution comprised a 20 wt% solids mixture of the sulfonated copolymer dissolved in toluene/1-hexanol (85%:15%). Approximately 20 linear yards of fabric (62-inch width) were coated using five separate applications (runs). Each run comprised pouring the rubber solution onto the moving fabric just prior to passing through a doctor blade, followed by a short in open air before entrance into a drying oven. This oven uses forced hot air flow to facilitate solvent removal. Following the 5th coating run, average coating weight (total minus starting weight) is 3.81 oz/yd^2. During role transfer, some fingering of the rubber was observed, indicative of the high tack of the rubber. The coated fabric was immersed in barium chloride solution for a period of 96 hours to insure complete exchange. The fabric was then rinsed with distilled water and air dried.

Characterization of the microstructure of the polymer film was performed through transmission electron microscopy (TEM). Bright –field TEM images were obtained using a Philips EM400T with an accelerating voltage of 100 kV. Single yarns of the coated fabric were embedded in an epoxy resin. In an effort to obtain better penetration between individual fibers the resin was heated to 50° C thereby lowering its viscosity. In addition, a vacuum was pulled on the samples for ten minutes to remove any air bubbles. Thinner sections were difficult to achieve as the coated material tends to tear apart during the cutting process. The quality, uniformity, and degree of adhesion of the polymer coating were characterized through scanning electron microscopy (SEM). Images were obtained using a JEOL 7401F field emission SEM with an accelerating voltage of 1 kV and an emission current of 10 μA.

Tensile testing of the coated and uncoated samples was performed according to ASTM standard D 5035-06. Cut strips 1 inch wide and 7 inches long were tested as it proved impossible to ravel the coated material. The strips were tested using an Instron 8511 equipped with 18,000 N load cell. Samples were held in place using flat faced metal grips. A crosshead displacement rate of 300 mm/min was used to obtain sample break at about 20 seconds. Coated and uncoated fabrics were tested in both the warp and weft directions.

The shear properties of the coated and uncoated materials were characterized through shear-frame testing [5]. In this test a cross shaped sample is clamped into a square "picture-frame" grip. The central square of the sample measures approximately 5 inches on each side. The picture frame apparatus is loaded into an Instron which is then set to provide a constant crosshead displacement of 2 mm/s. As the frame is elongated the yarns of the fabric are prompted to rotate/shear with respect to one another.

DISCUSSION

Microscopic Results

The polymer coated layers were characterized by using electron microscopes such as SEM and TEM. First, the SEM utilized to examine coating uniformity and coverage of coated area. The phase separation behavior of SIBS copolymer coating material was investigated by performing TEM analysis. Those morphology studies could be acquainted to verify both coating quality and phase separation of block copolymer on the fabrics.

Figure 1 presents representative SEM images of the plain and the cross-sectional views of the coated fabric. As shown in figure 1a, the coating appears to be free of gaps and voids, which is crucial for the coating's performance as a chemical barrier. Figure 1b gives evidence that the coated polymer has penetrated into the fabric material (i.e. some of the coating material can be identified within the fabric tows) and that the coating adheres to the fabric fibers (i.e. the coating survives sample preparation). This penetration in-between the yarns of the material and this level of adhesion to the fibers will ensure that the layer of coating stays attached to the fabric substrate.

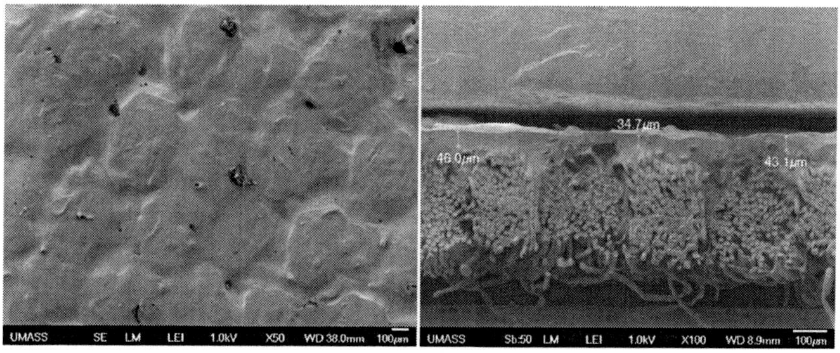

(a) (b)

Figure 1. (a) Plain view and (b) cross-sectional view of polymer coated fabric.

Measurements of the coating layer in both the warp and weft directions through the SEM analysis are gathered in Table I.

Table I. Thickness measurement of coating layers.

Coated Materials	Direction	
	Weft	Warp
Average Thickness (μm)	40.6	32.1
Standard Deviation (μm)	18.0	13.2

As illustrated in Figure 2, the TEM analysis reveals distinct organized domains present within the otherwise amorphous coating material. Literature [6] reports on SIBS thin films demonstrate the possible generation of cylindrical or lamellar domains in this block copolymer, with the dark bands attributed to sulfonated styrene. The structures observed so far appear to be mostly localized lamellar structures in nature (almost no cylinder cross-sectional morphology has been observed). One can note that the domain size of the present coating is on the order of 2 to 4 nm while phase separation domain sizes previously reported are typically about 10-20 nm.

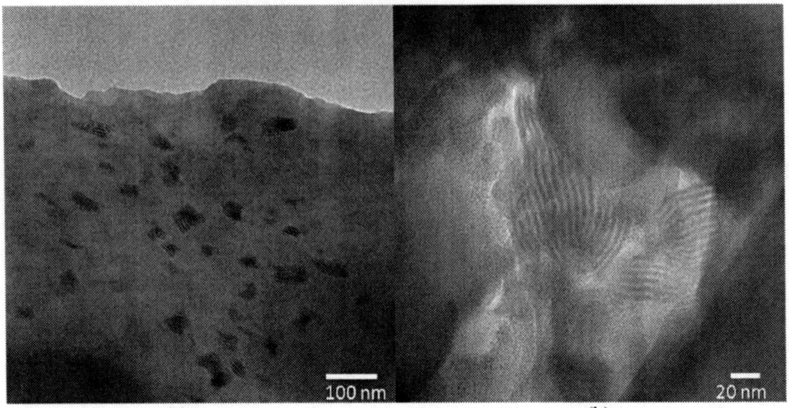

(a) (b)

Figure 2. TEM micrographs of coating material showing supposed localized phase separation at (a) whole coated layer and (b) the inside of coated layer.

Tensile Results

Figure 3 presents the results of tensile testing done in the warp direction of the fabric.

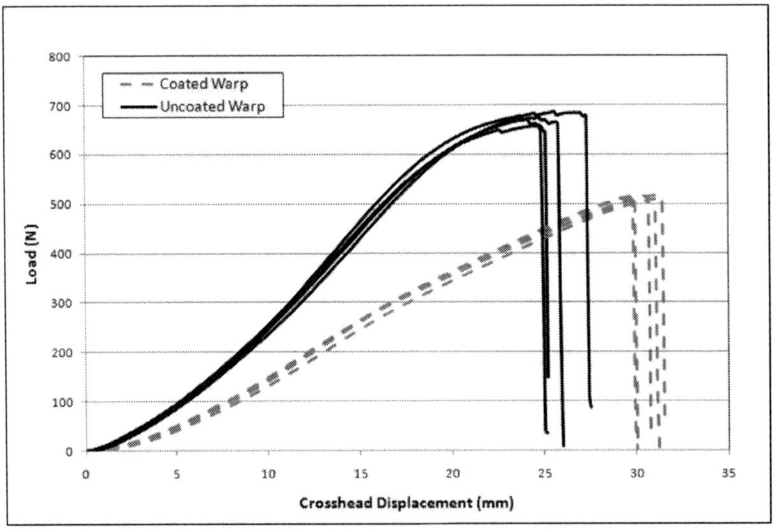

Figure 3. Tensile test results in the warp direction for coated and uncoated fabric.

The coated fabric experienced higher elongation than the uncoated fabric yet it supported a lower maximum load. Test results in the weft direction follow a similar trend. Table II compares the results for both directions. The uncoated and coated fabrics were stronger in the warp direction. This trend was expected as there is a higher density of yarns in the warp direction. The decrease in load at failure for the coated fabric could be due to the coating process. The fabric undergoes repeated thermal cycles which fatigues the fabric resulting in a decreased load bearing capacity. The presence of solvent may also be affecting the fabric by plasticizing the individual fibers, leading to a greater elongation during loading.

Table II. Tensile results for coated/uncoated fabrics in warp/weft directions.

Directions	Warp Direction		Weft Direction	
Specimen	Average Displacement At Break (mm)	Average Max Load (N)	Average Displacement At Break (mm)	Average Max Load (N)
Uncoated	25.654	671.7537	17.272	381.71
Coated	30.734	511.6005	20.574	307.34
% Difference	19.38	-23.8	18.6	-19.5

Shear Frame Results

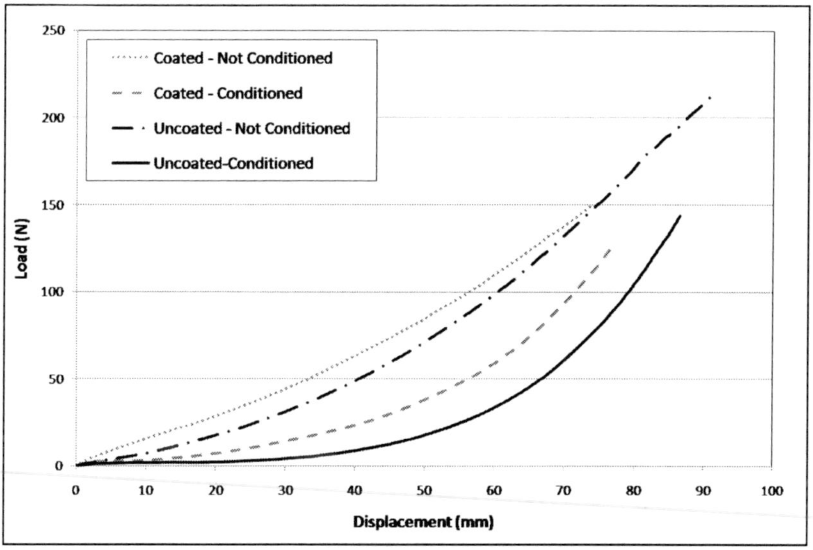

Figure 4. Shear frame results for conditioned and unconditioned coated/uncoated fabrics.

Figure 4 illustrates the shear frame test. Representative results are plotted for the initial testing of coated and uncoated samples as well as conditioned coated and uncoated samples. Conditioned samples were tested five times in a row, the fifth test is presented here.

One can observe in figure 4 a slight increase in the initial load experienced by the coated fabric when compared to the uncoated fabric. This is due to the coating material restricting the yarns from freely shearing relative to one another. After conditioning the difference between the coated and uncoated fabric behavior is more pronounced. The conditioned coated fabric experienced an increase in load sooner than the uncoated fabric. This indicates that the presence of the coating has decreased the ability of the yarns to shear relative to one another even after conditioning.

CONCLUSIONS

The microstructure and mechanical behavior of a fabric coated with sulfonated poly (styrene-isobutylene-styrene) block copolymer was examined. TEM imaging provided evidence of the presence of localized phase separated lamellar domains within the polymer coating. SEM analysis confirmed a uniform coating with apparent adhesion to the fabric substrate. The presence of the polymer coating affected the mechanical properties and behavior of the fabric. In particular the coated fabric exhibited a decrease in tensile strength and a decrease in the ability of the yarns to shear relative to one another. Future work will focus on the distribution of sulfonated SIBS phase separation and on identifying the reasons for the decrease in mechanical properties.

ACKNOWLEDGMENTS

This work was supported by the U.S. Army Research Lab under contract number, DoD-SBIR-A07-066, with EIC laboratories Inc.

REFERENCES

1. D. M. Crawford, J. Sloan, and E. Napadensky, *Chemical Agent Performance of Sulfonated Ionomeric Membranes For Chem/Bio Applications*, ADM002187, Army Research Laboratory (2008).
2. D. M. Crawford, E. Napadensky, N.C. Beck Tan, and J. Sloan, *Semipermeable Membrane From Ionomeric Self-Assembling Block Copolymer*, ARL-TR-2403, Army Research Laboratory (2001)
3. Y. Funaki, K. Kumano, T. Nakao, H. Jinnai, H. Yoshida, K. Kimishima, K. Tsutsumi, Y. Hirokawa, and T. Hashimoto, Polymer **40**, 7147 (1999).
4. G. Kim and M. Libera, Macromolecules **31**, 2569 (1998).
5. X.Q. Peng, J. Cao, J. Chen, P. Xue, D.S. Lussier, and L. Liu, Composites Science and Technology 64, 11 (2004).
6. D. M. Crawford, E. Napadensky, N. C. Beck Tan, D. A. Reuschle, D. A. Mountz, K.A. Mauritz, K.S. Laverdure, S.P. Gido, W. Liu, B. Hsiao, Thermochimica Acta **367**, 125 (2001).

Mater. Res. Soc. Symp. Proc. Vol. 1312 © 2011 Materials Research Society
DOI: 10.1557/opl.2011.136

Nano Patterning on Surfaces of 3-D Micro Structures by Means of Block Copolymer Self Assembly

Shin'ichi Warisawa[1] , Hiroyuki Mino[1], Reo Kometani[1] and Sunao Ishihara[1]
[1]Department of Mechanical Engineering, School of Engineering, The University of Tokyo, 7-3-1 Hongo Bunkyo-ku, Tokyo 113-8656, Japan

ABSTRACT

In this paper, we present the possibility of nano-scale patterning on surfaces of 3-D micro structures by means of block copolymer self assembly. We succeeded in forming vertical cylinders on the surfaces of the 3-D micro structures using dip-coat process with Au deposition instead of random copolymers. The material of 3-D structures fabricated by FIB-CVD was diamond-like carbon (DLC) due to selecting phenanthrene as a gas source. Since the surface energy of DLC is low, instead of PS-r-PMMA polymer brush, 100 nm-thick Au films were coated on the surfaces of 3-D structures. PS-b-PMMA (PS:PMMA = 50 kg/mol:21 kg/mol) was dip-coated on the Au coated surfaces and annealed. First, we prepared DLC 60° angled slope with size of 2.5 by 3 μm^2. As a result, the vertical cylinders were formed in a similar way regardless of the places, i.e., top, middle and bottom on the slopes. vertical cylinders were formed regularly and densely on the entire slope. In order to investigate the effect of slope angle, we prepared 3-D micro structures with different angled slopes such as 70, 80 and 90°. There was no slope angle specific difference, and vertical cylinders were also formed on each slope.

INTRODUCTION

Recently, three-dimensional (3-D) nano- and microstructures have been of great interest to realize innovative devices. However, such 3-D structures require very complicated and time-consuming processes for fabrication if based on semiconductor fabrication techniques. Focused ion beam etching and deposition techniques enable to fabricate complicated 3-D structures[1]. Such structures will need electric wiring with them to be actuated and smaller functional devices on them to sense something.

Recent block copolymer self-assembly researches have demonstrated the potential to form aligned line patterns[2], periodical sphere structures, and directed line and sphere patterns using chemical and physical templates at nano scale[3]. Many of these have been examined mainly on 2-D flat substrates. Ordered nano structures of block copolymer film are of great interest for lithographic applications such as patterned media, quantum dots and so on[4, 5]. For alignment control purpose, as well as chemically patterned substrates, topographic patterns such as groove[6], honeycomb, step-like[7], pillar array structures[8] have been investigated. Even in those cases, self-assembled copolymer structures are discussed from 2-D viewpoint. Enhancing the potential of nano patterning onto 3-D structures may extend further applications such as wiring for electro circuits on nano- and micro-electro mechanical systems, and 3-D structured optical devices and high-density memory devices. In this research, we demonstrate the possibility of nano-scale patterning on the surfaces of the 3-D micro structures by block copolymers. In our experiments, 3-D micro structures were fabricated by FIB-CVD with phenanthrene gas, and thus their material was diamond like carbon. In our preliminary investigation, because of uneven film thickness, we could not obtain self-assembled structures of block copolymer on the surfaces of DLC 3-D microstructures by spin-coating as is conducted on Si substrate. Therefore, this paper focuses on how to coat block copolymer on the surfaces of DLC 3-D microstructures in order to induce block copolymer self-assembly.

EXPERIMENT

Using a FIB system (SMI3050: SII NanoTechnology Inc.), Ga^+ FIB was irradiated at 30 keV acceleration voltages with 16 pA beam currents under a phenanthrene gas atomosphere to form micro slopes of a diamond-like carbon (DLC). We employed a polystyrene (PS)-block-polymethyl methacrylate (PMMA) diblock copolymer, which had a weight of 50 kg/mol for PS and 21 kg/mol for PMMA. Typically, a solution of a copolymer is spin coated onto a flat or topographically patterned surface and annealed. However, spin-coating is apt to embed 3-D micro structures in the copolymer film. Since we would like to coat a copolymer film on the surfaces of 3-D micro structures, we applied a dip-coating technique using a dip coater (DC4200: AIDEN CO., LTD). Adjusting the pull-up velocity and polymer solution concentration controlled the film thickness. After dip-coating, the block copolymer on the substrate was annealed in an oven at 175–210 °C for a number of hours or on a hot plate at 200 °C for a number of minutes. UV exposure and acetic acid immersion removed PMMA from the self-assembled copolymer film. The samples were observed by a field emission scanning electron microscope (FE-SEM) (JSM-7001F: JEOL Ltd.).

Self-assembly experiment on PS-r-PMMA treated micro slope

First, we fabricated DLC slope structures on a Si substrate. The height, angle and width of the slope were 2.5 μm, 60° and 3 μm, respectively (Figure 1(a)). A 1.5 wt-% toluene solution of PS-r-PMMA was dip-coated onto the DLC slope surfaces with a pull-up velocity of 4 mm/s. After a 12 hours 175 °C anneal, a 1 wt-% toluene solution of PS-b-PMMA was dip-coated onto the PS-r-PMMA pretreated DLC slope surfaces with a pull-up velocity of 4 mm/s, and annealed at 210 °C for 5 minutes on a hot plate. As shown Figure 1, vertical cylinder structures were observed on the DLC slope, but the structures on the slope surface were not so dense as expected (Figure 1(b)). The boundary between the substrate surface and the slope surface was dull (Figure 1(c)). Self-assembled structures were formed closely on the boundary. These micrographs suggest that PS-b-PMMA was not dip-coated uniformly on the slope surface.

Self-assembly experiment on Au coated micro slope

Thus, in order to obtain more dense structures, instead of PS-r-PMMA brush layer, we employed sputter coating to cover the DLC surfaces with a 100 nm-thick Au film. Figure 2 (a) shows a relationship between pull-up velocity and PS-b-PMMA film thickness in case of 1 wt-% toluene solution. Figures 2 (b) – (e) show AFM topography images of self-assembled structures with pull-up velocities at 2, 4, 6 and 8 mm/s, respectively. These images suggest that Au film provides non-preferential affinity to PS and PMMA, leading to forming vertical cylinders. In the following experiments, we dip-coated with pull-up velocity at 4 mm/s forming most dense vertical cylinders.

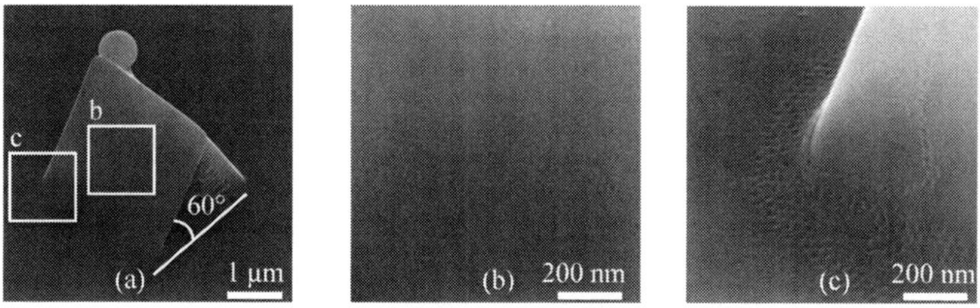

Figure 1. SEM micrographs of PS-b-PMMA film on a DLC micro slope pretreated with PS-r-PMMA. (a) A DLC micro slope structure dip-coated with PS-b-PMMA. (b) Disperse vertical cylinders were formed on the surface of the slope. (c) Vertical cylinders were formed densely on the boundary between the substrate surface and the slope surface.

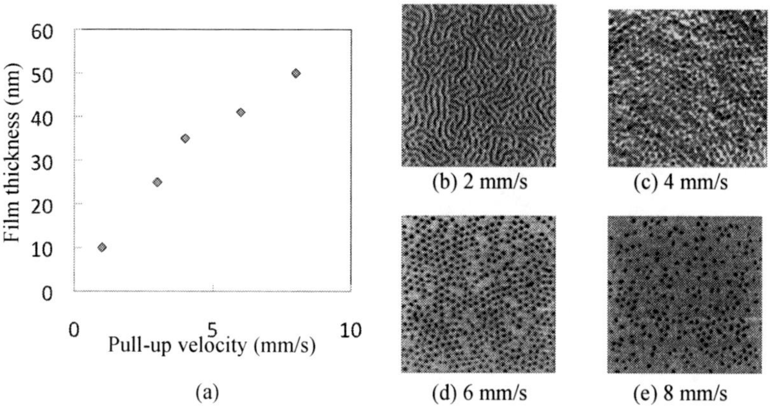

Figure 2. PS-b-PMMA (PS:PMMA=50kg/mol:21kg/mol) dip-coated on Au coated DLC surfaces. (a) Relationship between pull-up velocity and PS-b-PMMA film thickness. AFM topographical images of self-assembled structures with pull-up velocity at (b) 2mm/s, (c) 4mm/s (d) 6mm/s, and (e) 8mm/s.

After a 100-nm thick Au film deposition, a 1 wt-% toluene solution of PS-b-PMMA was dip-coated onto the DLC 60° angled slope. The substrate was subsequently annealed at 210 °C for 5 minutes on a hot plate. Figure 3 shows FE-SEM micrographs. Regular and dense vertical cylinders were successfully formed on both the flat (Figure 3 (e)) and sloped (Figures 3 (b) and (c)) surfaces. Furthermore, the cylindrical structures stand uniformly on both the sloped and flat surfaces around the boundary. The boundary clearly observed in Figure 3 (d) suggests that PS-b-PMMA dip-coating was done very well onto the micro 3-D structures.

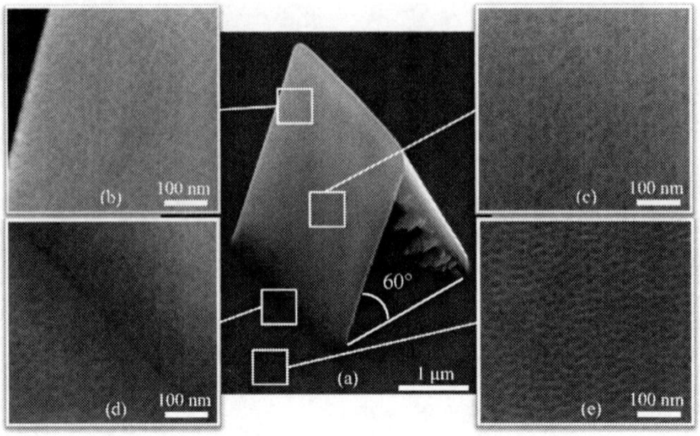

Figure 3. SEM micrographs of PS-b-PMMA self assembly on a DLC micro slope pretreated with Au. Angle of the slope was 60°. (a) A DLC micro slope structure dip-coated with PS-b-PMMA. Dense cylindrical structures were successfully formed (b) on the upper of the surface, (c) on the middle of the surface, (d) around the boundary, and (e) on the substrate surface.

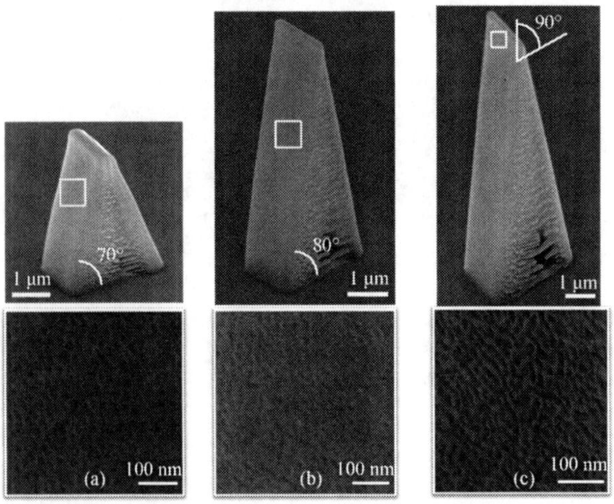

Figure 4. SEM micrographs of PS-b-PMMA film on a DLC micro slope pretreated with Au. Angles of the surfaces we evaluated were (a) 70°, (b) 80° and (c) 90°. Regardless of the surface angle, dense self-assembled structures were formed.

Figure 4 shows how the self-assembled structures change with the angle of the slope. DLC micro slopes with 70°, 80° and 90° were prepared. The samples were dip-coated at 4 mm/s of pull-up velocity and annealed at 210 °C for 5 minutes on a hot plate. Dense vertical cylinders were self-assembled regardless of the slope angle. The DLC micro slopes with 90° exhibited coexistence of vertical and horizontal cylinders, a texture of which was similar to that of the surface at 2 mm/s of pull-up velocity shown in Figure 2 (b).

Application to more complicated surfaces

We prepared a more complicated 3-D structure to demonstrate the possibility of our method. Figure 5 (a) shows the entire structure fabricated by FIB-CVD. The 100-nm thick Au film deposited DLC structure is like a pyramid having four surfaces, each of which consists of three slightly curved surfaces. A 1 wt-% toluene solution of PS-b-PMMA was dip-coated, and annealed at 210 °C for 5 minutes on a hot plate. Along the ridge of the pyramid, horizontal cylinders were densely formed (Figure 5(b)). On the contrary, vertical cylinders were densely formed on the sloped and curved surfaces (Figure 5(c)).

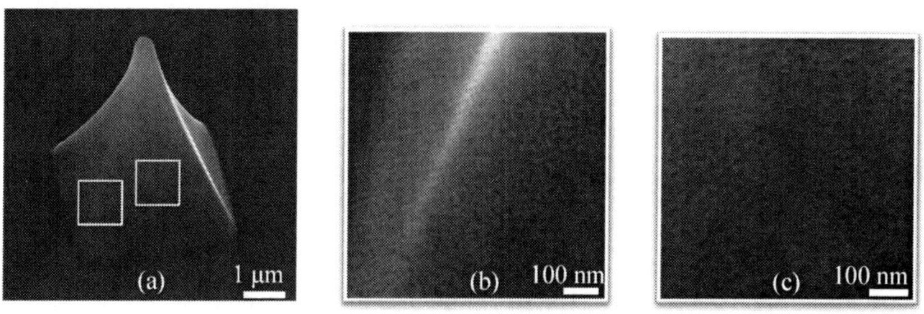

Figure 5. SEM micrographs of PS-b-PMMA self assembly on a DLC micro pyramid pretreated with Au. (a) Each surface of the micro pyramid consists of three slightly curved surfaces. (b) Horizontal cylinders were observed around the ridge. (c) Vertical cylinders were observed on the sloped and curved surface.

DISCUSSION

The difference between PS-r-PMMA on DLC and Au treatment is due to the difference in the surface free energy between DLC and Au. The surface free energy of DLC and Au were measured as 42.5 mJ/m^2 and 62.8 mJ/m^2, respectively. Eventually, Figures 1 (b) and (c) show that the copolymer was not coated evenly, especially around the edges. The clear edge of the intersection between the slope and the substrate shown in Figure 3 (d) also suggests that better coating result is due to the surface energy difference.

Horizontal cylinders were observed on the top of the slope in Figure 4 (c) and around the ridge of the pyramid in Figure 5 (b). The common situation between the two cases is that the area is close to the edge of the structure where the copolymer film seems coated thinner than other places. This is also suggested from the result that horizontal cylinders were formed in case of thinner film thickness as displayed in Figure 2 (b). This may be improved when coating films with larger surface free energy than Au.

Ordered block copolymer films are of great interest, and are also necessary for 3-D micro structures. Chemical or physical template methods that have been proposed are effective methods for that end. FIB-CVD technique enables to fabricate physical templates on 3-D micro structures. Furthermore, as presented in this paper, DLC can change the affinity of the surfaces to block copolymers, and thus can be used for physical and chemical templates on the 3-D micro structures as well as physical templates. Combination of DLC deposition on Au films would be very interesting.

CONCLUSIONS

In summary, we have demonstrated that dip-coating of PS-b-PMMA onto Au-film-deposited DLC offers a facile and promising way to produce vertical and horizontal cylindrical block copolymer domains. As shown in the body, this method could be useful for putting those domains on 3-D micro structures, especially DLC structures fabricated by FIB-CVD. The block copolymer domains that are ordered well in a way like straight lines or periodic dots, if further investigations will be proceeded, may lead to new functional devices and actuators.

ACKNOWLEDGMENTS

This work was partly conducted at the Center for Nano Lithography & Analysis, the University of Tokyo, which is supported by the Ministry of Education, Culture, Sports, Science and Technology (MEXT) of Japan. This work was also supported by a Grant-in-Aid for Scientific Research (B) No. 21360059 from the Japan Society for the Promotion of Science.

REFERENCES

1. Kometani, R., et al., *Performance of nanomanipulator fabricated on glass capillary by focused-ion-beam chemical vapor deposition.* Journal of Vacuum Science & Technology B: Microelectronics and Nanometer Structures, 2005. **23**: p. 298.
2. Stoykovich, M.P., et al., *Directed assembly of block copolymer blends into nonregular device-oriented structures.* Science (New York, N.Y.), 2005. **308**: p. 1442-6.
3. Yang, J.K.W., et al., *Complex self-assembled patterns using sparse commensurate templates with locally varying motifs.* Nature nanotechnology, 2010. **5**: p. 256-60.
4. Park, M., *Block Copolymer Lithography: Periodic Arrays of 1011 Holes in 1 Square Centimeter.* Science, 1997. **276**: p. 1401-1404.
5. Cheng, J.Y., et al., *Formation of a Cobalt Magnetic Dot Array via Block Copolymer Lithography.* Advanced Materials, 2001. **13**: p. 1174-1178.
6. Cheng, J.Y., et al., *Fabrication of nanostructures with long-range order using block copolymer lithography.* Applied Physics Letters, 2002. **81**: p. 3657.
7. Segalman, R.A., H. Yokoyama, and E.J. Kramer, *Graphoepitaxy of Spherical Domain Block Copolymer Films.* Advanced Materials, 2001. **13**: p. 1152-1155.
8. Bita, I., et al., *Graphoepitaxy of self-assembled block copolymers on two-dimensional periodic patterned templates.* Science (New York, N.Y.), 2008. **321**: p. 939-43.

AUTHOR INDEX

SUBJECT INDEX

CPSIA information can be obtained at www.ICGtesting.com
Printed in the USA
242025LV00001B/2/P

9 781605 112893